PRINCIPLES OF PHYSICAL OPTICS

PRINCIPLES OF PHYSICAL OPTICS

C. A. Bennett
University of North Carolina At Asheville

WILEY-INTERSCIENCE

A JOHN WILEY & SONS, INC., PUBLICATION

VICE PRESIDENT AND EXECUTIVE PUBLISHER	Kaye Pace
ASSOCIATE PUBLISHER	Petra Recter
EXECUTIVE EDITOR	Stuart Johnson
EXECUTIVE MARKETING MANAGER	Amanda Wainer
MANAGER, PHOTO DEPARTMENT	Hilary Newman
PRODUCTION EDITOR	Kerry Weinstein
SENIOR MEDIA EDITOR	Tom Kulesa
DESIGNER	Michael St. Martine
ASSISTANT EDITOR	Aly Rentrop
EDITORIAL ASSISTANT	Veronica Armour
PHOTO EDITOR	Sarah Ascione

To order books or for customer service,please call 1-800-CALL WILEY (225-5945).

ISBN-13 978-0-470-12212-9
ISBN-10 0-470-12212-9

Printed in the United States of America

10 9 8 7 6 5 4 3 2

PREFACE

The study of optics has much to offer to the STEM[1] curriculum. If taken prior to upper-level quantum mechanics, the classical waves of physical optics can provide context and an intuitive foundation for the more abstract study of quantum mechanical waves. Wave interference and coherence can be observed with interferometers and within diffraction patterns, giving context to more abstract notions of quantum coherence. Fourier methods become intuitive when the connection to diffraction is established and they can be displayed as patterns on the wall. Quantum uncertainty becomes less mysterious after the classical counterparts are studied. Single-photon interference and Fresnel diffraction can open windows into subtle and interesting outcomes of quantum electrodynamics. Ideas such as these are what attract majors to science in the first place, and optics can provide a conceptual foundation for more advanced study. Optics is also very useful. Students from across the STEM disciplines can benefit from the technology of optics. Scientists measure things, and measurement usually involves some aspect of optics. Engineers design and use optical sensors, and build communication infrastructures that are increasingly based in optics. Biologists and astronomers benefit from a fundamental understanding of resolution limits and coherence. A foundation in optics can facilitate scientific careers.

Despite the practical and conceptual benefits that the study of optics has to offer, it is often missing from undergraduate STEM curricula. Perhaps this is because in many cases, optics is offered as a senior-level physics course with advanced electricity and magnetism as a prerequisite. However, most of the content in an undergraduate optics course can be discussed very effectively without this level of preparation. The prerequisites assumed in this text are limited to the standard introductory sequences in calculus and calculus-based

---

[1]Science, Technology, Engineering and Mathematics.

physics. Principles of electricity and magnetism are developed as needed, beginning with integral forms of Maxwell's equations. A derivation of the electromagnetic wave equations is included, as is a thorough determination of all properties of transverse electromagnetic waves. This is a text that is accessible to sophomores and juniors, and the accessibility is achieved without compromise.

The text is designed to support a one-semester first course in optics. The material is organized to allow coverage at least through Chapter 7 on lasers within a three hour, one semester course. To facilitate this goal, the discussion of elliptical polarization, birefringence, and wave plates is postponed until Chapter 9; while important, this material is not more important than lasers, and it is not prerequisite for any of the earlier material. Other nonessential topics have been move to appendices, or included as starred sections. A starred section is not prerequisite for any discussions within Chapters 1-6. Chapter 8 (Optical Imaging) and Chapter 9 (Polarization and Nonlinear Optics) are available for courses that maintain a more lively pace, or perhaps do not include a complete coverage of earlier chapters.

For most STEM majors, the transition from sophomore to more advanced study is largely one of sophistication. In the standard undergraduate physics curriculum for example, this transition is typically facilitated by junior level courses such as thermodynamics. This text is designed to provide a similar transition for all STEM disciplines. Early chapters provide adequate scaffolding to help students with weaker math backgrounds. Numerous example problems illustrate both math and physics principles throughout the text. By the end of the course, students will have a firm foundation in physical optics, along with practical experience in a range of mathematical applications such as matrix methods, Fourier analysis, and complex algebra. On the other hand, it is certainly my goal to keep this text relevant and interesting to students with more advanced preparation. For example, the derivation of the differential forms of Maxwell's equations is contained in an appendix of Chapter 2, since it has the potential to be a bit hard for juniors and a bit boring for more advanced students. Instructors teaching a more advanced course might consider assigning the first two chapters and the associated on-line homework for independent study.

As a long-time user and developer of on-line homework, I have come to rely increasingly upon this resource in my teaching. The majority of on-line problems available for this text currently consist of medium-difficulty problems designed to encourage students to read the text carefully. There are other advantages; for example, less class time can be devoted to working example problems if on-line homework is assigned. I often assign the on-line homework for sections that I do not cover or only cover lightly in class; this increases the coverage, and leaves more class time for conceptual discussions.

Acknowledgments

The guidance and assistance of the following reviewers is gratefully acknowledged: John Ballato, *Clemson University*; Ralph Benbow, *Illinois University*; George Bissinger, *East Carolina University*; Dale Byrne, *University of Texas at Dallas*; K. Kelvin Cheng, *Texas Tech University*; Heidi Fearn, *California State University-Fullerton*; Enrique Galvez, *Colgate University*; Magnus Gustafsson, *University of Colorado*; Martin Hackworth, *Idaho State University*; L. Kent Morrison, *University of New Mexico*; Guido Mueller, *University of Florida*; Jay Newman, *Union College*; Michael Paesler, *North Carolina State University*; Oren Quist, *South Dakota State University*; Charles Sackett, *University of Virginia*; David Shiner, *University of North Texas*; Orven Swenson, *North Dakota State University*; David Voelz, *New Mexico State University*; William J.F. Wilson, *University of Calgary*; Lowell

Wood, *University of Houston*; Jens Zorn, *University of Michigan*. The text was carefully checked for accuracy by Lowell Wood (*University of Houston*). Special thanks and recognition go to Wiley editors Aly Rentrop and Stuart Johnson for their skilled advice, support and encouragement.

CHUCK BENNETT

Asheville, NC
November 15, 2007

Some Physical Constants

Constant	Symbol	Value
Permittivity of free space	ϵ_0	$8.854 \times 10^{-12} \frac{C_2}{Nm^2}$
Permeability of free space	μ_0	$4\pi \times 10^{-7} \frac{T \cdot m}{A}$
Speed of light	c	$2.9979 \times 10^8 \frac{m}{s}$
Electron charge	e	$1.602 \times 10^{-19} \, C$
Electron mass	m_e	$9.108 \times 10^{-31} \, kg$
Planck's constant	h	$6.626 \times 10^{-34} J \cdot s$
Boltzmann's constant	k_B	$1.380 \times 10^{-23} \frac{J}{K}$
Stefan's constant	σ	$5.6699 \times 10^{-8} \frac{W}{m^2 \cdot K^4}$

CONTENTS

CHAPTER 1

THE PHYSICS OF WAVES

The solution of the difficulty is that the two mental pictures which experiment lead us to form —
the one of the particles, the other of the waves — are both incomplete and have only the validity
of analogies which are accurate only in limiting cases.

—Heisenberg

Contents

1.1 INTRODUCTION

The properties of waves are central to the study of optics. As we will see, light (or more
properly, electromagnetic radiation) has both particle and wave properties. These *complementary* aspects are a result of quantum mechanics, and prior to the early 1900s there were

1

two schools of thought. Newton postulated that light consists of particles, while contemporaries Huygens and Hooke promoted a wave theory of light. The matter seemed settled with Young's important double-slit experiment that offered clear experimental evidence that light is a wave. Maxwell's sweeping theory of electromagnetism finally provided a deep and complete description of electromagnetic waves that we consider in detail in Chapter 2. Although current theories of optics include both wave and particle descriptions, the wave picture still forms the bedrock of most optical technology. In this chapter, we will outline some general properties that apply to traveling waves of all types.

1.2   ONE-DIMENSIONAL WAVE EQUATION

Mechanical waves travel within elastic media whose material properties provide restoring forces that result in oscillation. When a guitar string is plucked, it is displaced away from its equilibrium position, and the mechanical energy of this disturbance subsequently propagates along the string as traveling waves. In this case, the waves are *transverse*, meaning that the displacement of the medium (the string) is perpendicular to the direction of energy travel. Acoustic waves in a gas are *longitudinal*, meaning that the gas molecules are displaced back and forth along the direction of energy flow as regions of high and low pressure are created along the wave.

As we shall see in the next chapter, *electromagnetic waves* are transverse, but differ from mechanical waves in that they do not require an elastic medium. Rather, they propagate as disturbances in the *electromagnetic field*. Mechanical waves and electromagnetic waves are both examples of *classical waves*, waves that can be described with classical physics[1].

Consider a mechanical wave, and let $\Psi(x,t)$ describe a disturbance of the medium away from its equilibrium condition. For a transverse wave along a horizontal string, $\Psi(x,t)$ represents a vertical displacement along the y-axis. For a longitudinal acoustic wave traveling horizontally through air, $\Psi(x,t)$ might represent deviations away from ambient pressure along the x-axis. In any case, we refer to $\Psi(x,t)$ as the *wavefunction*. Since $\Psi(x,t)$ depends only on the single spatial coordinate x, it is a *one-dimensional wavefunction*.

All classical mechanical waves can be described by the same equation, the *differential wave equation*:

$$\frac{\partial^2 \Psi(x,t)}{\partial x^2} = \frac{1}{v^2} \frac{\partial^2 \Psi(x,t)}{\partial t^2} \tag{1.1}$$

where v is the wave speed and $\Psi(x,t)$ is the wave function. In order to demonstrate that a given function $\Psi(x,t)$ describes a classical wave, it is necessary only to show that this function satisfies Equation 1.1. Conversely, any physical system that can be shown to be described by Equation 1.1 must necessarily involve classical traveling waves.

Equation 1.1 is an example of a class of mathematical equations known as *differential equations*. When taking a partial derivative with respect to a given parameter, the remaining parameters are treated as constants.

■ **EXAMPLE 1.1**

Show that

$$\Psi(x,t) = (x + vt)^2$$

---

[1] The term classical refers to the physics that came before the modern theories of relativity and quantum mechanics.

is a solution to the differential wave equation. Assume that v is a constant.

Solution

We must show that $\Psi(x,t)$ solves Equation 1.1:

$$\frac{\partial^2 \Psi(x,t)}{\partial x^2} = \frac{1}{v^2} \frac{\partial^2 \Psi(x,t)}{\partial t^2}$$

To do this, we must take the indicated derivatives. Begin with the partial derivative with respect to x:

$$\frac{\partial \Psi(x,t)}{\partial x} = \frac{\partial}{\partial x} (x + vt)^2 = 2(x + vt)\frac{\partial}{\partial x}(x + vt) = 2(x + vt)(1) = 2(x + vt)$$

Now take the second partial derivative with respect to x:

$$\frac{\partial^2 \Psi}{\partial x^2} = \frac{\partial}{\partial x}\left[\frac{\partial \Psi}{\partial x}\right] = \frac{\partial}{\partial x}\left[2\left(x + vt\right)\right] = 2$$

Next, take the partial derivatives of $\Psi(x,t)$ with respect to t:

$$\frac{\partial \Psi}{\partial t} = 2\left(x + vt\right)\frac{\partial}{\partial t}\left(x + vt\right) = 2\left(x + vt\right)\left(v\right) = 2v\left(x + vt\right)$$

$$\frac{\partial^2 \Psi}{\partial t^2} = \frac{\partial}{\partial t}\left[2v\left(x + vt\right)\right] = \frac{\partial}{\partial t}\left[2vx + 2v^2 t\right] = 2v^2$$

Substitute these results into the differential wave equation.

$$\frac{\partial^2 \Psi(x,t)}{\partial x^2} = 2$$

$$\frac{1}{v^2} \frac{\partial^2 \Psi(x,t)}{\partial t^2} = \frac{1}{v^2}\left(2v^2\right) = 2$$

Thus

$$\frac{\partial^2 \Psi(x,t)}{\partial x^2} = \frac{1}{v^2} \frac{\partial^2 \Psi(x,t)}{\partial t^2}$$

$\Psi(x,t)$ solves the differential wave equation, and thus we have shown that it is a traveling wave.

1.2.1 Transverse Traveling Waves On A String

It is instructive to see how Equation 1.1 arises for a particular example from classical mechanics. Consider the case of a horizontal string supported at both ends and stretched to a tension of magnitude F. Figure 1.1 shows a small section of the string that has been displaced from its horizontal equilibrium position. The y coordinate of the string in the figure indicates the values of $\Psi(x,t)$ between the points x and $x + dx$. The force of tension acting on each end of the string segment is indicated in the figure along with the x and y components of theses forces.

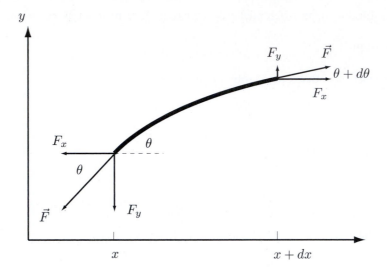

Figure 1.1   A segment of a tight string that has been displaced from equilibrium. Only the segment between x and $x + dx$ is shown.

It will simplify the analysis if we assume that the oscillations are small so that the angle shown in Figure 1.1 remains small across the string segment. To find the net force acting on the string segment, we sum the horizontal and vertical components. In the horizontal direction,

$$F_x(x, t) = -F\cos(\theta)$$

$$F_x(x + dx, t) = F\cos(\theta + d\theta)$$

For small angles, $\cos(\theta + d\theta) \approx \cos(\theta) \approx 1$, so

$$F_x(x, t) + F_x(x + dx, t) \approx 0$$

Thus the net horizontal force across the string segment is zero. In the vertical direction, we use the small angle approximation: $\sin(\theta) \approx \theta$

$$F_y(x, t) = -F\sin(\theta) \approx -F\theta$$

$$F_y(x + dx, t) = F\sin(\theta + d\theta) \approx F\theta + F\,d\theta$$

The net vertical force is

$$F_y(x + dx, t) + F_y(x, t) = F\,d\theta$$

To find $d\theta$, we first find θ by noting that for small angles, $\theta \approx \sin(\theta) \approx \tan(\theta)$. The tangent is given by the slope of $\Psi(x, t)$

$$\theta(x, t) \approx \frac{\partial \Psi(x, t)}{\partial x}$$

The differential $d\theta$ is given by

$$d\theta = \frac{\partial \theta}{\partial x}dx = \frac{\partial^2 \Psi(x, t)}{\partial x^2}dx$$

Thus, the net vertical force acting on the string segment is

$$F_y(x + dx, t) + F_y(x, t) = F \frac{\partial^2 \Psi(x, t)}{\partial x^2} dx$$

By Newton's second law, this must be equal to the mass of the string segment times its acceleration. Let μ be the mass per unit length of the string. The mass of the string segment is given by $(\mu \, dx)$. In the case at hand, $\Psi(x, t)$ represents a displacement, so its second time derivative is the segment's acceleration:

$$F_y(x + dx, t) + F_y(x, t) = F \frac{\partial^2 \psi(x, t)}{\partial x^2} dx = (\mu \, dx) \frac{\partial^2 \psi(x, t)}{\partial t^2}$$

Canceling the factor of dx and dividing by F gives

$$\frac{\partial^2 \psi(x, t)}{\partial x^2} = \frac{\mu}{F} \frac{\partial^2 \psi(x, t)}{\partial t^2}$$

This is the differential wave equation with a wave speed given by

$$v = \sqrt{\frac{F}{\mu}}$$

Notice that once the wave equation has been deduced, the wave speed may be determined by inspection.

In summary, we have used the laws of classical mechanics to demonstrate that the motion of a stretched string can be described by the differential wave equation, Equation 1.1. Thus, we now know that a disturbance of the string away from its equilibrium position will propagate along the string as traveling waves. In the next section, we will say more about the functions $\Psi(x, t)$ that describe these and other classical traveling waves.

Problem 1.1   Show that
$$\Psi(x, t) = (x - vt)^2$$
is a traveling wave.

Problem 1.2   A $1.00 \, m$ long string is stretched to a tension of $100 \, N$. Find the wave speed if it has a mass of $0.100 \, g$.

1.3   GENERAL SOLUTIONS TO THE 1-D WAVE EQUATION

Consider a plot of $\Psi(x, t)$ vs. x. Data for such a plot could be provided by an array of measuring devices, such as an array of pressure sensors arranged linearly to record the pressure amplitude of a passing acoustic wave. A plot of $\Psi(x, t)$ vs. x at a particular time t represents a *snap shot* of the wave as it passes by an array of measuring devices.

There are many possible shapes for this amplitude. Perhaps a sinusoidal function comes to mind, with distinct periodic crests and troughs. However, such waves are not the most general solution to Equation 1.1, as you can determine by comparing the sound of your voice as you hum or sing a specific musical note (if you can!) to the sound that your hands make when you clap them together. We will refer to the sound of a clap as a *pulse*.

We will show below that the most general solutions to Equation 1.1 may be expressed as follows:

$$\Psi(x, t) = f(x - vt) \tag{1.2}$$

$$\Psi(x, t) = g(x + vt) \tag{1.3}$$

where the functions f and g represent any function that has finite second derivatives, and where the parameters x, v, and t all occur explicitly within the function as $x - vt$ or $x + vt$.

As an example, consider the function

$$\Psi(x, t) = \frac{A}{1 + (x - vt)^2} \tag{1.4}$$

where A is a constant. Equation 1.4 represents a peaked function whose maximum is located at points given by $x = vt$. A plot of this wave functions at two different times is shown in Figure 1.2.

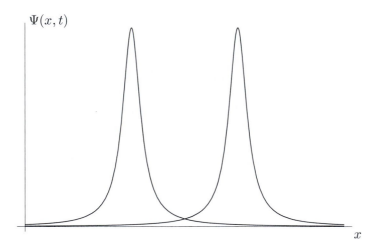

Figure 1.2   The traveling pulse of Equation 1.4, shown at two different times $t_1 < t_2$.

In order to show that Equation 1.4 represents a traveling wave, we could simply check to see if it solves the differential wave equation. It is more elegant, however, to show this for any differentiable functions given by Equations 1.2 and 1.3. Begin with the function $f(x - vt)$, and define a new variable u given by $u = x - vt$. Differentiate $f(u)$ using the *chain rule*:

$$\frac{\partial f(u)}{\partial x} = \frac{\partial f}{\partial u} \frac{\partial u}{\partial x}$$

In this case, $u = x - vt$, so the derivative of u with respect to x is just 1. Thus,

$$\frac{\partial f(u)}{\partial x} = \frac{\partial f}{\partial u}$$

and

$$\frac{\partial^2 f(u)}{\partial x^2} = \frac{\partial^2 f}{\partial u^2}\frac{\partial u}{\partial x} = \frac{\partial^2 f}{\partial u^2} \tag{1.5}$$

The time derivative of $f(u)$ is given by

$$\frac{\partial f(u)}{\partial t} = \frac{\partial f}{\partial u}\frac{\partial u}{\partial t} = -v\frac{\partial f}{\partial u}$$

and

$$\frac{\partial^2 f(u)}{\partial t^2} = \frac{\partial^2 f}{\partial u^2}\frac{\partial u}{\partial t} = v^2\frac{\partial^2 f}{\partial u^2} \tag{1.6}$$

Substitute the results of Equations 1.5 and 1.6 into the differential wave equation:

$$\frac{1}{v^2}\frac{\partial^2 f}{\partial t^2} = \frac{1}{v^2}\left(v^2\frac{\partial^2 f}{\partial u^2}\right) = \frac{\partial^2 f}{\partial u^2} = \frac{\partial^2 f}{\partial x^2}$$

Thus Equation 1.1 is satisfied by $f(u) = f(x - vt)$. In a similar way, you may show that $g(x + vt)$ also solves Equation 1.1(see Problem 1.4). Thus $f(x - vt)$ and $g(x + vt)$ both solve the differential wave equation, and therefore represent traveling waves.

In summary, we have shown that any function of x and t with finite second derivatives and with explicit occurrences of x and t that can be grouped as $x - vt$ or $x + vt$ are solutions to the differential wave equation and thus represent traveling waves. In particular, the function of Equation 1.4 fits this requirement, and is therefore a traveling wave. Of course, you can also demonstrate this by substituting Equation 1.4 directly into Equation 1.1 (see Example 1.2 below).

We now show that the function $f(x - vt)$ represents a *forward-traveling wave*. Consider values of x and t determined by $x - vt = constant$. In Equation 1.4, choosing the value zero for this constant locates the peak of the pulse. As time proceeds, the specific value of x that satisfies this equation changes according to

$$dx - v\,dt = 0$$

or

$$\frac{dx}{dt} = +v$$

Thus, $f(x - vt)$ propagates in the positive x direction with velocity $+v$, and is therefore a forward-traveling wave. In a similar way, you can show that $g(x + vt)$ represents a backward-traveling wave (see Problem 1.5).

■ **EXAMPLE 1.2**

Show explicitly that Equation 1.4 satisfies the differential wave equation.

Solution

We must substitute

$$\Psi(x, t) = \frac{A}{1 + (x - vt)^2}$$

into the differential wave equation

$$\frac{\partial^2 \Psi(x, t)}{\partial x^2} = \frac{1}{v^2}\frac{\partial^2 \Psi(x, t)}{\partial t^2}$$

Begin by differentiating with respect to x:

$$\frac{\partial \Psi}{\partial x} = \frac{\partial}{\partial x}\left(\frac{A}{1+(x-vt)^2}\right) = \frac{(-1)\,A\,[2\,(x-vt)]}{\left[1+(x-vt)^2\right]^2}$$

$$\frac{\partial^2 \Psi}{\partial x^2} = \frac{A\,(2)\,[2\,(x-vt)]^2}{\left[1+(x-vt)^2\right]^3} - \frac{A\,[2]}{\left[1+(x-vt)^2\right]^2}$$

Now differentiate with respect to t:

$$\frac{\partial \Psi}{\partial t} = -\frac{A\,[2\,(x-vt)\,(-v)]}{\left[1+(x-vt)^2\right]^2} = (-v)\left[-\frac{A\,[2\,(x-vt)]}{\left[1+(x-vt)^2\right]^2}\right]$$

$$\frac{\partial^2 \Psi}{\partial t^2} = (v^2)\left[\frac{A\,(2)\,[2\,(x-vt)]^2}{\left[1+(x-vt)^2\right]^3} - \frac{A\,[2]}{\left[1+(x-vt)^2\right]^2}\right]$$

Thus

$$\frac{1}{v^2}\frac{\partial^2 \Psi}{\partial t^2} = \frac{\partial^2 \Psi}{\partial x^2}$$

and $\Psi(x,t)$ is a traveling wave.

If you can cast $\Psi(x,t)$ in the form of Equation 1.2 or 1.3, you do not have to compute the derivatives explicitly in order to show that $\Psi(x,t)$ is a solution to Equation 1.1.

Problem 1.3　Show that

$$\Psi(x,t) = Ae^{-(a^2 x^2 + b^2 t^2 + 2\,a\,b\,x\,t)}$$

is a traveling wave, and find the wave speed and direction of propagation. Assume that A, a and b are all constants, and that a and b have units that make the quantity in the exponential function unitless.

Problem 1.4　Show that $g(x+vt)$ in Equation 1.3 is a solution of the one-dimensional differential wave equation.

Problem 1.5　Show that $g(x+vt)$ in Equation 1.3 represents a wave that travels in the negative-x direction.

1.4　HARMONIC TRAVELING WAVES

According to the results of the previous section, any function described by Equation 1.2 or 1.3 represents a traveling wave. In particular, *harmonic functions* (i.e., sines and cosines) with the appropriate arguments solve the differential wave equation. Thus, the following function represents a traveling wave:

$$\Psi(x,t) = A\sin\frac{2\pi}{\lambda}(x \mp vt) \tag{1.7}$$

where A is the wave amplitude, and λ is defined below.

Harmonic functions are periodic with a period of 2π radians:

$$\sin(\theta \pm 2\pi) = \sin(\theta)$$

$$\cos(\theta \pm 2\pi) = \cos(\theta)$$

The $\Psi(x,t)$ given in Equation 1.7 is periodic in both space and time coordinates. The term λ represents the *spatial period*:

$$\Psi(x+\lambda, t) = A\sin\frac{2\pi}{\lambda}(x + \lambda \mp vt) = A\sin\left[\frac{2\pi}{\lambda}(x \mp vt) + 2\pi\right] = \Psi(x,t)$$

The parameter λ is also called the *wavelength*. It has SI units of meters.

Let T represent the *temporal period*; i.e., the time required for one cycle. The SI unit of T is seconds (s), but it is convenient to use s/cycle as a reminder of what T represents. Since Equation 1.7 is periodic in T, we have

$$\Psi(x, t+T) = A\sin\frac{2\pi}{\lambda}[x \mp v(t+T)] = A\sin\left[\frac{2\pi}{\lambda}(x \mp vt) \mp \frac{2\pi}{\lambda}vT\right]$$

T represents the temporal period provided that

$$\frac{vT}{\lambda} = 1 \tag{1.8}$$

or

$$v = \frac{\lambda}{T} \tag{1.9}$$

Thus, a periodic classical wave travels one wavelength λ in one temporal period T. It is customary to define the wave *frequency* as

$$f = \frac{1}{T} \tag{1.10}$$

The units of f are *cycles/s* (SI unit: s^{-1}), often referred to as Hertz (Hz). In terms of frequency, Equation 1.9 becomes

$$f\lambda = v \tag{1.11}$$

A plot of $\Psi(x,t)$ vs. x is shown in Figure 1.3(a). Figure 1.3(b) shows a plot of $\Psi(x,t)$ vs. t. A plot such as this could be obtained from data provided by a measuring device located at a particular value of x.

It is customary to define the *propagation constant* as follows:

$$k = \frac{2\pi}{\lambda} \tag{1.12}$$

This quantity is also sometimes referred to as the *wave number*. Since k converts meters to radians, the units are rad/m (SI unit: m^{-1}). We may rewrite Equation 1.7 as

$$\Psi(x,t) = A\sin k(x \mp vt)$$

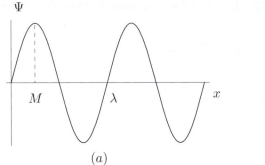
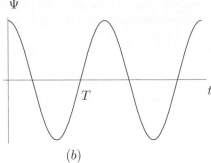

Figure 1.3   Plots of a harmonic wavefunction. (a) A plot of $\Psi(x,t)$ vs. position x. (b) A plot of $\Psi(x,t)$ vs. time using data recorded by a single measuring device located at M in Figure (a).

Similarly, we can define the *angular frequency*:

$$\omega = kv = \frac{2\pi}{\lambda}v = 2\pi f \tag{1.13}$$

Since ω converts time to radians, the units are rad/s (SI unit: s^{-1}). Note that according to the last result,

$$\frac{\omega}{k} = v \tag{1.14}$$

In terms of k and ω, Equation 1.7 becomes

$$\Psi(x,t) = A\sin(kx \mp \omega t)$$

Collectively, the terms $kx \mp \omega t$ are called the *phase* of the harmonic traveling wave. We may also include an explicit value of the phase when x and t are zero by specifying the *initial phase ϕ*:

$$\Psi(x,t) = A\sin(kx \mp \omega t + \phi) \tag{1.15}$$

The units of ϕ are radians (SI unit: dimensionless).

Problem 1.6   Show explicitly (by direct substitution) that the function

$$\Psi(x,t) = A\sin\frac{2\pi}{\lambda}(x \mp vt)$$

solves the one-dimensional wave equation.

Problem 1.7   Show explicitly that the function

$$\Psi(x,t) = A\cos(kx \mp \omega t + \phi)$$

solves the one-dimensional wave equation.

Problem 1.8   The speed of sound is $343\,m/s$ in air and $1493\,m/s$ in water. Find the wavelength emitted by a transducer that oscillates at 1000 Hz in both air and water. Assume that the frequency of the oscillator is the same in both media. In each case, find k and ω.

Problem 1.9   A harmonic traveling wave is given by $\Psi(z, t) = A\sin(50z + 3000t)$. Find the wave speed, frequency, angular frequency, and direction of propagation.

1.5   THE PRINCIPLE OF SUPERPOSITION

The differential wave equation is a linear differential equation. If a differential equation is linear, then the principle of superposition holds: if Ψ_1 and Ψ_2 individually satisfy the differential equation, then $\Psi_1 + \Psi_2$ is also a solution. We may demonstrate this explicitly for Equation 1.1 as follows:
given that

$$\frac{\partial^2 \Psi_1}{\partial x^2} = \frac{1}{v^2} \frac{\partial^2 \Psi_1}{\partial t^2}$$

and

$$\frac{\partial^2 \Psi_2}{\partial x^2} = \frac{1}{v^2} \frac{\partial^2 \Psi_2}{\partial t^2}$$

then

$$\frac{\partial^2 (\Psi_1 + \Psi_2)}{\partial x^2} = \frac{\partial^2 \Psi_1}{\partial x^2} + \frac{\partial^2 \Psi_2}{\partial x^2} = \frac{1}{v^2} \frac{\partial^2 \Psi_1}{\partial t^2} + \frac{1}{v^2} \frac{\partial^2 \Psi_2}{\partial t^2} = \frac{1}{v^2} \frac{\partial^2 (\Psi_1 + \Psi_2)}{\partial t^2}$$

The proof may be extended to any number of solutions.

The principle of superposition provides the basis for our subsequent study of interference and diffraction in Chapters 5 and 6.

1.5.1   Periodic Traveling Waves

It is certainly possible for a function to be periodic but not harmonic. In fact, it is very common for waves to be periodic, but it is physically impossible for any classical wave to be perfectly harmonic. We will gain more insights into this interesting fact in Chapter 5. For now, suffice it to say that no laser is perfectly monochromatic, nor does any tuning fork emit a perfectly precise musical note. An example of a periodic function that is not harmonic is illustrated in Figure 1.4.

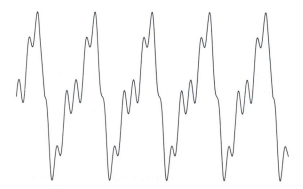

Figure 1.4   A periodic wave that is not harmonic.

1.5.2 Linear Independence

As we will discuss in Chapter 5, a periodic wave with arbitrary profile such as that shown in Figure 1.4 may be represented as a linear combination of traveling harmonic waves. However, a forward-traveling periodic wave must be constructed of forward-traveling harmonic waves, and backward-traveling periodic waves are represented as linear combinations of backward-traveling harmonic waves. Backward and forward-traveling waves are *linearly independent*, meaning that one cannot be obtained as a linear combination of the other. Thus, the most general solution of the one-dimensional differential wave equation is a linear combination of both solutions given by Equations 1.2 and 1.3.

Problem 1.10 Show explicitly that the function

$$\Psi(x, t) = A\sin(kx - \omega t + \phi) + B\sin(kx + \omega t + \phi)$$

solves the one-dimensional wave equation.

1.6 COMPLEX NUMBERS AND THE COMPLEX REPRESENTATION

Complex numbers can provide algebraic shortcuts that will prove very convenient as we continue our discussion of classical optics. In this section, we provide a quick overview of the properties of complex numbers, and a few of their algebraic features that we will find most useful.

Complex numbers include the concept of the *imaginary* number i:

$$i = \sqrt{-1} \tag{1.16}$$

Clearly, the square root of a negative number has no counterpart within the set of all *real* numbers. A complex number z has both a real part and an imaginary part:

$$z = x + iy \tag{1.17}$$

where x is the *real part* of z and y is the *imaginary part* of z. Complex numbers may be visualized in the *complex plane*, as shown in Figure 1.5.

In the *Cartesian representation*, a complex number is plotted with coordinates (x, y); thus the horizontal axis is called the *real axis*, and the vertical axis is called the *imaginary axis*.

Many of the features of complex numbers that we will find most useful result from the *Euler relation*:

$$e^{i\theta} = \cos\theta + i\sin\theta \tag{1.18}$$

See Example 1.3 for a derivation of this important formula. We may use the Euler relation to express any complex number z in *polar form*. In Figure 1.5, let

$$x = r\cos\theta \tag{1.19}$$

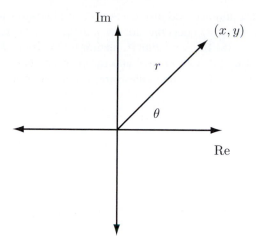

Figure 1.5   The complex plane.

$$y = r\sin\theta \tag{1.20}$$

where

$$r = \sqrt{x^2 + y^2} \tag{1.21}$$

$$\theta = \tan^{-1}\left(\frac{y}{x}\right) \tag{1.22}$$

Using these relations, we may represent z as

$$z = x + iy = r\cos\theta + i\left(r\sin\theta\right) \tag{1.23}$$

Thus, according to Equation 1.18,

$$z = r\,e^{i\theta} \tag{1.24}$$

In the polar representation, r is called the *magnitude* of z, and θ is called the *phase* of z. The following examples are often useful:

$$e^{\pm i(2\pi)} = 1; \quad e^{\pm i(\pi)} = -1; \quad e^{i\frac{\pi}{2}} = i; \quad e^{i\frac{3\pi}{2}} = -i \tag{1.25}$$

The *complex conjugate* z^* of a complex number z is obtained by inverting the sign on each occurrence of i. For example, in the Cartesian representation where $z = x + iy$, z^* is given by

$$z^* = x - iy \tag{1.26}$$

In the polar representation where $z = r\,e^{i\theta}$, z^* is given by

$$z^* = r\,e^{-i\theta} \tag{1.27}$$

1.6.1 Complex Algebra

Let $z_1 = x_1 + i\,y_1 = r_1\,e^{i\theta_1}$ and $z_2 = x_2 + i\,y_2 = r_2\,e^{i\theta_2}$ with r_1, r_2, θ_1 and θ_2 defined by Equations 1.21 and 1.22. Addition of complex numbers is easiest in the Cartesian representation:

$$z_1 + z_2 = (x_1 + iy_1) + (x_2 + iy_2) = x_1 + x_2 + i\left(y_1 + y_2\right) \tag{1.28}$$

Notice that complex numbers add just like vectors in a two-dimensional real space. To add or subtract two complex numbers that are given in the polar representation, first convert to Cartesian form using the Euler relation (Equation 1.23).

Let $\mathrm{Re}[z]$ and $\mathrm{Im}[z]$ denote the real and imaginary parts of z. In the Cartesian form, $\mathrm{Re}[z] = x$ and $\mathrm{Im}[z] = y$. In the polar form, $\mathrm{Re}[z] = r\cos\theta$ and $\mathrm{Im}[z] = r\sin\theta$. In either case,

$$\mathrm{Re}[z] = \frac{z + z^*}{2} \tag{1.29}$$

$$\mathrm{Im}[z] = \frac{z - z^*}{2i} \tag{1.30}$$

According to the Euler relation,

$$\cos\theta = \frac{e^{i\theta} + e^{-i\theta}}{2} \tag{1.31}$$

$$\sin\theta = \frac{e^{i\theta} - e^{-i\theta}}{2i} \tag{1.32}$$

Let λ be any real number. Scalar multiplication is defined by

$$\lambda z = \lambda(x + iy) = \lambda x + i(\lambda y) \tag{1.33}$$

To subtract two complex numbers, let $\lambda = -1$:

$$z_1 - z_2 = (x_1 + iy_1) - (x_2 + iy_2) = x_1 - x_2 + i(y_1 - y_2) \tag{1.34}$$

Complex multiplication is straightforward in either representation. Using the Cartesian representation gives

$$z_1 z_2 = (x_1 + iy_1)(x_2 + iy_2) = x_1 x_2 - y_1 y_2 + i(x_1 y_2 + x_2 y_1) \tag{1.35}$$

where we used the fact that $i^2 = -1$. In the polar representation,

$$z_1 z_2 = \left(r_1 e^{i\theta_1}\right)\left(r_2 e^{i\theta_2}\right) = r_1 r_2 e^{i(\theta_1 + \theta_2)} \tag{1.36}$$

Multiplication of a complex number by its complex conjugate gives the square of the magnitude. In the Cartesian form,

$$z\, z^* = (x + iy)(x - iy) = x^2 + y^2 \tag{1.37}$$

and in the polar form,

$$z\, z^* = \left(r\, e^{i\theta}\right)\left(r\, e^{-i\theta}\right) = r^2 \tag{1.38}$$

which, according to Equation 1.21, agrees with the previous result. We will also refer to the magnitude of z as $|z|$:

$$z\, z^* = |z|^2$$

Division is straightforward in the polar representation:

$$\frac{z_1}{z_2} = \frac{r_1 e^{i\theta_1}}{r_2 e^{i\theta_2}} = \frac{r_1}{r_2} e^{i(\theta_1 - \theta_2)} \tag{1.39}$$

In the Cartesian representation, division is a bit more involved if we wish to express the result in Cartesian form:

$$
\begin{aligned}
\frac{z_1}{z_2} = \frac{x_1 + iy_1}{x_2 + iy_2} &= \left(\frac{x_1 + iy_1}{x_2 + iy_2} \right) \left(\frac{x_2 - iy_2}{x_2 - iy_2} \right) \\
&= \left(\frac{x_1 x_2 - y_1 y_2}{x_2^2 + y_2^2} \right) + i \left(\frac{x_2 y_1 - x_1 y_2}{x_2^2 + y_2^2} \right)
\end{aligned} \tag{1.40}
$$

This can also be written as

$$
\frac{z_1}{z_2} = \frac{z_1 \, z_2^*}{z_2 \, z_2^*} = \frac{z_1 z_2^*}{|z_2|^2} \tag{1.41}
$$

■ **EXAMPLE 1.3**

Show that

$$
e^{i\theta} = \cos\theta + i\sin\theta
$$

Solution

We will demonstrate this relation using a *Taylor series*[2]:

$$
f(x) = f(x_0) + \frac{df}{dx}\bigg|_{x_0} (x - x_0) + \frac{1}{2!} \frac{d^2 f}{dx^2}\bigg|_{x_0} (x - x_0)^2 + \frac{1}{3!} \frac{d^3 f}{dx^3}\bigg|_{x_0} (x - x_0)^3 + \dots
$$

where x_0 is the *expansion point*. The accuracy of this expansion depends on how close x_0 is to the *evaluation point* x.

Let us expand the function $f(x) = \sin x$ in a Taylor series about the point $x_0 = 0$. Begin by computing each of the indicated derivatives evaluated at the expansion point:

$$
\frac{df}{dx} = \frac{d}{dx} \sin x = \cos x \qquad \frac{df}{dx}\bigg|_{x_0 = 0} = 1
$$

Similarly,

$$
\frac{d^2 f}{dx^2}\bigg|_0 = 0 \qquad \frac{d^3 f}{dx^3}\bigg|_0 = -1 \qquad \frac{d^4 f}{dx^4}\bigg|_0 = 0 \qquad \frac{d^3 f}{dx^3}\bigg|_0 = 1
$$

and so on. Substitution into the Taylor series formula gives

$$
\sin x = x - \frac{x^3}{3!} + \frac{x^5}{5!} - \frac{x^7}{7!} + \dots
$$

In a similar way, one finds

$$
\cos x = 1 - \frac{x^2}{2!} + \frac{x^4}{4!} - \frac{x^6}{6!} + \dots
$$

$$
e^x = 1 + x + \frac{x^2}{2!} + \frac{x^3}{3!} + \frac{x^4}{4!} + \dots
$$

---

[2] See, for example, Thomas et. al.[13].

as the reader should verify. Substitution of $x = i\theta$ with θ measured in radians gives

$$e^{i\theta} = 1 + (i\theta) + \frac{(i\theta)^2}{2!} + \frac{(i\theta)^3}{3!} + \frac{(i\theta)^4}{4!} + \frac{(i\theta)^5}{5!} + \frac{(i\theta)^6}{6!} + \frac{(i\theta)^7}{7!} + \ldots$$

$$= 1 + i\theta + \frac{i^2\theta^2}{2!} + \frac{i^3\theta^3}{3!} + \frac{i^4\theta^4}{4!} + \frac{i^5\theta^5}{5!} + \frac{i^6\theta^6}{6!} + \frac{i^7\theta^7}{7!} + \ldots$$

$$= \left(1 - \frac{\theta^2}{2!} + \frac{\theta^4}{4!} - \frac{\theta^6}{6!} + \ldots \right) + i \left(\theta - \frac{\theta^3}{3!} + \frac{\theta^5}{5!} - \frac{\theta^7}{7!} + \ldots \right)$$

$$= \cos\theta + i\sin\theta$$

Taking the real part or the imaginary part of $e^{i\theta}$ gives the corresponding harmonic function.

1.6.2   The Complex Representation of Harmonic Waves

We will often find it useful to represent harmonic traveling waves in the *complex representation*

$$\Psi(x,t) = Ae^{i(kx \mp \omega t + \theta)} \tag{1.42}$$

According to the Euler relation, this is equal to

$$\Psi(x,t) = Ae^{i(kx \mp \omega t + \theta)} = A\cos(kx \mp \omega t + \theta) + iA\sin(kx \mp \omega t + \theta) \tag{1.43}$$

The rules of complex algebra just outlined will often simplify the trigonometry of harmonic traveling waves. For classical traveling waves, the complex representation is not necessary, but only a convenience. Techniques for using this representation will be illustrated as we proceed.

Problem 1.11   Show the *small angle approximation*: for small θ, $\sin\theta \approx \theta$. Check the approximation for $\theta = 5°$, $10°$, and $20°$.

Problem 1.12   Show that $\cos x = 1 - \frac{x^2}{2!} + \frac{x^4}{4!} - \frac{x^6}{6!} + \ldots$.

Problem 1.13   Show that $e^x = 1 + x + \frac{x^2}{2!} + \frac{x^3}{3!} + \frac{x^4}{4!} + \ldots$.

Problem 1.14   Express the following numbers in polar form: $4 + 5i$, $-4 - 5i$, $4 - 5i$, $-4 + 5i$. In each case, graph the number on the complex plane.

Problem 1.15   Express the following numbers in polar form: $3 + 5i$, $-2 - 6i$, $5 - 4i$, $-3 + 8i$. In each case, graph the number on the complex plane.

Problem 1.16   Express the following in Cartesian form: $(3+5i)/(4-7i)$, $(-3+6i)/(3+9i)$.

Problem 1.17   Let $z_1 = 10\,e^{0.5\,i}$ and $z_2 = 20\,e^{-0.8\,i}$. Find the real part, imaginary part, and polar form of

    a) $z_1 + z_2$

b) $z_1 - z_2$

c) $z_1 z_2$

d) z_1/z_2

Problem 1.18   For each of the following complex numbers, find the real part and the imaginary part, and express the number in polar form.

    a) $z = (4 + 5i)^2$

    b) $z = 5(1 + i)e^{i\pi/6}$

1.7   THE THREE-DIMENSIONAL WAVE EQUATION

In Cartesian coordinates, the extension of Equation 1.1 to include three dimensions is made in the obvious way:

$$\frac{\partial^2 \Psi}{\partial x^2} + \frac{\partial^2 \Psi}{\partial y^2} + \frac{\partial^2 \Psi}{\partial z^2} = \frac{1}{v^2} \frac{\partial^2 \Psi}{\partial t^2} \tag{1.44}$$

where $\Psi = \Psi(x, y, z, t)$ is now understood to be a function of time and all three spatial coordinates.

Equation 1.44 can be written in a more general form using the Laplacian operator:

$$\nabla^2 \Psi = \frac{1}{v^2} \frac{\partial^2 \Psi}{\partial t^2} \tag{1.45}$$

where, in Cartesian coordinates, the *Laplacian* is given by

$$\nabla^2 = \frac{\partial^2}{\partial x^2} + \frac{\partial^2}{\partial y^2} + \frac{\partial^2}{\partial z^2} \tag{1.46}$$

Equation 1.45 is the *coordinate independent* representation of the three-dimensional wave equation. The wave equation in curvilinear coordinate systems is obtained by using the explicit form of the Laplacian for that system of coordinates.

In three dimensions, it is important to use a *right-handed* Cartesian coordinate system. According to the *right-hand rule*, $\hat{i} \times \hat{j} = \hat{k}$ where $\hat{i}$, $\hat{j}$, and $\hat{k}$ are the unit normal vectors along the x, y, and z axes[3]. A right-handed Cartesian coordinate system is illustrated in Figure 1.6.

Spherical Coordinates

Figure 1.6 illustrates the spherical coordinate system, where the three spatial coordinates are $(r, \theta, \phi)^4$. The coordinate r is the magnitude of the position vector $\vec{r}$, θ measures the angle between $\vec{r}$ and the positive z-axis, and ϕ measures the angle between the positive x-axis and the projection of $\vec{r}$ onto the x-y plane. The transformation equations between

---

[3]When used with vector quantities, the symbol $\times$ denotes a *vector cross product*. For details, see any introductory physics text (e.g. Halliday et.al. [11]).

[4]Mathematicians typically interchange the roles of θ and ϕ in spherical coordinates. The convention adopted here is more common in physics.

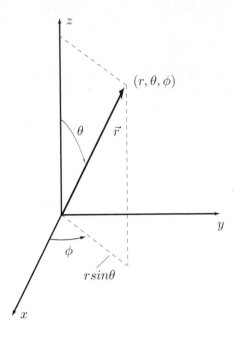

Figure 1.6 Spherical coordinates illustrated within a right-handed Cartesian coordinate system.

spherical and Cartesian coordinates are

$$x = r \sin \theta \cos \phi \qquad\qquad r = \sqrt{x^2 + y^2 + z^2}$$

$$y = r \sin \theta \sin \phi \qquad\qquad \theta = \tan^{-1}\left(\frac{x^2 + y^2}{z}\right) \qquad (1.47)$$

$$z = r \cos \theta \qquad\qquad \phi = \tan^{-1}\left(\frac{y}{x}\right)$$

The volume element is obtained by incrementing each coordinate

$$d\tau = (dr)\,(r\,d\theta)\,(r \sin \theta\, d\phi) = r^2 \sin \theta\, dr\, d\theta\, d\phi \qquad (1.48)$$

In spherical coordinates, the Laplacian is given by[5]

$$\nabla^2 = \frac{1}{r^2}\frac{\partial}{\partial r}\left(r^2 \frac{\partial}{\partial r}\right) + \frac{1}{r^2 \sin \theta}\frac{\partial}{\partial \theta}\left(\sin \theta \frac{\partial}{\partial \theta}\right) + \frac{1}{r^2 \sin^2 \theta}\frac{\partial^2}{\partial \phi^2} \qquad (1.49)$$

Substitution of the Laplacians given by Equation 1.46 or 1.49 into the coordinate independent version of the differential wave equation given by Equation 1.45 yield very different looking but entirely equivalent differential equations. The choice of which coordinate system to use will typically be dictated by the *symmetry* of the problem. Problems with rectangular symmetry are typically easier to solve in Cartesian coordinates, while problems with spherical symmetry are more easily solved in the spherical coordinate system. A traveling wave solution that satisfies the differential wave equation in any particular coordinate system will always solve the wave equation when expressed in any other coordinate system.

[5] See, for example, Boas[1].

As a final note, it is customary to use the notation $\Psi(\vec{r}, t)$ to represent a wavefunction in an arbitrary coordinate system. Thus, in Cartesian coordinates, $\Psi(\vec{r}, t)$ is interpreted as $\Psi(x, y, z, t)$, while in spherical coordinates, $\Psi(\vec{r}, t)$ represents $\Psi(r, \theta, \phi, t)$. Using this notation, the coordinate independent differential wave equation of Equation 1.45 becomes

$$\nabla^2 \Psi(\vec{r}, t) = \frac{1}{v^2} \frac{\partial^2 \Psi(\vec{r}, t)}{\partial t^2} \tag{1.50}$$

1.7.1 Three-Dimensional Plane Waves

Consider an acoustic wave traveling through air. Pressure is a property of space, and the pressure variations that define the acoustic wave are defined throughout a region. In a three-dimensional *plane wave*, the properties of the medium are constant over any *plane* oriented normal to the direction of propagation. To describe such a wave, we define the propagation vector $\vec{k}$ with magnitude $\frac{2\pi}{\lambda}$ and direction given by the wave propagation. The corresponding plane wave is given by

$$\Psi(x, y, z, t) = A \sin\left(\vec{k} \cdot \vec{r} - \omega t + \varphi\right) \tag{1.51}$$

A plane has Cartesian symmetry[6], and in Cartesian coordinates, $\vec{k}$ is given by

$$\vec{k} = k_x \hat{i} + k_y \hat{j} + k_z \hat{k} \tag{1.52}$$

The position vector in Cartesian coordinates is given by

$$\vec{r} = x\hat{i} + y\hat{j} + z\hat{k} \tag{1.53}$$

For example, in a plane harmonic sound wave traveling in the positive x-direction, the value of the air pressure is constant over any plane that is parallel to the y-z plane:

$$\Psi(x, y, z, t) = A \sin(kx - \omega t + \varphi) \tag{1.54}$$

This expression only *looks* one-dimensional; the value of $A \sin(kx - \omega t + \varphi)$ at any point on the x-axis determines the value of $\Psi(x, y, z, t)$ at all points on a plane parallel to the y-z plane located at x.

In Figure 1.7 it is seen that $\vec{k}$ is normal to planes defined by $\vec{k} \cdot \vec{r} = const$. It is convenient to define *wavefronts* located at points where the phase of Equation 1.51 is equal to integer multiples of 2π. Thus, a three-dimensional plane wave may be visualized as a train of wavefronts separated by one wavelength λ and moving with the wave speed v.

In the complex representation, harmonic plane waves are given by

$$\Psi(x, y, z, t) = A e^{i\left(\vec{k} \cdot \vec{r} - \omega t + \varphi\right)} \tag{1.55}$$

It is instructive to show explicitly that the plane wave in Equation 1.55 solves the differential wave equation. Plane waves have rectangular symmetry, so we will use Cartesian coordinates:

$$\nabla^2 \Psi = \frac{\partial^2 \Psi}{\partial x^2} + \frac{\partial^2 \Psi}{\partial y^2} + \frac{\partial^2 \Psi}{\partial z^2} = \frac{1}{v^2} \frac{\partial^2 \Psi}{\partial t^2} \tag{1.56}$$

---

[6]An object with Cartesian symmetry can be oriented so that it is unaffected by reflections about x, y, or z axes. It is always simpler to use a coordinates system with the same symmetry as the system being studied.

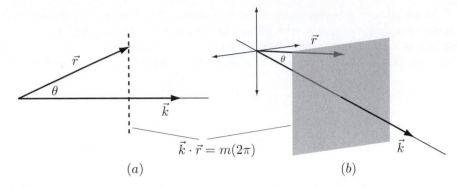

Figure 1.7 (a)—(b): A wavefront defined by $\vec{k} \cdot \vec{r} = m(2\pi)$, where m is an integer. Vectors $\vec{k}$ and $\vec{r}$ do not have the same units.

Using Equations 1.52 and 1.53 for $\vec{k}$ and $\vec{r}$, we find that

$$\Psi(x, y, z, t) = Ae^{i\left(\vec{k}\cdot\vec{r}-\omega t+\varphi\right)} = Ae^{i(k_x x + k_y y + k_z z - \omega t + \varphi)}$$

Thus,

$$\frac{\partial \Psi}{\partial x} = ik_x \Psi$$

$$\frac{\partial^2 \Psi}{\partial x^2} = -k_x^2 \Psi$$

with similar expressions for derivatives with respect to y and z. Thus

$$\nabla^2 \Psi = -\left(k_x^2 + k_y^2 + k_z^2\right)\Psi = -k^2\Psi$$

The time derivatives are given by

$$\frac{\partial \Psi}{\partial t} = -i\omega\Psi$$

$$\frac{\partial^2 \Psi}{\partial t^2} = -\omega^2\Psi$$

Substitution into the wave equation gives

$$\nabla^2\Psi = \frac{1}{v^2}\frac{\partial^2\Psi}{\partial t^2} \Rightarrow -k^2\Psi = \frac{1}{v^2}\left(-\omega^2\right)\Psi$$

This will be true provided

$$v = \frac{\omega}{k}$$

in agreement with Equation 1.14.

1.7.2 Spherical Waves

Spherical waves are waves with wavefronts that are spherical in shape. Examples include waves that emanate from an isotropic point source, or are converging to a point. An isotropic source emits waves symmetrically in all directions.

In spherical coordinates, spherical waves have no dependence on the angular coordinates θ and ϕ, so the derivatives with respect to these coordinates are zero. In the notation introduced by Equation 1.50, this means that

$$\Psi(\vec{r}, t) = \Psi(r, t)$$

In other words, $\Psi(r, t)$ depends only on the spherical coordinate r and the time t. Derivatives of such a function with respect to θ and ϕ give zero, leading to a simplified version of the Laplacian for spherical coordinates (Equation 1.49):

$$\nabla^2 = \frac{1}{r^2} \frac{\partial}{\partial r} \left(r^2 \frac{\partial}{\partial r} \right) \tag{1.57}$$

Using this Laplacian, we obtain the *spherically symmetric differential wave equation*:

$$\frac{1}{r^2} \frac{\partial}{\partial r} \left(r^2 \frac{\partial \Psi(r, t)}{\partial r} \right) = \frac{1}{v^2} \frac{\partial^2 \Psi(r, t)}{\partial t^2} \tag{1.58}$$

An important solution to this equation is the *harmonic spherical wave*

$$\Psi(r, t) = \frac{A}{r} \sin (kr \mp \omega t + \varphi) \tag{1.59}$$

If the minus sign is chosen, the wave emanates from an isotropic source located at $r = 0$. A plus sign represents a wave that is converging to the point $r = 0$. The amplitude $\frac{A}{r}$ of the wave is not constant, since it depends on the radial coordinate r. For a wave emanating from a point, the amplitude decreases as the wave travels. Waves converging to a point have amplitudes that increase as time increases.

In the complex representation, Equation 1.59 becomes

$$\psi(r, t) = \frac{A}{r} e^{i(kr \mp \omega t + \varphi)} \tag{1.60}$$

■ **EXAMPLE 1.4**

Show that Equation 1.59 solves the spherically symmetric differential wave equation.

Solution

We must show that Equation 1.58

$$\frac{1}{r^2} \frac{\partial}{\partial r} \left(r^2 \frac{\partial \Psi(r, t)}{\partial r} \right) = \frac{1}{v^2} \frac{\partial^2 \Psi(r, t)}{\partial t^2}$$

is solved by Equation 1.59

$$\Psi(r, t) = \frac{A}{r} \sin (kr \mp \omega t + \varphi)$$

Begin by taking the derivative of $\Psi(r, t)$ with respect to r:

$$\frac{\partial \Psi}{\partial r} = -\frac{A}{r^2} \sin (kr \mp \omega t + \varphi) + \frac{A}{r} (k) \cos (kr \mp \omega t + \varphi)$$

To evaluate the left-hand side of the wave equation, multiply this last result by r^2, take another derivative with respect to r, then divide by r^2:

$$\frac{1}{r^2}\frac{\partial}{\partial r}\left(r^2\frac{\partial\Psi(r,t)}{\partial r}\right) = \frac{1}{r^2}\frac{\partial}{\partial r}\left[r^2\left(-\frac{A}{r^2}\sin\left(kr\mp\omega t+\varphi\right)+\frac{A}{r}(k)\cos\left(kr\mp\omega t+\varphi\right)\right)\right]$$

$$= \frac{1}{r^2}\frac{\partial}{\partial r}\left[-A\sin\left(kr\mp\omega t+\varphi\right)+rAk\cos\left(kr\mp\omega t+\varphi\right)\right]$$

$$= \frac{1}{r^2}\left[-rAk^2\sin\left(kr\mp\omega t+\varphi\right)\right]$$

$$= -k^2\left[\frac{A}{r}\sin\left(kr\mp\omega t+\varphi\right)\right]$$

$$= -k^2\Psi$$

The time derivatives on the right-hand side of the wave equation are given by

$$\frac{\partial\Psi}{\partial t} = \frac{\partial}{\partial t}\left(\frac{A}{r}\sin\left(kr\mp\omega t+\varphi\right)\right) = \mp\omega\frac{A}{r}\cos\left(kr\mp\omega t+\varphi\right)$$

$$\frac{\partial^2\Psi}{\partial t^2} = -\omega^2\Psi$$

Substitution into the wave equation gives

$$-k^2\Psi = \frac{1}{v^2}\left(-\omega^2\Psi\right)$$

which is satisfied since $v = \frac{\omega}{k}$.

Problem 1.19   Show explicitly that the following functions are solutions to the spherically symmetric wave equation.

$$(a)\quad \Psi(r,t) = \frac{A}{r}\cos\left(kr\mp\omega t+\varphi\right)$$

$$(b)\quad \Psi(r,t) = \frac{A}{r}e^{i((kr\mp\omega t+\varphi)}$$

Problem 1.20   Find the equation for a plane electromagnetic wave whose propagation vector $\vec{k}$ is parallel to a line in the x-z plane that is 30° measured counter clockwise from the positive x-axis. Assume that $\vec{k}$ lies in the first quadrant of the x-z plane.

Additional Problems

Problem 1.21   Determine the direction of propagation of the following harmonic traveling waves:

a) $\Psi(z, t) = A\sin(kz - \omega t)$
b) $\Psi(y, t) = A\cos(\omega t - ky)$
c) $\Psi(x, t) = A\cos(\omega t + kx)$
d) $\Psi(x, t) = A\cos(-\omega t - kx)$

Problem 1.22   Show that the *Gaussian* wave $\Psi(x, t) = Ae^{-a(bx - ct)^2}$ is a solution to the one-dimensional wave equation. If $a = 5.00$, $b = 10.0$, $c = 100$, with $a(bx - ct)^2$ unitless, determine the wave speed.

Problem 1.23   Sketch or plot the following wavefunction at times $t = 0$, $t = 0.5$ s, and $t = 1.0$ s:

$$\Psi(x, t) = \frac{1.0}{1 + (x + 10t)^2}$$

Problem 1.24   A 1-D harmonic traveling wave that travels in the $-y$ direction has amplitude 10 (unitless), wavelength $10.0\,m$, period $2.0\,s$ and initial phase π. Using the complex representation, find an expression for this wave that uses angular frequency and propagation constant.

Problem 1.25   Light from a helium-neon laser has a wavelength of $633\,nm$ and a wave speed of $3.00 \times 10^8\,m/s$. Find the frequency, period, angular frequency, and wave number for this light.

Problem 1.26   Consider a harmonic wave given by

$$\Psi(x, t) = U(x, y, z)e^{-i\omega t}$$

where $U(x, y, z)$ is called the *complex amplitude*. Show that U satisfies the *Helmholz equation*:

$$\left(\nabla^2 + k^2\right) U(x, y, z) = 0$$

where k is the propagation constant.

Problem 1.27   Show that a complex number divided by its complex conjugate gives a result whose magnitude is one.

Problem 1.28   Show that $z = \frac{\sqrt{2}}{2}(1 + i)$ is a square root of i. Find another one.

Problem 1.29   Find the real and imaginary parts of

a) $z = \left(2e^{i\frac{\pi}{4}}\right)^3$
b) $z = (2 + 3i)^3$

Problem 1.30   Hyperbolic sines and cosines are defined as follows: $\sinh x = \frac{e^x - e^{-x}}{2}$ and $\cosh x = \frac{e^x + e^{-x}}{2}$. Show that $\sin(ix) = i\sinh(x)$ and $\cos(ix) = \cosh(x)$.

Problem 1.31   Find the Taylor series expansions for hyperbolic sine and hyperbolic cosine.

Problem 1.32   Consider a vector $\vec{v}$, and let a be the angle between $\vec{v}$ and the positive x-axis, b be the angle between $\vec{v}$ and the positive y-axis, and c be the angle between $\vec{v}$ and the positive z-axis. Define the *direction cosines* α, β, and γ as follows:

$$\alpha = \cos a = \frac{v_x}{|v|}$$

$$\beta = \cos b = \frac{v_y}{|v|}$$

$$\gamma = \cos c = \frac{v_z}{|v|}$$

a) Show that $\alpha^2 + \beta^2 + \gamma^2 = 1$.

b) Show that the function

$$\Psi(x, y, z, t) = A e^{i[k(\alpha x + \beta y + \gamma z) \mp \omega t]}$$

is a three-dimensional plane wave that solves the differential wave equation in Cartesian coordinates.

Problem 1.33   The Laplacian in cylindrical coordinates (ρ, φ, z) is given by

$$\nabla^2 = \frac{1}{\rho}\frac{\partial}{\partial \rho} + \frac{\partial^2}{\partial \rho^2} + \frac{1}{\rho^2}\frac{\partial^2}{\partial \varphi^2} + \frac{\partial^2}{\partial z^2}$$

A *cylindrical wave* has a wavefront that is constant on a cylinder. In other words, it does not depend upon φ or z. Show that

$$\Psi = \frac{A}{\sqrt{\rho}} e^{i(k\rho \mp \omega t)}$$

approximately solves the differential wave equation in clylindrical coordinates for values of ρ that are sufficiently large.

Problem 1.34   Show that the spherically symmetric wave equation can be written as

$$\frac{\partial^2}{\partial r^2}\left[r\Psi(r, t)\right] = \frac{1}{v^2}\frac{\partial^2}{\partial t^2}\left[r\Psi(r, t)\right]$$

which is a *linear* wave equation in the quantity $r\Psi(r, t)$. Show that Equations 1.59 and 1.60 are solutions.

Problem 1.35   *Photons* are particles of light with energy and momentum given by

$$E = hf$$

$$p = \frac{h}{\lambda}$$

where f is the light frequency, λ is the light wavelength, and h is *Planck's constant*: $h = 6.626 \times 10^{-34}\, J \cdot s$. Show that for photons, $E = cp$.

CHAPTER 2

ELECTROMAGNETIC WAVES AND PHOTONS

When I am judging a theory, I ask myself whether, if I were God, I would have arranged the world in such a way.

—Einstein

Now, let's look at the first basic action — a photon goes from [A to B]. I will draw this action as a wiggly line from A to B for no good reason.

—Feynman

Contents

2.1 INTRODUCTION

In this chapter, we attempt to answer the question: "What is light?" Interestingly, there seems to be more than one answer. The bedrock of our description will be the theory of electromagnetism, as summarized by Maxwell's equations. From this perspective, light is most certainly a wave — a transverse electromagnetic wave — with properties that we will obtain from Maxwell's equations. We will use the "wave picture" of electromagnetic radiation extensively in subsequent chapters to develop many optical concepts and applications. Examples include effects due to interference and diffraction, coherence, optical imaging and resolution, and many of the concepts relating to laser design.

However, it will often be necessary to utilize the concept of a light *particle*, or *photon*. Thus, electromagnetic radiation presents *complementary* aspects that are determined by the type of observation being made. From a fundamental point of view, the particle-wave dualism of electromagnetic radiation (and, as it turns out, of matter as well) offers many fascinating mysteries that still await resolution. From a practical perspective, both pictures will be useful as we develop the many aspects of modern optical technology discussed within this text.

2.2   ELECTROMAGNETISM

All known observations of classical electromagnetism can be explained with the set of equations collectively known as *Maxwell's equations*[1]. These equations are expressed using Faraday's concept of *electric and magnetic fields* that ultimately become the framework for our description of electromagnetic waves[2].

The *electric field* $\vec{E}$ at a point is defined as the force per unit charge experienced by a *small positive test charge* placed at that point. Thus, any real charge q will experience a force $\vec{F} = q\vec{E}$ when placed within an electric field $\vec{E}$. The *magnetic field* $\vec{B}$ is also defined in terms of a force; however, the charge q must be moving in order for a magnetic force to act: $\vec{F} = q\,\vec{v} \times \vec{B}$. The combination of the electric and magnetic force is called the *Lorentz force*:

$$\vec{F} = q\vec{E} + q\,\vec{v} \times \vec{B} \tag{2.1}$$

The Lorentz force can be taken as the definition of electric and magnetic fields.

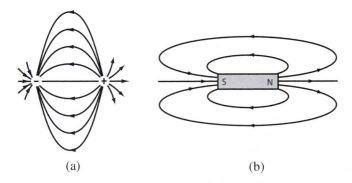

(a)   (b)

Figure 2.1   (a) Field lines for an electric dipole. (b) Field lines for a magnetic dipole.

Electric and magnetic *field lines* provide an important and intuitive visual aid. Field lines are drawn so that the tangent to the curve passing through a given point gives the field

---

[1] James Clerk Maxwell: 1831-1879. Scottish mathematician and physicist.
[2] This section constitutes a concise review of electromagnetism. For more detail, see any calculus-based introductory physics text (e.g. Halliday, Resnick, and Walker [11]).

direction at that point. Electric field lines begin on positive charges and end on negative charges, as shown in Figure 2.1(a). Magnetic field lines are produced by currents and bar magnets (apparently, there is no magnetic charge) as shown in Figure 2.1(b). In a solenoid or bar magnet, magnetic field lines enter into south poles and emerge out of north poles. Drawing a finite number of field lines allows one to visualize the field strength: regions where the field lines are more closely spaced indicate where the field magnitude is larger.

Maxwell's equations consist of four physical laws, each of which is summarized below. Together, they provide a deep and fundamental explanation of all electromagnetic phenomena, including electromagnetic waves.

Gauss's Law for Electric Fields: *the outward electric flux integrated over a closed surface is proportional to the net electric charge enclosed by the surface. Electric flux* through a *closed surface* is a mathematical concept that can be intuitively visualized with electric field lines. *Outward electric flux* is defined so that field lines *leaving* the closed surface contribute positively while field lines entering the closed surface contribute negatively. The contribution to the total electric flux over an infinitesimally small area is $d\Phi_E = \vec{E} \cdot \hat{n} \, dA$, where $\hat{n}$ is the *outward unit normal* to the surface at the position of dA. It is customary to define $d\vec{A} \equiv \hat{n} \, dA$. If the net charge enclosed by the surface is nonzero, there will be an imbalance between sources and sinks of electric field lines, and the net flux will be nonzero. According to Gauss,

$$\Phi_E = \oint_A \vec{E} \cdot d\vec{A} = \frac{q_{net}}{\epsilon_0} = \frac{1}{\epsilon_0} \int_V \rho \, dV \qquad (2.2)$$

where A is a closed surface, V is the volume enclosed by A, q_{net} is the algebraic sum of all charges contained in V, ρ is the charge per unit volume (charge distribution) within V, and ϵ_0 is the *permittivity of free space*

$$\epsilon_0 = 8.854 \times 10^{-12} \frac{C^2}{Nm^2} \qquad (2.3)$$

The term q_{net} in Gauss's law includes both *free* and *bound* charges. Free charges can consist of charge carriers placed onto or removed from conductors and insulators, or ions and electrons within a region of space. Bound charge distributions result from the molecular alignment of *dipole moments* (induced or permanent) within a *dielectric material*.

Effects related to dielectric materials may be included using the *relative permittivity*, also referred to as the *dielectric constant* K_E:

$$\epsilon = K_E \, \epsilon_0 \qquad (2.4)$$

An external electric field causes alignment of molecular dipole moments resulting in bound charge densities that tend to cancel the aligning field. A *linear isotropic homogeneous dielectric* is one for which K_E is a constant of the space coordinates, and whose value does not depend on direction[3]. Effects due to linear isotropic dielectrics (and their associated bound charge densities) are particularly easy to include within Maxwell's equations: simply replace every occurrence of ϵ_0 with ϵ in the equations that involve free charge. For such materials, Gauss's law becomes

$$\Phi_E = \oint_A \vec{E} \cdot d\vec{A} = \frac{q_{free}}{\epsilon} = \frac{1}{\epsilon} \int_V \rho_{free} \, dV \qquad (2.5)$$

---

[3]For linear dielectrics, the bound charge density at any point is a linear function of the applied electric field.

where q_{free} and ρ_{free} account for free charge and free charge densities.

Gauss's[4] Law for Magnetic Fields: *the outward magnetic flux integrated over a closed surface is zero*. This is a statement of the interesting empirical fact that magnetic charges (*magnetic monopoles*) have never been observed. Thus,

$$\Phi_M = \oint_A \vec{B} \cdot d\vec{A} = 0 \tag{2.6}$$

Faraday's[5] Law of Induction: *a changing magnetic field induces an electric field*. For a curve C that bounds an area A, Faraday's law is stated as follows:

$$\oint_C \vec{E} \cdot d\vec{\ell} = -\frac{d}{dt} \int_A \vec{B} \cdot d\vec{A} \tag{2.7}$$

The minus sign on the right-hand side of Faraday's law is an expression of *Lenz's Law*, which states that currents flowing in response to the induced *emf*[6] produce magnetic fields oriented so as to *oppose the change* in magnetic flux.

The magnetic flux integral on the right-hand side of Equation 2.7 removes all explicit space dependence from the integrand. The time derivative in Faraday's law may be moved inside the integral sign as a partial derivative:

$$\oint_C \vec{E} \cdot d\vec{\ell} = -\int_A \frac{\partial \vec{B}}{\partial t} \cdot d\vec{A} \tag{2.8}$$

■ **EXAMPLE 2.1**

A cylindrical region of space of radius R contains a uniform magnetic field $\vec{B}$ with direction into the page, as shown in Figure 2.2. If the magnitude changes in time, describe the induced electric field for points inside the cylinder ($r < R$).

Solution

Since $\vec{B}$ points into the page, the scalar product $\vec{B} \cdot d\vec{A}$ evaluates to a product of scalars $B\,dA$. The magnitude of $\vec{B}$ changes in time, so the right-hand side of Equation 2.8 is nonzero. From the symmetry of the magnetic field region, we choose a circular path C, and note that the tangential component of the induced electric field $\vec{E}$ must be constant along C. However, Equation 2.8 must hold for an *arbitrary* path C, thus we conclude that the direction of the induced electric field $\vec{E}$ for this very symmetrical case is *azimuthal* (tangent to the circular path). We find the magnitude of $\vec{E}$ with Equation 2.8:

$$\oint_C \vec{E} \cdot d\vec{\ell} = \oint_C E\,d\ell = E \oint_C d\ell = E\,(2\pi r) = -\frac{d}{dt} \int_A \vec{B} \cdot d\vec{A} = -\pi\,r^2\,\frac{dB}{dt}$$

---

[4]Johann Carl Friedrich Gauss: 1777-1855. German mathematician and physicist.
[5]Michael Faraday: 1791-1867. English physicist who made many remarkable discoveries in electromagnetism.
[6]This is an acronym that stands for *electromotive force* . For historical reasons, it is used to describe the induced *voltage*.

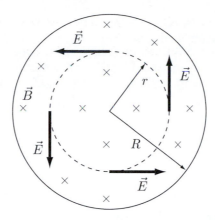

Figure 2.2   Induced azimuthal electric field $\vec{E}$ when magnetic field $\vec{B}$ points into the page and $\frac{dB}{dt} > 0$.

Thus,

$$E = -\frac{r}{2}\frac{dB}{dt}$$

The minus sign in the last expression results from Lenz's law; the magnitude of $\vec{E}$ is of course positive. If a circular wire loop of radius r is placed along the dashed integration path shown in Figure 2.2, the current induced in the loop must create a magnetic field tending to cancel the increase in $\vec{B}$. Since $\vec{B}$ points into the page and has increasing magnitude, the magnetic field induced by the current in the loop must point out of the page, so the induced current must flow *counter-clockwise*. Therefore, the induced $\vec{E}$ that causes this current is azimuthal in the counter-clockwise sense, as shown.

Note that the induced electric field $\vec{E}$ has a direction that is perpendicular to the changing magnetic field $\vec{B}$.

Ampere's[7] **Circuital Law**: *a changing electric field induces a magnetic field*. Ampere's original version of this law was restated by Maxwell to include a displacement current:

$$\oint_C \vec{B} \cdot d\vec{\ell} = \mu I + \mu \left(\int_A \epsilon \frac{\partial \vec{E}}{\partial t} \cdot d\vec{A} \right)$$

As in Faraday's law, C bounds A. The symbol I represents the flux of free charge through A:

$$\oint_C \vec{B} \cdot d\vec{\ell} = \mu \left(\int_A \vec{J} \cdot d\vec{A} \right) + \mu \left(\int_A \epsilon \frac{\partial \vec{E}}{\partial t} \cdot d\vec{A} \right)$$

For *linear, isotropic magnetic materials*, $\mu = K_M \mu_0$, where K_M is the *relative permeability* and μ_0 is the *permeability of free space*:

$$\mu_0 = 4\pi \times 10^{-7} \frac{Tm}{A} \tag{2.9}$$

[7] Andre-Marie Ampere: 1775-1836. French physicist.

For non-magnetic materials, $K_M = 1$. In this text, *we assume that all optical materials of interest are non-magnetic.* In this case, $K_M = 1$, $\mu = \mu_0$, and Ampere's law becomes

$$\oint_C \vec{B} \cdot d\vec{\ell} = \mu_0\, I + \mu_0 \left(\int_A \epsilon \frac{\partial \vec{E}}{\partial t} \cdot d\vec{A} \right) \tag{2.10}$$

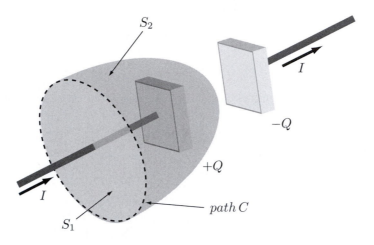

Figure 2.3   Ampere's law applied over two surfaces.

The *displacement current term* (in parentheses in Equation 2.10) is necessary from more than one perspective. Consider Figure 2.3, which shows a circular path C linked by a current I that charges a capacitor. The surfaces S_1 and S_2 are both bounded by the path C, but since S_2 extends between the plates of the capacitor, the current I passes through S_1 but not S_2. More importantly, Maxwell demanded that if a changing magnetic field induces an electric field as described by Faraday's law, then a changing electric field should induce a magnetic field. Thus, Maxwell proposed the *displacement current density* in the region between the plates of Figure 2.3:

$$\vec{J}_D = \epsilon \frac{\partial \vec{E}}{\partial t} \tag{2.11}$$

As the capacitor in Figure 2.3 charges, a magnetic field is induced between the plates by the displacement current density $\vec{J}_D$ that can be computed with Equation 2.10.

■ **EXAMPLE 2.2**

A cylindrical region of empty space of radius R contains a uniform electric field $\vec{E}$ with direction into the page, as shown in Figure 2.4. If the magnitude E changes in time, describe the induced magnetic field for points inside the cylinder $(r < R)$.

Solution

As in the previous example, the symmetry of the problem dictates an induced $\vec{B}$ with azimuthal direction and magnitude constant along a circle with the same center

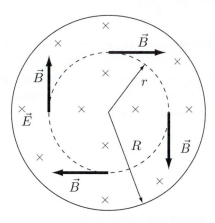

Figure 2.4   azimuthal induced magnetic field when $\vec{E}$ points into the page and $\frac{dE}{dt} > 0$

of symmetry as $\vec{E}$. The free current I is zero in this region, so Equation 2.10 gives

$$\oint_C \vec{B} \cdot d\vec{\ell} = \oint_C B \, d\ell = B \oint_C d\ell = B \, (2\pi r)$$

$$= \mu_0 \left(\int_A \epsilon_0 \frac{\partial \vec{E}}{\partial t} \cdot d\vec{A} \right) = \mu_0 \epsilon_0 \frac{d}{dt} \left(\int_A E \, dA \right)$$

$$= \mu_0 \epsilon_0 \, \pi \, r^2 \frac{dE}{dt}$$

Thus,

$$B = \mu_0 \epsilon_0 \frac{r}{2} \frac{dE}{dt}$$

The direction of $\vec{B}$ is given by the *right-hand rule*: point your thumb in the direction of $\frac{d\vec{E}}{dt}$, and your fingers curl in the azimuthal sense of $\vec{B}$. As in Example 2.1, the induced field is perpendicular to the changing field.

We will be primarily concerned with regions of space that contain no free charge distributions and no free currents. In this case, Maxwell's equations become:

$$\oint_A \vec{E} \cdot d\vec{A} = 0 \tag{2.12}$$

$$\oint_A \vec{B} \cdot d\vec{A} = 0 \tag{2.13}$$

$$\oint_C \vec{E} \cdot d\vec{\ell} = -\int_A \frac{\partial \vec{B}}{\partial t} \cdot d\vec{A} \tag{2.14}$$

$$\oint_C \vec{B} \cdot d\vec{\ell} = \mu_0 \epsilon \int_A \frac{\partial \vec{E}}{\partial t} \cdot d\vec{A} \tag{2.15}$$

Problem 2.1   Find the magnitude and direction of the induced electric field of Example 2.1 at $r = 5.00\,cm$ if the magnetic field changes at a constant rate from $0.500\,T$ to zero in $0.100\,s$. Assume that the cylindrical region has a radius of $10.0\,cm$.

Problem 2.2   A cylindrical region of space of radius R contains a uniform magnetic field $\vec{B}$ with direction into the page, as shown in Figure 2.2. If the magnitude B inside the cylinder changes in time and outside the cylinder it is zero, describe the induced electric field (magnitude and direction) for points outside the cylinder $(r > R)$. Find the magnitude and direction of the induced electric field at $r = 15.00\,cm$ if $R = 10.0\,cm$ and the magnetic field changes at a constant rate from zero to $0.500\,T$ in $0.100\,s$.

Problem 2.3   Find the magnitude and direction of the induced magnetic field of Example 2.2 at $r = 5.00\,cm$ if the electric field changes at a constant rate from $5000\,V/m$ to zero in $0.100\,s$. Assume that the cylindrical region has a radius of $10.0\,cm$.

Problem 2.4   A cylindrical region of empty space of radius R contains a uniform electric field $\vec{E}$ with direction into the page, as shown in Figure 2.4. If the magnitude E inside the cylinder changes in time and outside the cylinder it is zero, describe the induced magnetic field (magnitude and direction) for points outside the cylinder $(r > R)$. Find the magnitude and direction of the induced magnetic field at $r = 15.00\,cm$ if $R = 10.0\,cm$ and the electric field changes at a constant rate from zero to $5000\,V/m$ in $0.100\,s$.

2.3   ELECTROMAGNETIC WAVE EQUATIONS

In this section, we deduce the properties of electromagnetic waves from Maxwell's equations. Equations 2.12—2.15 express Maxwell's equations in *integral form*. To find the properties of electromagnetic waves, we must derive the *differential wave equation for electromagnetic waves*. This is a differential equation, so we must re-express Maxwell's equations in *differential form*; in other words, as a set of differential equations. This is done in the Appendix 2 at the end of this chapter[8]. The results are summarized below:

Faraday's law:

$$\frac{\partial E_y}{\partial x} - \frac{\partial E_x}{\partial y} = -\frac{\partial B_z}{\partial t} \tag{2.A.3}$$

$$\frac{\partial E_x}{\partial z} - \frac{\partial E_z}{\partial x} = -\frac{\partial B_y}{\partial t} \tag{2.A.4}$$

$$\frac{\partial E_z}{\partial y} - \frac{\partial E_y}{\partial z} = -\frac{\partial B_x}{\partial t} \tag{2.A.5}$$

Ampere's law:

$$\frac{\partial B_y}{\partial x} - \frac{\partial B_x}{\partial y} = \mu_0\,\epsilon\,\frac{\partial E_z}{\partial t} \tag{2.A.6}$$

$$\frac{\partial B_x}{\partial z} - \frac{\partial B_z}{\partial x} = \mu_0\,\epsilon\,\frac{\partial E_y}{\partial t} \tag{2.A.7}$$

---

[8] See also any intermediate text in electricity and magnetism; e.g. Griffiths [9].

$$\frac{\partial B_z}{\partial y} - \frac{\partial B_y}{\partial z} = \mu_0 \, \epsilon \, \frac{\partial E_x}{\partial t} \tag{2.A.8}$$

Gauss's law for $\vec{E}$ and $\vec{B}$:

$$\frac{\partial E_x}{\partial x} + \frac{\partial E_y}{\partial y} + \frac{\partial E_z}{\partial z} = 0 \tag{2.A.9}$$

$$\frac{\partial B_x}{\partial x} + \frac{\partial B_y}{\partial y} + \frac{\partial B_z}{\partial z} = 0 \tag{2.A.10}$$

The differential wave equation for $\vec{E}$ is obtained for Cartesian coordinates as follows. Begin by taking the time derivative of Equation 2.A.8, then use Equations 2.A.3 and 2.A.4:

$$\epsilon \, \mu_0 \frac{\partial^2 E_x}{\partial t^2} = \frac{\partial}{\partial t} \left(\frac{\partial B_z}{\partial y} - \frac{\partial B_y}{\partial z} \right) = \frac{\partial}{\partial y} \left(\frac{\partial B_z}{\partial t} \right) - \frac{\partial}{\partial z} \left(\frac{\partial B_y}{\partial t} \right)$$

$$= \frac{\partial}{\partial y} \left(\frac{\partial E_x}{\partial y} - \frac{\partial E_y}{\partial x} \right) - \frac{\partial}{\partial z} \left(\frac{\partial E_z}{\partial x} - \frac{\partial E_x}{\partial z} \right)$$

$$= \frac{\partial^2 E_x}{\partial y^2} + \frac{\partial^2 E_x}{\partial z^2} - \frac{\partial}{\partial x} \left(\frac{\partial E_y}{\partial y} + \frac{\partial E_z}{\partial z} \right)$$

Add and then subtract the term $\frac{\partial^2 E_x}{\partial x^2}$.

$$\epsilon \, \mu_0 \frac{\partial^2 E_x}{\partial t^2} = \frac{\partial^2 E_x}{\partial x^2} + \frac{\partial^2 E_x}{\partial y^2} + \frac{\partial^2 E_x}{\partial z^2} - \frac{\partial}{\partial x} \left(\frac{\partial E_y}{\partial y} + \frac{\partial E_z}{\partial z} \right) - \frac{\partial^2 E_x}{\partial x^2}$$

$$= \frac{\partial^2 E_x}{\partial x^2} + \frac{\partial^2 E_x}{\partial y^2} + \frac{\partial^2 E_x}{\partial z^2} - \frac{\partial}{\partial x} \left(\frac{\partial E_x}{\partial x} + \frac{\partial E_y}{\partial y} + \frac{\partial E_z}{\partial z} \right)$$

The last term is zero by virtue of Equation 2.A.9, giving the result we seek:

$$\frac{\partial^2 E_x}{\partial x^2} + \frac{\partial^2 E_x}{\partial y^2} + \frac{\partial^2 E_x}{\partial z^2} = \epsilon \, \mu_0 \frac{\partial^2 E_x}{\partial t^2}$$

We can express this last result using the Laplacian operator defined in Equation 1.46

$$\nabla^2 \equiv \frac{\partial^2}{\partial x^2} + \frac{\partial^2}{\partial y^2} + \frac{\partial^2}{\partial z^2}$$

giving

$$\nabla^2 E_x = \epsilon \, \mu_0 \frac{\partial^2 E_x}{\partial t^2}$$

Repeat this derivation with Equation 2.A.7 to give the wave equation for E_y, and with Equation 2.A.6 to give the wave equation for E_z. Thus each component of $\vec{E}$ satisfies the wave equation, and we have

$$\nabla^2 \vec{E} = \epsilon \, \mu_0 \frac{\partial^2 \vec{E}}{\partial t^2} \tag{2.16}$$

Equations 2.A.3 — 2.A.8 and 2.A.10 can be used in a similar way to find the wave equation for $\vec{B}$:

$$\nabla^2 \vec{B} = \epsilon \, \mu_0 \frac{\partial^2 \vec{B}}{\partial t^2} \tag{2.17}$$

At this point, it is worthwhile to reflect that, if it were not for the extra term added by Maxwell to Ampere's law, the right-hand sides of Equations 2.A.6 — 2.A.8 would all be zero, making it impossible to obtain Equation 2.17. Finally, although the differential wave equations 2.16 and 2.17 were obtained for the specific case of Cartesian coordinates, the result is quite general and valid for all commonly used coordinate systems.

Equations 2.16 and 2.17 have the general form of a differential wave equation as discussed in Chapter 1:

$$\nabla^2 \Psi = \frac{1}{v^2} \frac{\partial^2 \Psi}{\partial t^2}$$

We identify the wave speed by inspection:

$$v = \frac{1}{\sqrt{\epsilon \mu_0}}$$

The vacuum value of ϵ gives the speed of electromagnetic waves in vacuum.

$$c = \frac{1}{\sqrt{\epsilon_0 \mu_0}} = 2.998 \times 10^8 \, m/s \tag{2.18}$$

Since $\epsilon = K_E \epsilon_0$, the electromagnetic wave speed in a material becomes

$$v = \frac{1}{\sqrt{K_E \epsilon_0 \mu_0}} = \frac{c}{\sqrt{K_E}} = \frac{c}{n} \tag{2.19}$$

where n is the *index of refraction*:

$$n = \sqrt{K_E} \tag{2.20}$$

Thus, the index of refraction is given by the square root of the dielectric constant[9]. Finally, we note that K_E depends on frequency, so it is better to say that the index of refraction is the square root of the dielectric constant measured at the frequency of the electromagnetic wave (see Section 3.7).

2.3.1 Transverse Electromagnetic Waves

As discussed in Chapter 1, functions that solve the differential wave equation are quite general, needing only sufficient differentiability and the right form of argument. We will begin our discussion of the solutions to Equations 2.16 and 2.17 by considering an important subclass of solutions that have the following properties:

1. The solutions are harmonic.

2. They are three-dimensional plane waves of the form discussed in Section 1.7 of Chapter 1.

3. They are *linearly polarized*, meaning that the wave amplitude consists of field vectors that oscillate along definite directions.

4. Finally, they are electromagnetic waves, meaning that they must satisfy Maxwell's equations (Equations 2.A.3 — 2.A.10) in addition to the differential wave equations.

---

[9]Recall our convention in this text to assume a value of unity for the relative permeability K_m. If K_m is greater than one, $n = \sqrt{K_E K_M}$. For details, see Guenther [10].

We will utilize the complex representation of three-dimensional waves introduced in Chapter 1. According to item 2 above, the wave amplitude will consist of the field vectors $\vec{E}$ and $\vec{B}$. Begin with the electric field $\vec{E}$:

$$\vec{E}\,(x, y, z, t) = \vec{E}_0\, e^{i(\vec{k}\cdot\vec{r} \mp \omega t + \varphi)} \tag{2.21}$$

where $i = \sqrt{-1}$, $\omega = 2\pi f$, $k = \frac{2\pi}{\lambda}$, and in Cartesian coordinates, $\vec{k}\cdot\vec{r} = k_x x + k_y y + k_z z$ and $\vec{E}_0 = E_{0x}\,\hat{i} + E_{0y}\,\hat{j} + E_{0z}\,\hat{k}$. You should be careful not to confuse the unit vectors $\hat{i}$ and $\hat{k}$ with the imaginary number i and wavenumber k.

As in Chapter 1, we may demonstrate that 2.21 solves the wave equation 2.16. To do so, note that

$$\frac{\partial E_x}{\partial x} = ik_x\,\vec{E}_0\, e^{i(\vec{k}\cdot\vec{r} \mp \omega t + \varphi)}$$

$$\frac{\partial^2 E_x}{\partial x^2} = -k_x^2\,\vec{E}_0\, e^{i(\vec{k}\cdot\vec{r} \mp \omega t + \varphi)}$$

so that

$$\nabla^2 \vec{E} = -k^2 \vec{E}$$

The time derivatives give

$$\frac{\partial \vec{E}}{\partial t} = \mp i\omega\, \vec{E}$$

$$\frac{\partial^2 \vec{E}}{\partial t^2} = -\omega^2\, \vec{E}$$

Thus, Equation 2.16 is satisfied provided

$$k^2 = \frac{\omega^2}{v^2}$$

$$v = \mp\frac{\omega}{k} = \mp f\,\lambda \tag{2.22}$$

as found in Chapter 1.

According to Maxwell's equations, and Equations 2.16 and 2.17, an electromagnetic wave must have both $\vec{E}$ and $\vec{B}$ fields. Thus

$$\vec{B}\,(x, y, z, t) = \vec{B}_0\, e^{i(\vec{k}\cdot\vec{r} \mp \omega t + \varphi)} \tag{2.23}$$

By the same arguments given above, Equation 2.23 solves wave equation 2.17.

We now demonstrate that electromagnetic waves are transverse: that is, $\vec{E}$ and $\vec{B}$ must both point in directions that are perpendicular to the direction of propagation, determined by the direction of $\vec{k}$. Begin with the electric field and consider Equation 2.A.9:

$$\frac{\partial E_x}{\partial x} + \frac{\partial E_y}{\partial y} + \frac{\partial E_z}{\partial z} = i\,(k_x E_x + k_y E_y + k_z E_z) = i\,\vec{k}\cdot\vec{E} = 0$$

Since $\vec{E}\cdot\vec{k} = 0$, $\vec{E}$ and $\vec{k}$ must be perpendicular. In a similar way, Equation 2.A.10 shows that $\vec{B}$ is perpendicular to $\vec{k}$. Thus, according to Maxwell's equations, electromagnetic waves are transverse.

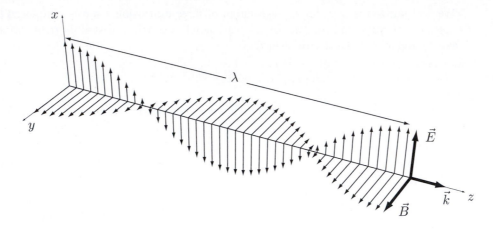

Figure 2.5 A transverse electromagnetic wave. The electric field $\vec{E}$ oscillates in the x-z plane, the magnetic field $\vec{B}$ oscillates in the y-z plane, and the wave propagates in the z-direction.

We now show that $\vec{E}$ and $\vec{B}$ are perpendicular to each other. To do so, we will use Equations 2.A.3 - 2.A.5. Begin with Equation 2.A.3 and assume a forward-traveling wave (see Section 1.3):

$$\frac{\partial E_y}{\partial x} - \frac{\partial E_x}{\partial y} = i\left(k_x E_y - k_y E_x\right) = -\frac{\partial B_z}{\partial t} = i\omega\, B_z$$

with similar relations for Equations 2.A.4 and 2.A.5. These results may be summarized using the cross-product:

$$\vec{k} \times \vec{E} = \omega\, \vec{B} \qquad (2.24)$$

From this, we conclude that $\vec{B}$ is perpendicular to both $\vec{E}$ and $\vec{k}$. Since the wave is transverse, $\vec{E}$ is also perpendicular to $\vec{k}$. Thus, $\vec{E}$, $\vec{B}$, and $\vec{k}$ define a mutually orthogonal right-handed triad as illustrated in Figure 2.5. The *directions* in Equation 2.24 may be cyclically permuted to see that the vector $\vec{E} \times \vec{B}$ points in the direction of $\vec{k}$.

Taking the magnitude of both sides of Equation 2.24 gives

$$E = \frac{\omega}{k}B = v\, B \qquad (2.25)$$

In vacuum, this becomes

$$E = c\, B \qquad (2.26)$$

Both fields $\vec{E}$ and $\vec{B}$ are harmonic functions of both space and time, and according to Equations 2.25 and 2.3.1, they oscillate *in phase*. The harmonic dependence can be canceled from both sides, so that Equations 2.25 and 2.3.1 are also satisfied by the wave amplitudes (E_0 and B_0 in Equations 2.21 and 2.23).

In summary, we have shown that functions of the form given in Equations 2.21 and 2.23 solve the differential wave equations that were deduced from Maxwell's equations. Both solutions combine to form a plane polarized transverse electromagnetic wave, with $\vec{E}$ and $\vec{B}$ oriented so that $\vec{E} \times \vec{B}$ points in the direction of energy flow, and where the magnitudes of $\vec{E}$ and $\vec{B}$ are such that $E = v\, B$ ($E = c\, B$ in vacuum). By plane polarized, we mean that $\vec{E}$ and $\vec{B}$ have magnitude and direction that are constant and uniform over an entire plane (e.g. the x-y plane in Figure 2.5). The planes normal to $\vec{k}$ located at points where $\vec{E} = \vec{E}_0$ and $\vec{B} = \vec{B}_0$ are called *wavefronts*.

The solutions 2.21 and 2.23 are unphysical in several respects. Plane waves have wavefronts that extend to infinity in two coordinate directions, whereas real waves rarely have wavefronts that are correlated over distances of over a meter. Real light beams can be *unpolarized*, meaning that the polarization states change randomly in time. Furthermore, real light beams never have precise values of frequency and wavelength. Even laser beams have ω and k values distributed over a spread of values as opposed to the precise values in implied by Equations 2.21 and 2.23. Nevertheless, the plane wave solutions will be very important to us as we use them to build more realistic models of electromagnetic waves.

Finally, we note that it is redundant to specify both fields (Equations 2.21 and 2.23) for a given electromagnetic wave. Once either is specified, the direction and magnitude of the other can be obtained from the electromagnetic wave properties just deduced. It is customary to specify the electric field: Equation 2.21.

■ **EXAMPLE 2.3**

A plane electromagnetic wave traveling in vacuum is described by

$$\vec{E}(x, y, z, t) = \left(E_0 \hat{i}\right) e^{i(ky+\omega t)}$$

Find an equation for the magnetic field of this wave.

Solution

When x and t are zero, $\vec{E}$ points along positive x and the wave propagates along negative y. A diagram similar to Figure 2.6 is helpful to find $\vec{B}$. Using this, we can see that when x and t are zero, $\vec{B}$ must point along positive z. Finally, the magnitude B must be given by E/c. Thus, the magnetic field of this wave is given by

$$\vec{B}(x, y, z, t) = \left(\frac{E_0}{c}\hat{k}\right) e^{i(ky+\omega t)}$$

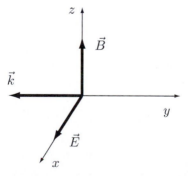

Figure 2.6   The fields $\vec{E}$ and $\vec{B}$ are oriented so that $\vec{E} \times \vec{B}$ has the same direction as $\vec{k}$.

Problem 2.5   A region of space has a permittivity of $1.9 \times 10^{-11} \frac{C^2}{Nm^2}$. What is the speed of electromagnetic radiation within this region? What is the index of refraction of this material?

Problem 2.6   Derive the differential wave equation for $\vec{B}$ (Equation 2.17).

Problem 2.7   Show that in a transverse electromagnetic wave, the magnetic field is perpendicular to the propagation direction.

Problem 2.8   A plane electromagnetic wave traveling in vacuum is described by

$$\vec{E}(x, y, z, t) = \left(E_0 \hat{j} \right) e^{i(kz - \omega t)}$$

Find an equation for the magnetic field of this wave.

Problem 2.9   A plane electromagnetic wave traveling in vacuum is described by

$$\vec{B}(x, y, z, t) = \left(B_0 \hat{k} \right) e^{i(kx + \omega t)}$$

Find an equation for the electric field of this wave.

Problem 2.10   A plane electromagnetic wave traveling in vacuum is described by

$$\vec{E}(x, y, z, t) = \left(-E_0 \hat{i} \right) e^{i(ky - \omega t)}$$

Find an equation for the magnetic field of this wave.

2.3.2   Energy Flow and the Poynting Vector

It takes work to establish electric and magnetic fields, and a region of space that contains such fields has an associated energy density. We begin by considering a vacuum, then include the effect of linear isotropic homogeneous materials by letting $\epsilon_0 \rightarrow \epsilon$.

Consider a parallel-plate capacitor of area A, plate separation d and capacitance $C = \frac{\epsilon_0 A}{d}$ charged to an electric potential V. The capacitor stores energy as an electric field. The energy density between the plates is

$$u_E = \frac{\frac{1}{2} C V^2}{Ad} = \frac{\frac{1}{2} \left(\frac{\epsilon_0 A}{d} \right) (E\,d)^2}{Ad} = \frac{\epsilon_0}{2} E^2 \tag{2.27}$$

Similarly, an inductor stores energy as a magnetic field. Consider a simple solenoid of length ℓ, current I, cross-sectional area A, and inductance $L = \mu_0 n^2 \ell A$. The magnetic field has value $B = \mu_0 n I$, and the energy density within the solenoid is

$$u_B = \frac{\frac{1}{2} L I^2}{A\ell} = \frac{\frac{1}{2} \left(\mu_0 n^2 \ell A \right) \left(\frac{B}{\mu_0 n} \right)^2}{A\ell} = \frac{B^2}{2\mu_0} \tag{2.28}$$

The derivations of 2.27 and 2.28 both ignore end effects but nonetheless give correct results.

An electromagnetic wave has both electric and magnetic fields. The energy density of the wave is the sum

$$u = u_E + u_B \qquad (2.29)$$

By Equation 2.26, u_E and u_B are *equal* for an electromagnetic wave:

$$u_B = \frac{B^2}{2\,\mu_0} = \frac{\left(\frac{E}{c}\right)^2}{2\,\mu_0} = \frac{\epsilon_0\,\mu_0\,E^2}{2\,\mu_0} = u_E \qquad (2.30)$$

Thus, according to Equation 2.29,

$$u = \epsilon_0\,E^2 = \frac{B^2}{\mu_0} \qquad (2.31)$$

Equation 2.29 expresses the energy per unit volume in a region where an electromagnetic wave exists. However, electromagnetic waves *travel*, so *energy flux* is a more useful quantity. Consider a time Δt during which the wave travels a distance $c\,\Delta t$. Let the wave move through an area A. The amount of energy that flows through A in time Δt is given by $u\,(c\,\Delta t\,A)$. The energy flux S is defined as the energy flux per unit area:

$$S = \frac{u\,(c\,\Delta t\,A)}{\Delta t\,A} = u\,c \qquad (2.32)$$

Using Equations 2.30 and 2.29, we find

$$S = \left(\epsilon_0 E^2\right)\,c = \epsilon_0 E\,(c\,B)\,c = \frac{\epsilon_0 E\,B}{\epsilon_0\,\mu_0} = \frac{1}{\mu_0}E\,B$$

We may use the fact that $\vec{E} \times \vec{B}$ points in the direction of energy flow to express S as a vector:

$$\vec{S} = \frac{1}{\mu_0}\vec{E} \times \vec{B} \qquad (2.33)$$

$\vec{S}$ is called the *Poynting vector*. It has units of W/m^2.

2.3.3 Irradiance

Optical frequencies for visible light are on the order of 10^{14} Hz, and a measurement of $\vec{S}$ in this case necessarily involves a *time average*. The time average of the magnitude of $\vec{S}$ is what we mean by *irradiance*.

Since $\vec{S}$ involves a *product* of harmonic terms, we must express these fields in their real rather than complex representations (see Problem 2.23). Thus, for a forward-traveling wave

$$\vec{E} = \vec{E}_0\,\cos(\vec{k}\cdot\vec{r} - \omega t + \varphi)$$

$$\vec{B} = \vec{B}_0\,\cos(\vec{k}\cdot\vec{r} - \omega t + \varphi)$$

giving

$$\vec{S} = \frac{1}{\mu_0}\vec{E}_0 \times \vec{B}_0\,\cos^2\left(\vec{k}\cdot\vec{r} - \omega t + \varphi\right) \qquad (2.34)$$

Irradiance is the time average of the Poynting vector magnitude:

$$I = \left\langle \left| \vec{S} \right| \right\rangle = \frac{1}{\mu_0} \left| \vec{E}_0 \times \vec{B}_0 \right| \left\langle \cos^2 \left(\vec{k} \cdot \vec{r} - \omega t + \varphi \right) \right\rangle \qquad (2.35)$$

This result requires some explanation. The time average is taken only over the harmonic term because it is presumed that the field amplitudes either do not vary in time, or that they do so slowly enough so that their time dependence can be measured.

The fields are perpendicular, giving

$$\frac{1}{\mu_0} \left| \vec{E}_0 \times \vec{B}_0 \right| = \frac{E_0 B_0}{\mu_0} = \epsilon_0 c \, E_0^2 \qquad (2.36)$$

The time average of $\cos^2 \left(\vec{k} \cdot \vec{r} - \omega t + \varphi \right)$ over many cycles is $\frac{1}{2}$, as shown in Example 2.4. Thus according to Equations 2.35 and 2.36, the irradiance is given by

$$I = \langle S \rangle = \frac{\epsilon_0 c}{2} E_0^2 \qquad (2.37)$$

Within a homogeneous linear isotropic material, this becomes

$$I = \frac{\epsilon \, v}{2} E_0^2 \qquad (2.38)$$

■ **EXAMPLE 2.4**

Show that $\left\langle \cos^2 \left(\vec{k} \cdot \vec{r} - \omega t + \varphi \right) \right\rangle$ over many cycles is $1/2$.

Solution

From elementary calculus, the time average of a time-varying function is

$$\langle f(t) \rangle = \frac{1}{T} \int_{t_0}^{t_0+T} f(t) \, dt$$

Thus,

$$\left\langle \cos^2 \left(\vec{k} \cdot \vec{r} - \omega t + \varphi \right) \right\rangle = \frac{1}{T} \int_{t_0}^{t_0+T} \cos^2 \left(\vec{k} \cdot \vec{r} - \omega t + \varphi \right) dt$$

To evaluate this expression, change variables by letting $\theta = \vec{k} \cdot \vec{r} - \omega t + \varphi$. We measure the average at a definite location, so $\vec{k} \cdot \vec{r} + \varphi$ is constant in this integral, giving $d\theta = -\omega \, dt$. To evaluate the integration limits, let $\theta_0 = \vec{k} \cdot \vec{r} - \omega t_0 + \varphi$ so that $t_0 + T \rightarrow \theta_0 - \omega T$. Thus

$$\left\langle \cos^2 \left(\vec{k} \cdot \vec{r} - \omega t + \varphi \right) \right\rangle = -\frac{1}{\omega T} \int_{\theta_0}^{\theta_0 - \omega T} \cos^2 (\theta) \, d\theta$$

Use the trigonometric identity

$$\cos^2\left(\theta\right) = \frac{1}{2}\left(1 + \cos 2\theta\right)$$

to give

$$\left\langle \cos^2\left(\vec{k}\cdot\vec{r} - \omega t + \varphi\right)\right\rangle = -\left[\frac{1}{2\omega T}\int_{\theta_0}^{\theta_0 - \omega T} d\theta + \frac{1}{2\omega T}\int_{\theta_0}^{\theta_0 - \omega T}\cos 2\theta\, d\theta\right]$$

$$= \frac{1}{2} - \frac{1}{4\omega T}\left[\sin 2\left(\theta_0 - \omega T\right) - \sin\left(2\theta_0\right)\right]$$

The second term becomes zero in the limit of large T. Thus,

$$\left\langle \cos^2\left(\vec{k}\cdot\vec{r} - \omega t + \varphi\right)\right\rangle = \frac{1}{2}$$

when integrated over many cycles.

The fastest detectors known have electrical bandwidths that are many orders of magnitude less than the frequency of visible light.

■ **EXAMPLE 2.5**

A certain electromagnetic wave has an electric field given by

$$\vec{E} = \left(-200\frac{V}{m}\right)\hat{j}\, e^{i(kz + \omega t)}$$

with $\omega = 2.00 \times 10^{15}\, rad/s$ and $k = 1.00 \times 10^7\, rad/m$. Find (a) the corresponding magnetic field, and (b) the irradiance of the wave.

Solution

(a) Find the corresponding magnetic field.

As in Example 2.3, we determine that $\vec{B}$ must point along positive y when z and t are zero. To find the field amplitude, we must first find the wave speed:

$$v = \frac{\omega}{k} = 2.00 \times 10^8\frac{m}{s}$$

Use $E = vB$ to give

$$\vec{B} = \left(10^{-6}T\right)\hat{k}\, e^{i(kz + \omega t)}$$

(b) Find the wave irradiance.

Begin with

$$I = \frac{\epsilon v}{2}E_0^2$$

From part (a), the wave speed $v = 2.00 \times 10^8 \frac{m}{s}$. Thus, the index of refraction is $n = 1.50$ and for optical frequencies $K_E = 2.25$. Thus

$$I = \frac{(2.25) \left(8.854 \times 10^{-12} \frac{C^2}{N \cdot m^2}\right) \left(2.00 \times 10^8 \frac{m}{s}\right)}{2} \left(200 \frac{N}{C}\right)^2 = 79.7 \frac{W}{m^2}$$

Problem 2.11   Show that average of $\sin^2 \left(\vec{k} \cdot \vec{r} - \omega t + \varphi\right)$ over many cycles is $1/2$.

Problem 2.12   Let

$$\vec{E} = \left(200 \frac{V}{m}\right) \hat{k} \, e^{i(ky - \omega t)}$$

with $\omega = 3.00 \times 10^{15} \, rad/s$ and $k = 1.50 \times 10^7 \, rad/m$. Find the corresponding magnetic field and the irradiance of the wave.

Problem 2.13   Let

$$\vec{B} = \left(9.00 \times 10^{-7} T\right) \hat{j} \, e^{i(kz - \omega t)}$$

with $\omega = 3.00 \times 10^{15} \, rad/s$ and $k = 1.50 \times 10^7 \, rad/m$. Find the corresponding electric field and the irradiance of the wave.

Problem 2.14   Find the irradiance of the following transverse electromagnetic waves, assuming that they travel in vacuum:

   a) $\vec{E} = \left(200 \frac{V}{m} \, \hat{i}\right) e^{i(kz - \omega t)}$

   b) $\vec{B} = \left(7.00 \times 10^{-7} T \, \hat{j}\right) e^{i(kz - \omega t)}$

Problem 2.15   The output of a green laser pointer has a beam power of $10.0 \, mW$ and a beam diameter of $1.00 \, mm$. Calculate the beam irradiance, and the maximum values of electric and magnetic fields within the beam. Assume uniform irradiance across the beam's cross section.

2.4 PHOTONS

The effort to understand light has a long and interesting history. Isaac Newton favored a particle (corpuscular) theory, as opposed to wave theories advanced by contemporaries Huygens and Hooke. Newton's arguments were temporarily more convincing, and the corpuscular theory dominated until the early 1800s, when Thomas Young's analysis of his double-slit experiment and Fresnel's wave theory of diffraction left little doubt that light was in fact a wave.

   We will discuss Young's double-slit experiment more completely in Chapters 5 and 6, but it will be helpful to give a brief overview here. Figure 2.7 shows a double-slit aperture illuminated by plane-parallel wavefronts. Light from each aperture diffracts to illuminate a wide area on the observing screen, combining *coherently*[10] with similar light from the other

[10]Coherence will be discussed in Chapter 5.

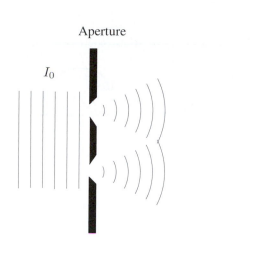

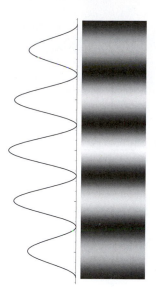

Figure 2.7   Plane wavefronts with irradiance I_0 illuminate a double-slit aperture in a Young's double-slit experiment. Interference fringes are indicated in profile on an observing screen. The figure is not drawn to scale; the observing screen should be very far away, and the width of the apertures should be larger than the illumination wavelength.

aperture. Bright fringes occur where the two beams combine constructively (in phase), and dark fringes locate areas of destructive interference. Effects such as constructive and destructive interference are easier to explain with a wave picture, and Young's experiment seemed to provide evidence that light is a wave. The full development of electromagnetism as expressed by Maxwell in 1865, and as outlined earlier in this chapter, gives a detailed model of the transverse electromagnetic wave. Such concise and convincing evidence for the wave picture of light set the stage for Einstein's analysis of the *photoelectric effect* in 1905.

Investigations into the photoelectric effect provided early evidence that was difficult to explain using classical physics and a wave picture of electromagnetic radiation. To see why, consider Figure 2.8, which shows a metallic plate (labeled E for emitter) illuminated by a beam of light with irradiance I. The illumination causes electrons to be ejected from the emitter surface, and collected by the collector plate (labeled C in the Figure). Ammeter A measures the resulting *photocurrent*.

A potential difference is applied across plates E and C. Charging C negatively provides a *stopping potential*, which can be used to determine the kinetic energy of the ejected electrons. According to Equation 2.37, one would expect an increase in irradiance to result in a larger electric field, causing the electron to be ejected with greater force and hence greater kinetic energy. Interestingly, this is not at all what is observed: increasing the irradiance increases the amount of current, but the stopping potential remains the same. Thus, increasing the irradiance increases the *number* of ejected electrons but not their initial kinetic energy. In order to resolve this, Einstein in 1905 reintroduced the concept of a light *particle* or *quanta*, subsequently termed a *photon*[11]. According to this picture, a photon

---

[11] The term photon was coined by American chemist G. N. Lewis in 1926.

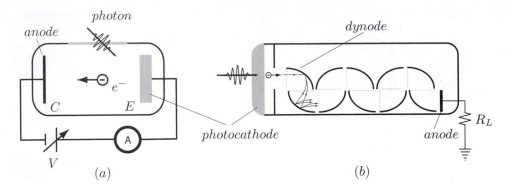

Figure 2.8   (a) Diagram of a typical photoelectric effect experiment. (b) A photomultiplier tube with semitransparent photocathode.

has energy given by

$$E = hf \tag{2.39}$$

where h is Planck's constant

$$h = 6.626 \times 10^{-34} \, J \cdot s = 4.136 \times 10^{-15} \, eV \cdot s$$

Thus, light is *quantized*. A beam of electromagnetic radiation consists of discrete quanta (photons) whose energy cannot be subdivided. In Einstein's analysis, increasing the illumination irradiance increases the flux of photons, increasing the flux of ejected electrons and thus increasing the photocurrent. Since each photon has a fixed energy, the kinetic energy of each ejected electron is fixed. All observations are thus accounted for, and Einstein was awarded the Nobel Prize in Physics in 1921 in part for his analysis of this effect.

According to the special theory of relativity, photons have zero mass, but nonzero momentum[12]. Photon momentum and energy are related by

$$p = \frac{E}{c} \tag{2.40}$$

By Equation 2.39

$$p = \frac{hf}{c} = \frac{h}{\lambda} = \hbar k \tag{2.41}$$

where $\hbar \equiv \frac{h}{2\pi}$. Since the momentum is in the direction of energy flow, we may write

$$\vec{p} = \hbar \vec{k} \tag{2.42}$$

Electromagnetic radiation reflecting from or absorbed by an object must transfer momentum to the object, resulting in *radiation pressure*. The momentum transfer per photon Δp is twice the photon momentum for complete reflection and equal to the photon momentum for complete absorption. A beam of light with irradiance I has a photon flux per unit area given by $\frac{I}{hf}$. Thus, for the case of complete reflection, the pressure exerted on the object becomes

$$P = \frac{I}{hf} \left(2\frac{h}{\lambda} \right) = \frac{2I}{c} \tag{2.43}$$

[12] See any introductory physics text that includes modern physics (eg. Halliday, Resnick, and Walker [11]).

For complete absorption, the pressure is

$$P = \frac{I}{c} \tag{2.44}$$

Equations 2.40, 2.43, and 2.44 were first obtained by Maxwell using electromagnetism and the wave picture. Briefly, the argument proceeds as follows. The field of an electromagnetic wave that is incident on a conductor imparts forces on charge carriers that subsequently place a force on the sides of the conductor. Conservation of momentum gives the momentum of the electromagnetic wave. Details of the analysis can be found in Guenther [10].

Similarly, there are *semi-classical* treatments of the photoelectric effect that yield the results obtained by Einstein without resorting to photons. In such analyses, matter is quantized but the electromagnetic radiation is not. Thus, experiments such as the photoelectric effect do not prove the existence of photons.

■ **APPLICATION NOTE 2.1**

Photodetectors and Photomultipliers

The photoelectric effect is the basis for these two commonly used light detectors. A *photodetector* (also called a *vacuum phototube*) is illustrated by Figure 2.8(a) with the stopping potential reversed to make the anode positive. Photoelectrons are ejected from the photocathode with maximum kinetic energy given by

$$E_{\max} = hf - W$$

where the *work function* W represents the work that must be done to liberate a conduction electron from the photocathode material. The value of W determines the *cutoff frequency* for a particular photocathode material:

$$f_c = \frac{W}{h}$$

Photomultipliers use a series of *dynodes* that are made from or coated with a material that provides *secondary electron emission*, as illustrated in Figure 2.8(b). Each dynode is maintained at a positive electric potential relative to the previous one. A photoelectron ejected from the photocathode is accelerated toward the first dynode, where two or more electrons are liberated by secondary emission. These travel on through the *dynode chain*, each liberating more secondary electrons with each dynode collision. The result is an avalanche that can create millions of electrons at the anode for each photoelectron ejected from the photocathode. Very sensitive photomultipliers that can easily detect individual photons are called *photon counters*.

As a practical matter, it is necessary to allow the *photocurrent* collected by the anode to pass through a load resistor R_L and then into electrical ground. If the secondary electrons are not allowed to escape the tube, they can create an "electron gas" that spoils the potential difference between dynodes and causes the sensitivity to suffer dramatically. The value used for R_L combines with the capacitance of the photomultiplier tube and the connecting signal cable to form an RC network whose RC time constant affects the time response of the detector. The time response is also affected by the *transit time* of photoelectrons through the detector itself.

The *responsivity* of a photodetector determines the amount of photocurrent generated:

$$\Re = \frac{\eta e}{h f}$$

where η is the *quantum efficiency* of the photodetector. It is customary to quote responsivity in units of A/W. The responsivity of a photomultiplier incorporates the detector *gain G*:

$$\Re = \frac{G \eta e}{h f}$$

Sensitive photon counters can have gains of several million.

The *quantum efficiency* η represents the probability that an incident photon will create a photoelectron. Even above the cutoff frequency, this is typically a fairly strong function of frequency. There are many available types of photodetectors and photomultiplier tubes, each having a different *spectral response curve* over the target spectral region.

Electrons thermally ejected from the photocathode and dynodes cause *dark current*, which as a noise source limits the minimum detectable light level. Cooling the detector can dramatically reduce dark current. Photocathode materials with low work functions are more susceptible to thermal noise, and this limits their applicability for detection of infrared photons.

There are many other detection strategies that utilize photon-electron interactions in both conductors and semiconductors. See, for example, *Photonics*, by Saleh and Teich [17].

Photon detectors do not conclusively prove the existence of photons. Detector elements contain matter with quantized energy levels that can only respond discretely. Critics of the photon picture interpret signals from a photon counter as arising from the interaction of non-quantized electromagnetic waves and the discrete quantum levels of the detector element.

However, other experiments have been reported that give more convincing evidence for the *necessity* of the photon picture[13]. In 1986, coworkers P. Grangier, A. Aspect and G. Roger placed photon counters on either side of a beam-splitter designed to reflect and transmit 50% of incident radiation. In the wave picture, the incident wave is divided, with half the wave energy transmitted and half reflected. In the photon picture, a photon is either reflected or transmitted with 50% probability. Thus, in the photon picture, the two photon counters will only rarely measure coincident photon pulses, and then only as an artifact of the detection electronics. The results of their experiment fell conclusively in favor of the photon picture by over 13 standard deviations. Improvements in detector and electronics technology have currently advanced to the point where this experiment can now be conducted (even more conclusively) within undergraduate laboratories[14].

---

[13]For a readable account of these and related issues, see *The Quantum Challenge* by Greenstein and Zajonc [8].
[14]J. J. Thorn, M. S. Neel, V. W. Donato, G. S. Bergreen, R. E. Davies and M. Beck, "Observing the quantum behavior of light in an undergraduate laboratory," Am. J. Phys. **72** (9), p. 1210, 2004.

2.4.1 Single-Photon Interference

Consider a double-slit interference experiment conducted with illumination irradiance so small that at most one photon interacts with the aperture at any one instant of time (e.g. one photon per second). The experimental details are unimportant here — suffice it to say that current technology allows such an experiment to be conducted as a classroom demonstration [15].

According to the photon picture, each photon has energy given by Equation 2.39 and may not be subdivided. Thus, any individual photon must pass through one of the available slits, but not both. However, when data are accumulated over time, the integrated irradiance builds toward exactly the same distribution as would be obtained with more intense illumination. Photographic film or a sensitive photon detector array placed at the observation screen records data similar to Figure 2.9, which shows a distribution of photon detection events that fall beneath an envelope given by the normal irradiance profile of a double-slit experiment conducted with more intense illumination. Over time, the randomness integrates away, and the familiar double-slit pattern emerges.

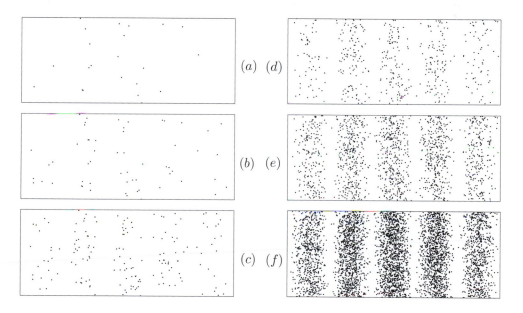

Figure 2.9 Typical data from a single-photon interference experiment. The number of photons detected increases by 10 in each figure.

Such results imply that photons propagate according to a *probability distribution* obtained using the irradiance profile of the associated electromagnetic waves. For example, in Figure 2.9, the probability for detecting a photon is low in regions where the electromagnetic waves interfere destructively, and is high where there is constructive interference.

In the modern view of Quantum Electrodynamics (QED), particle-wave duality disappears. Electromagnetic waves gives probabilities that allow one to *predict* the outcome of measurement, but the measurement itself typically involves random detection of photons.

[15] W. Rueckner and P. Titcomb, "A lecture demonstration of single-photon interference," Am. J. Phys. **64** (2), p. 184, 1996.

These ideas will be discussed in more detail in Chapter 6. For more information, see Section 6.5, and the elegant introduction by R. P. Feynman: *QED: The Strange Theory of Light and Matter* [5].

2.5   THE ELECTROMAGNETIC SPECTRUM

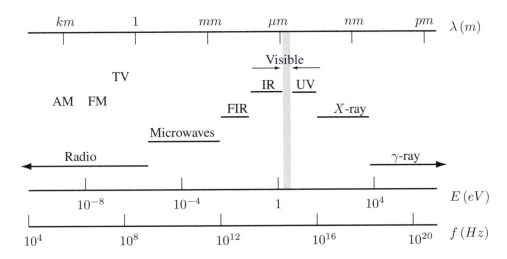

Figure 2.10   The electromagnetic spectrum.

Since (in vacuum) $f = c/\lambda$, electromagnetic waves with larger frequencies have smaller wavelengths. The range of possibilities for f and λ is roughly subdivided into the conceptual categories illustrated in Figure 2.10.

Radio waves include electromagnetic radiation that may be generated and detected with electric circuit elements such as capacitors and inductors. Examples include AM and FM radio, broadcast television, and microwaves. The field amplitudes of radio waves may be detected directly using antennas. For these relatively low frequencies $(0 - 10^{10} Hz)$, the photon energy is less than or on the order of thermal fluctuations for detectors cooled to even a few degrees above absolute zero, making photon detection impractical. Wavelengths in this region vary from many kilometers for long wave radio waves to millimeters for high-frequency microwaves. Photon energies range from zero to $10^{-3} eV$.

Infrared radiation (IR) includes electromagnetic radiation with wavelength only somewhat larger than visible (near IR, 1-10 μm) to wavelengths a bit smaller than microwaves (far IR, 10-1000 μm). Photon detection strategies become more useful in this region, even though such detectors must usually be cooled to well below room temperature to reduce thermal noise. Blackbody radiation from a room-temperature object has a wavelength maximum at around 10 μm. Photon energies range from $10^{-3} eV$ to about $1 eV$.

Visible radiation ranges from about 0.4 μm for violet to 0.7 μm for red. The term light is usually reserved for visible electromagnetic radiation. Photon energies $(1-3 eV)$ are large enough to allow routine single-photon detection.

Ultra-violet (UV) radiation has smaller wavelength and higher frequency than visible light. Photon energies of 4-100 eV are high enough to be damaging to living tissue. UV photons with energy greater than about $10 eV$ will not pass through Earth's atmosphere.

X-rays include photon energies of $100\,eV$ to about $10^4\,MeV$. For all practical purposes, particle characteristics completely dominate wave properties here. Photons in this region come mainly from inner-shell atomic transitions.

Gamma-rays (γ-rays) include photons emitted from nuclear transitions. Photon energies are greater than about $10^4\,MeV$. Both x-rays and γ-rays can be extremely damaging to living tissue, and can provide sources of genetic mutation.

■ **EXAMPLE 2.6**

A monochromatic beam of electromagnetic radiation has a beam power of $1.00\,mW$. How many photons per second pass by a given point if the beam radiation is (a) infrared with wavelength $10.0\,\mu m$, and (b) gamma-ray with wavelength $0.100\,nm$?

Solution

(a) The beam wavelength is in the infrared region, with wavelength $10.0 \times 10^{-6}\,m$. The energy of an individual photon is

$$E_{ph} = hf = \frac{hc}{\lambda} = \frac{\left(4.136 \times 10^{-15}\,eV \cdot s\right)\left(2.998 \times 10^8 m/s\right)}{1.00 \times 10^{-5}\,m} = 0.124\,eV$$

The number of photons per second that pass by any point in the beam is

$$N_s = \frac{P}{E_{ph}} = \frac{1.00 \times 10^{-3}\,J/s}{\left(1.602 \times 10^{-19}\,J/eV\right)\left(0.124\,eV\right)} = 5.03 \times 10^{16}\,photon/s$$

(b) The beam wavelength is in the gamma-ray region with wavelength $1.0 \times 10^{-10}\,m$. The energy of an individual photon is

$$E_{ph} = \frac{hc}{1.00 \times 10^{-10}\,m} = 1.24 \times 10^4\,eV$$

The number of photons per second that pass by any point in the beam is

$$N_s = \frac{P}{\left(1.602 \times 10^{-19}\,J/eV\right)\left(1.24 \times 10^4\,eV\right)} = 5.03 \times 10^{11}\,photon/s$$

Gamma-ray beams are composed of photons with energy that is far greater than photons in an infrared beam. Optical detectors that detect photons are affected by *shot noise* — a noise source that originates from the fact that the beam energy is not continuous, but consists of discrete photon packets. The name derives intuitively from the effect of projectiles fired from a shotgun and striking a surface. As a noise source, shot noise becomes more significant as the wavelength becomes shorter.

Problem 2.16 The output of a certain laser pointer consists of a beam with uniform irradiance across the beam's cross section. Assume a beam power of $1.00\,mW$ and a beam diameter of $1.00\,mm$. (a) Calculate the beam irradiance. (b) How many photons per second pass by any point in the beam if the wavelength is $633\,nm$? (c) Repeat part (b) if the wavelength is $530\,nm$.

Problem 2.17   Calculate the radiation pressure as the beams of Problem 2.16 impinge on perfectly reflecting and perfectly absorbing surfaces. Calculate the photon momentum in each case.

Additional Problems

Problem 2.18   An *elliptically polarized* electromagnetic wave has perpendicular components of $\vec{E}$ that are *out of phase*. For example,

$$E_x = E_{0x}e^{i(kz-\omega t)}$$

$$E_y = E_{0y}e^{i(kz-\omega t+\varphi)}$$

with E_{0x} and E_{0y} real. Find the magnetic field components of this wave, assuming that it travels in vacuum.

Problem 2.19   Show explicitly that $\vec{B}(x, y, z, t) = \vec{B}_0 \, e^{i(\vec{k}\cdot\vec{r} \mp \omega t+\varphi)}$ is a solution to the differential wave equation for $\vec{B}$.

Problem 2.20   For an electromagnetic wave, show that $\vec{k} \cdot \vec{B} = 0$, and thus that $\vec{B}$ is perpendicular to the direction of propagation.

Problem 2.21   For an electromagnetic wave, show that $\vec{k} \times \vec{E} = \omega \, \vec{B}$.

Problem 2.22   For an electromagnetic wave, show that $-\vec{k} \times \vec{B} = \epsilon \, \mu_0 \, \omega \, \vec{E}$.

Problem 2.23   Let $f(x, t) = f_0 \, e^{i(kx - \omega t)}$ with f_0 a real constant.
   a) Find $\mathrm{Re}[f^2]$.
   b) Find $(\mathrm{Re}[f])^2$.
   c) Show that $(\mathrm{Re}[f])^2 = \frac{1}{2}\left(f_0^2 + \mathrm{Re}[f^2]\right)$.

Problem 2.24   Show that the average of $\sin\left(\vec{k} \cdot \vec{r} - \omega t + \varphi\right)$ and $\cos\left(\vec{k} \cdot \vec{r} - \omega t + \varphi\right)$ over many cycles is zero.

Problem 2.25   Show that the average of

$$\sin\left(\vec{k} \cdot \vec{r} - \omega t + \varphi\right) \cos\left(\vec{k} \cdot \vec{r} - \omega t + \varphi\right)$$

over many cycles is zero.

Problem 2.26   Show that an electromagnetic wave with spherical wavefronts of radius r has an irradiance that varies as $1/r^2$.

Problem 2.27   Calculate the peak electric and magnetic fields $1.0 \, km$ from a $1.0 \, MW$ radio station, assuming that it radiates electromagnetic waves as an isotropic point source.

Problem 2.28   It is often convenient to define the *optical thickness* of a material as nd where n is the index of refraction and d is the physical thickness.
   a) Find the optical thickness for vacuum and glass ($n = 1.5$) for a physical thickness of one meter.
   b) Find the time it takes light to travel $d = 1.0 \, m$ in vacuum and in glass of index 1.50.

Problem 2.29   The maximum electric field sustainable in a material before electrical breakdown is called the *dielectric strength*. For dry air at STP, the dielectric strength is about $3.0 \times 10^6 \, V/m$.
   a) Use this to estimate the maximum irradiance of a laser beam that can propagate through air.

b) If the beam profile is uniform and the beam diameter is $10\,cm$, what is the maximum beam power if it is to travel through air?

c) What radiation pressure would this beam exert on absorbing and reflecting surfaces? In each case, what net force would be exerted by a $10\,cm$-diameter beam?

Problem 2.30   A certain pulsed laser has a maximum power output of $10^7\,W$.

a) Find the maximum values of $\vec{E}$ and $\vec{B}$ if the beam diameter is $10\,cm$. Assume a uniform beam profile.

b) Find the maximum values of $\vec{E}$ and $\vec{B}$ if the beam is focused to a spot of diameter $100\,\mu m$.

Problem 2.31   Calculate the electric and magnetic fields at the top of Earth's atmosphere where the solar irradiance is $1340\,W/m^2$.

Problem 2.32   Find the frequency of a photon whose momentum is that of a $0.300\,g$ BB traveling at $100\,m/s$.

Problem 2.33   Determine the responsivity at $550\,nm$ of a photomultiplier tube that has a gain of 10^6 and a quantum efficiency of 30%. Repeat for a phototube with unity gain and the same quantum efficiency.

Problem 2.34   A $100\,W$ laser beam with $2.00\,mm$ beam diameter illuminates a surface that absorbs 40% and reflects 60%. Find the net force on the surface due to radiation pressure.

Problem 2.35   What is the minimum detectable wavelength of a photodetector with a cathode workfunction of $2.26\,eV$?

Problem 2.36   Under ideal circumstances, the human eye can detect a photon flux of about $250\,photons/s$ of $550\,nm$ light. If the eye has a pupil diameter of $8.0\,mm$, what irradiance does this correspond to?

Appendix: Maxwell's Equations in Differential Form

The most straightforward and elegant way to deduce the differential form of Maxwell's equations is to use the tools of vector calculus; however, it is possible to deduce the differential form of Maxwell's equations without relying on these more advanced mathematical tools. We will use both approaches: Method 1 takes a simpler approach that is limited to Cartesian coordinates, and Method 2 uses vector calculus for a coordinate independent demonstration.

Method 1: Cartesian Coordinates

We begin with Faraday's law (Equation 2.14). In order to evaluate the left-hand side, we must evaluate the line integral for an arbitrary electric field $\vec{E}$. Consider the path illustrated in Figure 2.A.1 which shows a small rectangle situated in the x-y plane. As we proceed around this path, we orient the vector $d\vec{A}$ according the rules of vector calculus: a person in the x-y plane standing in the direction of $d\vec{A}$ will be walking around the path positively if her left shoulder faces the inside of the curve. Thus, traversing the rectangle shown in Figure 2.A.1 in the counter-clockwise sense corresponds to a $d\vec{A}$ that points along positive z.

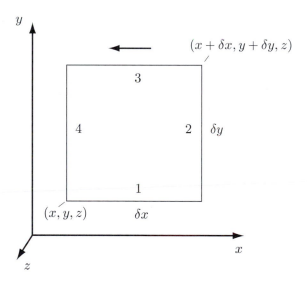

Figure 2.A.1 Path integration in the x-y plane.

As we proceed around the rectangle, the integrand $\vec{E} \cdot d\vec{\ell}$ selects the component of $\vec{E}$ that is tangent to the curve. So, for example, along the bottom of the rectangle, the x-component of $\vec{E}$ is selected, while along the top $-E_x$ is selected. Similarly, the right-hand side of the rectangle selects $+E_y$ and the left-hand side selects $-E_y$.

To complete the path integration, we use a Taylor series expansion to express the values of E_x and E_y at all points along the path:

$$E_x(x', y', z', t) = E_x(P, t) + \left.\frac{\partial E_x}{\partial x}\right|_P \Delta x + \left.\frac{\partial E_x}{\partial y}\right|_P \Delta y + \left.\frac{\partial E_x}{\partial z}\right|_P \Delta z + \ldots$$

$$E_y(x', y', z', t) = E_y(P, t) + \left.\frac{\partial E_y}{\partial x}\right|_P \Delta x + \left.\frac{\partial E_y}{\partial y}\right|_P \Delta y + \left.\frac{\partial E_y}{\partial z}\right|_P \Delta z + \ldots$$

In these equations, we are expanding each component E_x and E_y about the point $P = (x, y, z)$ located at the lower left corner of the rectangle in Figure 2.A.1. All partial derivatives and the leading term on the right-hand side are evaluated at the single point P. The terms Δx, Δy, and Δz each measure displacements from P; for example, $\Delta x = x' - x$ where x' corresponds to an evaluation point, with $x \leq x' \leq x + \delta x$. The sides δx of the rectangle are taken to be small enough so that higher-order terms in each series can be neglected.

We are now in a position to complete the path integration. Begin by breaking the path up into four segments (see Figure 2.A.1).

$$\oint \vec{E} \cdot d\vec{\ell} = \int_1 \vec{E} \cdot d\vec{\ell} + \int_2 \vec{E} \cdot d\vec{\ell} + \int_3 \vec{E} \cdot d\vec{\ell} + \int_4 \vec{E} \cdot d\vec{\ell}$$

Horizontal segments 1 and 3 involve x-components of $\vec{E}$ that differ only in the value of y. Along segment 3, x decreases as ℓ increases:

$$\int_1 \vec{E} \cdot d\vec{\ell} + \int_3 \vec{E} \cdot d\vec{\ell} = \int_x^{x+\delta x} \left(E_x(P, t) + \left.\frac{\partial E_x}{\partial x}\right|_P \Delta x \right) dx'$$

$$+ \int_{x+\delta x}^x -\left(E_x(P, t) + \left.\frac{\partial E_x}{\partial x}\right|_P \Delta x + \left.\frac{\partial E_x}{\partial y}\right|_P \delta y \right) (-dx')$$

$$= \int_x^{x+\delta x} \left(E_x(P, t) + \left.\frac{\partial E_x}{\partial x}\right|_P \Delta x \right) dx'$$

$$- \int_x^{x+\delta x} \left(E_x(P, t) + \left.\frac{\partial E_x}{\partial x}\right|_P \Delta x \right) dx' - \int_x^{x+\delta x} \left(\left.\frac{\partial E_x}{\partial y}\right|_P \delta y \right) dx'$$

$$= - \left.\frac{\partial E_x}{\partial y}\right|_P \int_x^{x+\delta x} (\delta y) \, dx'$$

$$= - \left.\frac{\partial E_x}{\partial y}\right|_P \delta y \, \delta x$$

Similarly, vertical segments 2 and 4 involve y-components of $\vec{E}$ that differ only in the value of x.

$$\int_2 \vec{E} \cdot d\vec{\ell} + \int_4 \vec{E} \cdot d\vec{\ell} = \int_y^{y+\delta y} \left(E_y + \left.\frac{\partial E_y}{\partial x}\right|_P \delta x + \left.\frac{\partial E_y}{\partial y}\right|_P \Delta y \right) dy' +$$

$$+ \int_{y+\delta y}^y -\left(E_y + \left.\frac{\partial E_y}{\partial y}\right|_P \Delta y \right) (-dy')$$

$$= \left.\frac{\partial E_y}{\partial x}\right|_P \delta x \, \delta y$$

Thus,

$$\oint \vec{E} \cdot d\vec{\ell} = \left(\frac{\partial E_y}{\partial x} - \frac{\partial E_x}{\partial y} \right)\Bigg|_P \delta x \, \delta y \qquad (2.\text{A}.1)$$

The right-hand side of Faraday's law involves an integral over the surface enclosed by the path of Figure 2.A.1.

$$-\frac{d}{dt} \int_A \vec{B} \cdot d\vec{A} = -\frac{d}{dt} \int_A B_z \, dA = -\int_A \frac{\partial B_z}{\partial t} dA$$

We assume that the rectangle is sufficiently small so that $\frac{\partial B_z}{\partial t}$ may be considered constant in the space coordinates (x, y, z). Thus

$$-\frac{d}{dt} \int_A \vec{B} \cdot d\vec{A} = -\frac{\partial B_z}{\partial t}\Bigg|_P \int_A dA = -\frac{\partial B_z}{\partial t}\Bigg|_P \delta x \, \delta y \qquad (2.\text{A}.2)$$

Thus, for the path shown in Figure 2.A.1, Faraday's law 2.14 along with Equations 2.A.1 and 2.A.2 gives

$$\frac{\partial E_y}{\partial x} - \frac{\partial E_x}{\partial y} = -\frac{\partial B_z}{\partial t} \qquad (2.\text{A}.3)$$

where all derivatives are understood to be evaluated at the point (x, y, z). The reader may show that similar rectangles situated in the x-z and z-y planes give

$$\frac{\partial E_x}{\partial z} - \frac{\partial E_z}{\partial x} = -\frac{\partial B_y}{\partial t} \qquad (2.\text{A}.4)$$

$$\frac{\partial E_z}{\partial y} - \frac{\partial E_y}{\partial z} = -\frac{\partial B_x}{\partial t} \qquad (2.\text{A}.5)$$

Equations 2.A.3 — 2.A.5 are the differential form of Faraday's law.

Ampere's law (Equation 2.15) may be evaluated in precisely the same way to give:

$$\frac{\partial B_y}{\partial x} - \frac{\partial B_x}{\partial y} = \mu_0 \epsilon \frac{\partial E_z}{\partial t} \qquad (2.\text{A}.6)$$

$$\frac{\partial B_x}{\partial z} - \frac{\partial B_z}{\partial x} = \mu_0 \epsilon \frac{\partial E_y}{\partial t} \qquad (2.\text{A}.7)$$

$$\frac{\partial B_z}{\partial y} - \frac{\partial B_y}{\partial z} = \mu_0 \epsilon \frac{\partial E_x}{\partial t} \qquad (2.\text{A}.8)$$

Equations 2.A.6 — 2.A.8 are the differential form of Ampere's law.

To evaluate Gauss's law for electric fields (Equation 2.12), we consider a small cube with sides parallel to each coordinate axis, as shown in Figure 2.A.2. In order to evaluate the surface integral over the entire cube, we integrate separately over the six cube faces.

Consider the top and bottom surfaces of the cube, located at y and $y + \delta y$. The direction of $d\vec{A}$ over the top surface points along $+y$ and over the bottom surface points in the $-y$ direction. The contribution to the total electric flux over these two faces can be expressed

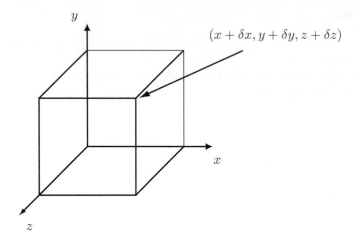

Figure 2.A.2   Surface integration over a cube. The origin of the coordinate axes shown are at (x, y, z), and the corner in the first quadrant is located at $(x + \delta x, y + \delta y, z + \delta z)$.

as follows:

$$\int_A \vec{E} \cdot d\vec{A} = \int_A E_y\,(x,\,y + \delta y,\,z,\,t)\,dx\,dz - \int_A E_y(x, y, z, t)\,dx\,dz$$

$$= \int_A \left(E_y(x, y + \delta y, z, t) - E_y(x, y, z, t) \right) dx\,dz$$

$$= \int_A \left(\int_y^{y+\delta y} \frac{\partial E_y}{\partial y}\,dy \right) dx\,dz$$

$$= \int_V \frac{\partial E_y}{\partial y}\,dx\,dy\,dz$$

where in the last result, the integration is over the entire volume of the cube. Similar results are obtained over the other two sets of cube faces:

$$\oint_A \vec{E} \cdot d\vec{A} = \int_V \left(\frac{\partial E_x}{\partial x} + \frac{\partial E_y}{\partial y} + \frac{\partial E_z}{\partial z} \right) dx\,dy\,dz = 0$$

Since the volume is arbitrary, the integrand of the volume integral must be zero.

$$\frac{\partial E_x}{\partial x} + \frac{\partial E_y}{\partial y} + \frac{\partial E_z}{\partial z} = 0 \qquad (2.A.9)$$

Equation 2.A.9 is the differential form of Gauss's law for electric fields.

Similar remarks hold for Gauss's law for magnetic fields (Equation 2.13). Thus,

$$\frac{\partial B_x}{\partial x} + \frac{\partial B_y}{\partial y} + \frac{\partial B_z}{\partial z} = 0 \qquad (2.A.10)$$

Equation 2.A.10 is the differential form of Gauss's law for magnetic fields.

Method 2: Vector Calculus

The results of the last subsection may be obtained more elegantly using the resources of multivariable calculus. Readers whose background does not include these resources may prefer to scan this subsection for the essential ideas.

We begin by defining the *gradient operator* in Cartesian coordinates:

$$\vec{\nabla} = \frac{\partial}{\partial x}\hat{i} + \frac{\partial}{\partial y}\hat{j} + \frac{\partial}{\partial z}\hat{k} \tag{2.A.11}$$

Using the gradient operator, we may define two new differential operations: the *divergence* and *curl*. Let $\vec{F}$ represent an arbitrary vector field. The divergence is given by the dot product of $\vec{\nabla}$ and $\vec{F}$. In Cartesian coordinates,

$$\vec{\nabla} \cdot \vec{F} = \left(\frac{\partial}{\partial x}\hat{i} + \frac{\partial}{\partial y}\hat{j} + \frac{\partial}{\partial z}\hat{k} \right) \cdot \left(F_x\hat{i} + F_y\hat{j} + F_z\hat{k} \right)$$
$$= \frac{\partial F_x}{\partial x} + \frac{\partial F_y}{\partial y} + \frac{\partial F_z}{\partial z} \tag{2.A.12}$$

The curl is given by the cross product of $\vec{\nabla}$ and $\vec{F}$:

$$\vec{\nabla} \times \vec{F} = \begin{vmatrix} \hat{i} & \hat{j} & \hat{k} \\ \frac{\partial}{\partial x} & \frac{\partial}{\partial y} & \frac{\partial}{\partial z} \\ F_x & F_y & F_z \end{vmatrix}$$
$$= \left(\frac{\partial F_z}{\partial y} - \frac{\partial F_y}{\partial z} \right)\hat{i} + \left(\frac{\partial F_x}{\partial z} - \frac{\partial F_z}{\partial x} \right)\hat{j} + \left(\frac{\partial F_y}{\partial x} - \frac{\partial F_x}{\partial y} \right)\hat{k} \tag{2.A.13}$$

These differential operations lead to the following two theorems, stated here without proof [16]: the *divergence theorem* and *Stokes' theorem*.

Divergence Theorem: Let V be a volume enclosed by area A and let $\vec{F}$ be a differentiable vector field. Then

$$\int_V \left(\vec{\nabla} \cdot \vec{F} \right) d\tau = \oint_A \vec{F} \cdot d\vec{A} \tag{2.A.14}$$

where in Cartesian coordinates, $d\tau = dx\,dy\,dz$. The divergence theorem states that the integral of the divergence of a vector field over a region V is determined by the value of the field on the boundary A that encloses the region.

Stokes' Theorem: Let the area A be bounded by a curve C, and let $\vec{F}$ be a differentiable vector field. Then

$$\int_A \left(\vec{\nabla} \times \vec{F} \right) \cdot d\vec{A} = \oint_C \vec{F} \cdot d\vec{\ell} \tag{2.A.15}$$

Stokes' theorem states that the surface integral of the curl of a vector field is determined by the magnitude and direction of the vector field on the boundary of the area.

We may use the divergence theorem and Stokes' theorem to express Maxwell's equations in differential form. Begin with Gauss's law for electric fields given by Equation 2.12 and

---

[16]See, for example, Boas [1].

apply the divergence theorem:

$$\oint_A \vec{E} \cdot d\vec{A} = \int_V \left(\vec{\nabla} \cdot \vec{E} \right) d\tau = 0$$

Since the volume V is arbitrary, the integrand of the right-hand side must be zero:

$$\vec{\nabla} \cdot \vec{E} = 0 \tag{2.A.16}$$

This is Gauss's law for electric fields (free space) in differential form. In Cartesian coordinates, it is identical to Equation 2.A.9.

Similarly, Gauss's law for magnetic fields (Equation 2.13 gives

$$\oint_A \vec{B} \cdot d\vec{A} = \int_V \left(\vec{\nabla} \cdot \vec{B} \right) d\tau = 0$$

As above, this implies

$$\vec{\nabla} \cdot \vec{B} = 0 \tag{2.A.17}$$

This is Gauss's law for magnetic fields (free space) in differential form. In Cartesian coordinates, it is identical to Equation 2.A.10.

For Faraday's law (Equation 2.14) use Stokes' theorem:

$$\oint_C \vec{E} \cdot d\vec{\ell} = \int_A \left(\vec{\nabla} \times \vec{E} \right) \cdot d\vec{A} = - \int_A \frac{\partial \vec{B}}{\partial t} \cdot d\vec{A}$$

Since the area A is arbitrary,

$$\vec{\nabla} \times \vec{E} = -\frac{\partial \vec{B}}{\partial t} \tag{2.A.18}$$

This is Faraday's law for free space in differential form. In Cartesian coordinates, it is equivalent to Equations 2.A.3 — 2.A.5.

Similarly, Ampere's law (Equation 2.15) gives

$$\oint_C \vec{B} \cdot d\vec{\ell} = \int_A \left(\vec{\nabla} \times \vec{B} \right) \cdot d\vec{A} = \mu_0 \epsilon \int_A \frac{\partial \vec{E}}{\partial t} \cdot d\vec{A}$$

Since the area A is arbitrary,

$$\vec{\nabla} \times \vec{B} = \mu_0 \epsilon \frac{\partial \vec{E}}{\partial t} \tag{2.A.19}$$

This is Ampere's law for free space in differential form. In Cartesian coordinates, it is equivalent to Equations 2.A.6 — 2.A.8.

CHAPTER 3

REFLECTION AND REFRACTION

And perhaps, posterity will thank me for having shown it that the ancients did not know everything.

—Fermat

Contents

3.1 INTRODUCTION

This chapter investigates the propagation of electromagnetic waves at interfaces between different optical media. The direction of refracted beams was first described by Snell in the early 1600s; reflection was described much earlier by the Greeks. Both phenomena were explained first variationally in 1657 by Fermat, then later by Maxwell with his field equations of electromagnetism. Fresnel was first to describe the *quantity* of light reflected and refracted at a boundary with the equations that bear his name. His accomplishment was all the more remarkable in that he derived these results well before the availability of Maxwell's equations. The chapter concludes with an introduction to *dispersion* in optical materials that provides a conceptual foundation for refraction and absorption.

3.2   OVERVIEW OF REFLECTION AND REFRACTION

Reflection and *refraction* occur whenever light enters into a new medium where its speed changes value. From the previous chapter, we found that the speed of light within a material is given by

$$v = \frac{1}{\sqrt{\epsilon \mu_0}} = \frac{1}{\sqrt{(K_E \epsilon_0)\, \mu_0}} = \frac{1}{\sqrt{K_E}} \frac{1}{\sqrt{\epsilon_0 \mu_0}} = \frac{c}{n} \tag{3.1}$$

where c is the speed of light in vacuum, K_E is the dielectric constant, and n is the index of refraction

$$n = \sqrt{K_E} \tag{3.2}$$

Thus, reflection and refraction occur whenever light enters into a new medium where the value of n changes value. Table 3.1 gives the indices of refraction for some common optical materials.

The *law of reflection* has been known experimentally since antiquity: *the angle of incidence is equal to the angle of reflection.* In order to define these angles, we make use of the concept of a *light ray*. For a plane wave with plane-parallel wavefronts, the associated ray is a directed line segment pointing perpendicular to the wavefronts in the direction of propagation of the wave.

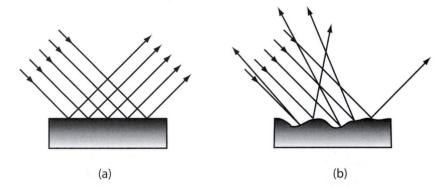

(a)   (b)

Figure 3.1   (a) Specular reflection. (b) Diffuse reflection.

An entire beam will reflect according to the law of reflection only if the surface of the interface between the two media is smooth. Reflection from a smooth surface is called *specular reflection*. Reflection from a rough surface is called *diffuse reflection*. Specular and diffuse reflection are illustrated in Figure 3.1.

Figure 3.2 shows a beam of light reflected from an interface separating two media. The incident beam is indicated by the incident ray, and the reflected beam is indicated by the reflected ray. The angle of incidence and angle of reflection are measured from the *normal to the interface*. Using these angles, the law of reflection is stated as follows:

$$\theta_i = \theta_r \tag{3.3}$$

Table 3.1. Range of transparency and indices of refraction for selected optical materials.

Material	Nominal Range of Transparency (μm)	n_r [a]
Vacuum	—	1
Air	$0.2 - 5.0, 9.0 - 14.0$	1.00029
Water	—	1.33
Ice	—	1.31
BK-7 Glass	$0.35 - 2.0$	1.50
Optical Crown Glass	$0.35 - 2.5$	1.52
Fused Silica (SiO_2) UV Grade	$0.175 - 2.2, 2.9 - 3.6$	1.46
Fused Silica (SiO_2) IR Grade	$0.2 - 3.9$	1.46
Quartz Crystal (SiO_2)	$0.2 - 2.4$	1.46
Zirconium Dioxide (ZrO_2)	$0.36 - 7.0$	2.40
Sapphire (Al_2O_3)	$0.15 - 5.0$	1.76
Diamond (C)	$0.22 - 4.3, 5.4 - FIR$ [b]	2.42
Calcium Fluoride (CaF_2)	$0.18 - 8.0$	1.40
Magnesium Oxide (MgO)	$0.4 - 8.0$	1.72
Barium Fluoride (BaF_2)	$0.2 - 11$	1.47
Strontium Fluoride (SrF_2)	$0.15 - 11$	1.44
Strontium Titanate ($SrTiO_3$)	$0.4 - 6$	2.31
Sodium Chloride ($NaCl$)	$0.35 - 15$	1.53
Potassium Bromide (KBr)	$0.3 - 25$	1.54
Potassium Chloride (KCl)	$0.3 - 20$	1.48
Cesium Iodide (CsI)	$0.5 - 70$	1.76
Potassium Iodide (KI)	$0.4 - 40$	1.64
Zinc Selenide ($ZnSe$)	$0.5 - 20$	2.49
Zinc Sulfide (ZnS)	$0.45 - 14$	2.27
Silicon (Si)	$1.0 - 100$	3.42
Germanium (Ge)	$2 - 17$	4.02
Gallium Arsenide ($GaAs$)	$1.0 - 17$	3.14
Cadmium Telluride ($CdTe$)	$1.0 - 28$	2.56
Polyethylene, High Density	$16 - FIR$	1.55

[a] Nominal, over the range of transparency. See Section 3.7.

[b] Far-infrared.

The *law of refraction*, also called *Snell's law*, was found experimentally by Snell early in the seventeenth century. In Figure 3.2, the light ray is incident within the *incident medium* and the light beam transmits into the *transmitted medium*. The index of refraction of the incident medium is n_i and the index of refraction of the transmitted medium is n_t. According to Snell's law

$$n_i \sin \theta_i = n_t \sin \theta_t \tag{3.4}$$

Finally, it is implicit in Figure 3.2 that the incident, reflected, and refracted rays all lie within a single plane. This plane is called the *plane of incidence*.

We will refer to the case where $n_i < n_t$ as *external incidence*. An example of external incidence occurs at an air-glass interface. *Internal incidence* occurs when $n_i > n_t$; for example, when a light beam passes from glass to air.

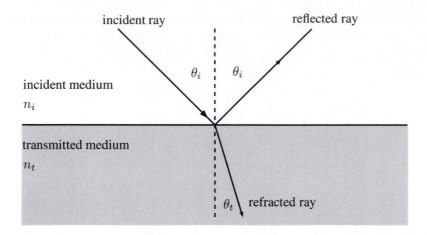

incident ray

reflected ray

θ_i θ_i

incident medium

n_i

transmitted medium

n_t

θ_t refracted ray

Figure 3.2   An interface between two media, showing the incident, reflected, and refracted rays.

■ **EXAMPLE 3.1**

Show that a light beam passing through a plane slab emerges parallel to its incident direction displaced laterally by a distance d given by

$$d = \frac{t}{\cos\theta_2} \sin(\theta_1 - \theta_2) \tag{3.5}$$

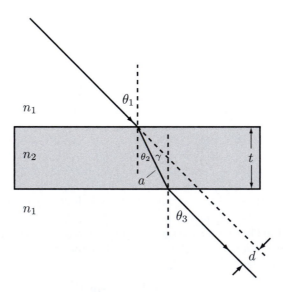

n_1

θ_1

n_2

θ_2 γ

a

t

n_1

θ_3

d

Figure 3.3   Light rays refracting through a slab of material with parallel sides.

Solution

Let the slab have index of refraction n_2 and the incident and transmitted medium have index of refraction n_1 as shown in Figure 3.3. To show that the incident and transmitted rays are parallel, begin with Snell's law at the first interface:

$$\sin \theta_2 = \frac{n_1}{n_2} \sin \theta_1$$

The beam transmitted by the first interface is incident internally on the second interface. Since the two interfaces are parallel, the angle of incidence for the second interface is θ_2:

$$\sin \theta_3 = \frac{n_2}{n_1} \sin \theta_2$$

Combining these results gives

$$\sin \theta_3 = \frac{n_2}{n_1} \left(\frac{n_1}{n_2} \sin \theta_1 \right) = \sin \theta_1$$

Since θ_1 and θ_3 are both positive and less than $90°$, $\theta_1 = \theta_3$, and thus the incident and transmitted beams are parallel.

To find the offset distance d, refer to Figure 3.3, and note that

$$d = a \sin \gamma$$

where

$$a = \frac{t}{\cos \theta_2}$$

and $\gamma = \theta_1 - \theta_2$. Combining these results gives Equation 3.5:

$$d = \frac{t}{\cos \theta_2} \sin (\theta_1 - \theta_2)$$

The amount of offset is a linear function of the slab thickness.

The laws of reflection and refraction may be derived from Maxwell's equations, as we show in Section 3.3.2. Before doing so, it will be instructive to derive these laws using a variational approach first described by Fermat.

3.2.1   Fermat's Principle of Least Time

Hero of Alexander discovered that a light ray traveling from an object to a mirror and then to an observer's eye takes a path that is *shorter* than any other available path. In 1657, the French mathematician Fermat deduced the law of refraction from a similar principle: that light travels between two points along the path that requires the smallest *transit time*[1].

Consider first the case of reflection as illustrated in Figure 3.4, where light from a source S travels to an observer at D. The total distance from S to D is

$$L = \sqrt{s^2 + x^2} + \sqrt{d^2 + (\rho - x)^2}$$

Since the entire path lies within the same medium, the light speed is constant along L. Thus, the minimum physical path length will give the minimum transit time. Different

---

[1]For a modern derivation of this principle from a completely different perspective, see Section 6.5. See also: *QED: The Strange Theory of Light and Matter*, by R. P. Feynman [5]

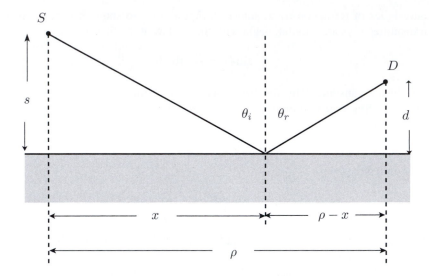

Figure 3.4   A light ray originates at S, reflects from the interface, then travels to D.

values for θ_i and θ_r correspond to different values of the parameter x in Figure 3.4. To find the minimum value of L, take the derivative of L with respect to x, and set this to zero:

$$\frac{dL}{dx} = \frac{1}{2} \frac{2x\,(1)}{\sqrt{s^2 + x^2}} + \frac{1}{2} \frac{2\,(\rho - x)\,(-1)}{\sqrt{d^2 + (\rho - x)^2}}$$

$$= \frac{x}{\sqrt{s^2 + x^2}} - \frac{(\rho - x)}{\sqrt{d^2 + (\rho - x)^2}} = 0$$

From Figure 3.4, we identify the two terms above as $\sin \theta_i$ and $\sin \theta_r$. Since

$$\sin \theta_i = \sin \theta_r$$

θ_i and θ_r are both acute,

$$\theta_i = \theta_r$$

To find Snell's law from the principle of least time, we must minimize the light travel time over a path that contains segments within two different media, as shown in Figure 3.5. Along segment $\overline{SO}$ the light speed is $v_i = \frac{c}{n_i}$, and along the segment $\overline{OD}$ the light speed is $v_t = \frac{c}{n_t}$. The transit time is

$$t = \frac{\overline{SO}}{v_i} + \frac{\overline{OD}}{v_t} = \frac{\sqrt{s^2 + x^2}}{v_i} + \frac{\sqrt{d^2 + (\rho - x)^2}}{v_t}$$

All paths must begin at S and end at D; however different paths will enter the transmitted medium at different points O. To find the path that gives the least time, take the derivative of t with respect to x and set this equal to zero:

$$\frac{dt}{dx} = \frac{1}{v_i} \frac{x}{\sqrt{s^2 + x^2}} - \frac{1}{v_t} \frac{\rho - x}{\sqrt{d^2 + (\rho - x)^2}} = 0$$

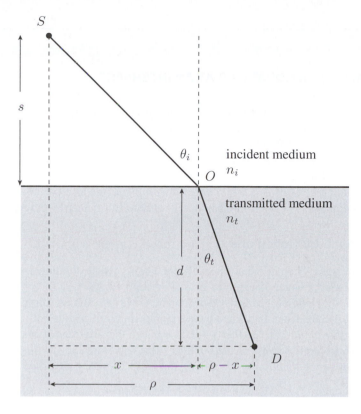

Figure 3.5 A light ray originates at S, refracts at the interface, then travels to D.

substituting $v = \frac{c}{n}$ in each medium, and identifying $\sin \theta_i$ and $\sin \theta_t$ in Figure 3.5, we obtain

$$\frac{n_i}{c} \sin \theta_i - \frac{n_t}{c} \sin \theta_t = 0$$

or

$$n_i \sin \theta_i = n_t \sin \theta_t$$

Fermat's principle of least time is an example of a *variational principle*. There are many other cases where Nature can be shown to behave according to maximum or minimum values of some physical quantity. It is fascinating and beautiful to discover examples such as Fermat's principle; however, Maxwell's equations provide a deeper explanation by describing all features of classical optics and electromagnetism, including Snell's law and the law of reflection.

Problem 3.1   Find the value of the offset d of Example 3.1 for a beam of light incident in air at $30°$ on a $6.35\,mm$-thick slab of glass with index of refraction equal to 1.52. Repeat for a slab of diamond with index 2.42.

Problem 3.2   Find θ_t at an air-glass and glass-air interface for $\theta_i = 40°$. Use 1.50 for the index of refraction of glass. Repeat for water-glass and glass-water interfaces.

3.3   MAXWELL'S EQUATIONS AT AN INTERFACE

In this section, we use Maxwell's equations to determine how electromagnetic waves reflect and refract from a plane interface separating two optical mediums. In doing so, we determine not only the direction of the reflected and refracted rays as described by the laws of reflection and refraction, but also how much of the incident electromagnetic energy is reflected and refracted.

We will approach this as a *boundary value problem*, and as a preliminary step, we use Maxwell's equations to determine *boundary conditions* for electric and magnetic fields.

3.3.1   Boundary Conditions

We have shown in Chapter 2 that Maxwell's equations can be combined to give differential wave equations for electromagnetic waves. Furthermore, we have found solutions to these equations in the form of traveling electromagnetic waves. To determine how electromagnetic waves behave when they encounter a change in their medium of travel, we utilize a *boundary value approach*. In other words, we match the solutions of Maxwell's equations (i.e., electromagnetic waves) at the boundary that separates the two media using the appropriate *boundary conditions*. We will obtain these boundary conditions from Maxwell's equations.

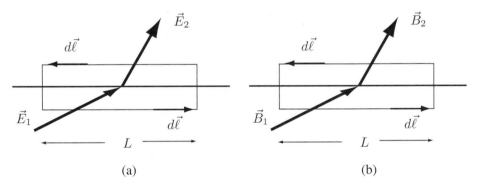

Figure 3.6   (a) Electric field vectors on either side of an interface separating two media. The rectangle is an integration path for Faraday's law. (b) Repeat of Figure (a) for the case of magnetic fields.

We begin with Faraday's law of induction:

$$\oint_C \vec{E} \cdot d\vec{\ell} = - \int_A \frac{\partial \vec{B}}{\partial t} \cdot d\vec{A} \tag{3.6}$$

where C is a path that encloses the area A. For a path C, we utilize an infinitely thin rectangle, as shown in Figure 3.6(a). The area enclosed by C is negligible, so the right hand side of Equation 3.6 is zero:

$$\oint_C \vec{E} \cdot d\vec{\ell} = 0$$

The bottom of the rectangle lies entirely within Region 1, and the dot product $\vec{E} \cdot d\vec{\ell}$ along this portion of the path selects the tangential component of the electric field in this region: E_{t1}. The top portion of the rectangle selects the tangential component of $\vec{E}_2$, this time with a minus sign since the orientation of $d\vec{\ell}$ reverses here. The sides of the rectangle are small enough so that their contribution to the line integral can be neglected. Finally, the length L of the rectangle is small enough so that the values of E_{t1} and E_{t2} remain constant over the integration. Faraday's law gives

$$\oint_C \vec{E} \cdot d\vec{\ell} = \oint_C (E_{t1} - E_{t2}) \, d\ell = (E_{t1} - E_{t2}) \, L = 0$$

By assumption, L is not zero, so

$$E_{t1} = E_{t2} \tag{3.7}$$

In words, *the tangential component of the electric field is continuous at a boundary.*

The corresponding boundary condition for magnetic fields is given by Ampere's law

$$\oint_C \vec{B} \cdot d\vec{\ell} = \mu_0 I + \mu_0 \left(\oint_A \epsilon \frac{\partial \vec{E}}{\partial t} \cdot d\vec{A} \right)$$

The small rectangular path in Figure 3.6(b) gives zero for the flux integral. If the current I enclosed by the path is negligible or zero, Ampere's law becomes

$$\oint_C \vec{B} \cdot d\vec{\ell} = 0$$

By the same arguments used above, this implies

$$B_{t1} = B_{t2} \tag{3.8}$$

In words, *in non-magnetic materials, the tangential component of the magnetic field $\vec{B}$ is continuous at a boundary.*[2]

3.3.2   Electromagnetic Waves at an Interface

We now derive the laws of reflection and refraction using the laws of electromagnetism. According to Maxwell's equations, solutions to the differential wave equations are transverse electromagnetic waves. Assume a plane-polarized harmonic electromagnetic wave incident upon an interface that separates two media, as shown in Figure 3.2. As shown in Chapter 2, we may specify this wave completely by giving its electric field

$$\vec{E}_i (\vec{r}, t) = \vec{E}_{0i} \exp i \left(\vec{k}_i \cdot \vec{r} - \omega_i t \right) \tag{3.9}$$

where $\vec{r}$ locates any point on the interface relative to an arbitrary origin of coordinates, $\vec{k}_i$ is the propagation vector of the incident wave, and ω_i is the angular frequency of the incident wave. In a similar way, we define the reflected and transmitted waves:

$$\vec{E}_r (\vec{r}, t) = \vec{E}_{0r} \exp i \left(\vec{k}_r \cdot \vec{r} - \omega_r t \right) \tag{3.10}$$

---

[2]For the corresponding boundary condition appropriate for materials with magnetic properties, see Guenther [10].

$$\vec{E}_t\left(\vec{r}, t\right) = \vec{E}_{0t} \exp i \left(\vec{k}_t \cdot \vec{r} - \omega_t t\right) \tag{3.11}$$

The amplitudes $\vec{E}_{0i}$, $\vec{E}_{0r}$ and $\vec{E}_{0t}$ may all be complex to account for any relative phase shifts in the reflected and transmitted waves.

According to the boundary conditions derived in Section 3.3.1, the tangential components of $\vec{E}$ and $\vec{B}$ must be continuous at the interface. The fields on both sides of the interface are given by Equations 3.9 — 3.11. These solutions are vector quantities that depend on both space and time coordinates. Thus, in order for the boundary conditions to hold, Equations 3.9 — 3.11 must

1. be identical functions of time,

2. be identical functions of the space coordinates, and

3. have amplitudes $\vec{E}_{0i}$, $\vec{E}_{0r}$ and $\vec{E}_{0t}$ that satisfy the boundary conditions specified by Equations 3.7 and 3.8.

According to item 1, the boundary conditions must be satisfied for all values of the time t. This can only be true if the frequencies of all three electromagnetic waves are identical

$$\omega_i = \omega_r = \omega_t \tag{3.12}$$

This result has an interesting and important consequence. Since the wave speed within a medium must equal the product of frequency and wavelength, the *wavelength* must change as it enters the new medium. Consider a wave traveling within a medium whose index of refraction is n:

$$f\lambda = v = \frac{c}{n} = \frac{f_0\lambda_0}{n} = \frac{f\lambda_0}{n}$$

cancellation gives

$$\lambda = \frac{\lambda_0}{n} \tag{3.13}$$

where λ_0 is the vacuum wavelength. Thus, the wavelength becomes *smaller* in a region of higher index of refraction.

According to item 2 above, the boundary conditions must hold for all points on the interface. Thus

$$\vec{k}_i \cdot \vec{r} = \vec{k}_r \cdot \vec{r} = \vec{k}_t \cdot \vec{r} \tag{3.14}$$

This implies that the propagation vectors $\vec{k}_i$, $\vec{k}_r$, and $\vec{k}_t$ are all coplanar. To see why, consider the plane defined by $\vec{k}_i$ and $\vec{k}_r$ and suppose that $\vec{k}_t$ has a component lying outside of that plane. Choose the origin 0 so that the position vector points normal to this plane. Thus $\vec{k}_i \cdot \vec{r} = \vec{k}_r \cdot \vec{r} = 0$ but $\vec{k}_t \cdot \vec{r} \neq 0$, so Equation 3.14 can only be satisfied for all points on the interface if $\vec{k}_i$, $\vec{k}_r$, and $\vec{k}_t$ all lie in the same plane. We will refer to this plane as the *plane of incidence*.

Now consider an origin O located at an arbitrary point on the interface in the plane of incidence, as shown in Figure 3.7. For this choice of origin, the scalar products indicated in Equation 3.14 are given by

$$\vec{k}_i \cdot \vec{r} = r k_i \sin \theta_i \tag{3.15}$$

$$\vec{k}_r \cdot \vec{r} = r k_r \sin \theta_r \tag{3.16}$$

$$\vec{k}_t \cdot \vec{r} = r k_t \sin \theta_t \tag{3.17}$$

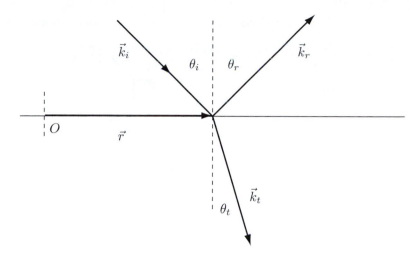

Figure 3.7   Origin of coordinates located at a point of intersection between the plane of incidence and the plane of the interface.

Since the incident and reflected waves travel within the same medium, $k_i = k_r$, and the first identity of Equation 3.14 gives

$$\sin \theta_i = \sin \theta_r$$

Since the incident and reflected angles are acute, this gives the law of reflection Equation 3.3:

$$\theta_i = \theta_r$$

Also, according to Equations 3.14, 3.15, and 3.17,

$$k_i \sin \theta_i = k_t \sin \theta_t$$

By Equation 3.13, $k = \frac{2\pi}{\lambda} = n \left(\frac{2\pi}{\lambda_0} \right)$. Substitution into the result above gives the law of refraction (Equation 3.4):

$$n_i \sin \theta_i = n_t \sin \theta_t$$

To summarize our results so far, the requirement that Equations 3.9 — 3.11 be identical functions of time and space (items 1 and 2) leads to the laws of reflection and refraction. These laws allow us to specify the *directions* of the refracted and reflected rays.

We must still apply the boundary conditions found in Section 3.3.1. The vector amplitudes of Equations 3.9 - 3.11, $\vec{E}_{0i}$, $\vec{E}_{0r}$ and $\vec{E}_{0t}$, must be oriented so that these boundary conditions are satisfied. We will examine this in the next section, where we will deduce the exact values of $\vec{E}_{0r}$ and $\vec{E}_{0t}$ for arbitrary incident wave with amplitude $\vec{E}_{0i}$.

3.4   THE FRESNEL EQUATIONS

We wish to apply the boundary conditions found in Section 3.3.1 to the wave amplitudes of Equations 3.9 — 3.11 for arbitrary angle of incidence and arbitrary incident polarization. To specify the polarizations of $\vec{E}_{0i}$, $\vec{E}_{0r}$, and $\vec{E}_{0t}$, we will use Cartesian coordinates where the x-y plane corresponds to the plane of incidence and the x-z plane corresponds to the

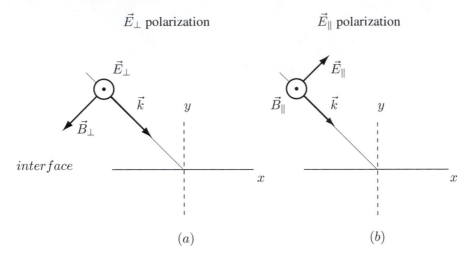

$\vec{E}_\perp$ polarization $\vec{E}_\parallel$ polarization

Figure 3.8   Orthogonal polarization states: (a) $\vec{E}_\perp$ and (b) $\vec{E}_\parallel$.

plane of the interface, as shown in Figure 3.8. An arbitrary state of linear polarization can be resolved along the two orthogonal directions indicated by $\vec{E}_\perp$ and $\vec{E}_\parallel$ in the diagram. The notation $\vec{E}_\perp$ corresponds to a state of linear polarization where the electric field is perpendicular to the plane of incidence, or parallel to the interface itself (in the z-direction in Figure 3.8). A $\vec{E}_\parallel$ polarization state has an electric field vector that lies entirely within the plane of incidence[3].

3.4.1   Incident Wave Polarized Normal to the Plane of Incidence

We first analyze the case where the incident wave has $\vec{E}_\perp$ polarization, as shown in Figure 3.9. The propagation vectors and magnetic field vectors for the incident, reflected, and transmitted rays are drawn so that $\vec{E} \times \vec{B}$ points in the direction of propagation (i.e. in the direction of $\vec{k}$) as the reader should verify. The vector triads of $\vec{E}$, $\vec{B}$ and $\vec{k}$ for each ray are drawn displaced from the point on the interface where the three rays meet. This is done for visual clarity; the reader should realize that we apply the boundary conditions at the intersection point on the interface.

According to boundary condition 3.7, the tangential component of the electric field is continuous at the interface. This gives:

$$E_{0i} + E_{0r} = E_{0t} \tag{3.18}$$

By boundary condition 3.8, the tangential component of the magnetic field is continuous across the interface. The tangential components of all magnetic field vectors shown in Figure 3.9 are x-components. The tangential component of $\vec{B}_{0i}$ is negative: $-B_{0i}\cos\theta_i$. $\vec{B}_{0r}$ has a positive tangential component equal to $B_{0r}\cos\theta_i$, and the tangential component

---

[3]The notation used to refer to these two polarization states is not standard. Some texts use s and p to refer to $\vec{E}_\perp$ and $\vec{E}_\parallel$, respectively. In this notation, s derives from the German word "senkrecht" which means "perpendicular", and p stands for parallel. In other texts, $\vec{E}_\perp$ is called "transverse electric" or TE, with the term transverse referring to the plane of the interface. In this notation, $\vec{E}_\parallel$ is called TM for transverse magnetic.

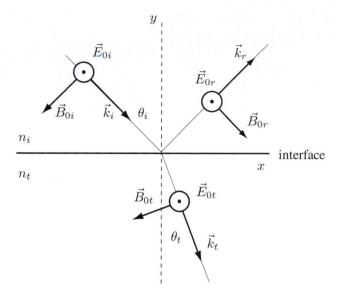

Figure 3.9   Incident $\vec{E}_\perp$ polarization.

of $\vec{B}_{0t}$ is $-B_{0t}\cos\theta_t$. Boundary condition 3.8 gives

$$-B_{0i}\cos\theta_i + B_{0r}\cos\theta_i = -B_{0t}\cos\theta_t \tag{3.19}$$

According to Equations 2.25 and 3.1

$$B = \frac{E}{v} = \frac{n}{c}E \tag{3.20}$$

The last two equations can be combined to give

$$n_i(E_{0i} - E_{0r})\cos\theta_i = n_t E_{0t}\cos\theta_t \tag{3.21}$$

Our goal is to solve for E_{0r} and E_{0t} for any given incident wave. Thus, divide both sides of Equation 3.18 by E_{0i} and rearrange to give

$$\frac{E_{0t}}{E_{0i}} - \frac{E_{0r}}{E_{0i}} = 1 \tag{3.22}$$

Similarly, dividing Equation 3.21 by E_{0i} and rearranging gives

$$(n_t\cos\theta_t)\frac{E_{0t}}{E_{0i}} + (n_i\cos\theta_i)\frac{E_{0r}}{E_{0i}} = n_i\cos\theta_i \tag{3.23}$$

Solving Equations 3.22 and 3.23 for the two unknowns $\frac{E_{0r}}{E_{0i}}$ and $\frac{E_{0t}}{E_{0i}}$ gives (see Example 3.4)

$$r_\perp \equiv \left(\frac{E_{0r}}{E_{0i}}\right)_\perp = \frac{n_i\cos\theta_i - n_t\cos\theta_t}{n_i\cos\theta_i + n_t\cos\theta_t} \tag{3.24}$$

$$t_\perp \equiv \left(\frac{E_{0t}}{E_{0i}}\right)_\perp = \frac{2n_i\cos\theta_i}{n_i\cos\theta_i + n_t\cos\theta_t} \tag{3.25}$$

where the $\perp$ subscript means "normal to the plane of incidence." For notational brevity we define $r_\perp = \left(\frac{E_{0r}}{E_{0i}}\right)_\perp$ and $t_\perp = \left(\frac{E_{0t}}{E_{0i}}\right)_\perp$.

We refer to $r_\perp$ and $t_\perp$ as the *Fresnel amplitude ratios*. It is important to note that they do not represent fractions of reflected and transmitted power. Rather, they are simply ratios of the electric field amplitudes of the reflected and refracted waves relative to the incident wave. The power ratios are called reflectivity and transmissivity, which we discuss in Section 3.6.

■ **EXAMPLE 3.2**

Find values of $r_\perp$ and $t_\perp$ for normal incidence when the incident medium is air ($n_i = 1.00$) and the transmitted medium is glass with $n_t = 1.50$. Repeat the calculation when the incident medium is glass and the transmitted medium is air.

Solution

If $\theta_i = 0$, then by Snell's law, $\theta_i = 0$. Substitution into Equations 3.24 and 3.25 gives

$$r_\perp(0) = \frac{n_i - n_t}{n_i + n_t}$$

$$t_\perp(0) = \frac{2n_i}{n_i + n_t}$$

When the incident medium is air:

$$r_\perp(0) = \frac{1.00 - 1.50}{1.00 + 1.50} = -0.200$$

$$t_\perp(0) = \frac{2.00}{2.50} = 0.800$$

If the incident medium is glass, the results are

$$r_\perp(0) = \frac{1.50 - 1.00}{2.50} = +0.200$$

$$t_\perp(0) = \frac{2(1.50)}{2.50} = 1.20$$

Notice that $r_\perp + t_\perp \neq 1$, consistent with our earlier remarks that these do not represent reflected powers (See Example 3.8). This is also evident from the fact that $t_\perp(0) > 1$. Note also that $r_\perp$ is negative for external incidence and positive for internal incidence. This will be discussed further in Section 3.5.5.

3.4.2 Incident Wave Polarized Parallel to the Plane of Incidence

In this case, the incident wave is polarized with an electric field that oscillates entirely within the plane of incidence. Figure 3.10 shows the incident, reflected and transmitted electromagnetic waves with $\vec{B}$ oriented so that $\vec{E} \times \vec{B}$ points in the direction of $\vec{k}$ as the reader should verify. The field orientations in Figure 3.10 are consistent with those of Figure 3.9 in that the electric fields all have positive tangential components.

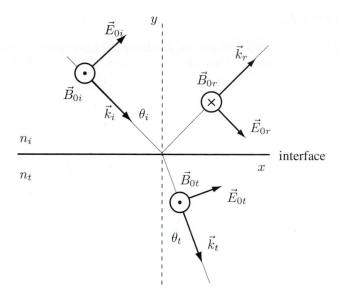

Figure 3.10   Incident $\vec{E}_\parallel$ polarization.

The boundary condition 3.7 that the tangential component of $\vec{E}$ be continuous across the interface leads to

$$E_{0i} \cos\theta_i + E_{0r} \cos\theta_i = E_{0t} \cos\theta_t \tag{3.26}$$

The tangential component of $\vec{B}$ must also be continuous (boundary condition 3.8)

$$B_{0i} - B_{0r} = B_{0t} \tag{3.27}$$

or by Equation 3.20,

$$n_i E_{0i} - n_i E_{0r} = n_t E_{0t} \tag{3.28}$$

Dividing Equations 3.26 and 3.28 by E_{0i} and rearranging gives

$$\cos\theta_t \left(\frac{E_{0t}}{E_{0i}}\right) - \cos\theta_i \left(\frac{E_{0r}}{E_{0i}}\right) = \cos\theta_i \tag{3.29}$$

$$n_t \left(\frac{E_{0t}}{E_{0i}}\right) + n_i \left(\frac{E_{0r}}{E_{0i}}\right) = n_i \tag{3.30}$$

Solving Equations 3.29 and 3.30 for the two unknowns $\frac{E_{0r}}{E_{0i}}$ and $\frac{E_{0t}}{E_{0i}}$ gives (see Problem 3.7)

$$r_\parallel \equiv \left(\frac{E_{0r}}{E_{0i}}\right)_\parallel = \frac{n_i \cos\theta_t - n_t \cos\theta_i}{n_i \cos\theta_t + n_t \cos\theta_i} \tag{3.31}$$

$$t_\parallel \equiv \left(\frac{E_{0t}}{E_{0i}}\right)_\parallel = \frac{2 n_i \cos\theta_i}{n_i \cos\theta_t + n_t \cos\theta_i} \tag{3.32}$$

where the $\parallel$ subscript means "parallel to the plane of incidence." Again, for notational brevity we define $r_\parallel = \left(\frac{E_{0r}}{E_{0i}}\right)_\parallel$ and $t_\parallel = \left(\frac{E_{0t}}{E_{0i}}\right)_\parallel$.

■ EXAMPLE 3.3

Find values of $r_\parallel$ and $t_\parallel$ for normal incidence when the incident medium is air and the transmitted medium is glass with $n = 1.5$. Repeat the calculation when the incident medium is glass and the transmitted medium is air.

Solution

Equations 3.31 and 3.32 give

$$r_\parallel(0) = \frac{n_i - n_t}{n_i + n_t}$$

$$t_\parallel(0) = \frac{2n_i}{n_i + n_t}$$

When the incident medium is air,

$$r_\parallel(0) = \frac{1.00 - 1.50}{1.00 + 1.50} = -0.200$$

$$t_\parallel(0) = \frac{2.00}{2.50} = 0.800$$

If the incident medium is glass, the results are

$$r_\parallel(0) = \frac{1.50 - 1.00}{2.50} = +0.200$$

$$t_\parallel(0) = \frac{2(1.50)}{2.50} = 1.20$$

These results are identical to those in Example 3.2 for $r_\perp$ and $t_\perp$ at normal incidence.

The field orientations of Figures 3.9 and 3.10 give identical values for $r_\perp$ and $r_\parallel$ at normal incidence. This is intuitively correct, since for normal incidence, there is no physical difference between the two orthogonal polarization states $\vec{E}_\perp$ and $\vec{E}_\parallel$. For angles other than normal incidence, the amplitude ratios r and t will depend on polarization.

■ EXAMPLE 3.4

Show that Equations 3.24 and 3.25 are the solutions to Equations 3.22 and 3.23.

Solution

We must solve Equations 3.22 and 3.23:

$$(1)\frac{E_{0t}}{E_{0i}} + (-1)\frac{E_{0r}}{E_{0i}} = 1$$

$$(n_t \cos\theta_t)\frac{E_{0t}}{E_{0i}} + (n_i \cos\theta_i)\frac{E_{0r}}{E_{0i}} = n_i \cos\theta_i$$

It is expedient to use *Cramer's rule*[4] to solve this set of simultaneous non-homogenous equations. Begin by forming a matrix of the coefficients that multiply the unknown quantities:

$$\begin{pmatrix} 1 & -1 \\ n_t \cos \theta_t & n_i \cos \theta_i \end{pmatrix}$$

We will also need the column vector of the non-homogeneous terms:

$$\begin{pmatrix} 1 \\ n_i \cos \theta_i \end{pmatrix}$$

The value of each unknown quantity is a ratio of determinants. The numerator for each ratio is obtained by modifying the above matrix by replacing the column that corresponds to the unknown being solved for with the non-homogeneous column vector, then taking its determinant. The denominator for both ratios the determinant of the coefficient matrix given above. Thus,

$$r_\perp \equiv \left(\frac{E_{0r}}{E_{0t}} \right)_\perp = \frac{\begin{vmatrix} 1 & 1 \\ n_t \cos \theta_t & n_i \cos \theta_i \end{vmatrix}}{\begin{vmatrix} 1 & -1 \\ n_t \cos \theta_t & n_i \cos \theta_i \end{vmatrix}} = \frac{n_i \cos \theta_i - n_t \cos \theta_t}{n_i \cos \theta_i + n_t \cos \theta_t}$$

$$t_\perp \equiv \left(\frac{E_{0t}}{E_{0t}} \right)_\perp = \frac{\begin{vmatrix} 1 & -1 \\ n_i \cos \theta_i & n_i \cos \theta_i \end{vmatrix}}{\begin{vmatrix} 1 & -1 \\ n_t \cos \theta_t & n_i \cos \theta_i \end{vmatrix}} = \frac{2 n_i \cos \theta_i}{n_i \cos \theta_i + n_t \cos \theta_t}$$

Cramer's rule is of course not necessary (especially for only two simultaneous equations), it is just convenient. We will use Cramer's rule again with systems of equations with more than two unknowns.

Problem 3.3   Find the wavelength, frequency, and speed of light with vacuum wavelength $633\,nm$ when it enters a region where the index of refraction is equal to 1.52.

Problem 3.4   Find values of $r_\perp$, $t_\perp$, $r_\parallel$, and $t_\parallel$ for normal incidence when the incident medium is air and the transmitted medium is diamond with $n_t = 2.42$. Repeat the calculation for internal incidence.

Problem 3.5   Find values of $r_\perp$, $t_\perp$, $r_\parallel$, and $t_\parallel$ for normal incidence when the incident medium is water ($n_i = 1.33$) and the transmitted medium is glass with $n_t = 1.55$. Repeat the calculation for internal incidence.

Problem 3.6   Find $r_\perp$, $t_\perp$, $r_\parallel$, and $t_\parallel$ at grazing incidence: $\theta_i = 90°$. Include values for both external and internal incidence using air and glass with $n = 1.50$.

[4]See, for example, Thomas *et al.* [13].

Problem 3.7   Derive Equations 3.31 and 3.32.

Problem 3.8   Find $r_\perp$, $t_\perp$, $r_\parallel$, and $t_\parallel$ for $\theta_i = 30°$ when the incident medium is air and the transmitted medium has index of refraction 1.50. Repeat the calculation for internal incidence.

Problem 3.9   Find $r_\perp$, $t_\perp$, $r_\parallel$, and $t_\parallel$ for $\theta_i = 30°$ when the incident medium is water ($n = 1.33$) and the transmitted medium has index of refraction 1.50. Repeat the calculation for internal incidence.

3.5   INTERPRETATION OF THE FRESNEL EQUATIONS

Before plotting the formulas obtained above for $r_\perp$, $t_\perp$, $r_\parallel$, and $t_\parallel$, we examine some limiting cases.

3.5.1   Normal Incidence

As shown in Examples 3.2 and 3.3 above, the values of $r_\perp$ and $r_\parallel$ for normal incidence are identical:

$$r_\perp(0) = r_\parallel(0) = \frac{n_i - n_t}{n_i + n_t} \tag{3.33}$$

Similarly, the values of $t_\perp$ and $t_\parallel$ are also identical for normal incidence:

$$t_\perp(0) = t_\parallel(0) = \frac{2n_i}{n_i + n_t} \tag{3.34}$$

For external incidence, $r_\perp$ and $r_\parallel$ are negative. This will be interpreted in Section 3.5.5.

3.5.2   Brewster's Angle

A close examination of the equation for $r_\parallel$ shows that it can be zero:

$$r_\parallel = \frac{n_i \cos\theta_t - n_t \cos\theta_i}{n_i \cos\theta_t + n_t \cos\theta_i}$$

This ratio will be zero when

$$n_i \cos\theta_t = n_t \cos\theta_i$$

By Snell's law, $n_i = n_t \frac{\sin\theta_t}{\sin\theta_i}$, giving

$$\sin\theta_t \cos\theta_t = \sin\theta_i \cos\theta_i$$

From trigonometry, $\sin 2\alpha = 2\sin\alpha\cos\alpha$, giving

$$\sin 2\theta_t = \sin 2\theta_i$$

The solution $\theta_t = \theta_i$ is not allowed by Snell's law, so

$$2\theta_t = \pi - 2\theta_i$$

or

$$\theta_i + \theta_t = \frac{\pi}{2} \tag{3.35}$$

Substituting this result into Snell's law gives

$$n_i \sin \theta_i = n_t \sin \left(\frac{\pi}{2} - \theta_i \right) = n_t \cos \theta_i$$

or

$$\tan \theta_B = \frac{n_t}{n_i} \tag{3.36}$$

where θ_B is called *Brewster's angle*.

It is only $r_\parallel$ that goes to zero when $\theta_i = \theta_B$; $r_\perp$ is nonzero for all incident angles when $n_i \neq n_t$. An electromagnetic wave with electric field parallel to the plane of incidence has a magnetic field that oscillates *parallel to the plane of the interface*. This situation is sometimes referred to as *transverse magnetic*, meaning that the magnetic field is *transverse to the plane of the interface*.

■ **APPLICATION NOTE 3.1**

Brewster Windows

A window oriented at Brewster's angle gives *zero* reflection for the $\vec{E}_\parallel$ polarization. *Brewster windows* are often used in laser cavities. In a typical laser, the beam traverses the laser cavity tens or perhaps hundreds of times before exiting the laser, and even a small amount of reflection from the windows can reduce the laser gain below the threshold required for operation. A laser that uses Brewster windows emits a *polarized* beam, as shown in Figure 3.11.

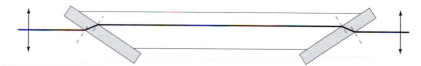

Figure 3.11   A cylindrical tube with Brewster windows attached to the ends.   The vertical polarization is $\vec{E}_\parallel$ to the window interfaces.

Brewster's angle is often called the *polarization angle*. A beam of light with arbitrary polarization can be resolved into two component beams with the $\vec{E}_\perp$ and $\vec{E}_\parallel$ polarizations illustrated in Figure 3.8. For a beam incident on a surface at $\theta_i = \theta_B$, the $\vec{E}_\parallel$ component will not be reflected, so the reflected beam will be completely polarized as $\vec{E}_\perp$. For example, specular reflection from horizontal surfaces typically occurs close enough to Brewster's angle so that the glare from the surface is almost completely polarized along $\vec{E}_\perp$. An electromagnetic wave with electric field oriented perpendicular to the plane of incidence oscillates parallel to the interface. Thus, glare from horizontal surfaces is highly polarized *horizontally*.

A *polarizer* is a device that transmits only the component of incident polarization that is parallel to its *transmission axis*. Polaroid© material is a common example of a polarizer, and this material is often used in sunglasses. Orienting the transmission axis of the Polaroid material *vertically* preferentially absorbs glare from horizontal surfaces.

■ **EXAMPLE 3.5**

Find Brewster's angle for internal and external incidence at an air-glass interface. Assume a index of refraction of 1.50 for the glass.

Solution

For external incidence, $n_i = 1.00$ and $n_t = 1.50$, giving

$$\theta_B = \tan^{-1}\left(\frac{1.50}{1.00}\right) = 56.3°$$

For internal incidence, the incident index is that of glass. Let θ_B' be the Brewster's angle for internal incidence:

$$\theta_B' = \tan^{-1}\left(\frac{1.00}{1.50}\right) = 33.7°$$

Since the role of n_i and n_t are interchanged between external and internal incidence, the Brewster's angles for these two cases are *complements*.

When light is incident externally on a slab at Brewster's angle, the beam that transmits into the slab is also incident internally on the second surface at the internal Brewster's angle θ_B'. To see that this is so, use Equation 3.35. As noted in the Remarks for the previous Example, the roles of n_i and n_t simply interchange in Equation 3.36, so θ_B and θ_B' are complements. According to Equation 3.35, the transmitted and incident angles are also complements when $\theta_i = \theta_B$. Thus, the beam transmitted into the slab is incident internally at θ_B' at the second parallel surface.

3.5.3 Total Internal Reflection

For *internal incidence*, Snell's law gives an interesting result:

$$\sin\theta_t = \left(\frac{n_i}{n_t}\right)\sin\theta_i$$

Since $(n_i > n_t)$, the term in parentheses is greater than unity, so $\theta_t > \theta_i$. The incident angle that gives a transmitted angle of 90° is called the critical angle θ_c. Since $\sin(90°) = 1$,

$$\sin\theta_c = \frac{n_t}{n_i} \tag{3.37}$$

Angles of incidence that are greater than θ_c are problematic with regards to Snell's law. In such a case Equation 3.37 appears to imply that $\sin\theta_t > 1$.

To investigate this, consider $r_\perp$ (Equation 3.24)

$$r_\perp = \frac{\frac{n_i}{n_t}\cos\theta_i - \cos\theta_t}{\frac{n_i}{n_t}\cos\theta_i + \cos\theta_t}$$

Use Snell's law to express $\cos\theta_t$ in terms of the incident angle

$$\cos\theta_t = \sqrt{1 - \sin^2\theta_t} = \sqrt{1 - \left(\frac{n_i}{n_t}\sin\theta_i\right)^2}$$

According to Snell's law and Equation 3.37, when $\theta_i > \theta_c$ the quantity under the square root sign becomes negative. In this case, $\cos \theta_t$ becomes *imaginary*

$$\cos \theta_t = i \sqrt{\left(\frac{n_i}{n_t} \sin \theta_i \right)^2 - 1}$$

Substituting this into the expression for $r_\perp$ gives

$$r_\perp = \left(\frac{E_{0r}}{E_{0i}} \right)_\perp = \frac{\frac{n_i}{n_t} \cos \theta_i - i \sqrt{\left(\frac{n_i}{n_t} \sin \theta_i \right)^2 - 1}}{\frac{n_i}{n_t} \cos \theta_i + i \sqrt{\left(\frac{n_i}{n_t} \sin \theta_i \right)^2 - 1}} \tag{3.38}$$

Notice that this expression for $r_\perp$ has the form of a ratio of a complex number and its complex conjugate. Such a complex number always has a magnitude of one:

$$r = \left| \frac{z}{z^*} \right| = \left(\frac{z}{z^*} \right) \left(\frac{z}{z^*} \right)^* = \left(\frac{z}{z^*} \right) \left(\frac{z^*}{z} \right) = 1 \tag{3.39}$$

Thus, E_{0r} and E_{0i} differ only in phase

$$r_\perp = \left(\frac{E_{0r}}{E_{0i}} \right)_\perp = e^{i\varphi_\perp}$$

or

$$E_{0r} = E_{0i} e^{i\varphi_\perp} \tag{3.40}$$

Since E_{0r} and E_{0i} have the same magnitude, the beam is *totally reflected*. By conservation of energy, no energy is transmitted across the interface[5]. The phase shift $\varphi_\perp$ is given by (see Example 3.6)

$$\varphi_\perp = 2 \tan^{-1} \left(\frac{n_t}{n_i} \alpha \right) \tag{3.41}$$

where

$$\alpha = \frac{\sqrt{\left(\frac{n_i}{n_t} \sin \theta_i \right)^2 - 1}}{\cos \theta_i} \tag{3.42}$$

Similarly, by Equation 3.31

$$r_\parallel \equiv \left(\frac{E_{0r}}{E_{0i}} \right)_\parallel = \frac{-\frac{n_t}{n_i} \cos \theta_i + \cos \theta_t}{\frac{n_t}{n_i} \cos \theta_i + \cos \theta_t}$$

$$= -\frac{\frac{n_t}{n_i} \cos \theta_i - i \sqrt{\left(\frac{n_i}{n_t} \sin \theta_i \right)^2 - 1}}{\frac{n_t}{n_i} \cos \theta_i + i \sqrt{\left(\frac{n_i}{n_t} \sin \theta_i \right)^2 - 1}} \tag{3.43}$$

This also has the form of $\frac{z}{z^*}$, so it also has unity magnitude. Consequently, for the parallel polarization,

$$E_{0r} = E_{0i} e^{i\varphi_\parallel} \tag{3.44}$$

[5]We will show this formally in Section 3.6.

The phase shift $\varphi_{\parallel}$ is given by (see Problem 3.12)

$$\varphi_{\parallel} = 2 \tan^{-1} \left(\frac{n_i}{n_t} \alpha \right)$$ (3.45)

where α is again given by Equation 3.42.

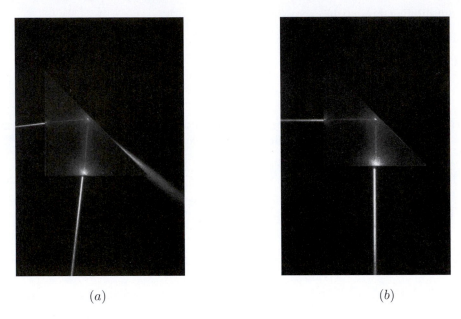

(a) (b)

Figure 3.12 (a) The beam incident from the left has an internal incident angle that is slightly less that the critical angle. The transmitted beam just grazes the interface. (b) Total internal reflection. (GIPhotoStock/Photo Researchers, Inc.)

In summary, for angles of incidence $\theta_i > \theta_c$, there is total internal reflection for both $\vec{E}_{\perp}$ and $\vec{E}_{\parallel}$. The phase shifts $\varphi_{\perp}$ and $\varphi_{\parallel}$ depend upon angle, and for each value of $\theta_i > \theta_c$ are in general not equal.

■ **EXAMPLE 3.6**

Show that

$$\varphi_{\perp} = 2 \tan^{-1} \left(\frac{n_t}{n_i} \alpha \right)$$

where

$$\alpha = \frac{\sqrt{\left(\frac{n_i}{n_t} \sin \theta_i \right)^2 - 1}}{\cos \theta_i}$$

Solution

Being careful here can save a lot of work. Recall that $r_{\perp}$ can be expressed as a complex ratio of the form

$$r_{\perp} = \frac{z}{z^*}$$

where, according to Equation 3.38

$$z = \frac{n_i}{n_t} \cos\theta_i - i\sqrt{\left(\frac{n_i}{n_t} \sin\theta_i\right)^2 - 1} \qquad (3.46)$$

Multiplying the numerator and denominator of $r_\perp$ by z makes the denominator real:

$$r_\perp = \frac{z}{z^*} = \frac{z}{z^*}\frac{z}{z} = \frac{z^2}{|z|^2} = \frac{\left(|z|\,e^{i\varphi}\right)^2}{|z|^2} = \frac{|z|^2\,e^{i(2\varphi)}}{|z|^2} = e^{i(2\varphi)}$$

Consequently, the phase angle for $r_\perp$ is twice the phase angle for z in Equation 3.46.

$$\varphi_\perp = 2\varphi = 2\tan^{-1}\left(\frac{\mathrm{Im}[z]}{\mathrm{Re}[z]}\right) = 2\tan^{-1}\left(\frac{\sqrt{\left(\frac{n_i}{n_t}\sin\theta_i\right)^2 - 1}}{\frac{n_i}{n_t}\cos\theta_i}\right)$$

$$= 2\tan^{-1}\left(\frac{n_t}{n_i}\frac{\sqrt{\left(\frac{n_i}{n_t}\sin\theta_i\right)^2 - 1}}{\cos\theta_i}\right)$$

$$= 2\tan^{-1}\left(\frac{n_t}{n_i}\alpha\right)$$

This result only has meaning for $\theta_i > \theta_c$. This phase angle will be interpreted as a phase shift in Section 3.5.5.

In order for total internal reflection to remain consistent with the boundary conditions found in Section 3.3.1, there must be electric and magnetic fields on the transmission side of the interface. These fields belong to the *evanescent wave*, which will be discussed in more detail in Section 3.6.2.

3.5.4 Plots of the Fresnel Equations vs. Incident Angle

Figure 3.13 shows plots of reflection and transmission ratios for both external and internal incidence at an air-glass interface ($n_g = 1.50$). The symbol $r_{x\perp}$ refers to the reflection ratio for $\vec{E}_\perp$ polarization and external incidence. Similarly, $r_{x\|}$ gives the external incidence reflection ratio for $\vec{E}_\|$ polarization. The corresponding ratios for internal incidence are indicated by $r_{i\perp}$ and $r_{i\|}$. The values of $r_{x\|}$ and $r_{i\|}$ both pass through zero at Brewster's angles θ_B and θ'_B, respectively. The internal incidence reflection ratios approach an absolute value of one at the critical angle for total internal reflection, indicated by θ_c. The transmission ratios for external incidence are given by $t_{x\perp}$ for $\vec{E}_\perp$ polarization and by $t_{x\|}$ for $\vec{E}_\|$ polarization. The transmission ratios for internal incidence are not shown, but they are positive and greater than one for all angles of incidence less than the critical angle.

■ **EXAMPLE 3.7**

Generate the plots of Figures 3.13.

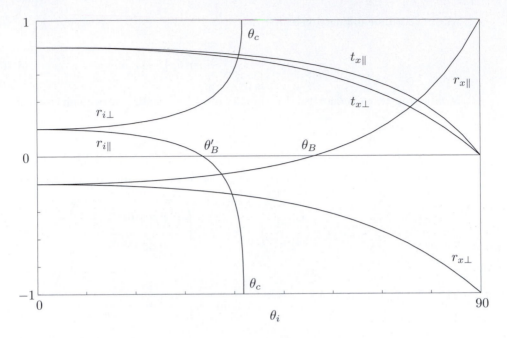

Figure 3.13   Fresnel amplitude ratios for internal and external incidence for an air-glass ($n = 1.50$) interface. Ratios subscripted with x are for external incidence, and ratios subscripts with i are for internal incidence. Transmission ratios for internal incidence are not shown.

Solution

Plotting the Fresnel equations will be easier if we use Snell's law to express them as a function of only the incident angle θ_i. Thus

$$\sin \theta_t = \frac{n_i}{n_t} \sin \theta_i$$

$$\cos \theta_t = \sqrt{1 - \sin^2 \theta_t} = \sqrt{1 - \left(\frac{n_i}{n_t} \sin \theta_i \right)^2}$$

For $\vec{E}_\perp$ polarization,

$$r_\perp = \frac{n_i \cos \theta_i - n_t \sqrt{1 - \left(\frac{n_i}{n_t} \sin \theta_i \right)^2}}{n_i \cos \theta_i + n_t \sqrt{1 - \left(\frac{n_i}{n_t} \sin \theta_i \right)^2}} \tag{3.47}$$

$$t_\perp = \frac{2 n_i \cos \theta_i}{n_i \cos \theta_i + n_t \sqrt{1 - \left(\frac{n_i}{n_t} \sin \theta_i \right)^2}} \tag{3.48}$$

and for $\vec{E}_\parallel$ polarization,

$$r_\parallel = \frac{n_i \sqrt{1 - \left(\frac{n_i}{n_t}\sin\theta_i\right)^2} - n_t\cos\theta_i}{n_i \sqrt{1 - \left(\frac{n_i}{n_t}\sin\theta_i\right)^2} + n_t\cos\theta_i} \qquad (3.49)$$

$$t_\parallel = \frac{2n_i\cos\theta_i}{n_i \sqrt{1 - \left(\frac{n_i}{n_t}\sin\theta_i\right)^2} + n_t\cos\theta_i} \qquad (3.50)$$

For external incidence, you can plot from $0 \leq \theta_i \leq 90°$. For internal incidence, you should limit θ_i to values less than the critical angle. Plot Equations 3.47 — 3.50 and compare your results to Figure 3.13.

There are many options for plotting these functions. Your favorite spreadsheet will do a fine job.

3.5.5 Phase Changes on Reflection

Unless there is total internal reflection, the amplitude ratios given by the Fresnel equations for non-absorbing dielectrics are real, so the only possible *phase shifts* for the transmitted and reflected waves relative to the incident wave are zero and π. The angles θ_i and θ_t are always acute, so the values given by Equations 3.25 and 3.32 for $t_\perp$ and $t_\parallel$ are positive for all incident angles. This means that our *initial guess* for the field directions of the transmitted rays indicated in Figures 3.9 and 3.10 were correct. For normal incidence, it is also clear that the transmitted beams are not phase shifted, since in this case the relative orientations of $\vec{E}$ and $\vec{B}$ are identical for the incident and transmitted beams. For nonzero angles of incidence, the direction of propagation for the transmitted rays changes according to Snell's law, and we may define the *deviation angle* between the incident and transmitted beams, which according to Figure 3.2 is $\theta_d = \theta_i - \theta_t$. A *propagation reversal* occurs whenever $|\theta_d|$ exceeds $90°$. According to Snell's law, there is never a propagation reversal between the incident and transmitted rays. We interpret a π phase shift for rays with no propagation reversal according to Figure 3.14(a). According to this convention, the transmitted ray is never phase-shifted for all angles of external incidence. For internal incidence, $t_\perp$ and $t_\parallel$ are both positive for incident angles less than the critical angle (see Problem 3.39), $|\theta_d|$ is again less than $90°$, and again there is no phase change. To summarize, the transmitted ray is never phase-shifted for any polarization or any angle of incidence.

From the plots shown in Figures 3.13, it is evident that $r_\perp$ and $r_\parallel$ can be positive or negative depending on the angle of incidence. If these amplitude ratios are negative, then we did not guess the field directions correctly in Figures 3.9 and 3.10. Since $e^{i\pi} = -1$, it is tempting to also interpret a negative value as a phase shift of π. However, since the propagation may reverse for the reflected ray, this must be done with care.

Figure 3.14(b) is similar to our initial guess in Figure 3.9, except that it has been modified to show the true field directions before and after external reflection for the case of $\vec{E}_\perp$ polarization. According to our calculations, $r_{x\perp}$ is negative for all incident angles. Thus, $\vec{E}_{0i}$ and $\vec{E}_{0r}$ point in *opposite* directions, as shown. It is also clear from the figure that for small angles of incidence the propagation of the reflected rays reverses. In order to

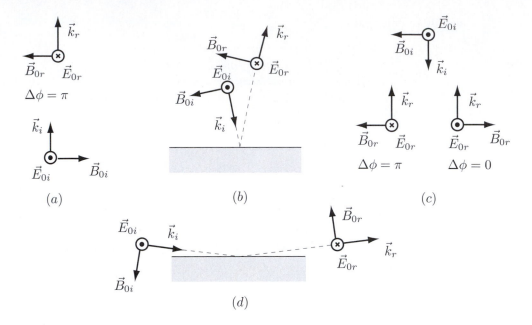

Figure 3.14   (a) Definition of a π phase shift for the case where there is no propagation reversal. (b) External incidence of the $\vec{E}_\perp$ polarization showing the true orientations of all fields when the angle of incidence is small. In this case, there is a propagation reversal between the incident and reflected rays. (c) Definition of a π phase shift for the case where there is a propagation reversal. (d) External incidence of the $\vec{E}_\perp$ polarization showing the true orientations of all fields when the angle of incidence is large. In this case, there is no propagation reversal between the incident and reflected rays.

reverse the propagation of a transverse electromagnetic wave, it is necessary that one (but not both) of the field directions reverse. Whether or not to interpret the situation of Figure 3.14(b) as a phase change should be based on observation; namely, the phase changes for normal incidence that determine whether constructive or destructive interference is observed for reflection from thin films[6]:

1. For external incidence ($n_i < n_t$), the reflected wave is phase-shifted by π.

2. For internal incidence ($n_i > n_t$), the reflected wave is not phase-shifted.

3. The transmitted wave is never phase-shifted.

These observations are consistent with an interpretation of Figure 3.14(b) as a π phase shift. Figure 3.14(c) illustrates the corresponding convention for the case of normal incidence when the propagation reverses. Notice that the $\Delta\phi = 0$ case (no phase change) is identical to the incident wave if you invert the page, so this convention is both observationally and intuitively correct. For large angles of incidence, there is no longer any propagation reversal, as illustrated in Figure 3.14(d). According to the convention of Figure 3.14(a), there is still a phase change of π. Thus, the phase of a $\vec{E}_\perp$ polarized beam is phase-shifted by π for all angles of external incidence.

[6]See any introductory physics text (e.g Halliday, Resnick and Walker [11]).

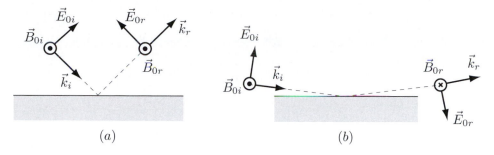

Figure 3.15 (a) External incidence of the $\vec{E}_{\parallel}$ polarization showing the true orientations of all fields for angle of incidence that is less than Brewster's angle. (b) Same as (a), but for a large angle of incidence.

To determine the phase shifts for external incidence of $\vec{E}_{\parallel}$ polarization, consider Figure 3.15. Figure 3.15(a) shows the true orientation of all fields for external incidence of $\vec{E}_{\parallel}$ polarization for an angle of incidence that is less than Brewster's angle. When $\theta_i > \theta_B$, $r_{\parallel}$ is positive, giving the field directions shown in Figure 3.15(b).

The reflected ray in Figure 3.15(a) undergoes a propagation reversal, and according to the convention of Figure 3.14(c), we interpret this as a phase shift of π. On the other hand, the reflected ray of Figure 3.15(b) does not undergo a propagation reversal, and by Figure 3.14(a) is also phase-shifted by π even though $r_{\parallel}$ for this case is *positive*. The sign transition at Brewster's angle is due to Equation 3.35; here, the reflected and transmitted rays are 90° apart. For $\theta_i < \theta_B$, the reflected ray has a propagation reversal relative to the *transmitted* ray, and for $\theta_i > \theta_B$, there is no such propagation reversal. The conventions of Figures 3.14(a) and 3.14(c) indicate for both cases a phase shift of π relative to the transmitted ray, which as we have shown is never phase-shifted relative to the incident wave. Thus, the phase of a $\vec{E}_{\parallel}$ polarized beam is phase-shifted for all angles of external incidence. To summarize this and the previous result, we conclude that an externally reflected wave is always phase-shifted by π regardless of polarization[7].

It remains to investigate internally reflected rays. For internal incidence we must consider three cases: $\theta_i < \theta_{Bi}$, $\theta_{Bi} \leq \theta_i \leq \theta_c$ and $\theta_i > \theta_c$. Comparing the reflection ratios for internal versus external incidence in Figure 3.13 shows a reversal in sign in each of the first two cases, so by the same arguments used above, the phase shifts for internal incidence are opposite to those for external incidence when $\theta_i < \theta_c$. Therefore, for internal incidence and $\theta_i < \theta_c$, the reflection phase shift is zero for all beam polarizations.

When $\theta_i > \theta_c$, there is total internal reflection. The phase shifts $\Delta\phi$ are then given by the phase angles $\varphi_{\perp}$ and $\varphi_{\parallel}$ found in Section 3.5.3. For a given angle of incidence, the phase shift is larger for the $\vec{E}_{\parallel}$ polarization, and in both cases varies continuously from zero to π (see Problem 3.18). Figure 3.16 shows the phase shift difference $\Delta\varphi = \varphi_{\parallel} - \varphi_{\perp}$ versus incident angle for two values of n_i/n_t. For $n_i/n_t = 1.50$, the maximum value of $\Delta\varphi$ is about 45°; in Chapter 5, we will see how to use this effect to prepare light with *circular* polarization.

Summary We may summarize our conclusions as follows:

---

[7]Evidence for the π phase shift for grazing incidence includes the adjacent *dark* fringe in interference from Lloyd's mirror. For normal incidence, evidence includes the central dark fringe in Newton's rings. Both topics are discussed in Chapter 6.

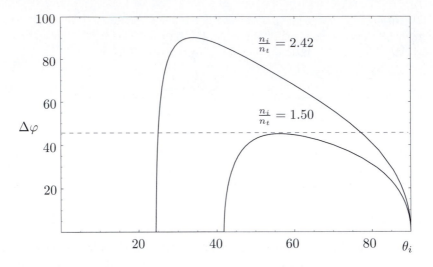

Figure 3.16   Relative phase shift $\Delta\varphi = \varphi_\| - \varphi_\perp$ for total internal reflection.

1. For all angles of external incidence ($n_i < n_t$), the reflected wave is phase-shifted by π.

2. For internal incidence ($n_i > n_t$), the reflected wave is never phase-shifted when $\theta_i < \theta_c$.

3. For internal incidence with $\theta_i > \theta_c$, the $\vec{E}_\perp$ polarization is phase-shifted according to Equation 3.41, and the $\vec{E}_\|$ polarization is phase-shifted according to Equation 3.45.

4. The transmitted wave is never phase-shifted.

Problem 3.10   Find Brewster's angle for internal and external incidence upon an air-diamond interface. Assume an index of refraction of 2.42 for diamond.

Problem 3.11   Find the critical angle for total internal reflection at an air-glass ($n = 1.50$) interface. Repeat for an air-diamond ($n = 2.42$) interface.

Problem 3.12   Show that the phase shift for total internal reflection of the $\vec{E}_\|$ polarization is given by

$$\varphi_\| = 2\tan^{-1}\left(\frac{n_i}{n_t}\alpha\right)$$

where α is defined in Equation 3.42.

Problem 3.13   Repeat the plots of Example 3.7 when the glass is replaced by diamond ($n = 2.42$).

Problem 3.14   Using a block of transparent, unknown material, it is found that a beam of light inside the material is totally internally reflected at the air-block interface at a minimum angle of 42.6 degrees. What is the index of refraction of this material?

Problem 3.15   A glass block having an index of 1.68 is covered with a layer of plastic whose index is 1.46. What is the critical angle for total internal reflection?

Problem 3.16   For glass of index 1.52 immersed in air,
  a) calculate the critical angle for total internal reflection,
  b) find the phase shift for total internal reflection when the incident beam is polarized perpendicular to the plane of incidence and the angle of incidence is $60°$.
  c) Repeat (b) for $E_\parallel$ polarization.

Problem 3.17   Glass of index 1.50 is immersed in air. If the angle of incidence is $55°$, find the total internal reflection phase shift difference between $E_\parallel$ and $E_\perp$ polarization.

Problem 3.18   Plot the reflection phase shifts $\varphi_\perp$ and $\varphi_\parallel$ for total internal reflection at glass ($n = 1.50$) and diamond ($n = 2.42$) interfaces immersed in air. Note that they vary continuously from zero to π for both polarizations. Use these to reproduce the plots of Figure 3.16.

Problem 3.19   Plot the phase difference $\Delta\varphi = \varphi_\parallel - \varphi_\perp$ vs. $\theta_i > \theta_c$ for internal incidence on a water-air interface.

3.6 REFLECTIVITY AND TRANSMISSIVITY

In this section, we use the amplitude ratios found in Section 3.4 to determine the fraction of power in the incident beam that is reflected from and transmitted across an interface that separates two media. The fraction of incident power that is reflected is called the *reflectivity* R, and the fraction of transmitted power is called the *transmissivity* T[8].

Consider an incident beam with circular cross section that is incident on an interface with incident angle θ_i as shown in Figure 3.17. For simplicity, assume that the beam irradiance is uniform over its cross section. The incident, reflected and transmitted beams all intersect within a "spot" on the interface whose area we define as A. In terms of the spot area, the incident and reflected beams have cross-sectional area equal to

$$A_i = A_r = A \cos \theta_i \tag{3.51}$$

However, since the transmitted beam leaves the interface at the refracted angle θ_t, it has a different cross-sectional area:

$$A_t = A \cos \theta_t \tag{3.52}$$

In Chapter 2, we defined the irradiance as the time average of the Poynting vector

$$I \equiv \langle S \rangle = \frac{\epsilon v}{2} E_0^2 = \frac{K_E \epsilon_0 c}{2n} E_0^2 = \frac{n \epsilon_0 c}{2} E_0^2 \tag{3.53}$$

---

[8]Reflectivity and transmissivity depend only on material properties. The related terms *reflectance* and *transmittance* can depend on the material dimensions. For example, thin films have reflectance and transmittance that are affected by interference.

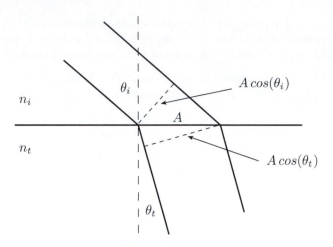

Figure 3.17   Cylindrical beam (shown in cross section) incident on an interface at incident angle θ_i. The incident beam illuminates a spot of area A on the interface. The incident beam has a cross-sectional area $A\cos(\theta_i)$ and the transmitted beam has cross-sectional area $A\cos(\theta_t)$.

where we have used the fact that $n = \sqrt{K_E}$. The irradiance of each beam in Figure 3.17 is equal to the beam power divided by its cross-sectional area.

The reflectivity is the ratio of reflected to incident power. Since each beam power is the product of the irradiance and the cross-sectional area, this gives

$$R \equiv \frac{reflected\ power}{incident\ power} = \frac{I_r A \cos\theta_i}{I_i A \cos\theta_i} = \frac{I_r}{I_i} \tag{3.54}$$

The incident and reflected beam both travel within the incident medium, so according to Equation 3.53,

$$R = \frac{E_{0r}^2}{E_{0i}^2} = r^2 \tag{3.55}$$

Similarly, the transmissivity is defined by

$$T \equiv \frac{transmitted\ power}{incident\ power} = \frac{I_t A \cos\theta_t}{I_i A \cos\theta_i} = \frac{I_t \cos\theta_t}{I_i \cos\theta_i} \tag{3.56}$$

The transmitted beam travels within a region where the index of refraction is n_t, and Equation 3.53 gives

$$T = \frac{I_t \cos\theta_t}{I_i \cos\theta_i} = \frac{n_t E_{0t}^2 \cos\theta_t}{n_i E_{0i}^2 \cos\theta_i} = \left(\frac{n_t \cos\theta_t}{n_i \cos\theta_i}\right) t^2 \tag{3.57}$$

The explicit form of R and T may be given for the two polarization states illustrated in Figure 3.8. For the $\vec{E}_\perp$ polarization,

$$R_\perp = r_\perp^2 = \left(\frac{n_i \cos\theta_i - n_t \cos\theta_t}{n_i \cos\theta_i + n_t \cos\theta_t}\right)^2 \tag{3.58}$$

$$T_\perp = \left(\frac{n_t \cos\theta_t}{n_i \cos\theta_i}\right) t_\perp^2 = \frac{4 n_i n_t \cos\theta_i \ \cos\theta_t}{(n_i \cos\theta_i + n_t \cos\theta_t)^2} \tag{3.59}$$

For the $\vec{E}_\parallel$ polarization state,

$$R_\parallel = \left(\frac{n_i \cos\theta_t - n_t \cos\theta_i}{n_i \cos\theta_t + n_t \cos\theta_i} \right)^2 \tag{3.60}$$

$$T_\parallel = \left(\frac{n_t \cos\theta_t}{n_i \cos\theta_i} \right) t_\parallel^2 = \frac{4 n_i n_t \cos\theta_i \, \cos\theta_t}{(n_i \cos\theta_t + n_t \cos\theta_i)^2} \tag{3.61}$$

■ **EXAMPLE 3.8**

Show that $R_\perp + T_\perp = 1$ for arbitrary angle of incidence.

Solution

From Equations 3.58 and 3.59, we have

$$
\begin{aligned}
R_\perp + T_\perp &= \left(\frac{n_i \cos\theta_i - n_t \cos\theta_t}{n_i \cos\theta_i + n_t \cos\theta_t} \right)^2 + \frac{4 n_i n_t \cos\theta_i \, \cos\theta_t}{(n_i \cos\theta_i + n_t \cos\theta_t)^2} \\
&= \frac{(n_i \cos\theta_i)^2 - 2 n_i n_t \cos\theta_i \, \cos\theta_i + (n_t \, \theta_t)^2 + 4 n_i n_t \cos\theta_i \, \cos\theta_t}{(n_i \cos\theta_i + n_t \cos\theta_t)^2} \\
&= \frac{(n_i \cos\theta_i)^2 + 2 n_i n_t \cos\theta_i \, \cos\theta_i + (n_t \cos\theta_t)^2}{(n_i \cos\theta_i + n_t \cos\theta_t)^2} \\
&= \frac{(n_i \cos\theta_i + n_t \cos\theta_t)^2}{(n_i \cos\theta_i + n_t \cos\theta_t)^2} \\
&= 1
\end{aligned}
$$

Notice that this result is valid for both external and internal incidence.

If r is complex, as is the case for total internal reflection, then R is equal to the *magnitude-squared* of r:

$$R = |r|^2 = r \, r^* \tag{3.62}$$

From Equations 3.39, 3.44, and 3.55, we see that for internal incidence, $R_\perp = R_\parallel = 1$ when the incident angle exceeds the critical angle.

■ **EXAMPLE 3.9**

Find values of the reflectivity and transmissivity for normal incidence when the incident medium is air and the transmitted medium is glass with $n = 1.50$. Repeat the calculation when the incident medium is glass and the transmitted medium is air.

Solution

The polarization states $\vec{E}_\perp$ and $\vec{E}_\parallel$ are physically identical at normal incidence, and

$$R_\perp = R_\parallel = \left(\frac{n_i - n_t}{n_i + n_t} \right)^2 \tag{3.63}$$

$$T_\perp = T_\parallel = \frac{4n_i n_t}{(n_i + n_t)^2} \tag{3.64}$$

Substituting $n_i = 1.00$ and $n_t = 1.50$ for the indices of air and glass gives $R = 0.0400$ and $T = 0.960$. Clearly, the roles of n_i and n_t can be reversed above without affecting the results, so R and T have the same values when the incident medium is glass and the transmitted medium is air. Thus, 4% of the incident beam power is reflected and 96% is transmitted for both external and internal incidence. At normal incidence, a glass slab (e.g. windowpane) reflects about 4% of externally incident light. About 4% of the *remaining* light is reflected internally. For a glass windowpane or uncoated glass lens, about 8% of incident light is removed by reflection. These percentages are of course higher for materials with a larger index of refraction.

3.6.1   Plots of Reflectivity and Transmissivity vs. Incident Angle

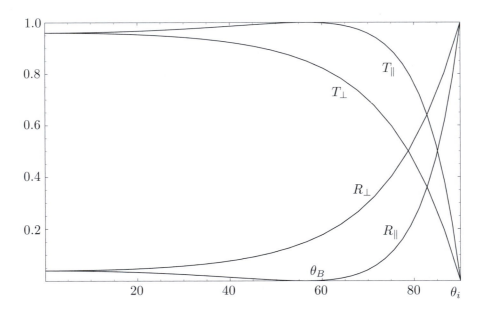

Figure 3.18   Reflectivity and Transmissivity for external incidence when $n_i = 1.00$ and $n_t = 1.50$.

Figure 3.18 shows plots of $R_\perp$, $R_\parallel$, $T_\perp$, and $T_\parallel$ vs. θ_i for external incidence when $n_i = 1.00$ and $n_t = 1.50$. The value of each quantity at normal incidence was computed in Examples 3.9 and the value of Brewster's angle was calculated in Example 3.5. $R_\perp$ and $R_\parallel$ are plotted for internal incidence ($n_i = 1.50$, $n_t = 1.00$) in Figure 3.19; in this case, $T_\perp$ and $T_\parallel$ are not shown, but can be determined from the relation that $R + T = 1$ as shown in Example 3.8.

Note that $R_\parallel = 0$ and $T_\parallel = 1$ at Brewster's angle for both external and internal incidence. For internal incidence, $R_\parallel$ rises sharply from zero at θ'_B to one at θ_c. The steepness of the curve in this region depends on the ratio n_i/n_t.

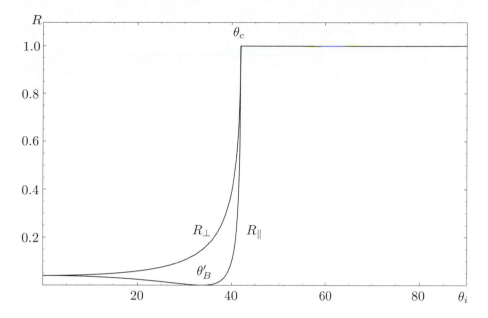

Figure 3.19   Reflectivity for internal incidence when $n_i = 1.00$ and $n_t = 1.50$.

■ **EXAMPLE 3.10**

Generate plots of reflectivity vs. incident angle for external and internal incidence on an air-diamond interface.

Solution

As in Example 3.7, we use Snell's law to express R and T in terms of θ_i:

$$R_\perp = \left(\frac{n_i \cos\theta_i - n_t \sqrt{1 - \left(\frac{n_i}{n_t}\sin\theta_i\right)^2}}{n_i \cos\theta_i + n_t \sqrt{1 - \left(\frac{n_i}{n_t}\sin\theta_i\right)^2}} \right)^2$$

$$T_\perp = \frac{4 n_i n_t \cos\theta_i \sqrt{1 - \left(\frac{n_i}{n_t}\sin\theta_i\right)^2}}{\left(n_i \cos\theta_i + n_t \sqrt{1 - \left(\frac{n_i}{n_t}\sin\theta_i\right)^2} \right)^2}$$

$$R_\parallel = \left(\frac{n_i \sqrt{1 - \left(\frac{n_i}{n_t}\sin\theta_i\right)^2} - n_t \cos\theta_i}{n_i \sqrt{1 - \left(\frac{n_i}{n_t}\sin\theta_i\right)^2} + n_t \cos\theta_i} \right)^2$$

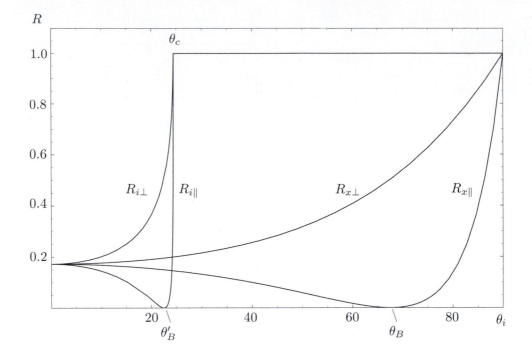

Figure 3.20   Reflectivity for diamond.

$$T_{\parallel} = \frac{4 n_i n_t \cos\theta_i \sqrt{1 - \left(\frac{n_i}{n_t}\sin\theta_i\right)^2}}{\left(n_i \sqrt{1 - \left(\frac{n_i}{n_t}\sin\theta_i\right)^2} + n_t \cos\theta_i\right)^2}$$

For external incidence, use $n_i = 1.00$ and $n_t = 2.42$. Finish the Example by plotting the above functions, and compare your results to Figure 3.20.

3.6.2   The Evanescent Wave

When discussing total internal reflection in Section 3.5.3 we noted that the boundary conditions determined in Section 3.3.1 require a wave to be transmitted across the boundary even when the incident ray is totally reflected. However, no *energy* is transmitted across the interface when $\theta_i > \theta_c$. In this section we see how this can occur, and we determine the properties of this transmitted wave, commonly called the *evanescent wave*.

Let the plane of incidence be the x-y plane as shown in Figure 3.21, and let the wave transmitted across the interface be given by

$$\vec{E}_t = \vec{E}_{0t} \exp i(\vec{k}_t \cdot \vec{r} - \omega t)$$

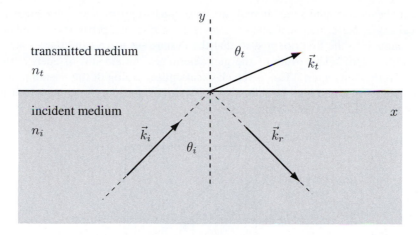

Figure 3.21 Internal incidence ($n_i > n_t$). Note that the incident beam arrives from below.

where $\vec{E}_{0t}$ represents an arbitrary state of polarization. In Figure 3.21, the propagation vector of the transmitted wave lies in the first quadrant, giving

$$\vec{k}_t \cdot \vec{r} = k_{tx}x + k_{ty}y$$

From Figure 3.21, $k_{tx} = k_t \sin \theta_t$ and $k_{ty} = k_t \cos \theta_t$. As in Section 3.5.3, $\cos \theta_t$ is imaginary when $\theta_i > \theta_c$:

$$\cos \theta_t = \sqrt{1 - \sin^2 \theta_t} = i\sqrt{\left(\frac{n_i}{n_t} \sin \theta_i\right)^2 - 1}$$

Thus,

$$k_{ty} = ik_t \sqrt{\left(\frac{n_i}{n_t} \sin \theta_i\right)^2 - 1}$$

$$k_{tx} = k_t \left(\frac{n_i}{n_t} \sin \theta_i\right)$$

The transmitted wave for $\theta_i > \theta_c$ is then given by

$$\vec{E}_t = \vec{E}_{0t} \exp\left[-k_t \left(\left[\frac{n_i}{n_t} \sin \theta_i\right]^2 - 1\right)^{\frac{1}{2}} y\right] \exp i\left[k_t \left(\frac{n_i}{n_t} \sin \theta_i\right) x - \omega t\right] \quad (3.65)$$

The first exponential term rapidly damps the evanescent wave to zero in the direction *normal to the interface*. Even for incident angles only slightly larger than the critical angle, the evanescent wave has negligible amplitude after just a few wavelengths of penetration into the transmitted medium. So far, we have assumed non-absorbing dielectrics, so by energy conservation, the wave is totally reflected, and $T = 0$ for all polarizations. In the next section, we consider materials where the absorption is not zero.

If the transmitted medium is modified before the evanescent wave damps to zero, energy *can* be transmitted across the interface. Figure 3.22 illustrates *frustrated total reflection* produced by two prisms separated by a gap that is a fraction of a wavelength wide. The incident beam encounters the internal surface of the first prism at an angle greater than the critical angle. Nevertheless, a beam transmits into the second prism if it can be positioned close enough to the first prism such that the evanescent wave can couple into it before damping to zero. The width of the gap determines the amount of energy transmitted across "forbidden region." This is the classical optical analog of the quantum mechanical phenomenon of *tunneling*. By adjusting the width of the gap, the arrangement in Figure 3.22 can be used as a variable attenuator.

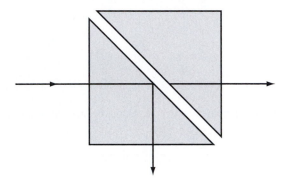

Figure 3.22   Frustrated total reflection. The gap between the two prisms is a fraction of a wavelength wide.

Problem 3.20   A slab of glass ($n = 1.52$) is immersed in water ($n = 1.33$).
 a) Find the critical angle for total internal reflection.
 b) Find Brewster's angle for both external and internal incidence.
 c) Find the reflectivity R for both internal and external incidence at normal incidence.

Problem 3.21   As a gemstone, diamond is valued not only for its hardness and dispersion, but also for the *brilliance* that results from its high reflectivity. Find the reflectivity of an air-diamond interface at normal incidence.

Problem 3.22   A beam of light is incident in air normally on a pane of glass with index of refraction equal to 1.52. Neglecting interference and absorption effects, what fraction of the incident beam irradiance is transmitted by the slab? Repeat for a slab of diamond with index of refraction equal to 2.42.

Problem 3.23   Show that $R_\parallel + T_\parallel = 1$ for arbitrary angle of incidence.

Problem 3.24   Find values of the reflectivity and transmissivity for an angle of incidence $\theta_i = 20°$ when the incident medium is air and the transmitted medium is glass with $n = 1.50$. Repeat the calculation for internal incidence.

Problem 3.25   Find values of the reflectivity and transmissivity for an angle of incidence $\theta_i = 20°$ when the incident medium is water ($n = 1.33$) and the transmitted medium is glass with $n = 1.50$. Repeat the calculation for internal incidence.

Problem 3.26   Find values of the reflectivity and transmissivity for an angle of incidence $\theta_i = 80°$ when the incident medium is air and the transmitted medium is glass with $n = 1.50$. Repeat the calculation for internal incidence.

Problem 3.27   Generate plots of reflectivity vs. incident angle for external and internal incidence on an air-water interface.

Problem 3.28   Fill in all algebraic details leading to equations 3.58 — 3.61 for $R_\perp$, $T_\perp$, $R_\parallel$, and $T_\parallel$.

3.7   *DISPERSION

In this section, we take a closer look at the index of refraction, and how it depends on the material properties of the medium. It is usually the case that the index of refraction of a material varies with frequency, and if this is the case, the material is said to exhibit *dispersion*. For most optical materials (i.e., materials that are transparent in the visible), the index of refraction is larger in the blue than for the red. As we will examine in more detail in Chapter 4, dispersion causes a prism to spread a white light beam into a rainbow of colors. Also, many optical materials exhibit *absorption* over some or all of the optical spectrum, and as we will see below, a material's absorptive and dispersive properties are very much related.

A quantitative discussion of dispersion and absorption requires the quantum theory of solids. However, we can obtain a qualitative understanding and some useful intuitions of these effects based on some very simple models based on classical physics.

3.7.1   Dispersion in Dielectric Media

Dielectric materials are composed of molecules that have or can have *dipole moments*. A *dipole* is a charge distribution with charges $-Q$ and $+Q$ separated by a displacement vector $\vec{s}$. The dipole moment is a vector with magnitude Qs and direction pointing $-Q$ to $+Q$, as shown in Figure 3.23(a):

$$\vec{p} = Q\vec{s} \tag{3.66}$$

Individual molecules can have *dipole moments* that are either *permanent* or *induced*. Water molecules, for example, have permanent dipole moments that exist in the absence of any external electric field. Normally, the permanent dipole moments are randomly oriented, but if an electric field is present, they align with the field. Induced dipole moments occur when the electrons and ions interact with an external electric field to create molecular dipole moments in otherwise symmetric molecules, and these are again aligned with the external field. In either case, placing a dielectric material within an external electric field results in *bound charge* distributions, as illustrated in Figure 3.23(b). Here, an external electric field $\vec{E}_f$ results from free conduction electrons with area charge density σ_f on flat metallic plates of area A. This field induces or orients molecular dipoles as indicated in the figure. This results in a net accumulation of charge along the dielectric surface only, since within the dielectric volume, the charge accumulations tend to cancel. This bound charge, indicated by the area charge distribution σ_b in Figure 3.23(b), combines to partially *cancel* the free charge distribution σ_f, or equivalently, produces an electric field $\vec{E}_b$ that orients to partially cancel the free charge field $\vec{E}_f$.

It is customary to define a new field called the *electric polarization*[9] $\vec{P}$, defined as the dipole moment per unit volume:

$$\vec{P} = N\vec{p} \tag{3.67}$$

where N is the number of dipole moments per unit volume (units: m^{-3}). In Figure 3.23(b), we may define the dipole moment of the entire dielectric volume according to Equation 3.66: $p = (\sigma_b A)\, d$. Thus, the magnitude of the electric polarization $\vec{P}$ is obtained by dividing $(\sigma_b A)\, d$ by the volume of the dielectric Ad, giving $P = \sigma_b$. Noting that on either surface, the bound charge has the opposite sign as the free charge that polarizes the material, it appears that

$$\sigma_b = \vec{P} \cdot \hat{n} \tag{3.68}$$

where $\hat{n}$ is the *outward* unit normal to the dielectric surface. This is in fact the case, and holds for any shape dielectric surface of arbitrary polarization $\vec{P}$.

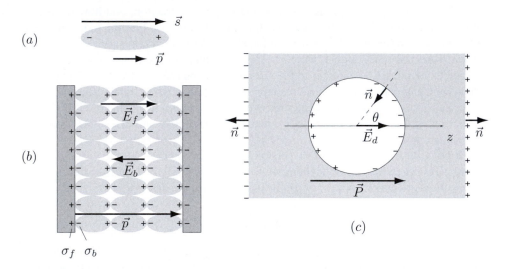

Figure 3.23   (a)A molecule with dipole moment $\vec{p}$. (b) A dielectric slab between two conducting plates. (c) The local electric field in a small spherical cavity inside a dielectric.

The total electric field $\vec{E}$ within the dielectric is due to the sum of both free and bound charge distributions:

$$E = \frac{\sigma_f - \sigma_b}{\epsilon_0} = E_f - \frac{P}{\epsilon_0} \tag{3.69}$$

where the minus sign on σ_b acknowledges that it tends to cancel σ_f.

The electric polarization $\vec{P}$ is due to the molecular orientation caused by the total electric field $\vec{E}$. For *linear dielectrics*, the polarization is proportional to $\vec{E}$:

$$\vec{P} = \epsilon_0 \chi_e \vec{E} \tag{3.70}$$

where χ_e is the *electric susceptibility* (unitless). Combining the last two results gives

$$E_f = (1 + \chi_e)\, E \tag{3.71}$$

---

[9]The reader should be careful not to confuse electric polarization as discussed in this section with the polarization of an electromagnetic wave.

As in Chapter 2, the total electric field $\vec{E}$ for linear isotropic dielectric materials may be obtained from $\vec{E}_f$ by multiplying ϵ_0 by the *dielectric constant K_e*:

$$E = \frac{\sigma_f}{K_e \epsilon_0} = \frac{E_f}{K_e}$$

Combining this with the previous result determines the relationship between the dielectric constant K_e and the electric susceptibility:

$$K_e = 1 + \chi_e \tag{3.72}$$

As shown in Chapter 2, the index of refraction n given by the square root of the dielectric constant:

$$n = \sqrt{K_e} \tag{3.73}$$

As we proceed to model the optical properties of dielectric materials, we shift our emphasis from the static external electric field $\vec{E}_f$ produced by the free charge distribution of Figure 3.23, to consider instead the rapidly oscillating electric field of a passing electromagnetic wave. As we will see, the ability of such a rapidly changing field to induce a dielectric polarization $\vec{P}$ is a function of frequency, and this makes χ_e, K_e, and n all functions of frequency as well. Our goal is to find the dependence of index of refraction n on the frequency of the electromagnetic wave.

To model a dielectric on the microscopic scale, we imagine that electrons are bound to nuclei by *linear springs* that obey Hooke's law. In this model, the electric field of a passing electromagnetic wave places a force on the electron that stretches the spring, thereby inducing a dipole moment. However, the electric field of the passing wave is not the only field that acts on the electron; we must also account for the effects of nearby dielectric molecules. We will refer to this total electric field as the *local field*.

Let $\vec{E}$ be the electric field of the passing electromagnetic wave, and let $\vec{E}_p$ be the field that results from nearby polarized dielectric molecules. To find $\vec{E}_p$, we imagine the molecule to be enclosed within a tiny spherical volume of radius R, as illustrated in Figure 3.23. Since $\sigma_b = \vec{P} \cdot \hat{n}$, the left half of the sphere has positive bound charge and the right half is charged negatively. The electric field $\vec{E}_p$ results from this bound charge. We note that the charge distribution has cylindrical symmetry about the horizontal axis (z-axis in Figure 3.23), so the transverse components of the local field vanish after integration. Integrating $dE_z = dE \cos \theta$ in spherical coordinates, we obtain:

$$E_p = \int_{\varphi=0}^{2\pi} \int_{\theta=0}^{\pi} \left[\frac{(P \cos \theta)}{4\pi \epsilon_0 R^2} \right] \cos \theta R^2 \sin \theta \, d\theta \, d\varphi$$

Integration over φ gives 2π. Cancellation gives

$$E_p = \frac{P}{2\epsilon_0} \int_0^{\pi} \cos^2 \theta \sin \theta \, d\theta = \frac{P}{2\epsilon_0} \left(-\frac{\cos^3 \theta}{3} \Big|_0^{\pi} \right) = \frac{P}{3\epsilon_0} \tag{3.74}$$

Strictly, we have found E_p at the center of the spherical volume. Remarkably, E_p is *uniform* throughout the spherical volume[10]. The local field at the position of the electron is thus

$$\vec{E}_L = \vec{E} + \frac{\vec{P}}{3\epsilon_0} \tag{3.75}$$

---

[10]See Example 2 in Chapter 4 of Griffiths [9] for a demonstration that a bound charge of $\vec{P} \cdot \hat{n}$ on the surface of a sphere due to a uniform $\vec{P}$ gives a uniform electric field of magnitude $P/3\epsilon_0$ inside the sphere.

The local electric field causes the spring to stretch, thereby inducing a dipole moment. We assume a dipole moment that is proportional to the local field:

$$\vec{p} = \epsilon_0 \alpha \vec{E}_L$$

where α is the *atomic polarizability*. By Equations 3.67 and 3.75, we have

$$P = \epsilon_0 N \alpha E_L = \epsilon_0 N \alpha \left(E + \frac{P}{3\epsilon_0} \right)$$

where, as above, N is the number per unit volume of induced dipole moments. Solving for P gives

$$P = \frac{N\alpha}{1 - (N\alpha/3)} \epsilon_0 E$$

By Equation 3.70, the electric susceptibility is given by

$$\chi_e = \frac{N\alpha}{1 - (N\alpha/3)} \tag{3.76}$$

By Equations 3.72 and 3.73,

$$n^2 = 1 + \frac{N\alpha}{1 - (N\alpha/3)}$$

Solving this for $N\alpha$ gives

$$3\frac{n^2 - 1}{n^2 + 2} = N\alpha \tag{3.77}$$

This is known as the Clausius-Mossotti equation.

To model the frequency dependence of n, it remains to model the atomic polarizability α in the Clausius-Mossotti equation. To do this, we will treat the electron-spring system as a driven harmonic oscillator with a forcing function provided by the electric field in the electromagnetic wave. In addition to this driving force, we will include a damping force proportional to the electron velocity to account for dissipative effects such as radiation. Finally, there is the spring restoring force, which by Hooke's law is $-kx$, where the spring constant k is given in terms of the natural frequency of oscillation ω_0 as $m\omega_0^2$, where m is the electron mass. Newton's second law becomes

$$m\frac{d^2 x}{dt^2} = -m\gamma\frac{dx}{dt} - m\omega_0^2 x + eE_0 e^{-i\omega t} \tag{3.78}$$

where the first term on the right is the resistive damping force whose value depends on the constant γ (units: s^{-1}). The second term is the Hooke's law restoring force, and the third is the driving force provided by the *local* electric field (for simplicity, at $x = 0$) due to the electromagnetic wave. Rearrangement gives

$$m\frac{d^2 x}{dt^2} + m\gamma\frac{dx}{dt} + m\omega_0^2 x = eE_0 e^{-i\omega t} \tag{3.79}$$

We assume an oscillation of the electron position that is at the same frequency as the driving force: $x = x_0 e^{-i\omega t}$. Substituting this into Equation 3.79 gives

$$\left(m\left(-\omega^2\right) + m\gamma\left(-i\omega\right) + m\omega_0^2 \right) x = eE_0$$

or, solving for x_0,

$$x_0 = \frac{\frac{e}{m}E_0}{\omega_0^2 - \omega^2 - i\gamma\omega} \tag{3.80}$$

This displacement determines the dipole moment of the electron-spring system: $p = ex$. Since $P = Np$,

$$\vec{P} = \frac{\frac{Ne^2}{m}\vec{E}}{\omega_0^2 - \omega^2 - i\gamma\omega}$$

Finally, since $\vec{P} = N(\epsilon_0\alpha\vec{E})$, $N\alpha$ is given by

$$N\alpha = \frac{N\frac{e^2}{m\epsilon_0}}{\omega_0^2 - \omega^2 - i\gamma\omega}$$

So far, our model is a bit unrealistic in that we have only allowed for one value of Hooke's law spring constant. In a typical dielectric, there may be any number of available electronic transitions, each characterized by a different spring constant. Let ω_j be the resonant frequency of the j-th Hooke's law spring, and let N_j be the number of electrons per unit volume that are bound by such a spring. Combining the last result with the Clausius-Mossotti equation gives

$$3\frac{n^2 - 1}{n^2 + 2} = \sum_j \frac{N_j\frac{e^2}{m\epsilon_0}}{\omega_j^2 - \omega^2 - i\gamma_j\omega}$$

It is customary to define the *oscillator strength* f_j such that $N_j = f_j N$. Substituting this into the last result and rearranging gives

$$\frac{n^2 - 1}{n^2 + 2} = \frac{Ne^2}{3m\epsilon_0}\sum_j \frac{f_j}{\omega_j^2 - \omega^2 - i\gamma_j\omega} \tag{3.81}$$

Equation 3.81 determines the frequency dependence of n. Because of the damping factor γ, n is *complex*. Denoting the real and imaginary parts of n as n_r and n_{im}, we have

$$n = n_r + in_{im} \tag{3.82}$$

An electromagnetic wave traveling within the dielectric is described by

$$\vec{E} = \vec{E}_0 e^{i(\tilde{k}z - \omega t)}$$

where the symbol $\tilde{k}$ denotes that the index of refraction is now also complex:

$$\tilde{k} = \frac{2\pi}{\lambda} = \frac{\omega}{c}n = \frac{\omega}{c}(n_r + in_i) = \frac{\omega n_r}{c} + i\frac{\omega n_{im}}{c} = k + i\frac{\beta}{2} \tag{3.83}$$

where β is the *absorption coefficient* (units: m^{-1}).

$$\beta = \frac{2\omega n_{im}}{c} \tag{3.84}$$

Note that we have used the previous symbol k to represent the real part of the complex $\tilde{k}$. The traveling electromagnetic wave is now given by

$$\vec{E} = \vec{E}_0 e^{-\frac{\beta}{2}z}e^{i(kz - \omega t)} \tag{3.85}$$

or, in terms of irradiance[11],

$$I = I_0 e^{-\beta z} \qquad (3.86)$$

For optical frequencies, k is on the order of $10^7 \, m^{-1}$. For transparent materials, $\beta << k$ and hence $n_i << n_r$.

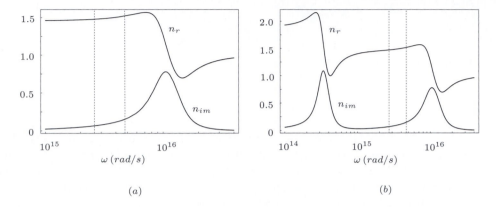

(a) (b)

Figure 3.24   Equation 3.81 applied to silicon dioxide. The dashed vertical lines indicate the visible spectrum. (a) An absorption band at $160 \, nm$ (b) Two absorption bands: $160 \, nm$ and $5.0 \, \mu m$.

■ EXAMPLE 3.11

Use Equation 3.81 to investigate dispersion in fused silica glass. Use $60 \, g/mole$ and a density of 2.55 for SiO$_2$. Assume a UV absorption line at $160 \, nm$, which is a known absorption band for quartz.

Solution

Let N be the number of SiO$_2$ atoms per unit volume. First find the volume of a mole:

$$V = \left(\frac{60 \, g/mole}{2.55 \, g/cm^3} \right) \left(\frac{1 \, m}{100 \, cm} \right)^3 = 2.35 \times 10^{-5} \, \frac{m^3}{mole}$$

Use this to find N:

$$N = \frac{6.023 \times 10^{23} \, molecules/mole}{V} = 2.56 \times 10^{28} \, m^{-3}$$

Using $e = 1.6 \times 10^{-19} \, C$, $m = 9.11 \times 10^{-31} \, kg$ and $\epsilon_0 = 8.854 \times 10^{-12} \, C^2/Nm^2$ in Equation 3.81 gives

$$N\alpha = \left(2.71 \, x \, 10^{31} \, s^{-2} \right) \left(\frac{f}{\omega_0^2 - \omega^2 - i\gamma\omega} \right)$$

Solving Equation 3.81 for the index of refraction gives

$$n = \sqrt{\frac{3 + 2N\alpha}{3 - N\alpha}}$$

[11] Most optical detectors measure *power* rather than field amplitude.

An absorption band at $\lambda = 160 \times 10^{-9}\,m$ gives $\omega_0 = 1.18 \times 10^{16}\,rad/s$. As a starting point, use a damping factor $\gamma = \omega_0/2$, and let $f = 0.5$. Most symbolic programming environments can compute the real and imaginary parts of the above expression for n using built-in functions. Figure 3.24(a) shows the results. Figure 3.24(b) includes an infrared absorption feature.

The real part of n is in the right range of values. At $700\,nm$, $n_r = 1.478$ and at $400\,nm$, $n_r = 1.519$. The difference is 0.041, compared to 0.013 for optical quartz, so the model gives a bit too much dispersion. Also, the imaginary part n_{im} is much too large across the visible spectrum, where pure quartz has a very low absorption. Still, the results are remarkably close for such a simple model.

For most transparent materials, n_r increases with frequency. The sudden decrease of n_r in the vicinity of an absorption feature (see Figure 3.24) is called *anomalous dispersion*.

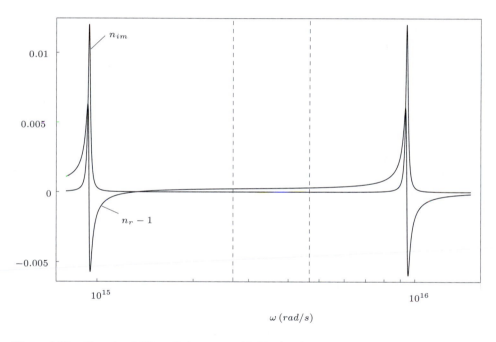

Figure 3.25   Equation 3.87 applied to a gas with $30\,g/mole$ at standard temperature and pressure, assuming absorption features at $200\,nm$ and $2.00\,\mu m$. The $200\,nm$ feature uses $f_1 = 0.5$ and $\gamma_1 = \omega_1/50$. The $2.00\,\mu m$ feature uses $f_2 = 0.005$ and $\gamma_2 = \omega_2/50$. Smaller values of γ give sharper resonances with larger peak values of n_{im}. Vertical dashed lines indicate the visible spectrum.

3.7.1.1 *Non-conducting Gases*

Molecules in a gas are separated by larger distances, so we may neglect the contribution of $\vec{E}_d$ in Equation 3.74. Thus, $\chi_e = \sum_j N_j \alpha_j$, and

$$n^2 = 1 + \frac{Ne^2}{m\epsilon_0} \sum_j \frac{f_j}{\omega_j^2 - \omega^2 - i\gamma_j \omega} \tag{3.87}$$

Figure 3.25 shows a plot of Equation 3.87 that models air at STP. Absorption features at $200\,nm$ and $2.00\,\mu m$ roughly bracket the range where air is transparent. This model gives

$n_r = 1.00024$ at $650\,nm$ and $n_r = 1.00029$ at $450\,nm$, compared to experimental values of $n_r = 1.00029188$ at $656.4\,nm$ and $n_r = 1.00029713$ at $435.8\,nm$[12].

It is useful to separate the quantity n^2 into real and imaginary parts. Begin with Equation 3.87, and multiply numerator and denominator of the term under the sum sign by the complex conjugate of the denominator to give

$$n^2 = \left(1 + \frac{Ne^2}{m\epsilon_0} \sum_j \frac{f_j\left(\omega_j^2 - \omega^2\right)}{\left(\omega_j^2 - \omega^2\right)^2 + \gamma_j^2\omega^2}\right) + i\left(\frac{Ne^2}{m\epsilon_0} \sum_j \frac{f_j\gamma\omega}{\left(\omega_j^2 - \omega^2\right)^2 + \gamma_j^2\omega^2}\right)$$

$$= (n_r + in_{im})^2 = n_r^2 - n_{im}^2 + i\,(2n_r n_{im})$$

The real and imaginary parts of n^2 are

$$n_r^2 - n_{im}^2 = 1 + \frac{Ne^2}{m\epsilon_0} \sum_j \frac{f_j\left(\omega_j^2 - \omega^2\right)}{\left(\omega_j^2 - \omega^2\right)^2 + \gamma_j^2\omega^2} \tag{3.88}$$

$$2n_r n_{im} = \frac{Ne^2}{m\epsilon_0} \sum_j \frac{f_j\gamma\omega}{\left(\omega_j^2 - \omega^2\right)^2 + \gamma_j^2\omega^2} \tag{3.89}$$

For transparent gases, $n_r \cong 1$ and $n_r \gg n_i$. Under these circumstances, we may develop some approximate formulas. First consider Equation 3.88, and note that since $n_r \gg n_{im}$, we may ignore the term n_{im}^2. Since $n_r \cong 1$, we may use the binomial expansion: $(1 + x)^n \approx 1 + nx$ for $|x| \ll 1$. Taking the square root of both sides and applying this approximation with $n = \frac{1}{2}$ gives

$$n_r \cong 1 + \frac{Ne^2}{2m\epsilon_0} \sum_j \frac{f_j\left(\omega_j^2 - \omega^2\right)}{\left(\omega_j^2 - \omega^2\right)^2 + \gamma_j^2\omega^2} \tag{3.90}$$

In Equation 3.89, we may substitute $n_r \cong 1$ and solve for n_{im} to give

$$n_{im} \cong \frac{Ne^2}{2m\epsilon_0} \sum_j \frac{f_j\gamma_j\omega}{\left(\omega_j^2 - \omega^2\right)^2 + \gamma_j^2\omega^2} \tag{3.91}$$

For frequencies sufficiently far from resonance, we may ignore the damping term in Equation 3.90, giving

$$n_r = 1 + \frac{Ne^2}{2m\epsilon_0} \sum_j \frac{f_j}{\left(\omega_j^2 - \omega^2\right)} \tag{3.92}$$

Rearranging, and using $(1 - x)^{-1} \approx 1 + x$ gives

$$n_r = 1 + \frac{Ne^2}{2m\epsilon_0} \sum_j \frac{f_j}{\omega_j^2}\left(1 - \frac{\omega^2}{\omega_j^2}\right)^{-1} \approx 1 + \frac{Ne^2}{2m\epsilon_0} \sum_j \frac{f_j}{\omega_j^2}\left(1 + \frac{\omega^2}{\omega_j^2}\right)$$

which simplifies to

$$n_r = 1 + \left(\frac{Ne^2}{2m\epsilon_0} \sum_j \frac{f_j}{\omega_j^2}\right) + \left(\frac{Ne^2}{2m\epsilon_0} \sum_j \frac{f_j}{\omega_j^4}\right)\omega^2 \tag{3.93}$$

---

[12] See S. A. Korff and G. Breit, "Optical Dispersion," *Rev. Mod. Phys.*, 4: 471—503 (1932).

Grouping constant terms and expressing in terms of wavelength gives:

$$n_r = 1 + A + \frac{B}{\lambda^2} \tag{3.94}$$

Thus, away from resonance, the dispersion varies as one over the wavelength-squared.

■ **EXAMPLE 3.12**

Estimate the values of n_r and n_{im} at resonance using the parameter values of Figure 3.25.

Solution

The approximations used above may not be appropriate close to resonance. For example, let the $j = 1$ resonance be the $2.00\,\mu m$ feature of Figure 3.25, and let $\gamma_1 = \delta_1 \omega_1$. At resonance, n_{im} approaches

$$n_{im} \rightarrow \frac{Ne^2}{2m\epsilon_0 \omega_1^2} \frac{f_1}{\delta_1}$$

For an ideal gas, we use the fact that at STP, one mole occupies $22.4 \times 10^{-3}\,m^3$ to give $N = 2.69 \times 10^{25}\,m^{-3}$. Define ω_p such that

$$\omega_p = \sqrt{\frac{Ne^2}{m\epsilon_0}} = \frac{(2.69 \times 10^{25}\,m^3)\,(1.6 \times 10^{-19}C)}{(9.11 \times 10^{-31}kg)\,(8.854 \times 10^{-12}Nm^2/C^2)} = 2.92 \times 10^{14} s^{-1}$$

Thus, at resonance,

$$n_{im} \rightarrow \frac{1}{2} \frac{\omega_p^2}{\omega_1^2} \frac{f_1}{\delta_1}$$

Figure 3.25 uses $\delta_1 = 1/50$, $f_1 = 0.005$, and $\omega_1 = 9.42 \times 10^{14}\,s^{-1}$. Thus $n_{im} \rightarrow 0.0120$ as indicated in the figure. Note, however, that increasing the oscillator strength or decreasing δ can easily give peak values of n_{im} that differ substantially from zero, and hence values of n_r that differ substantially from one (see Problem 3.47). At atmospheric pressure, atomic absorption lines have values of δ in the range of 10^{-6}.

Notice that at a given temperature and pressure, N and the quantity ω_p have the same values for any ideal gas.

Problem 3.29   Add an infrared absorption line to the glass modeled in Example 3.11. Figure 3.24(b) uses a γ factor of one third the infrared resonant angular frequency, and an oscillator strength of 0.001 with an infrared absorption line centered at $\lambda = 5.0\,\mu m$.

Problem 3.30   (a) Repeat Example 3.12 for the $200\,nm$ absorption feature of Figure 3.25. (b) Estimate n_r and n_{im} for air at $550\,nm$ using the parameter values of Figure 3.25.

Problem 3.31   Write a computer program that uses the model of Equation 3.87 to investigate the index of refraction for air. Use the parameter values of Figure 3.25 as a starting point.

3.7.2 Dispersion in Conducting Media

Conductors have *free conduction electrons* that flow in response to external electric fields. To account for the induced currents within conducting materials, we need a general form of Ohm's law. Consider a conducting material in the shape of a cylinder or parallelepiped with cross-sectional area A, length d, and connected to a voltage source V, as shown in Figure 3.26. In terms of the *resistivity* ρ (units: Ωm), the resistance of the piece of material is given by

$$R = \rho \frac{d}{A}$$

The resistance R determines the current I that flows through the material according to Ohm's law expressed as $V = IR$. This current flows because of the electric field $\vec{E}$ that exists within the material, due to the potential difference placed across it by the voltage source V. Assuming a uniform electric field, we obtain

$$V = Ed = I\left(\rho\frac{d}{A}\right) \quad \Rightarrow \quad E = \rho\left(\frac{I}{A}\right)$$

We define the *current density* $\vec{J}$ as a vector with magnitude I/A (units: C/m^2) and direction given by the direction of $\vec{E}$. We also define the *conductivity* $\sigma = 1/\rho$ (units: $(\Omega m)^{-1}$), giving[13]

$$\vec{J} = \sigma\vec{E}$$

This expresses Ohm's law in terms of the electric field, current density, and the conductivity of the material.

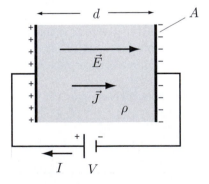

Figure 3.26   A conducting slab connected to a battery.

We may model the conductivity for static fields by defining a damping term γ that causes individual conduction electrons to reach a terminal velocity. As in Equation 3.78,

---

[13] Be careful not to confuse electrical conductivity with surface charge density, which also uses the symbol σ.

we assume a damping force that is proportional to the velocity:

$$m\gamma \frac{dx}{dt} = eE$$

giving a terminal velocity of

$$v_t = \frac{eE}{m\gamma}$$

The resulting current density is due to the collective movement of the free electrons

$$J = Nev_t = \frac{Ne^2}{m\gamma} E$$

giving the conductivity for steady fields:

$$\sigma = \frac{Ne^2}{m\gamma} \tag{3.95}$$

We may include the effects of free conduction electrons using an equation similar to Equation 3.78 with the restoring force removed:

$$m\frac{d^2x}{dt^2} + m\gamma\frac{dx}{dt} = eE_0 e^{-i\omega t} \tag{3.96}$$

We could also include the effects of bound electrons, as in Equation 3.78, which in gold and copper give the metal a distinctive color. In many cases, however, the absorption effects of bound electrons can be ignored for conducting media. We also ignore any contribution to the local field by bound charge distributions, assuming such effects to be reduced by available free charges. Since unbound charges have a resonance frequency of zero, we find

$$n^2 = 1 + \frac{Ne^2}{m\epsilon_0}\left(\frac{f}{-\omega^2 + i\gamma\omega}\right) \tag{3.97}$$

where f is the number of conduction electrons per atom, and γ is defined above. Define the *plasma frequency* ω_p such that

$$\omega_p = e\sqrt{\frac{Nf}{m\epsilon_0}} \tag{3.98}$$

Using this, Equation 3.97 becomes

$$n^2 = 1 + \frac{\omega_p^2}{-\omega^2 + i\gamma\omega} \tag{3.99}$$

For good conductors at higher frequencies, we may neglect the damping term:

$$n^2 = 1 - \frac{\omega_p^2}{\omega^2} \tag{3.100}$$

■ **EXAMPLE 3.13**

(a) Find γ for copper with density $8.95\ g/cm^3$, $63.5\ g/mole$, one conduction electron per atom, and a conductivity of $5.88 \times 10^7\ (\Omega m)^{-1}$. (b) Find plasma frequencies for copper and for the ionosphere when $N = 10^{11}\ m^3$.

Solution

(a) Following Example 3.11, we find that for copper, $N = 8.49 \times 10^{28}\ atoms/m^3$. Equation 3.95 gives

$$\gamma = \frac{Ne^2}{m\sigma} = \frac{\left(8.49 \times 10^{28}m^{-3}\right)\left(1.6 \times 10^{-19}C\right)^2}{\left(9.11 \times 10^{-31}kg\right)\left(5.88 \times 10^7\ (\Omega m)^{-1}\right)} = 4.06 \times 10^{13}s^{-1}$$

or, in terms of wavelength,

$$\lambda_\gamma = \frac{2\pi c}{\gamma} = 46.4\ \mu m$$

(b) For copper,

$$\omega_p = e\sqrt{\frac{Nf}{m\epsilon_0}} = \left(1.6 \times 10^{-19}C\right)\sqrt{\frac{\left(8.49 \times 10^{28}m^{-3}\right)(1)}{\left(9.11 \times 10^{-31}kg\right)\left(8.854 \times 10^{-12}C^2/(Nm^2)\right)}}$$

$$= 1.64 \times 10^{16}rad/s$$

or, in terms of wavelength,

$$\lambda_p = \frac{2\pi c}{\omega_p} = 115\ nm$$

For the ionosphere, with $N = 10^{11}\ m^3$,

$$\omega_p = \left(1.6 \times 10^{-19}C\right)\sqrt{\frac{\left(1.00 \times 10^{11}m^{-3}\right)(1)}{\left(9.11 \times 10^{-31}kg\right)\left(8.854 \times 10^{-12}C^2/(Nm^2)\right)}}$$

$$= 1.78 \times 10^7 rad/s$$

or, in terms of frequency,

$$f_p = \frac{\omega_p}{2\pi} = 2.84 \times 10^6\ Hz$$

The number density N in the ionosphere fluctuates, reaching peak values during periods of high solar illumination. Shortwave radio signals propagate worldwide due to reflections from the ionosphere. The upper limit of the shortwave band is 30 MHz.

―――――――――――――――――――

Effects due to γ can be neglected when $\omega \gg \gamma$. In this case, Equation 3.100 becomes

$$n = \sqrt{1 - \frac{\omega_p^2}{\omega^2}} \tag{3.101}$$

For frequencies less than the plasma frequency, the index of refraction becomes *imaginary*, with $n_r = 0$ and n_{im} given by

$$n_{im} = \sqrt{\frac{\omega_p^2}{\omega^2} - 1} \tag{3.102}$$

As the frequency becomes lower (i.e. $\lambda \to \lambda_\gamma$ in the example above), effects due to γ become non-negligible, and n_r becomes nonzero.

For frequencies greater than the plasma frequency, n becomes real, with n_r given by Equation 3.101. When this happens, the conductor becomes *transparent*, provided that there are no nearby bound-charge resonances. Note that in this case according to Equation 3.101, $n < 1$, suggesting that $v > c$. This is in fact the case, and we shall come back to this point in Chapter 5, where we will define a new velocity called the *group velocity* v_g. There, we will show that *information* travels at v_g, and that in all cases (including the one we are considering), $v_g < c$. The velocity given by c/n will be called the *phase velocity*.

3.7.2.1 Reflection from Conductors

According to Equation 3.62, $R = |r|^2 = rr^*$, with $r_\perp$ given by Equation 3.24 and $r_\parallel$ given by Equation 3.31. If the transmitted index of refraction is imaginary ($n = in_{im}$), Equations 3.24 and 3.24 become

$$r_\perp = \frac{n_i \cos\theta_i - (in_{im})\cos\theta_t}{n_i \cos\theta_i + (in_{im})\cos\theta_t} \tag{3.103}$$

$$r_\parallel = \frac{n_i \cos\theta_t - (in_{im})\cos\theta_i}{n_i \cos\theta_t + (in_{im})\cos\theta_i} \tag{3.104}$$

In this case, both $r_\perp$ and $r_\parallel$ are ratios of a complex number z and its complex conjugate z^*, and by Equation 3.39, both have unity magnitude. Thus for a conductor with infinite conductivity, $R = 1$ and $T = 0$ for all incident angles and all polarizations. In other words, a perfect conductor is perfectly reflecting.

If the effects of γ are not negligible, then there are dissipative losses that result in absorption. As noted above, effects due to γ are more important at lower frequencies. Because copper has a smaller value of γ than aluminum, it makes a better reflector for infrared wavelengths. Applications such as low-gain lasers that require nearly perfect reflectors commonly utilize multi layer dielectric films rather than conducting surfaces for reflection (see Chapter 5).

Both conductors and absorbing dielectrics are characterized by indices of refraction with both real and imaginary parts. In either case, the transmitted beam is eventually completely attenuated.

■ EXAMPLE 3.14

Use the Fresnel equations to find the normal-incidence reflectivity and absorptivity of a material with $n = n_r + i\,n_{im}$. Evaluate numerically at $560\,nm$ for aluminum ($n = 0.92 + i6.79$) and silver ($n = 0.12 + i3.45$). Assume an incident medium with $n_i = 1.00$.

Solution

From Example 3.9, at normal incidence

$$R = \left|\frac{n-1}{n+1}\right|^2 = \frac{(n_r - 1)^2 + n_{im}^2}{(n_r + 1)^2 + n_{im}^2} \tag{3.105}$$

Since

$$1 - R = 1 - \frac{(n_r - 1)^2 + n_{im}^2}{(n_r + 1)^2 + n_{im}^2} = \frac{(n_r + 1)^2 - (n_r - 1)^2}{(n_r + 1)^2 + n_{im}^2} = \frac{4n_r}{(n_r + 1)^2 + n_{im}^2}$$

we identify the absorptivity A:

$$A = \frac{4n_r}{(n_r + 1)^2 + n_{im}^2} \tag{3.106}$$

For silver,

$$A = \frac{4\,(0.12)}{(1.12)^2 + (3.45)^2} = 0.0365$$

and the reflectivity is $96.4\,\%$. For aluminum,

$$A = \frac{4\,(0.92)}{(1.92)^2 + (6.79)^2} = 0.0739$$

with a reflectivity of $92.6\,\%$.

For good conductors with $\gamma \ll \omega_p$, there is a large range of frequencies for which $n_r \ll n_{im}$. From Snell's law,

$$\sin \theta_t = \frac{n_i}{n_t} \sin \theta_i$$

giving

$$\cos \theta_t = \sqrt{1 - \left(\frac{n_i}{n_t}\right)^2 \sin^2 \theta_i}$$

For good conductors and frequencies well under the plasma frequency, n_r is small and n_{im} is large. Under these circumstances, the second term under the square root in the last equation can be neglected to give $\cos \theta_t \approx 1$. This means that the transmitted beam travels approximately normal to the conductor surface for all angles of incidence. Since n_{im} is large, the absorption coefficient is also large (see Equation 3.84), and the beam transmitted into the conductor is rapidly attenuated. Figure 3.27 shows $R_\perp$ and $R_\parallel$ for gold at $10\,\mu m$ where it is a good reflector, and at $450\,nm$ where its reflectivity is poor.

■ **APPLICATION NOTE 3.2**

Photoconducting Detectors

In a *photoconductor*, charge carriers are created by the absorption of photons. According to the band theory of solids, charge carriers reside in the *conduction band* which is at a higher energy than the *valence band*. When a photon is absorbed, an electron makes the transition to the conduction band, leaving a positively-charged *hole* in the valence band. The hole can also move in response to an external electric field (although not necessarily with the same *mobility*), and can also contribute to the *photocurrent* I_p (see Figure 3.28). The photocurrent is proportional to the photon flux, and constitutes a detection signal. Photoconductors are often used as light detectors. For more information, see Saleh and Teich [17].

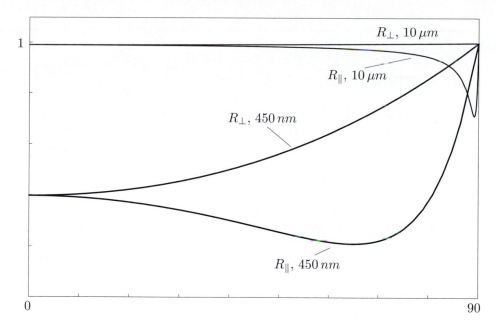

Figure 3.27 Reflectivity for gold at $10\,\mu m$ where $n = 11.5 + i67.5$, and at $450\,nm$ where $n = 1.40 + i1.88$.

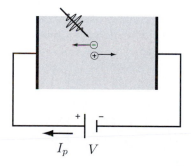

Figure 3.28 A photon is absorbed by a photoconducting material, creating an electron-hole pair.

Problem 3.32 Find γ and the plasma frequency for

    a) aluminum with density $2.70\,g/cm^3$, $27.0\,g/mole$, one conduction electron per atom, and a conductivity of $3.77 \times 10^7\,(\Omega m)^{-1}$.

    b) silver with density $10.49\,g/cm^3$, $108\,g/mole$, one conduction electron per atom, and a conductivity of $6.30 \times 10^7\,(\Omega m)^{-1}$.

Problem 3.33 The index of refraction of gold is $n = 0.31 + i2.88$ at $560\,nm$, and $n = 12.2 + i54.7$ at $9.92\,\mu m$. Find the normal incidence absorptivity and reflectivity at both wavelengths.

3.8 SCATTERING

Scattering occurs when electromagnetic waves interact with particles. If the particles have diameters much larger than the wavelength, the scattering can be understood mostly on a basis of reflection and refraction. Examples of large-particle scattering include rainbows and scattering from clouds. Scattering from all sizes of spherical particles can be calculated rigorously using *Mie scattering* theory[14], which takes Maxwell's equations as a starting point, and thus includes effects of *diffraction*. In *Rayleigh scattering*[15], the particle size is much smaller than the scattering wavelength. Both Rayleigh scattering and large-particle scattering can utilize simplifying assumptions, but when the particle size is on the order of the scattering wavelength, Mie scattering must be used. It just so happens that Earth's atmosphere has just the right buoyancy to suspend particles that are about equal to the wavelength of visible light.

3.8.1 Atmospheric Scattering

Rayleigh scattering makes the daytime sky blue. Equation 3.80 gives the periodic displacement amplitude of a Hooke's law electron interacting with a passing electromagnetic wave. Clean air has very little absorption, and well away from resonances, this amplitude is inversely proportional to ω^2, or directly proportional to λ^2. The scattering results from radiation by the accelerating electron, which is proportional to the square of the oscillation amplitude. Hence, the Rayleigh scattering is inversely proportional to λ^4, and blue is scattered more effectively than red.

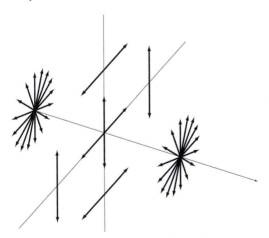

Figure 3.29 Polarization by scattering. Scattering at right angles is completely polarized; at other angles the polarization is partial.

In most cases, light scattered from particles is partially polarized. As illustrated in Figure 3.29, transverse electromagnetic waves scattered at right angles cannot have a polarization

[14]Gustav Mie (1869-1957).
[15]John Strutt, Lord Rayleigh (1842-1919).

component in the direction of propagation. The polarization of Rayleigh scattering in the atmosphere is at its maximum when viewed at a point 90° from the Sun along the solar path (ecliptic). Holding the transmission axis of a polarizer parallel to the ecliptic gives minimum transmission. Light scattered from clouds and haze has usually been multiply scattered and so is less polarized.

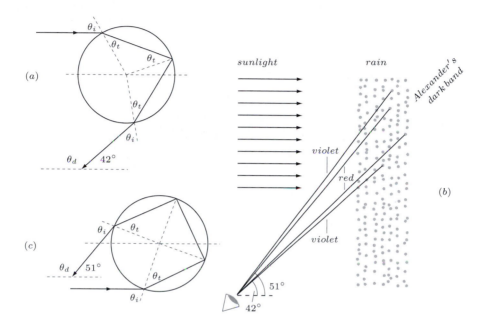

Figure 3.30   (a) Ray path for the primary rainbow. (b) Geometry that determines the observation angles for primary and secondary rainbows. (c) Ray path for the secondary rainbow.

3.8.1.1 *Rainbows*   Rainbows are formed when sunlight is backscattered from rain-drops. Surface tension causes raindrops to be almost perfectly spherical, and they are often large enough to allow diffraction effects to be neglected. Figure 3.30(a) shows a light ray incident on a drop at angle θ_i, refracting into the drop at angle θ_t. Because the two radii form an isosceles triangle, this ray reflects off the back surface with an incident angle of θ_t, and refracts back into air at the original angle θ_i. We determine the deviation angle θ_d between the incident and final backscattered ray as follows. The first refraction gives a deviation of $\theta_i - \theta_t$. The reflection contributes a deviation of $180° - 2\theta_t$. Finally, the last refraction deviates the beam another factor of $\theta_i - \theta_t$. The net deviation for a single internal reflection is

$$\theta_d = (\theta_i - \theta_t) + (180 - 2\theta_t) + (\theta_i - \theta_t)$$

or,

$$\theta_d = 180 + 2\theta_i - 4\theta_t \tag{3.107}$$

As we show below, the deviation has a *minimum* when θ_d is about 42°. The backscattered rays cluster at this *minimum deviation angle*, creating an enhanced brightness that consti-tutes the rainbow. The spectral colors result from dispersion within the raindrop that causes the minimum deviation angle to depend upon wavelength.

To find the minimum deviation angle $\theta_{d,min}$, define a new parameter $u \equiv \sin\theta_i$. Note that as u varies between zero and one, θ_i varies between zero and 90°. According to Snell's law, Equation 3.107 can be expressed as follows:

$$\theta_d = 180 + 2\sin^{-1}(u) - 4\sin^{-1}\left(\frac{u}{n}\right) \tag{3.108}$$

Since

$$\frac{d}{du}\sin^{-1}(\alpha u) = \frac{\alpha}{\sqrt{1 - (\alpha u)^2}}$$

the derivative of θ_d with respect to u is

$$\frac{d\theta_d}{du} = \frac{2}{\sqrt{1 - u^2}} - \frac{4\frac{1}{n}}{\sqrt{1 - \left(\frac{u}{n}\right)^2}}$$

To find the minimum deviation angle, set the last result to zero and solve for u:

$$n^2 - u^2 = 4 - 4u^2$$

or

$$u = \sqrt{\frac{4 - n^2}{3}} \tag{3.109}$$

■ **EXAMPLE 3.15**

Describe the rainbow that results if the index of refraction of water is 1.331 in the red, and 1.344 in the violet.

Solution

For red, Equation 3.109 gives $u = 0.86187$. This gives $\theta_i = 59.53°$, $\theta_t = 40.36°$, and a minimum deviation angle of $\theta_{d,min} = 137.63°$. An observer sees this angle at $180 - \theta_d = 42.4°$ above a line passing from the Sun to the eye. This line extends to any point that is opposite the Sun (an antisolar point). For example, such a line passes from the tip of your head to the tip of your shadow, in the direction away from the Sun.

For violet, $u = 0.8551$, $\theta_i = 58.77°$, $\theta_t = 39.51°$, $\theta_{d,min} = 139.49°$. The observer sees this angle at $40.5°$ above the antisolar point.

The entire rainbow is about 2° wide. It is circular because of the circular symmetry about the antisolar point. Colors range from red at the outer circumference to violet at the inner circumference. Scattering at angles for which θ_d exceeds the minimum value creates an increase in brightness *below* the rainbow, but because these angles can receive contributions from many wavelengths, the brightness here appears white. Likewise, there is no scattering at angles greater than 42.4° from the antisolar point, and this area appears dark, giving *Alexander's dark band* (see Figure 3.30(b)).

To observe a rainbow, the Sun must be low in the sky and reasonably unobstructed by clouds. The atmosphere in the direction of the antisolar point must contain suspended rain drops, ideally with diameters of 1—2 mm. Your shadow points the way!

Figure 3.30(c) shows a ray path with an additional reflection. This causes a *secondary rainbow* that is formed a bit higher in the sky. The deviation angle for this case is

$$\theta_d = 360 + 2\theta_i - 6\theta_t \tag{3.110}$$

The minimum deviation angle for the secondary rainbow occurs at $\theta_i \approx 72°$ with the minimum value of $\theta_{d,min}$ greater for violet than for red, so the inside of the secondary rainbow is red, and the outer circumference is violet (see Problem 3.34). From Figure 3.30(c), it is seen that the secondary rainbow contributes no scattered rays *below* the arc of the bow; thus, *Alexander's band* remains dark (see Figure 3.30(b)). See Figure 3.31.

Figure 3.31   Double rainbow over the Very Large Array (VLA) radio telescope at Socorrow, New Mexico. The dark region between the two rainbows is Alexander's dark band. (Doug Johnson/Photo Researchers, Inc.)

3.8.1.2 Parhelia   *Parhelia* (false suns), otherwise known as *Sun dogs*, arise from atmospheric scattering from hexagonal ice crystals. The ice crystals are usually formed high in the cold troposphere, but occasionally are located much closer to ground level. A hexagonal ice crystal is illustrated in Figure 3.32(a). Rays traversing through such a crystal have a deviation minimum that occurs for rays that transmit *symmetrically*. To see that this is so, suppose for a moment that a non-symmetric traversal gives a deviation minimum. Reversing that ray then gives the equal deviation at another incident angle. By contradiction, the minimum must occur only for the symmetric case. Since the index of refraction for ice is $n = 1.31$ (see Table 3.1.), the minimum deviation occurs for $\theta_d \approx 22°$, as shown in Figure 3.32(a). As with the rainbow, there is a concentration of rays at the deviation minimum, and this causes an increase in brightness at this deflection angle.

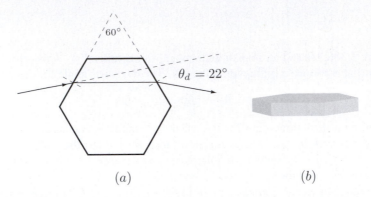

$$(a) \qquad\qquad (b)$$

Figure 3.32 (a) A ray that passes symmetrically through the effective prism of a hexagonal ice crystal has a minimum deviation at $22°$. (b) Thin plate-like ice crystals often orient horizontally as they fall through the atmosphere, resulting in two bright parhelia diametrically located $22°$ from the Sun along a horizontal line.

Randomly oriented crystals result in a *halo* with an angular radius of $22°$, centered on the Sun. Flat crystals tend to orient horizontally as they fall, as shown in Figure 3.32(b), and these create two bright parhelia each diametrically $22°$ from the Sun along a horizontal line. Each parhelia often contains spectral colors that result from dispersion within the ice.

Figure 3.33 Parhelia with halo. (Pekka Parviainen/Photo Researchers, Inc)

Problem 3.34   Derive Equation 3.110, then show that the deviation angle for a secondary rainbow has a minimum when

$$\sin \theta_i = \sqrt{\frac{9 - n^2}{8}}$$

Describe this rainbow using the data in Example 3.15.

Problem 3.35   Show that the ray that passes symmetrically through the hexagonal ice crystal of Figure 3.32 has a deviation angle of $21.8°$. Assume an index of 1.31 for ice.

3.8.2   Optical Materials

Optical materials have properties that are useful in optical applications. The most obvious property required of such a material is that it be transparent. Table 3.1. gives the nominal range of transparency for a variety of commonly used substances. Transparency requirements depend on the application. A percent or so of absorption might be acceptable in a lens used at low power levels, but if the same material is used in the cavity of a high-power laser, the heat deposited could distort and even destroy the material. Because of dispersion, the index of refraction and range of transparency depends upon wavelength. Making lenses from material with low dispersion can reduce *chromatic aberration*, as can using materials with different dispersive properties in achromat lens designs as described in Section 4.6.1.

Physical properties such as hardness are also important. For example, sodium chloride (common table salt) is quite transparent from the visible to middle infrared. In the early days of laser development, sodium chloride Brewster windows were commonly used in carbon dioxide and other infrared laser systems. However, sodium chloride is soft, and it is soluble in water; hence, the slightest abrasion or water vapor condensation will destroy a carefully polished optical surface. Zinc selenide is now more commonly used. It is hard, it also has very low absorption across the middle infrared, and has enough transparency in the visible to allow helium-neon lasers to be used as alignment aids. Zinc selenide is toxic if ingested, and care must be taken if it is shattered by dropping.

Other important properties of optical materials will be discussed in later chapters. Crystalline and stressed materials often exhibit *birefringence*, where the index of refraction depends on direction. Such materials have important applications that are discussed in Chapter 9. *Nonlinear materials* have optical properties that are affected by external fields, including fields in electrogmagnetic waves. Such materials can be used to enable many modern optical applications such as fast amplitude modulators and frequency doublers. Applications of nonlinear optics are also discussed in Chapter 9.

Additional Problems

Problem 3.36   The *optical thickness* is the product of physical thickness d and the refractive index n (see Problem 2.28). Show that within a material, the number of wavelengths in a physical thickness equals the number of vacuum wavelengths in the corresponding optical thickness.

Problem 3.37   A fish is at a depth d under the water. What is the apparent depth if it is viewed at an angle θ_i?

Problem 3.38   Show that when a beam is reflected from a flat mirror rotated by θ about an axis passing through its center, the beam direction is changed by 2θ.

Problem 3.39   Plot $t_\perp$ and $t_\parallel$ for internal incidence with $\theta_i < \theta_c$ for a glass-air interface. Assume $n = 1.50$ for glass.

Problem 3.40   For a water-glass interface ($n_g = 1.52$, $n_w = 1.33$),
   a) Plot $r_\perp$, $r_\parallel$, $t_\perp$, and $t_\parallel$ vs. θ_i for internal and external incidence.
   b) Plot $R_\perp$, $R_\parallel$, $T_\perp$, and $T_\parallel$ vs. θ_i for internal and external incidence.

Problem 3.41   Design a cell with Brewster windows similar to the one shown in Figure 3.11 using window material with index of refraction 2.42. Label all angles carefully, especially the angles for the ends of the cylinder.

Problem 3.42   Suppose you find a linear polarizer with no indicator mark for the transmission axis (this is very common). Describe a method that uses Brewster's angle to identify the transmission axis.

Problem 3.43   A photographer wishes to use a linear polarizer in front of the camera lens to reduce glare from a surface with normal vector inclined at 30° to the vertical. How should the polarizer be oriented? Include a diagram.

Problem 3.44   What percentage of irradiance transmits completely through a slab of zirconium dioxide (index of refraction 2.40) at normal incidence? Ignore any effects due to interference or absorption.

Problem 3.45   Using Snell's law, the Fresnel amplitude ratios may be expressed as

$$r_\perp = -\frac{\sin\left(\theta_i - \theta_t\right)}{\sin\left(\theta_i + \theta_t\right)}$$

$$r_\parallel = -\frac{\tan\left(\theta_i - \theta_t\right)}{\tan\left(\theta_i + \theta_t\right)}$$

$$t_\perp = +\frac{2\sin\theta_t\cos\theta_i}{\sin\left(\theta_i + \theta_t\right)}$$

$$t_\parallel = +\frac{2\sin\theta_t\cos\theta_i}{\sin\left(\theta_i + \theta_t\right)\cos\left(\theta_i - \theta_t\right)}$$

Use these formulas to reproduce the graphs of Figure 3.13.

Problem 3.46   Use experimental values of n_r for air: $n_r = 1.00029188$ at $656.4\,nm$ and $n_r = 1.00029713$ at $435.8\,nm$, to determine values of A and B for air in Equation 3.94. Using these constants, compare the result of Equation 3.94 to the experimental value of $n_r = 1.00029391$ at $546.1\,nm$.

Problem 3.47   Consider a resonance at ω_1 in a non-conducting gas with $\gamma_1 \ll \omega_1$. (a) Show that the width of an anomalous dispersion region, as determined by the width $\Delta\omega$ between maximum and minimum points of n_r, is equal to γ. (b) Show that the maximum and minimum values of n_r in an anomalous dispersion region occur where n_{im} is at half-maximum.

Problem 3.48   Use Equation 3.100 and the fact the phase velocity $v = \omega/k$ to derive a *dispersion relation* for an electromagnetic wave traveling in a conductor when $\omega > \omega_p$:

$$\omega = \omega_p \sqrt{1 + \left(\frac{c^2}{\omega_p^2}\right) k^2}$$

Problem 3.49   Starting with Equation 3.99, show that

$$n_r^2 - n_{im}^2 = 1 - \left(\frac{\omega_p^2}{\omega^2 + \gamma^2}\right)$$

$$2n_r n_{im} = \frac{\gamma}{\omega}\left(\frac{\omega_p^2}{\omega^2 + \gamma^2}\right)$$

Using this, show that $n_r = n_{im}$ when

$$\omega = \sqrt{\omega_p^2 - \gamma^2}$$

For small γ, this occurs with $\omega = \omega_p$.

Problem 3.50   Show that an electromagnetic wave propagating through a conducting gas (i.e., a plasma) with $\omega \gg \omega_p$ propagates with a dispersion that is proportional to the square of the wavelength. In particular, show that

$$n = 1 - \frac{\lambda^2}{2\lambda_p^2}$$

where λ_p is the vacuum wavelength at the plasma frequency.

CHAPTER 4

GEOMETRIC OPTICS

My candle burns at both ends; it will not last the night;
But ah, my foes, and oh, my friends — it gives a lovely light.

—Edna St. Vincent Millay

Contents

4.1   INTRODUCTION

In this chapter, we explore applications of the laws of reflection and refraction. When discussing these laws in Chapter 3, we found it convenient to introduce the notion of a *light ray*, and we will continue to rely heavily on this idea. In fact, the concept of a light ray is so central to geometrical optics that it is sometimes called *ray optics*. We begin by discussing reflection and refraction from aspheric surfaces, then simplify the analysis with the paraxial and thin-lens approximations to allow the application of spherical refracting and reflecting surfaces. Lens combinations are discussed, along with computational techniques provided by principal points and effective focal lengths that are also later applied to thick lenses. Aberrations are discussed, and techniques for reducing them are discussed within the context of specific optical instruments. Techniques of radiometry are introduced and illustrated for measurement of blackbody radiation, which provided much of the impetus for the transition from classical to modern theories of physics. A discussion of optical fiber technology provides an introduction to this important and current optical application. Finally, matrix methods are introduced that can facilitate the analysis of more complicated optical systems.

4.2   REFLECTION AND REFRACTION AT ASPHERIC SURFACES

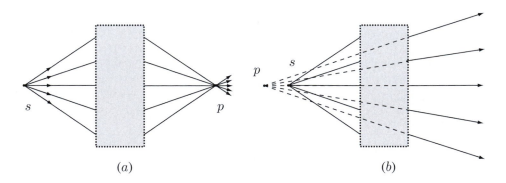

(a) (b)

Figure 4.1   (a) An optical element collects a bundle of rays leaving the source point s, and redirects them so that they pass through the collection point p. The time required for light to propagate along all paths is equal. (b) The optical element redirects rays that leave the source s so that they appear to emanate from a virtual point p.

A reflecting or refracting surface forms an image by collecting light rays that emanate from a source point s, and reassembling them so that they emanate or appear to emanate from another point p, as illustrated in Figure 4.1. This can happen only if it is possible for *all* rays leaving s to converge to p in a way that is consistent with Fermat's principle of least time discussed in Chapter 3. Since all rays must propagate from s to p in minimal time, and since all rays leaving s that encounter the reflective or refractive surface converge to p, it then follows that all such rays propagate from s to p in the *same* time. The points s and p of Figure 4.1 are called *conjugate points*.

The problem of finding surfaces that satisfy these conditions was first solved by Descartes[1] in 1637. According to this analysis, many of the surfaces that reflect or refract rays as in Figure 4.1 have cross-sections given by one of the *conic sections*[2], first studied by Euclid[3] and then more fully by the Greek geometer Apollonius[4] around the second century BC. Conic sections include the ellipse, parabola, hyperbola and circle. The corresponding surfaces are obtained by rotating these conic sections about an axis of symmetry to give ellipsoids, paraboloids, hyperboloids and spheres. In general, surfaces that refract or reflect rays leaving a source point s to a conjugate point p are known as *Cartesian surfaces*. It is instructive to examine a few important cases.

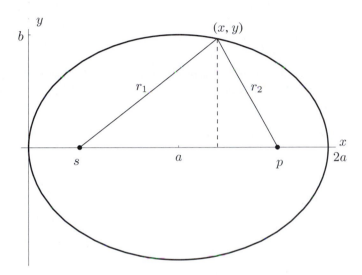

Figure 4.2   An ellipse with source s at the left focus and point p at the right focus. The semimajor axis is a and the semiminor axis is b. The ellipse is defined by the equation $r_1 + r_2 = 2a$.

We begin with an example of reflection from a concave surface. Figure 4.2 shows an *ellipse* with source s at one focus and point p at the other focus. An ellipse is defined by the relation

$$r_1 + r_2 = 2a \tag{4.1}$$

where r_1 and r_2 are the distances from any point on the ellipse to the two foci and a is the semimajor axis, as shown in Figure 4.2. To show that all rays leaving s reflect through p, we show that rays following r_1 and r_2 in Figure 4.2 obey the law of reflection. Consider Figure 4.3, which shows two such trajectories that reflect from closely spaced points on the ellipse. Since the points are spaced by an infinitesimal amount, the secant line connecting them is the tangent to the surface, so the law of reflection holds if $\theta_1 = \theta_2$. To see that this is the case, note that the two paths labeled r_1 and $r_1 + dr_1$ are parallel, as are the other two paths. Thus, the two angles labeled θ_1 are equal, as are the two angles labeled θ_2. According to Equation 4.1, in going from the left path to the right path, r_1 increases by precisely the same amount as r_2 decreases. Thus, the bold triangle is isosceles, and

---

[1]Rene Descartes: 1596-1650. French philosopher and mathematician.
[2]See, for example, Thomas [13].
[3]Euclid of Alexandria: 325-265 BC. Greek mathematician best known for his text on geometry.
[4]Apollonius of Perga: 262-190 BC. Greek geometer.

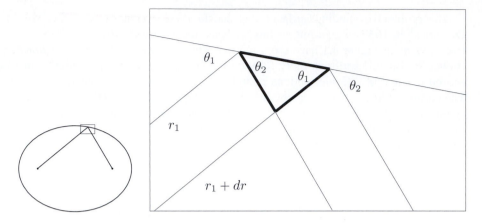

Figure 4.3   Construction to show that the law of reflection holds for a ray that leaves one focus and reflects to pass through the other focus.

$\theta_1 = \theta_2$ in accord with the law of reflection. Furthermore, by Equation 4.1, all paths from s to p have the same length, so the travel time from s to p is the same for all paths, in accord with Fermat's principle. This is often referred to as the *reflection property of the ellipse*. There are many proofs; see, for example, Stein [20].

When the separation between s and p approaches infinity, the Cartesian surface for reflection becomes a *parabola*. To see that this is so, refer to Figure 4.2. According to Equation 4.1,

$$\sqrt{(x-s)^2 + y^2} + \sqrt{(p-x)^2 + y^2} = 2a = p + s$$

Taking the limit as $p \to \infty$ gives

$$\sqrt{(x-s)^2 + y^2} + (p - x) = p + s$$

giving

$$\sqrt{(x-s)^2 + y^2} = x + s$$

Squaring both sides and simplifying gives

$$y^2 = 4sx \tag{4.2}$$

Equation 4.2 describes a parabola with center of symmetry (vertex) at $(0, 0)$ that opens to the right with the source s located at the *focus* f, as shown in Figure 4.4(a). By reversing the directions of the rays, it is seen that parallel rays arriving parallel to the x-axis as in Figure 4.4(b) are reflected through f. For this reason, parabolic mirrors are used to form precise images of astronomical objects; however, this is strictly only true for images formed at the focus. Images formed away from the axis of symmetry are not formed at a conjugate point, and thus suffer from imperfections referred to as *aberrations*. Aberrations are discussed in more detail in Section 4.6.

A *lens* is an object that focuses rays from s to p by refraction. To investigate this, consider a curved boundary that separates two regions of indices n_1 and n_2, as shown in Figure 4.5. Rays that leave the source s travel in the medium with index n_1, and rays that arrive at p travel in the medium of index n_2. We seek a boundary that gives equal

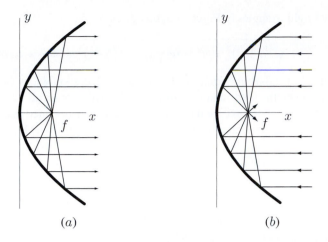

Figure 4.4 (a) A parabolic surface reflects rays emanating from s located at the focus toward infinity. (b) Conversely, rays arriving from infinity are reflected through the focus.

propagation time for all paths from s to p. To facilitate a graphical analysis, we introduce the notion of *optical path length*. In a region of higher index, the propagation time is longer since the light speed decreases. Since the transit time is given by $t = {d_0}/{v} = {nd_0}/{c}$ we can account for the increase in transit time by letting

$$d = nd_0 \qquad (4.3)$$

where d_0 is the *physical distance*.

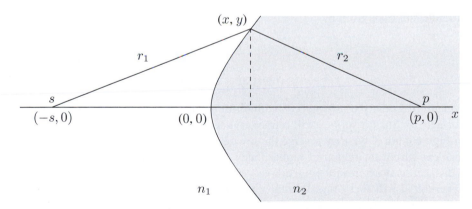

Figure 4.5 A refracting boundary separating two media with indices of refraction n_1 and n_2.

Figure 4.5 shows a refracting boundary with rays leaving source s located[5] at $(-s, 0)$ and arriving at p located at $(p, 0)$. The condition that all transit times be equal means that any arbitrary path has an optical path length equal to that of the straight-line path from s to p

$$n_1 r_1 + n_2 r_2 = n_1 s + n_2 p$$

---

[5]For notational brevity, we use the symbols s and p to denote point names, point locations, and distances to the points.

Using the two right triangles in Figure 4.5, this gives

$$n_1\sqrt{(s+x)^2 + y^2} + n_2\sqrt{(p-x)^2 + y^2} = n_1 s + n_2 p \tag{4.4}$$

The set of all (x, y) satisfying Equation 4.4 defines surfaces known as *Cartesian ovals*.

Let us investigate the Cartesian oval interface that will refract all rays leaving s so that they travel parallel to the x-axis in the second medium, as illustrated in Figure 4.6(a). Use Equation 4.4 and take the limit as p approaches infinity. In this case, Equation 4.4 becomes

$$n_1\sqrt{(s+x)^2 + y^2} + n_2(p-x) = n_1 s + n_2 p$$

giving

$$n_1\sqrt{(s+x)^2 + y^2} = n_1 s + n_2 x$$

Squaring both sides and rearranging gives

$$n_1^2 y^2 = \left(n_2^2 - n_1^2\right) x^2 + 2n_1\left(n_2 - n_1\right) sx$$

which, after completing a square, gives

$$\left(\frac{n_1^2}{n_2^2 - n_1^2}\right) y^2 = \left[x + \left(\frac{n_1 s}{n_2 + n_1}\right)\right]^2 - \left(\frac{n_1 s}{n_2 + n_1}\right)^2$$

This can be rearranged to give

$$\frac{(x+a)^2}{a^2} - \frac{y^2}{b^2} = 1 \tag{4.5}$$

where a and b are given by

$$a = \frac{n_1 s}{n_2 + n_1} \tag{4.6}$$

$$b = \left(\sqrt{\frac{n_2 - n_1}{n_2 + n_1}}\right) s \tag{4.7}$$

Equation 4.5 describes a *hyperbola* with vertex at the origin, semimajor axis a, and semiminor axis b. Such a surface will refract rays leaving s into a parallel beam as illustrated in Figure 4.6(a). We may reverse the ray direction in Figure 4.6(a); hence, parallel rays incident within the region of higher index will refract to pass through a single focus at a distance s away from the hyperbola vertex. By the same reasoning, a double-convex hyperboloid will focus rays leaving s to another point located at p, as shown in Figure 4.6(b).

In practice, there are difficulties associated with applying the aspheric Cartesian surfaces discussed in this section. Each surface is designed for a specific pair of conjugate points, and using them for different locations of s and p can mean that rays leaving s fail to converge to a single point p. Such a failure is referred to as an *aberration*, discussed more fully in Section 4.6. Finally, Cartesian surfaces can be difficult to manufacture with the accuracy needed for optical applications.

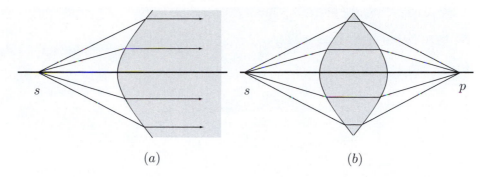

(a) (b)

Figure 4.6   (a) A hyperbolic surface that refracts rays leaving s toward infinity. (b) A double-convex hyperboloid that focuses rays leaving s to p.

Problem 4.1   The *sagitta* of a mirror is the depth of the surface curve, measured relative to the mirror edge. What is the sagitta of a $25.4\,cm$-diameter parabolic mirror that has a $25.4\,cm$ focal length? Repeat if the focal length is $2.54\,m$.

Problem 4.2   Find the semimajor and semiminor axes of the hyperbolic surface that will refract rays leaving s located $10.0\,cm$ from the vertex into parallel rays. Assume an incident medium of air and a transmitted medium of glass with index 1.50.

4.3   REFLECTION AND REFRACTION AT A SPHERICAL SURFACE

Compared to Cartesian surfaces, *spherical* surfaces are relatively easy to prepare. Furthermore, by utilizing the appropriate approximations, use of spherical refracting and reflecting surfaces will lead us to relatively simple formulas that describe the location, orientation and size of images formed from extended objects with a continuous range of location.

We will locate all positions relative to the *optical axis*. Many optical elements have cylindrical symmetry, so for a single optical element, the optical axis passes through this axis of symmetry. Multiple optical elements must be arranged coaxially for there to be a single axis of symmetry and corresponding optical axis. It is almost always important to arrange optical elements in this way to achieve an optimal image.

4.3.0.1   The Paraxial Approximation   We will often find it useful to apply the *paraxial approximation*. For this approximation to be valid, all rays passing through the optical system must make angles with the optical axis that satisfy the *small angle approximation*:

$$\sin \theta \approx \theta \qquad\qquad (4.8)$$

This will be the case if the nonlinear terms in the Taylor series approximation for $\sin \theta$ can be neglected (see Example 1.3); in other words, that $\frac{\theta^3}{3!} \ll \theta$. For most purposes, $\theta = 0.1\,rad$ or about $5°$ is small enough.

4.3.1   Spherical Reflecting Surfaces

In this section, we use the paraxial approximation to locate images formed by spherical reflecting surfaces. Figure 4.8 shows a concave spherical reflecting surface with center of

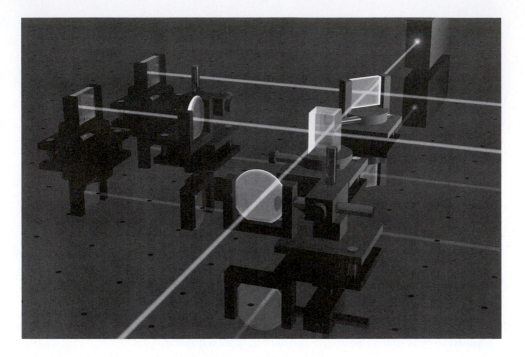

Figure 4.7 An *optics table* uses precisely aligned holes to mount optical elements along a well-defined optical axis. (Boris Starosta/Photo Researchers, Inc.)

curvature located at the point R. The source point s is located on the *optical axis*, which coincides with the axis of symmetry for the spherical surface. An arbitrary ray leaving s intersects the spherical surface at M with an angle of incidence given by θ_i, measured relative to the radius line that defines the normal to the surface. The reflected ray passes through a point p that is also located on the optical axis. The surface can form an image if all rays it receives from s are reflected through p. Let the distance along the optical axis from s to the reflecting surface be the *object distance* s_o, and the distance along the optical axis from p to the reflecting surface be the *image distance* s_i.

The *law of sines*[6] is illustrated in Figure 4.8(b):

$$\frac{\sin \theta_a}{A} = \frac{\sin \theta_b}{B} = \frac{\sin \theta_c}{C} \tag{4.9}$$

Applying the law of sines to triangle sRM in Figure 4.8 gives

$$\frac{\sin \theta_i}{s_o - R} = \frac{\sin \theta_1}{R} \tag{4.10}$$

Similarly, for triangle RpM, we have

$$\frac{\sin \theta_i}{R - s_i} = \frac{\sin (180 - \theta_2)}{R} \tag{4.11}$$

[6] See any introductory calculus text: e.g. Thomas [13].

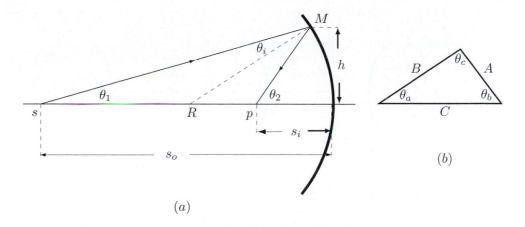

Figure 4.8 (a) A light ray emanating from source s reflects at a spherical surface and passes through point P. (b) A triangle illustrating the law of sines. Side a is opposite angle θ_a, and so on.

Combining Equations 4.10 and 4.11 gives

$$\frac{\sin \theta_1}{\sin \theta_2} = \frac{R - s_i}{s_o - R} \tag{4.12}$$

According to this result, the value of s_o would depend on the values of θ_1 and θ_2, so it would not be true that all reflected rays pass through a *single* point p. However, we may make the *paraxial approximation* that reflected rays have angles θ_1 and θ_2 small enough for the small angle approximation to apply: $\sin \theta \approx \theta$[7]. In this case,

$$\theta_1 \approx \frac{h}{s_o}$$

$$\theta_2 \approx \frac{h}{s_i}$$

After these approximations, Equation 4.12 becomes

$$\frac{s_i}{s_o} = \frac{R - s_i}{s_o - R}$$

This may be rearranged to give

$$\frac{1}{s_i} + \frac{1}{s_o} = \frac{2}{R} \tag{4.13}$$

An object at infinity gives parallel rays that reflect through the point $s_i = f$, as illustrated in Figure 4.9(a). This point is referred to as the *focal point*, and the distance from the mirror to f is called the *focal length*. According to Equation 4.13,

$$f = \frac{R}{2} \tag{4.14}$$

In terms of the focal length, Equation 4.13 becomes

$$\frac{1}{s_i} + \frac{1}{s_o} = \frac{1}{f} \tag{4.15}$$

---

[7]For purposes of illustration, the rays used in Figure 4.8 are inclined at angles that far exceed those appropriate for the paraxial approximation.

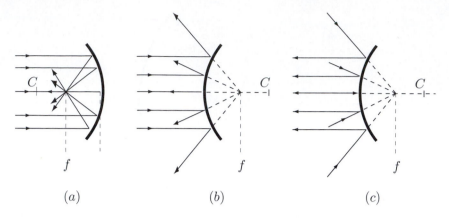

(a) (b) (c)

Figure 4.9   (a) A converging mirror formed from a concave reflecting surface. A source point at infinity produces parallel rays that reflect through a *real* focus located at f. (b) Parallel rays reflecting from a diverging mirror formed from a convex reflecting surface. The reflected rays appear to diverge from a *virtual* focus located behind the mirror. (c) A time reversal of (b). In this case, the incoming rays converge towards a *virtual object* located at the virtual focus f; hence, the reflected rays are parallel.

4.3.1.1   *Sign Conventions for Reflecting Surfaces*   The source point s in Figure 4.8 is located to the left of the concave reflecting surface to give incoming rays diagrammed intuitively as arriving from left to right. The corresponding object distance s_o shown in this figure is *positive* relative to an origin located at the center of the reflecting surface. Likewise, the radius of curvature R, focal point f, and image distance s_i are also all positive as illustrated in Figure 4.8. A *convex* reflecting surface thus has a negative radius of curvature, so according to Equation 4.14, it also has a *negative* focal length f, as illustrated in Figure 4.9(b). In this case, parallel rays are reflected so that they appear to emanate from a point located a distance f *behind* the mirror. Such a point is referred to as a *virtual focus*. A *virtual object* is illustrated in Figure 4.9(c). In this case, incoming rays are reflected *before* converging to a point located behind the mirror.

We may summarize the sign conventions used with Equation 4.13 for spherical reflecting surfaces in the following intuitive way. Mirrors *reflect* light, so images and objects to the left of the reflecting surface are located by positive values of s_i and s_o. The sign conventions for f and R are best remembered by recalling that focal lengths for converging mirrors are positive, and those for diverging mirrors are negative; hence, values of f and R located to the left of the reflecting surface are positive, and are otherwise negative.

4.3.2   Spherical Refracting Surfaces

We may also use the paraxial approximation to locate images formed by spherical refracting surfaces. Figure 4.10 shows a convex spherical refracting surface with radius of curvature R that separates regions with index of refraction n_m and $n_\ell > n_m$. A ray leaves the optical axis at point s and angle θ_1, and is refracted back through the optical axis at p with angle θ_2, as illustrated. According the law of sines,

$$\frac{\sin\left(180° - \theta_i\right)}{s_o + R} = \frac{\sin\left(\theta_1\right)}{R} = \frac{\sin\left(\theta_i\right)}{s_o + R} \tag{4.16}$$

Similarly,

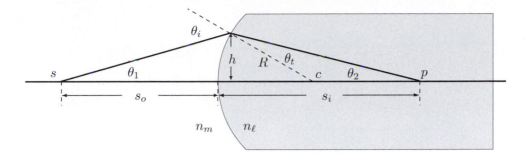

Figure 4.10   A light ray emanating from source s refracts from a spherical surface and passes through point P.

$$\frac{\sin(\theta_t)}{s_i - R} = \frac{\sin(\theta_2)}{R} \tag{4.17}$$

Divide Equations 4.16 by 4.17 to give

$$\frac{\sin(\theta_1)}{\sin(\theta_2)} = \left(\frac{s_i - R}{s_o + R}\right)\left(\frac{\sin(\theta_i)}{\sin(\theta_t)}\right) = \left(\frac{s_i - R}{s_o + R}\right)\left(\frac{n_\ell}{n_m}\right) \tag{4.18}$$

Solving Equation 4.18 s_i gives an expression whose value depends on the angles θ_1 and θ_2. Thus, there is no unique image point p. However, we may again apply the paraxial approximation:

$$\left(\frac{s_i - R}{s_o + R}\right)\left(\frac{n_\ell}{n_m}\right) = \frac{\sin(\theta_1)}{\sin(\theta_2)} \approx \frac{\theta_1}{\theta_2} \approx \frac{s_i}{s_o} \tag{4.19}$$

This may be rearranged to give

$$\frac{n_m}{s_o} + \frac{n_\ell}{s_i} = \frac{(n_\ell - n_m)}{R} \tag{4.20}$$

In this equation, the object distance s_o is positive for a source point s located to the left of the surface, relative to an origin located at the center of the refracting surface. The radius of curvature R is also positive for the convex refracting surface shown in Figure 4.8.

Most lenses have two refracting surfaces, as shown in Figure 4.11. Light rays traveling from left to right encounter surface 1 first, which has radius of curvature R_1. The second surface has radius of curvature R_2. Note that as diagrammed in Figure 4.11, curvatures R_1 and R_2 are both positive. The medium that surrounds the lens has index of refraction n_m, and the lens has index of refraction n_ℓ. The source point s is located a distance s_{o1} from surface 1. By Equation 4.20,

$$\frac{n_m}{s_{o1}} + \frac{n_\ell}{s_{i1}} = \frac{(n_\ell - n_m)}{R_1} \tag{4.21}$$

The ray that transmits into the lens propagates toward a point a distance s_{i1}, as illustrated in Figure 4.11. This point becomes the object for surface 2 with object distance equal to $d - s_{i1}$. If this difference is negative, the object for surface 2 is *virtual*. Also, for surface 2, the incident medium has index of refraction n_ℓ. Equation 4.20 gives

$$\frac{n_\ell}{(d - s_{i1})} + \frac{n_m}{s_{i2}} = \frac{(n_m - n_\ell)}{R_2} \tag{4.22}$$

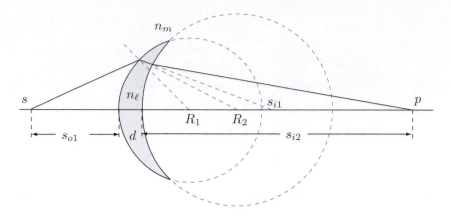

Figure 4.11 A lens formed by two spherical refracting surfaces. The radius of curvature of both spherical surfaces is positive.

Adding Equations 4.21 and 4.22 and rearranging gives

$$\frac{n_m}{s_{o1}} + \frac{n_m}{s_{i2}} = (n_\ell - n_m)\left(\frac{1}{R_1} - \frac{1}{R_2}\right) - \frac{n_\ell\, d}{(s_{i1})(s_{i1} - d)} \qquad (4.23)$$

In addition to the paraxial approximation, we must make the additional approximation that the lens is *thin*. Thus, we may ignore the last term in Equation 4.23. Furthermore, since d is small, we may let $s_{o1} \to s_o$ and $s_{i2} \to s_i$, giving

$$\frac{1}{s_o} + \frac{1}{s_i} = (n - 1)\left(\frac{1}{R_1} - \frac{1}{R_2}\right) \qquad (4.24)$$

where n is the relative index of refraction: $n = n_\ell/n_m$. As for reflection, the focal length is defined as the image distance as the object distance approaches infinity

$$\frac{1}{f} = (n - 1)\left(\frac{1}{R_1} - \frac{1}{R_2}\right) \qquad (4.25)$$

With this definition, Equation 4.24 becomes

$$\frac{1}{s_i} + \frac{1}{s_o} = \frac{1}{f}$$

which is identical to Equation 4.15 for a spherical reflecting surface. We will refer to this equation as the *thin-lens equation*, keeping in mind that the same equation describes spherical paraxial reflecting surfaces and thin spherical lenses.

It is common to specify the *f-number* of a lens or mirror:

$$f/\# = \frac{f}{D} \qquad (4.26)$$

where D is the lens diameter and f is the focal length. This is usually represented by the symbols $f/\#$; so for example, a lens with a focal length of four times the diameter would be described as an $f/4$ lens. As the f-number becomes small, the paraxial approximation becomes less valid.

■ EXAMPLE 4.1

A typical camcorder forms images with an $f/4$ converging lens. Estimate the largest angular deviation from paraxial for this lens, and evaluate the error from the paraxial approximation in this case. Assume a distant object.

Solution

The largest angular deviation from paraxial occurs for plane parallel rays that are refracted at the edge of the lens through the focus, as illustrated below. Thus,

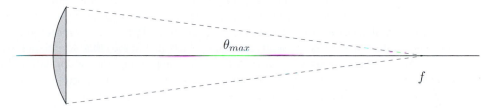

the maximum angle for any ray collected by this lens has a tangent of 1/8 giving $\theta_{max} = 0.124\,rad\,(7.125°)$. In this case, the error in the paraxial approximation is

$$\left|\frac{\sin\theta - \theta}{\sin\theta}\right| \times 100\% \approx 0.26\%$$

Whether or not this amount of error is acceptable is dictated by the desired image quality. For inexpensive camcorders, this may be fine. Using an aspheric lens can reduce this paraxial error, resulting in a more accurate image.

4.3.3 Sign Conventions and Ray Diagrams

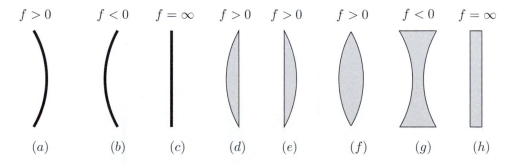

Figure 4.12   Sign conventions for focal lengths as determined by Equations 4.14 and 4.25 for light incident from left to right. Optics with positive focal lengths are converging and optics with negative focal lengths are diverging. (a) Concave reflecting surface: $R > 0$. (b) Convex reflecting surface: $R < 0$. (c) Plane reflecting surface: $R = \infty$. (d) Plano-convex lens: $R_1 > 0$, $R_2 = \infty$. (e) Plano-convex lens: $R_1 = \infty$, $R_2 < 0$. (f)Biconvex lens: $R_1 > 0$, $R_2 < 0$. (g) Bi-concave lens: $R_1 < 0$, $R_2 > 0$. (h) Plane slab: $R_1 = R_2 = \infty$.

Sign conventions for focal lengths of reflecting and refracting optical elements are illustrated in Figure 4.12. For brevity, we refer to an optical element as an *optic*. Converging optics have positive focal lengths and diverging optics have negative focal lengths. As with reflecting surfaces, sign conventions for the radii R_1 and R_2 of spherical refracting surfaces are best remembered for the double-convex lens of Figure 4.12(f). In this case, $R_1 > 0$ and $R_2 < 0$, so the focal length obtained with Equation 4.25 is positive; thus, a radius is positive if located to the right of the lens and otherwise it is negative. A lens *transmits* light, so a real image is located to the right of the lens, which must also be the positive side for s_i. As always, real objects are located to the left of the optic by a positive object distance s_o.

The location and size of images formed by converging and diverging optics may be obtained intuitively using a *ray diagram*, as illustrated for a converging lens in Figure 4.13 when the object distance s_o is larger than the focal length f. There are two representations of the lens in Figure 4.13: a conceptual representation shaded in grey that, for the purposes of illustration, is thicker than allowed by the thin-lens approximation, and a dashed line that represents the actual thin lens for purposes of drawing the refracted rays. Three rays are shown, each beginning at the tip of the arrow that represents the object:

1. The ray that begins traveling parallel to the optical axis is refracted to cross the optical axis at a distance f behind the lens.

2. The ray that begins in a direction that crosses the optical axis at a distance f in front of the lens is refracted parallel to the optical axis.

3. The ray that passes through the center of the lens is undeflected.

At its center, the lens is approximately a slab, and we ignore the transverse displacement of the third ray for a truly thin lens (see Example 3.1). The tip of the image is located where the transmitted rays cross. Similar rays could be drawn originating from every point on the object; thus this image could be displayed on an observing screen located at the point s_i, indicating that this image is *real*. From the diagram, we also see that the image is *inverted* relative to the object.

The *transverse magnification* (also sometimes referred to as *lateral magnification*) is defined as image size divided by object size. Using similar triangles in Figure 4.13, we

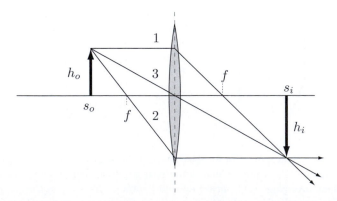

Figure 4.13   Ray diagram for a converging biconvex lens when the image distance is $1.75f$. The image is inverted and real. Rays are numbered as in the text.

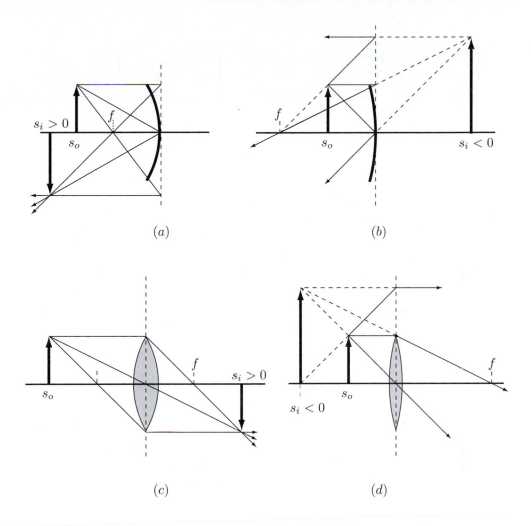

(a) (b)

(c) (d)

Figure 4.14   Images located by ray diagrams. (a) Converging mirror: real, inverted image when $s_o > f$. (b) Converging mirror: virtual upright image when $s_o < f$. (c) Converging lens: real inverted image when $s_o > f$. (d) Converging lens: virtual upright image when $s_o < f$.

observe that the transverse magnification is given by

$$M = -\frac{s_i}{s_o} \tag{4.27}$$

In this formula, a negative magnification indicates an image that is inverted, and if $|M| > 1$ then the image size is larger than the object size.

Figures 4.14 and 4.15 illustrate ray diagrams for reflecting and refracting optics for a variety of object locations. *Virtual images* are illustrated in Figures 4.14(b) and (d), and again in Figures 4.15. In such cases, an observer *perceives* the image because the lens of the eye refracts rays that *appear* to cross at the virtual image location to another real image formed on the retina of the eye. There are no reflected or refracted rays that really cross at this point; thus the image is not "real." In systems with a single optical element, virtual images are located "behind" the optical element (see the dashed lines in Figures 4.14 (b),

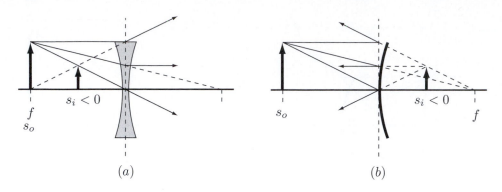

Figure 4.15   Images located by ray diagrams. (a) Diverging lens: virtual upright image. (b) Diverging mirror: virtual upright image.

(d), and 4.15). By behind, we mean the side opposite to the transmitted light for a lens and the reflected light for a mirror. Objects may also be virtual; in this case, rays converging toward a crossing point are intercepted by the optical element before crossing occurs (see Figure 4.9(c)). Virtual objects occur often in lens combinations, which are discussed in Section 4.4.

The image location s_i may be obtained analytically using the thin-lens equation. Solving Equation 4.15 for s_i gives

$$s_i = \frac{f\,s_o}{s_o - f} \tag{4.28}$$

When $f > 0$ and $s_o > f$, $s_i > 0$. This situation is illustrated for the real image diagrammed in Figure 4.13. When $s_i < 0$, the image is virtual. For light incident from left to right, real objects are on the left of the optical element ($s_o > 0$), and virtual objects are on the right. The positive side for image distances is different for lenses and mirrors. Since lenses transmit, real images are formed to the right of the lens. Mirrors reflect, so real images are formed to the left of the mirror.

■ **EXAMPLE 4.2**

A $1.00\ cm$-high object is located a distance s_o in front of an imaging optic. Determine the image orientation, size, and location for the following cases: (a) converging mirror when $s_o = 1.75f$; (b) converging mirror when $s_o = 0.500f$; (c) converging lens when $s_o = 2.00f$; (d) converging lens when $s_o = 0.500f$; (e) diverging lens when $s_0 = |f|$; (f) diverging mirror when $s_0 = |f|$. In each case, draw a ray diagram.

Solution

For light incident from left to right, a *real* object is located on the left of the optic. The object distance s_o for a real object is positive.

(a) Let the optical element be a converging mirror, and let the object be located at $s_o = 1.75f$. According to Equation 4.28

$$s_i = \frac{f\,s_o}{s_o - f} = \frac{1.75f^2}{0.75f} = 2.33f$$

Since the image distance is positive, the image is real. By Equation 4.27, the magnification is

$$M = -\frac{s_i}{s_o} = -1.33$$

The negative value for M means that the image is inverted. The image height is $1.33\,cm$. The ray diagram for this case is shown in Figure 4.14(a).

(b) Converging mirror with $s_o = 0.500f$. Equation 4.28 gives an image distance of $s_i = -f$. Since the image distance is negative, the image is virtual. Equation 4.27 gives a magnification of $+2.00$, indicating that the image is upright, and the image size is $2.00\,cm$ high. The ray diagram for this case is shown in Figure 4.14(b).

(c) Converging lens with $s_o = 2.00f$. Equation 4.28 gives a real image distance of $s_i = +2.00f$. This case is interesting since $s_i = s_o$ and thus $M = -1$, so the image is also $1.00\,cm$ high. This is the minimum possible separation $(4f)$ between a real object and its real image. The ray diagram for this case is shown in Figure 4.14(c).

(d) Converging lens with $s_o = 0.500f$. In this case, $s_i = -2.00f$ and the image is virtual. The magnification is $+2.00$, and the image is $2.00\,cm$ high. The ray diagram for this case is shown in Figure 4.14(d).

(e) Diverging lens when $s_0 = |f|$. Since this is a diverging optic, $f < 0$. Thus $(s_o - f) = +2.00\,|f|$, and Equation 4.28 gives a negative image distance: $s_i = -0.500\,|f|$. The image is virtual, and is $0.500\,cm$ high $(M = +0.500)$. The ray diagram for this case is shown in Figure 4.15(a).

(f) Diverging mirror when $s_0 = |f|$. The analysis here is the same as that for part (e), except for the ray diagram, which is shown in Figure 4.15(b).

Problem 4.3   An adjustable iris (called an *f-stop*) is placed in front of the lens so that its $f/\#$ can be varied. If the focal length of the lens is $5.00\,cm$, what diameter f-stop will limit the paraxial error to 0.100%?

Problem 4.4   A $1.00\,cm$-high object is located a distance s_o in front of a converging mirror of focal length f. Determine the image orientation, size, and location if $s_o = 2.75f$. Repeat for $s_o = 0.750f$. Include ray diagrams for each case.

Problem 4.5   A $1.00\,cm$-high object is located a distance s_o in front of a diverging mirror of focal length f. Determine the image orientation, size, and location if $s_0 = |3.00f|$. Repeat for $s_0 = |0.500f|$. Include ray diagrams for each case.

Problem 4.6   A $1.00\,cm$-high object is located a distance s_o in front of a converging lens of focal length f. Determine the image orientation, size, and location if $s_o = 2.00f$. Repeat for $s_o = 0.250f$. Include ray diagrams for each case.

Problem 4.7   A $1.00\,cm$-high object is located a distance s_o in front of a diverging lens of focal length f. Determine the image orientation, size, and location if $s_0 = |2.00f|$. Repeat for $s_0 = |0.500f|$. Include ray diagrams for each case.

Problem 4.8   A luminous object and observing screen are separated by a fixed distance L. Show that there are two locations separated by a distance d where the lens can be placed to give a focused image on the observing screen, provided that $L > 4f$, where f is the focal length of the lens. Show that f is given by

$$f = \frac{L^2 - d^2}{4L}$$

where $d = \sqrt{L(L - 4f)}$. Note that a measurement of d and L gives a quick and accurate way to determine the focal length of an unknown lens. Show that the two images formed on the observing screen have image sizes in the ratio

$$\frac{h_{large}}{h_{small}} = \left[\frac{L + d}{L - d}\right]^2$$

where h_{large} is the larger of the two image sizes.

4.4   LENS COMBINATIONS

Consider two lenses separated by a distance d, as illustrated in Figure 4.16(a). We may analyze the image formed by the combination by taking the image formed by lens 1 as the object for lens 2.

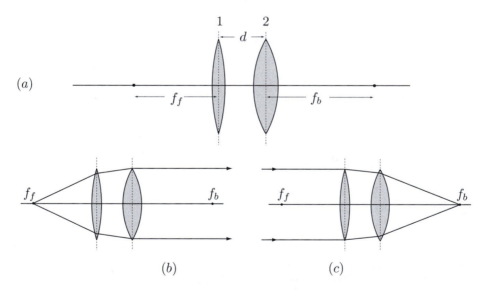

Figure 4.16   (a) A combination of two thin biconvex lenses with corresponding front and back focal points. (b) A point source at f_f gives parallel rays out 2. (c) Parallel rays in are focused at f_b.

Let s_{o1} be the object distance for lens 1, which forms an image located at

$$s_{i1} = \frac{s_{o1} f_1}{s_{o1} - f_1} \tag{4.29}$$

This image becomes the object for lens 2, with object distance given by $s_{o2} = d - s_{i1}$. Note that the object for lens 2 is virtual whenever $s_{o2} < 0$. The image distance for lens 2 is given by

$$s_{i2} = \frac{s_{o2}\, f_2}{s_{o2} - f_2} = \frac{(d - s_{i1})\, f_2}{d - s_{i1} - f_2} = \frac{f_2 d - \frac{f_2 s_{o1} f_1}{s_{o1} - f_1}}{d - f_2 - \frac{s_{o1} f_1}{s_{o1} - f_1}} \tag{4.30}$$

Letting $s_{o1} \rightarrow \infty$ gives plane-parallel illumination. In this case, s_{i2} is located at the *back focal point*. Taking this limit in Equation 4.30 gives

$$f_b = \frac{f_2 (d - f_1)}{d - (f_1 + f_2)} \tag{4.31}$$

Similarly, letting $s_{i2} \rightarrow \infty$ defines the *front focal point*. By Equation 4.30, this will be the case when

$$d - f_2 = \frac{s_{o1}\, f_1}{d - (f_1 + f_2)}$$

In this equation, we solve for s_{o1}, and identify this with the front focal point:

$$f_f = \frac{f_1 (d - f_2)}{d - (f_1 + f_2)} \tag{4.32}$$

When both lenses are thin, a positive front focal length is measured f_f to the left of the center of lens 1, and a positive back focal length is measured f_b to the right of the center of lens 2, as illustrated in Figure 4.16. Figure 4.16(a) shows two converging lenses, but the result of Equations 4.31 and 4.32 are valid for any combination of converging and diverging lenses.

The magnification of the combination is the product of the magnifications for each lens:

$$m = \left(-\frac{s_{i1}}{s_{o1}}\right)\left(-\frac{s_{i2}}{s_{o2}}\right) = \frac{s_{i1}\, s_{i2}}{s_{o1}\, s_{o2}} \tag{4.33}$$

4.4.1 Thin-Lenses in Close Combination

If both lenses in the combination of Figure 4.16 are thin, and we let the separation $d = 0$, then we have two thin lenses in *close contact*. Equations 4.31 and 4.32 give

$$f_f = f_b = f = \frac{f_1 f_2}{f_1 + f_2}$$

or

$$\frac{1}{f} = \frac{1}{f_1} + \frac{1}{f_2}$$

This formula may be extended to any number of thin lenses in close contact:

$$\frac{1}{f} = \frac{1}{f_1} + \frac{1}{f_2} + \dots \tag{4.34}$$

The reciprocal of focal length is often referred to as the *refracting power* P of the lens.

$$P = P_1 + P_2 + \dots \tag{4.35}$$

In SI units, P is expressed in inverse meters, commonly referred to as *diopters*.

■ **EXAMPLE 4.3**

Suppose that in Figure 4.16, $f_1 = 20.0\,cm$ and $f_2 = 10.0\,cm$. Find the front and back focal lengths when (a) $d = 5.00\,cm$, and (b) the lenses are in close contact.

Solution

(a) Let $d = 5.00\,cm$. Equation 4.31 and 4.32 give

$$f_f = \frac{f_1\,(d - f_2)}{d - (f_1 + f_2)} = \frac{20\,(5 - 10)}{5 - (10 + 20)} = \frac{100}{25} = 4.00\,cm$$
$$f_b = \frac{f_2\,(d - f_1)}{d - (f_1 + f_2)} = \frac{150}{25} = 6.00\,cm$$

Since both f_f and f_b are positive, they are located as illustrated in Figure 4.16.
(b) When the lenses are in close contact, Equation 4.34 gives

$$\frac{1}{f} = \frac{1}{20} + \frac{1}{10} = \frac{3}{20}$$

giving $f_f = f_b = 6.67\,cm$.

Problem 4.9   Two thin converging lenses with $f_1 = -20.0\,cm$ and $f_2 = 30.0\,cm$ are separated by $25.0\,cm$ on a common optical axis. Find the front and back focal lengths. Repeat if the lenses are in close contact.

Problem 4.10   Two thin converging lenses with $f_1 = -20.0\,cm$ and $f_2 = 40.0\,cm$ are separated by $5.00\,cm$ on a common optical axis. Find the front and back focal lengths. Repeat if the lenses are in close contact.

Problem 4.11   In microscopes and telescopes, d is close to $f_1 + f_2$. What are f_b and f_f in this case?

4.5   *PRINCIPAL POINTS AND EFFECTIVE FOCAL LENGTHS

In Equations 4.30—4.32, the distances s_{i2} and f_b are both measured relative to lens 2, while the distances s_{o1} and f_f are measured relative to lens 1. Let us define new points of reference that allow the entire combination to be used as a single optical element described by the thin-lens equation

$$\frac{1}{s_o} + \frac{1}{s_i} = \frac{1}{f_e} \tag{4.36}$$

where f_e is the *effective focal length* for the combination. For this equation to be valid, the *effective object distance* s_o and *effective image distance* s_i must be measured relative to the *principal points* of the combination, as illustrated in Figure 4.17. In Equation 4.36, s_o

is measured relative to the front principal point h_f, and s_i is measured relative to the back principal point h_b. If the effective focal length is positive, then the front focal length f_f is located a distance f_e to the left of h_f, and the back focal length is located f_e to the right of h_b, as illustrated. In order to use Equation 4.36 with the lens combination, we must find values for h_f, h_b, and f_e.

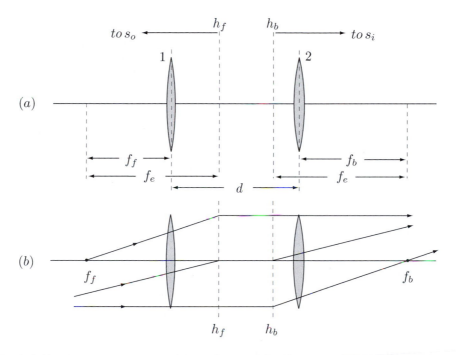

Figure 4.17 (a) A two-lens combination. The effective object distance s_o and front focal length f_f are measured relative to the front principal point h_f. The effective image distance s_i and back focal point f_b are measured relative to the back principal point h_b. (b) A ray that first passes through the front focal point f_f is redirected parallel to the optical axis at the front principal plane that passes through h_f. Similarly, an incident ray that is parallel to the optical axis is redirected at the back principal plane that contains h_b so that it passes through the back focal point f_b. The center ray is drawn to the optical axis at the front principle point, and then from the optical axis at the back principle point.

According to Equation 4.36, the magnification of the combination is given by $-s_i/s_o$ in terms of the effective image and object distances. Combining this with Equation 4.33 gives

$$m = \frac{s_{i1}}{s_{o1}} \frac{s_{i2}}{s_{o2}} = -\frac{s_i}{s_o} \qquad (4.37)$$

To find the effective focal length, we let $s_{o1} \to \infty$. In this case, $s_{i1} \to f_1$, $s_{o1} \to s_o$ and $s_i \to f_e$. The object distance for lens 2 is then $s_{o2} = d - f_1$, giving

$$\frac{f_1}{d - f_1} s_{i2} = -f_e$$

The image distance s_{i2} is given by

$$s_{i2} = \frac{f_2 s_{o2}}{s_{o2} - f_2} = \frac{f_2 (d - f_1)}{d - (f_1 + f_2)}$$

Combining the last two results gives the effective focal length:

$$f_e = - \left(\frac{f_1}{d - f_1} \right) \frac{f_2 \left(d - f_1 \right)}{d - \left(f_1 + f_2 \right)} = \frac{f_1 f_2}{f_1 + f_2 - d} \tag{4.38}$$

or, equivalently,

$$\frac{1}{f_e} = \frac{1}{f_1} + \frac{1}{f_2} - \frac{d}{f_1 f_2} \tag{4.39}$$

From Figure 4.17, we see that $h_f = f_e - f_f$. Combining Equations 4.32 and 4.38 gives

$$h_f = f_e - f_f = \frac{f_1 f_2}{f_1 + f_2 - d} - \frac{f_1 \left(d - f_2 \right)}{d - \left(f_1 + f_2 \right)}$$

$$= \frac{f_1 d}{f_1 + f_2 - d} \left(\frac{f_2}{f_2} \right)$$

$$= \left(\frac{f_1 f_2}{f_1 + f_2 - d} \right) \frac{d}{f_2}$$

or

$$h_f = \frac{f_e d}{f_2} \tag{4.40}$$

Principal points that lie to the left of the corresponding lens are negative. According to this sign convention, the value of h_b in Figure 4.17 is

$$h_b = - \left(f_e - f_b \right) = - \frac{f_e d}{f_1} \tag{4.41}$$

■ **EXAMPLE 4.4**

A Huygens eyepiece is designed with two plano-convex lenses separated by the average of their focal lengths. Let $f_1 = 30.0\,cm$, $f_2 = 10.0\,cm$ and $d = 20.0\,cm$. Find the effective focal length, the location of the principal points, and the front and back focal points. Typically, an optical instrument creates an object at the front focal point of the eyepiece. Describe the image in such a circumstance.

Solution

Begin by finding the effective focal length of the combination:

$$f_e = \frac{f_1 f_2}{f_1 + f_2 - d} = \frac{\left(30.0\,cm \right) \left(10.0\,cm \right)}{\left(30.0 + 10.0 - 20.0 \right)\,cm} = 15.0\,cm$$

The principal points are located at

$$h_f = \frac{f_e d}{f_2} = \frac{\left(15.0\,cm \right) \left(20.0\,cm \right)}{10.0\,cm} = 30.0\,cm$$

$$h_b = - \frac{f_e d}{f_1} = - \frac{\left(15.0\,cm \right) \left(20.0\,cm \right)}{30.0\,cm} = -10.0\,cm$$

As a check, calculate the front and back focal lengths two ways:

$$f_f = \frac{f_1 \left(d - f_2 \right)}{d - \left(f_1 + f_2 \right)} = -15.00\,cm = f_e - h_f$$

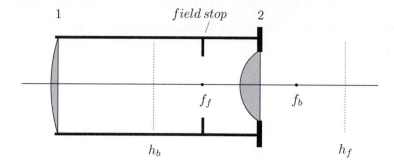

Figure 4.18 Huygens eyepiece. The field stop limits the field of view to the central area of best image quality.

$$f_b = \frac{f_2 \, (d - f_1)}{d - (f_1 + f_2)} = +5.00 \, cm = f_e + h_b$$

If the eyepiece is used in an optical instrument that creates an object at f_f, then this must be a *virtual object* for lens 1, with $s_{o1} = -15.0 \, cm$. This gives an effective object distance of $s_o = h_f + s_{o1} = +15.0 \, cm$, and an effective image distance of

$$s_i = \frac{f_e s_o}{s_o - f_e} = \infty$$

The normal eye is most comfortable viewing objects at infinity (distant stars, for example); in this case, the eye is said to be *relaxed*.

Images formed by lens combinations may be located with Equation 4.36 using the principal points and effective focal length. We may also use ray diagrams to locate images. Figure 4.17 illustrates three useful rays, the first two of which were used to derive formulas for the principal points.

1. ray that first passes through the front focal point f_f is redirected parallel to the optical axis at the *front principal plane* that contains h_f.

2. A ray traveling parallel to the optical axis is redirected through the back focal point f_b. The redirection occurs at the *back principal plane* that contains h_b.

3. A ray passing through h_f re-emerges at h_b in a parallel direction.

Both principal planes are oriented normal to the optical axis. In contrast to a single thin lens, Rays 1 and 2 change direction at the principal planes, and so may be thought of as effective refraction. Ray 3 corresponds to the ray that passes through a thin lens undeflected.

■ EXAMPLE 4.5

In a two-lens combination, let $f_1 = 20.0 \, cm$, $f_2 = 10.0 \, cm$ and $d = 5.00 \, cm$, as in Example 4.3(a). Find the image distance and image size of a $3.00 \, cm$-high object located $8.00 \, cm$ in front of lens 1.

Solution

Begin by finding the effective focal length and principal points:

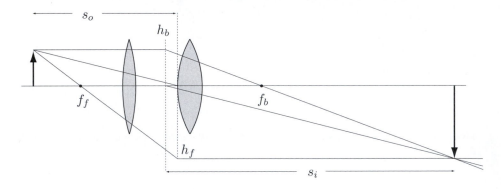

Figure 4.19   An image formed by two thin, symmetric biconvex lenses.

$$f_e = \frac{f_1 f_2}{f_1 + f_2 - d} = 8.00 \, cm$$

$$h_f = \frac{f_e d}{f_2} = 4.00 \, cm$$

$$h_b = -\frac{f_e d}{f_1} = -2.00 \, cm$$

Using these values, we find that f_f is located a distance f_e in front of h_f, or $4.00 \, cm$ in front of lens 1, as in Example 4.3. Similarly, f_b is located f_e to the right of h_b, or $6.00 \, cm$ behind lens 2, as before. The effective object distance s_o must locate the object relative to h_f; thus, $s_o = 12.0 \, cm$. The thin-lens equation with effective quantities gives

$$s_i = \frac{f_e s_o}{s_o - f_e} = \frac{(8.00 \, cm)\,(12.0 \, cm)}{(12 - 8) \, cm} = 24.0 \, cm$$

or $22.0 \, cm$ to the right of lens 2, as illustrated in Figure 4.19. The magnification is

$$m = -\frac{s_i}{s_o} = -2.00$$

Thus the image size is $6.00 \, cm$ and it is inverted.

Problem 4.12   Design a Huygens eyepiece that uses a $3.00 \, cm$ lens separation and a $1.00 \, cm$ focal length lens for f_2. Find the effective focal length, the front and back principal points, and the front and back focal points for this eyepiece.

Problem 4.13   In a two-lens combination, let $f_1 = -20.0 \, cm$, $f_2 = 10.0 \, cm$, and $d = 5.00 \, cm$. Find the effective focal length, the principal points, and the image distance and image size of a $3.00 \, cm$-high object located $20.0 \, cm$ in front of lens 1.

Problem 4.14   In a two-lens combination, let $f_1 = -20.0\,cm$, $f_2 = 10.0\,cm$, and $d = 5.00\,cm$. Find the effective focal length, the principal points, and the image distance and image size of a $3.00\,cm$-high object located $8.00\,cm$ in front of lens 1.

Problem 4.15   In a two-lens combination, let $f_1 = 20.0\,cm$, $f_2 = 30.0\,cm$, and $d = 25.0\,cm$. Find the effective focal length, the principal points, and the image distance and image size of a $3.00\,cm$-high object located $8.00\,cm$ in front of lens 1.

Problem 4.16   Derive Equation 4.41 for the location of the back principal point h_b of a two-lens combination.

4.6 ABERRATIONS

An *aberration* is any effect that prevents a lens from forming a perfect image. In this section, we present an overview of a few common aberrations and what may be done to correct them.

4.6.1 Chromatic Aberration

As noted in Chapter 3, *dispersion* causes the index of refraction of most materials to vary with frequency. The *dispersive* power of a material is defined as

$$\delta_d = \frac{n_b - n_r}{n_i - 1} \tag{4.42}$$

where n_b is the index measured at the blue end of the target spectrum, n_r is the index measured at the red end of the target spectrum, and n_i is measured at an intermediate wavelength. Traditionally, n_b is measured with the blue emission line of hydrogen (486.1 nm), n_r at the red line of hydrogen (656.3 nm), and n_i at the yellow line of helium (587.6 nm). For most optical materials, δ_d is positive indicating a larger index for blue than for red. In a *prism*, this causes more deviation for blue than for red, as illustrated in Figure 4.20(a). In a lens, this causes *chromatic aberration* where the focus depends on wavelength, as illustrated in Figure 4.20(b). Source points on the optical axis are imaged to collection points also on the optical axis but which depend on wavelength, leading to *longitudinal chromatic aberration*. Points off the optical axis are also dispersed transverse to the optical axis, leading also to *transverse chromatic aberration*.

Chromatic aberration is reduced in an *achromat* that consists of converging and diverging lenses in combination. Figure 4.20(c) illustrates an achromat *doublet* that consists of a converging and diverging lens in close contact. Each lens in the doublet is made of glass with different values of index n_i and dispersive power δ_d. If the diverging lens can be made of a glass with higher dispersive power and lower n_i, the chromatic aberration introduced by the converging lens can be reduced by the diverging lens without removing all of the refractive power of the combination.

Let us design an achromat doublet consisting of thin lenses that gives an identical focus at the red and blue wavelengths defined above. The focal length of each element in the doublet is given by Equation 4.25:

$$\frac{1}{f} = (n - 1) \left[\frac{1}{R_1} - \frac{1}{R_2} \right]$$

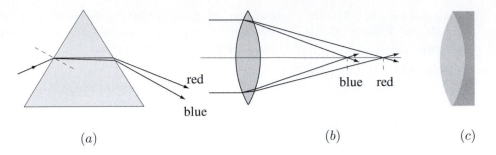

Figure 4.20   (a) A prism refracts different colors through different angles. The red ray passes through the prism symmetrically. (b) Dispersion causes the focus to vary with wavelength. (c) Achromat doublet.

For brevity, we replace the curvature terms with new parameters C_1 and C_2:

$$\frac{1}{f_1} = (n_1 - 1)\, C_1 \qquad\qquad \frac{1}{f_2} = (n_2 - 1)\, C_2$$

Since the lenses are in close contact, the focal length of the combination is given by Equation 4.34

$$\frac{1}{f} = \frac{1}{f_1} + \frac{1}{f_2} \tag{4.34}$$

We wish to find a combination such that

$$\frac{1}{f_r} = \frac{1}{f_b}$$

where f_r and f_b are the focal lengths of the combination measured at the red and blue design wavelengths defined above. Thus

$$(n_{1r} - 1)\, C_1 + (n_{2r} - 1)\, C_2 = (n_{1b} - 1)\, C_1 + (n_{2b} - 1)\, C_2 \tag{4.43}$$

This may be rearranged to give

$$\frac{C_1}{C_2} = -\frac{n_{b2} - n_{r2}}{n_{b1} - n_{r1}} \tag{4.44}$$

where we have grouped terms on the right according to our observation that refraction is higher for blue. At the intermediate wavelength, we have

$$\frac{1}{f_{1i}} = (n_{1i} - 1)\, C_1$$
$$\frac{1}{f_{2i}} = (n_{2i} - 1)\, C_2 \tag{4.45}$$

giving

$$\frac{C_1}{C_2} = \frac{(n_{2i} - 1)\, f_{2i}}{(n_{1i} - 1)\, f_{1i}} \tag{4.46}$$

Equations 4.44 and 4.46 may be combined to give

$$\frac{f_{2i}}{f_{1i}} = -\frac{(n_{2b} - n_{2r})/(n_{2i} - 1)}{(n_{1b} - n_{1r})/(n_{1i} - 1)} = -\frac{\delta_{2i}}{\delta_{1i}} = -\frac{V_{1i}}{V_{2i}} \tag{4.47}$$

Glass	n_i	Abbe Number V_i
Borosilicate BK7	1.517	64.2
Crown K5	1.522	59.5
Dense flint SF5	1.673	32.3

Table 4.1.   Indices of refraction for selected materials. The intermediate wavelength is the yellow line of helium (587.6 nm).

where V_i is the *Abbe number* and is defined as the reciprocal of the dispersive power δ_d. The subscript i in both parameters usually indicates a wavelength near the peak of the solar spectrum. Rearranging Equation 4.47 gives

$$f_{1i}V_{1i} + f_{2i}V_{2i} = 0 \tag{4.48}$$

Table 4.1. gives values of n_i and V_i for some commonly used optical glasses.

■ **EXAMPLE 4.6**

Design a converging achromat doublet using a 20.0 cm focal length BK7 symmetric biconvex converging lens and SF5 for the diverging lens, with identical inner curvatures as shown in Figure 4.20(c). Find the focal length of the doublet.

Solution

The symmetric biconvex lens has radii given by

$$\frac{1}{20} = (1.517 - 1)\left(\frac{2}{R}\right)$$

giving $R = 20.68\,cm$. Thus, the diverging lens will have $R_1 = -20.68\,cm$.

Using the data in Table 4.1., Equation 4.48 gives

$$f_{2i} = -\frac{V_{1i}}{V_{2i}}f_{1i} = -\frac{64.2}{32.3}(20.0\,cm) = -39.75\,cm$$

The remaining radius of the diverging lens is found using Equation 4.25:

$$\frac{1}{f_{2i}} = (1.673 - 1)\left(\frac{1}{(-20.68\,cm)} - \frac{1}{R_2}\right) = \frac{1}{-39.75\,cm}$$

giving $R_2 = -91.1\,cm$. Estimate the focal length of the doublet using the formula for lenses in close contact:

$$\frac{1}{f} = \frac{1}{f_{1i}} + \frac{1}{f_{2i}}$$

$$f = 40.3\,cm$$

The chromatic aberration is removed only for the red and blue design wavelengths. There could still be aberration at other wavelengths.

It is possible to reduce chromatic aberration with other lens combinations. Consider two thin converging lenses made of the same material and separated by distance d, as illustrated in Figure 4.21. The effective focal length of the combination is given by Equation 4.39

$$\frac{1}{f_e} = \frac{1}{f_1} + \frac{1}{f_2} - \frac{d}{f_1 f_2}$$

Chromatic aberration will be reduced whenever

$$\frac{\partial f_e}{\partial n} = 0 \tag{4.49}$$

Equivalently, we may find an extremum of $\frac{1}{f_e}$ (see Problem 4.17). Again, we replace the curvature terms for f_1 and f_2 with C_1 and C_2, respectively. Computing the derivative and solving for d gives

$$\frac{\partial}{\partial n} \frac{1}{f_e} = \frac{\partial}{\partial n} \left[(n-1) C_1 + (n-1) C_2 - d (n-1)^2 C_1 C_2 \right]$$
$$= C_1 + C_2 - 2d (n-1) C_1 C_2 = 0$$
$$\Rightarrow d = \frac{1}{2} \left[\frac{1}{(n-1) C_1} + \frac{1}{(n-1) C_2} \right] = \frac{f_1 + f_2}{2}$$

Thus, chromatic aberration is reduced when the two converging lenses are separated by the average of the individual focal lengths. The *Huygens eyepiece* (see Example 4.4) uses this effect to reduce chromatic aberration. In a typical design, two plano-convex lenses with $f_2 \approx 3 f_1$ are separated by $d = (f_1 + f_2) / 2$. In this orientation, the front focal point is located between the lenses, as shown in Figure 4.21. The eyepiece is positioned (focused) so that the lens 1 (called the *field lens*) forms a real image close to f_f. Lens 2 (called the *eye lens*) forms a virtual image with rays that are collected by the observer's eye. There are many other eyepiece designs, each typically designed to reduce one or more aberrations.

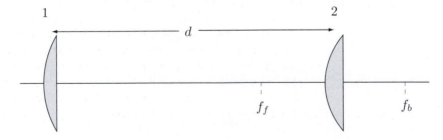

Figure 4.21   A combination of two thin plano-convex lenses separated the average focal length.

4.6.2 Spherical Aberration

Our derivation of the thin-lens equation for reflecting and refracting surfaces required the paraxial approximation which becomes less valid as the $f/\#$ of the lens decreases.

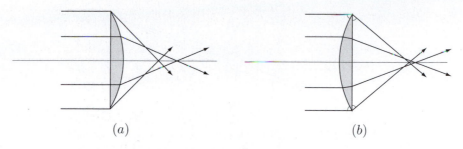

(a)   *(b)*

Figure 4.22   (a) Spherical aberration from a plano-convex lens when rays parallel to the optical axis first encounter the plane surface. (b) Reversing the orientation of the plano-convex lens reduces the spherical aberration.

When effects due to this approximation are no longer negligible, they produce *spherical aberrations*.

Since a circle is closed, spherical lenses typically have more curvature at the edges than the corresponding aspheric lens. Thus, rays passing through the outer regions of a spherical lens focus too closely, as shown in Figure 4.22(a).

Simply reversing the orientation of the lens can reduce the spherical aberration, as shown in Figure 4.22(b). To see why, consider the prism illustrated in Figure 4.20(a). The ray that passes through the prism *symmetrically* has the *minimum deviation* from the original direction[8]. The edges of the lens act as a prism, so orienting the lens so that rays pass through it symmetrically can reduce the deviation from the edges and hence can reduce the spherical aberration. Since the rays enter and leave the lens orientation of Figure 4.22(b) with incident and transmitted angles that are closer in value, it produces less spherical aberration than the orientation of Figure 4.22(a). Orienting the lens "flat to the focus" yields the best results.

Optical elements that are corrected for spherical aberration are called *aplanatic*. The corresponding conjugate points are called *aplanatic points*.

4.6.3   Astigmatism and Coma

Aspheric optics designed to image perfectly between conjugate points will nevertheless produce image distortions for source points located off the optical axis. In such a case, the edges of the lens tend to focus farther from optical axis than the center, giving an image that resembles a comet (*Coma* derives from the Greek word for comet). *Astigmatism* results from the fact that waves encountering an optics obliquely are imaged as if the reflecting or refracting surfaces have different curvatures in two orthogonal directions. Cylindrical lenses illustrate the extreme case of an astigmatic optic.

4.6.4   Field Curvature

A single thin lens not possessing any of the above aberrations would still suffer from *field curvature*. Off-axis source and collection points equidistant from the thin lens center are located on spherical rather than plane surfaces, as illustrated in Figure 4.23. This is

---

[8]Suppose that some incident angle giving a non-symmetric path gives the minimum deviation. Reversing the path gives another incident angle with the same deviation. Thus, the deviation for the non-symmetric path cannot be the minimum.

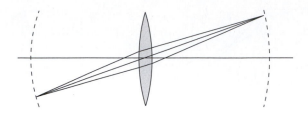

Figure 4.23   A single thin lens produces a spherical image field.

especially troublesome in photography, where it is desired to form a perfect image over a flat film plane. As with chromatic aberration, lens combinations can reduce the effects of this aberration.

4.6.5   Diffraction

All aberrations discussed so far can be largely eliminated by precision optical elements combined in clever ways. However, *diffraction* is the one aberration that can never be avoided. Physically, diffraction causes a light beam to spread whenever it passes through an aperture. The finite size of an objective lens provides just such an aperture, and no other apertures are necessary for diffraction to occur. Diffraction limits the resolution of, for example, telescopes to resolve stars with a small angular separation. We will defer further discussion of this important optical effect until Chapter 6. For more on optical resolution, see Section 6.3.4.

Problem 4.17   Show that an extremum found with Equation 4.49 is also an extremum of Equation 4.39.

Problem 4.18   Repeat Example 4.6 if the $20.0\,cm$ focal length symmetric biconvex lens is made from K5 crown glass.

Problem 4.19   Modify the design of Example 4.6 to give a *diverging* achromat doublet, again using a $20.0\,cm$ focal length symmetric biconvex lens.

Problem 4.20   The lens combination illustrated in Figure 4.21 is designed to reduce at least two sources of aberration. Identify these, and describe the design elements that reduce them.

4.7   OPTICAL INSTRUMENTS

4.7.1   The Camera

A camera is designed to place a real image on a film plane or detector array. The simplest such device is the *pinhole camera* illustrated in Figure 4.24. Each point on the object passes through the pinhole as a cone of light that illuminates a finite-sized image element on the image plane. The resolution of the eye (i.e., the ability of the eye to determine fine detail) is limited, and is only sufficient to distinguish detail that subtends an angle of greater than about $1/30°$. Image elements that are not much bigger than this combine to form a reasonably precise image. As the pinhole diameter is made even smaller, diffraction effects begin to increase the image element size. A pinhole of any size to produce a recognizable image gathers very little light.

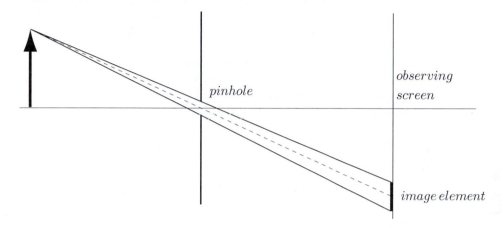

Figure 4.24   A pinhole camera. Each point on the object expands to an image element whose size is determined by the diameter of the pinhole.

From Figure 4.24 it is evident that both image and image element increase in proportion as the distance from the film to the pinhole is increased. Thus, the image of a pinhole camera is focused equally well no matter where the film is located. Because of this, a pinhole camera is said to have an infinite *depth of focus*.

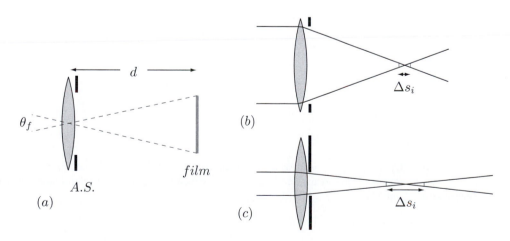

Figure 4.25   (a) A camera with aperture stop A. S. and field of view θ_f determined by the film size. The distance d is adjusted to coincide with the image distance s_i. (b) A large aperture stop opening (small $f/\#$) with a small depth of focus. (c) A small aperture stop opening (large $f/\#$) with a large depth of focus.

Figure 4.25(a) illustrates a *camera* whose light-gathering ability is greatly enhanced by a refracting objective lens. A real image is formed on the film plane by adjusting the distance d between the lens and the film. A *stop* is an aperture or other feature of the camera that limits the light delivered to the image plane by the objective lens. The size of the film, or a *field-stop* aperture placed just before the film, determines the field of view of the camera, as illustrated in the figure. An *aperture-stop A.S.* is an adjustable iris diaphragm placed just after the lens, as shown. Adjusting the diameter of the $A.S.$ allows the depth of focus to be controlled, as illustrated in Figures 4.25(b) and (c). An image will appear defocused if image detail that is larger than the eye's resolution limit is blurred. For the wide $A.S$ opening of Figure 4.25(b), the light-gathering ability is large but the range of image locations Δs_i that produce an acceptable image is relatively small. A small $A.S.$ opening such as in Figure 4.25(c) has a much larger depth of focus with correspondingly small light-gathering ability. An aperture-stop is also known as an *f-stop* and is usually calibrated according to the effective $f/\#$ of the lens, $A.S.$ combination.

4.7.2   The Eye

Figure 4.26 shows a simplified optical diagram of the human eye. The image is formed on the *retina* that consists of a dense array of light detectors. The detector density is more dense in the central region of this array, known as the *macula* (with center *fovea*), giving more resolution for objects viewed straight ahead as opposed to peripherally. The image is focused precisely on the retina by muscles that control the *shape* of the lens instead of the lens-detector distance d as in a camera. The ability of the eye to change the curvature of its lens is called its *accommodation*.

To form images on the retina of distance objects, the eye must *relax* so that the focal length is equal to the distance from the lens to the retina. If the focal length of the relaxed eye is shorter than this distance (i.e., the relaxed lens is too converging), then the eye is said to be *myopic* or *nearsighted*. To correct myopia, a *diverging* spectacle lens is used.

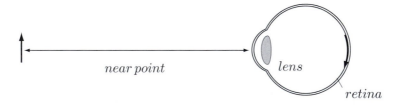

Figure 4.26   Optical diagram of the eye. The focal length of the lens is adjusted to form a real image on the retina of objects located no closer to the eye than its near point. The normal eye has a near point of about 25 *cm*.

The focal length of the eye-lens must be reduced to form retinal images of nearby objects. The normal eye can form images of objects as close as about 25 *cm* away; thus this distance is known as the *near point (N.P.)* of the eye. If the accommodation of the eye is such that this amount of curvature cannot be achieved, then the eye is said to be *hyperopic* or *farsighted*. To correct hyperopia, a *converging* spectacle lens is used.

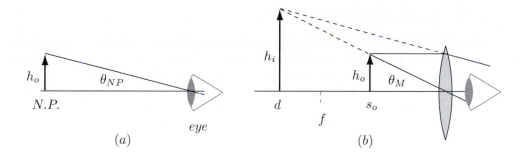

Figure 4.27 (a) The eye forms an image of an object located at the near point $N.P.$. (b) A magnifying glass forms a virtual image I at $d \geq N.P.$. The object is now much closer.

4.7.3 The Magnifying Glass

A magnifying glass is a converging lens that increases the curvature of the eye-lens combination so that objects closer than the $N.P.$ can be viewed, as illustrated in Figure 4.27. The *magnifying power* M is the ratio of the angular size of the image viewed through the magnifying glass (Figure 4.27(b)) relative to the same object viewed from the $N.P.$ (Figure 4.27(a)).

$$M = \frac{\theta_M}{\theta_{NP}} \tag{4.50}$$

We may approximate each angle with its tangent, giving $\theta_M \approx h_o/s_o$ and $\theta_{NP} \approx h_o/N.P.$.

$$M = \frac{N.P.}{s_o} \tag{4.51}$$

In Figure 4.27(b), the image distance is negative, giving $s_i = -d$. Thus

$$\frac{1}{s_o} = \frac{1}{f} + \frac{1}{d}$$

Combining this with Equation 4.51 and substituting $N.P. = 25\,cm$ gives

$$M = \frac{25}{f} + \frac{25}{d} \tag{4.52}$$

When $d = N.P. = 25\,cm$, this becomes

$$M = \frac{25}{f} + 1 \tag{4.53}$$

The eye must achieve maximum accommodation to view an image located at its $N.P.$. Objects located at infinity (such as stars) may be viewed more comfortably by the *relaxed eye*. In this case, $d \to \infty$ and the magnifying power is

$$M = \frac{25}{f} \tag{4.54}$$

The magnifying power takes into account that virtual images with large values of $m = -s_i/s_o$ have correspondingly large image distances. In practice, the paraxial approximation limits M to values less than about 8 for a single lens. Larger magnifications require lens combinations, such as in a microscope.

4.7.4 The Compound Microscope

A microscope is designed to produce accurate images with large magnification. A simplified diagram is shown in Figure 4.28. The objective lens forms a real image which is further magnified by the eyepiece. The image distance for the objective is determined by the length of the microscope body, indicated by L in the figure. Ignoring the minus sign in the formula for magnification, the real image formed by the objective is magnified by $s_i/s_o = L/f_{ob}$, where f_{ob} is the focal length of the objective. The eyepiece magnifies according to Equation 4.54. The final magnification is the product:

$$M = M_{ob}M_e \tag{4.55}$$

For small angles,

$$M = \frac{L}{f_{ob}}\frac{25}{f_e} \tag{4.56}$$

where f_e is the focal length of the eyepiece.

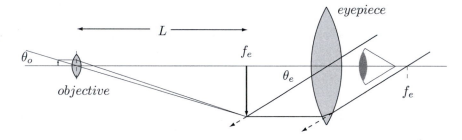

Figure 4.28 Optical diagram of a simple microscope. The objective lens forms a real image that is magnified by the eyepiece.

In a typical microscope, both objective and eyepiece consist of multiple optical elements, each designed to reduce aberrations. The length L in Equation 4.56 affects such designs, and typically standardized to $L = 16.0\,cm$.

With even the best optical designs, aberrations limit the magnification to about $M = 500$. Magnifications greater than this are practical with the *oil immersion* microscope objective. This design makes use of the conjugate points for a *spherical* refracting surface. To see that such points exist, consider Figure 4.29(a), which shows a sphere of radius r with index of refraction n. Shown also are two concentric circles, one with radius nr and the other with radius r/n. The conjugate points are located at the intersection of a line passing through the center of symmetry with these two circles. The object is located at the conjugate point labeled o, and a virtual image is formed at the conjugate point labeled i. The ray leaving o and intersecting the refracting surface at s has angle of incidence θ_i and transmits with angle θ_t, as shown. To show that this happens, we must show that θ_i and θ_t obey the law of refraction. Consider the two triangles sci and ocs shown in Figures 4.29(a) and (b). By construction,

$$\frac{\overline{sc}}{\overline{oc}} = \frac{\overline{ci}}{\overline{cs}} = n$$

and since the angle at c is common to both triangles, $\overline{si}$ and $\overline{os}$ have the same ratio[9]. Thus, the two triangles shown in Figures 4.29(a) and (b) are similar, and so each contains both

[9]You may show this formally with the law of cosines.

angles θ_i and θ_t. The law of sines with either triangle demonstrates the result we seek. For example, triangle sci gives

$$\frac{\sin \theta_t}{\overline{ci}} = \frac{\sin \theta_t}{nr} = \frac{\sin \theta_i}{\overline{sc}} = \frac{\sin \theta_i}{r}$$

giving

$$n \sin \theta_i = \sin \theta_t$$

The specimen is located within the region of higher index. This is achieved with *immersion oil* which must have an index of refraction that is very close to that of the objective lens and cover plate, as shown in Figure 4.29(c).

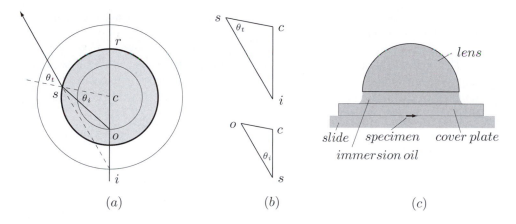

(a)   (b)   (c)

Figure 4.29   (a) Aplanatic conjugate points of a spherical refracting surface. (b) Construction triangles. (c) Oil-immersion microscope objective.

By using conjugate points, the oil immersion objective can utilize angles that far exceed the paraxial approximation without the effects of spherical aberration. As we will see in Chapter 8, increasing this acceptance angle is the only way to reduce aberrations due to diffraction. From Figure 4.29(a), rays are transmitted by the objective through a smaller angular spread, and additional optical components can provide magnification along with a reduction of the remaining aberrations. Well-designed oil immersion microscopes can provide detail that approaches the *diffraction limit*, which as we will see in Chapter 8 is on the order of the wavelength of light used for observation.

4.7.5   The Telescope

Astronomical objects have object distances that are effectively infinite. Figure 4.30 shows a *telescope*, which is designed to magnify the angular size of distant objects. The objective forms a real image at its focal point f_{ob}, and this image is magnified by the eyepiece, forming a virtual image at infinity when the image is located at the eyepiece focal point f_e. The object subtends an angular size of θ_o to the unaided eye, and the final image subtends an angular size of θ_e when viewed through the eyepiece. We again approximate each of these angles by its tangent. The magnification is defined as the ratio of θ_e and θ_o

$$M = \frac{\theta_e}{\theta_o} = \frac{h_i/f_e}{h_i/f_{ob}} = \frac{f_{ob}}{f_e} \tag{4.57}$$

A large magnification is achieved when f_{ob} is large and f_e is small.

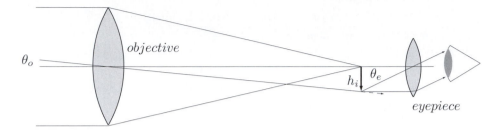

Figure 4.30   Optical diagram of a simple telescope. The objective lens forms a real image that is magnified by the eyepiece.

The objective should have one conjugate point at infinity. As shown in Section 4.2, a parabolic reflector or hyperbolic refracting lens fulfills this requirement when the objective consists of a single element. Both eyepiece and objective typically consist of multiple elements to correct aberrations.

4.7.6   The Exit Pupil

The eye has an aperture stop called the *iris diaphragm* that is, on the average, about $5\,mm$ wide. The bundle of rays produced by an optical instrument must ideally pass through a waist that is no larger than $5\,mm$; otherwise, some of the light gathered by the instrument will fail to enter the eye and is therefore wasted. This waist is called the *exit pupil*.

Figure 4.31 illustrates the exit pupil for a simple telescope that consists of a single-element objective of focal length f_{ob} and a single-element eyepiece of focal length f_e. Rays that pass from points on the object through the center of the objective are called *principal rays* (also sometimes referred to as *chief rays*). Such rays define the angular extent of the object, and when propagated through the eyepiece define the angular extent of the image, as shown by the bolded rays in Figure 4.31(a).

The exit pupil is most intuitively defined as the image of the objective aperture (or aperture stop) formed by the eyepiece. Since the exit pupil is an image of the input aperture, all rays passing through the instrument by definition pass through this image, as illustrated in Figure 4.31(b). The distance from the eyepiece to the exit pupil is called the *eye relief*. For the single-element eyepiece illustrated, the exit pupil is located just beyond f_e (see Problem 4.23).

The rays in Figure 4.31(b) that pass through f_{ob} define similar triangles that give

$$M = \frac{f_{ob}}{f_e} = \frac{D_{ob}}{D_e} \tag{4.58}$$

where D_{ob} is the diameter of the objective aperture, and D_e is the diameter of the exit pupil. The performance of an optical instrument can often be thought of intuitively in terms of an *antenna beam*, formed by the reverse projection of the light that actually enters the eye. If the iris diaphragm is smaller than the exit pupil, then the resulting antenna beam underfills the objective aperture, and aperture is wasted. If the iris diaphragm is larger than the exit pupil, the antenna beam overfills the objective aperture. When the iris and exit pupil are perfectly matched, the antenna beam optimally fills the object plane, and the maximum signal is collected with the minimum objective aperture.

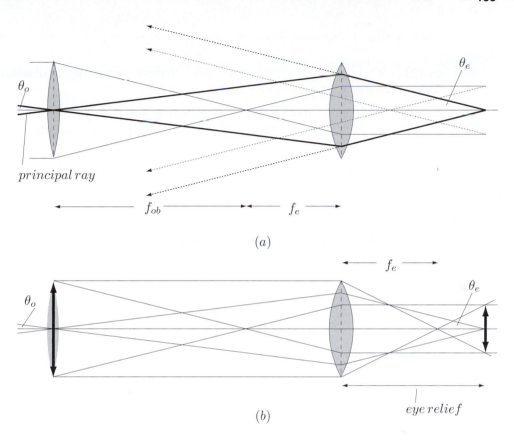

Figure 4.31 The exit pupil. (a) Ray diagram for a telescope. The principal rays are bolded. (b) The image of the objective formed by the eyepiece defines the exit pupil.

Binoculars are typically described by a figure of merit that includes both objective diameter in mm, and magnification power in the format: $M \times D_{ob}(mm)$. For example, a 7×35 binocular has a magnifying power of 7, an objective aperture of $35\,mm$, and by Equation 4.58, an exit pupil of $5\,mm$. Increasing the objective diameter of this binocular would increase both cost and weight without any increase in performance, unless it was used in low-light conditions when the iris diaphragm could be expected to be larger. In general, larger objective diameters require larger magnifications: e.g., 8×40 and 10×50. Binoculars designed for bright conditions where the expected iris opening is smaller can use smaller objectives: e.g., 8×25. The maximum exit pupil tends to decrease with age, so depending on the individual, 8×25 might be optimal even at night.

The exit pupil for telescopes is obtained in a similar way: divide the objective diameter by the working magnification.

The *entrance pupil* is defined as the area at the input of an optical system that can accept rays that can be transmitted through the exit pupil. The entrance pupil is the aperture stop if it is the first optical element; otherwise, it is determined by the image of the aperture stop when propagated backwards through the optical elements that precede it.

Problem 4.21   What pinhole diameter will give image elements with an angular diameter of $\frac{1}{30}^{\circ}$ using point-source objects located $1.00\,m$ away?

Problem 4.22   What is the focal length of a converging lens that can be used as a magnifying glass to give a magnification of 5.0? Assume an image formed at a near point of $25\,cm$.

Problem 4.23   Find an expression for eye relief of a telescope in terms of the eyepiece and objective focal lengths. You may assume a single-element objective and eyepiece, as illustrated in Figure 4.31.

Problem 4.24   Find the magnification of a $160\,cm$ microscope that uses a $4.0\,mm$ focal length objective and eyepiece that provides a magnification of 100.

Problem 4.25   Microscope objectives are sometimes characterized by the *numerical aperture NA*, defined by

$$NA = n \sin \theta$$

where n is the index of refraction of the medium that contains the object being observed, and θ is the half-angle of the collection cone of the objective lens. Show that the $f/\#$ of the objective is approximately given by

$$f/\# \approx \frac{n}{2NA}$$

Problem 4.26   A pair of binoculars are made with objective lenses that have a $50\,mm$ diameter. What magnifications will result in exit pupils of $3.00\,mm$ and $6.00\,mm$? Which design might give better results for daytime use, and why?

4.8   *RADIOMETRY

Radiometry concerns collection of electromagnetic energy from an arbitrary extended source. The related field of *photometry* is specific to the use of the human eye as a detector. In both cases, there are a variety of parameters that prove useful.

We begin by considering a *point source*, obtained when an extended source of finite size is located so far away that it cannot be resolved with the optical system used for collection. The total energy radiated by the source in a given time is the *radiant energy Q* with SI units of Joules. The rate of energy release is the *radiant flux Φ* with SI units of Watts.

To account for the fraction of radiant flux intercepted by a detector of area A_d, we define the *solid angle*

$$\Omega = \frac{A}{r^2} \tag{4.59}$$

where r is the distance from the detector to the source, and A is the projection of the detector area onto the area of a concentric sphere of radius r, as illustrated in Figure 4.33(a). Solid angle is measured in units of *steradians (sr)*. According to Equation 4.59, a spherical surface subtends a total solid angle of $4\pi\,sr$.

■ EXAMPLE 4.7

Find the solid angle subtended by lens of diameter D located a distance s_o from a point source p. Assume a lens of circular cross section coaxial with the line passing from its center through p.

Solution

To find the solid angle subtended by the lens about the point p, we must find

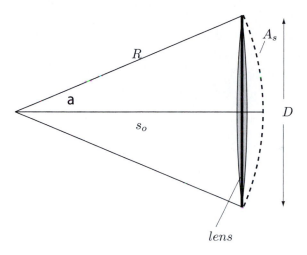

Figure 4.32 Solid angle projected by a thin lens of diameter D (bold line) located a distance s_0 from a point source p. The lens is coaxial with the line connecting its center with p. The angle θ is the half angle subtended by the lens at p.

the spherical area A_s bounded by the area of the lens (dotted line in Figure 4.32). Using spherical coordinates (Section 1.7.2) and taking the line of symmetry as the z-axis, we have

$$A_s = \int_0^\alpha 2\pi R^2 \sin\theta \, d\theta = 2\pi R^2 \left[1 - \cos\alpha\right]$$

The solid angle subtended at p is thus

$$\Omega = \frac{A_s}{R^2} = 2\pi \left[1 - \cos\alpha\right]$$

In terms of the lens diameter, the cosine of the half angle is given by

$$\cos\alpha = \frac{s_0}{\sqrt{s_0^2 + \frac{D^2}{4}}}$$

A point source is *isotropic* if it radiates equally in all directions. Most sources are not isotropic, and to account for directionality it is useful to define *radiant intensity* as the radiant flux per unit solid angle

$$J = \frac{\Phi}{\Omega} \tag{4.60}$$

with units W/sr. It can also be useful to define the source *irradiance*[10], defined in Section 2.3.3 as the flux that crosses a unit area A:

$$I = \frac{\Phi}{A} \tag{4.61}$$

with units W/m^2. Irradiance and radiant intensity are obviously very different quantities. For example, the irradiance from an isotropic point source follows an inverse-square law, but the radiant intensity is independent of distance (see Figure 4.33).

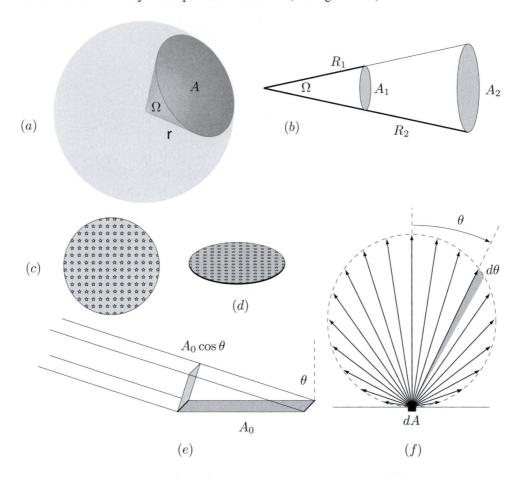

Figure 4.33   (a) Solid angle subtended by an area A that is a portion of a sphere of radius R. (b) Two areas A_1 and A_2 that subtend the same solid angle Ω. (c) Idealized point radiators spaced in a grid array. (d) The grid array of radiators in (c) viewed from a different perspective. (e) Projected area. (f) A Lambertian radiator.

---

[10]It is standard in radiometry to assign the symbol E to irradiance. We do not do so here to avoid confusion with electric field.

4.8.1 Extended Sources

We model extended sources as a collection of point radiators. Figure 4.33(c) represents a uniform radiating surface of finite area A, with idealized identical point radiators spaced in a grid array. For measurement angles inclined at θ with respect to the normal, the source area subtends a *projected area* that varies as

$$A = A_0 \cos \theta \tag{4.62}$$

where A_0 is the solid angle subtended by the source when the measurement angle is zero. The *radiance*, sometimes referred to as *brightness*, is defined as the flux emitted into unit solid angle per unit projected area:

$$L = \frac{\Phi}{A\Omega} \tag{4.63}$$

with units $W/m^2\ sr$.

For nonzero measurement angles, the same number of point radiators are contained within a smaller projected area, as illustrated in Figures 4.33(c) and (d). As measurement angles approach $90°$, the model of an isotropic point radiator gives results that are unphysical. If the extended source is a semi-infinite plane, then increasing the measurement angle increases the number of radiators sampled, and the radiance goes to infinity (see Figure 4.33(e)). A more realistic model for point radiators is that of the *Lambertian radiator*[11] illustrated in Figure 4.33(f). An extended source of Lambertian radiators emits a uniform radiance at all measurement angles. Consider a single Lambertian point radiator analyzed by a detector positioned at a constant distance from the source so that it subtends a constant solid angle Ω at all measurement angles. For a Lambertian point source, the flux received by this detector is independent of the measurement angle θ. Since the projected area of the source varies as $\cos \theta$, the flux emitted at angle θ must vary as $\cos \theta$ as well

$$L = \frac{d\Phi}{(\cos \theta dA)\ \Omega} = \frac{d\Phi_0}{dA\,\Omega}$$

where $d\Phi_0$ is the flux received when $\theta = 0$. Thus, $d\Phi = d\Phi_0 \cos \theta$, as indicated by the arrows in Figure 4.33(f). Examples of extended sources that are approximately Lambertian include the surface of the Sun and walls painted with diffusely reflecting paint.

4.8.1.1 Spectral Distributions

A *spectrometer* is an instrument designed to measure radiance over a specific portion of the electromagnetic spectrum. If the spectrometer is designed as a tunable wavelength filter[12], then the corresponding *spectral distribution* can be described with radiometric quantities that are also functions of wavelength. The *spectral radiant intensity* (sometimes referred to as the *specific radiant intensity*) is given by

$$J_\lambda = \frac{\partial J}{\partial \lambda} \tag{4.64}$$

with units W per steradian per meter. The *spectral irradiance* (sometimes referred to as the *specific irradiance*) is given by

$$I_\lambda = \frac{\partial I}{\partial \lambda} \tag{4.65}$$

[11] J. H. Lambert: 1728-1777. German mathematician, physicist, and astronomer.
[12] Spectrometers that utilize *diffraction gratings* (see Section 6.3.8) give output as a function of wavelength.

with units of W per cubic meter. The *spectral radiance* (sometimes referred to as the *specific radiance*) is given by

$$L_\lambda = \frac{\partial L}{\partial \lambda} \tag{4.66}$$

with units W per cubic meter per steradian.

Spectrometers can also be designed to give output as a function of other parameters such as frequency, photon energy, and wavenumber. In each case, there are corresponding spectral radiometric parameters that describe the associated spectral distribution.

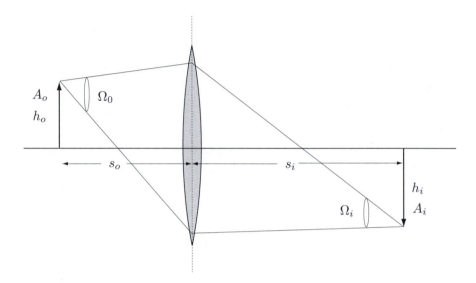

Figure 4.34   Conservation of radiance.

4.8.1.2 *Conservation of Radiance*

An image formed by an optical system has a radiance that is less than or at best equal to the radiance of the source. To see that this is so, consider the thin converging lens illustrated in Figure 4.34. The lens forms an image located at image distance s_i of an object located at object distance s_0 with magnification given by

$$\frac{h_i}{h_o} = \frac{s_i}{s_o} \tag{4.67}$$

The ratio of image area to source area is thus

$$\frac{A_i}{A_o} = \left(\frac{h_i}{h_o}\right)^2 = \frac{s_i^2}{s_o^2} \tag{4.68}$$

The refracted rays define an object solid angle given by $\Omega_0 = A_o/s_o^2$ and image solid angle given by $\Omega_i = A_i/s_i^2$, with ratio

$$\frac{\Omega_i}{\Omega_o} = \frac{s_0^2}{s_i^2} \tag{4.69}$$

The *optical throughput*, sometimes referred to as *entendu*, is defined as the product $A\,\Omega$. From Equations 4.68—4.69, we see that

$$A_o\,\Omega_0 = A_i\,\Omega_i \tag{4.70}$$

Thus, the radiance is conserved:

$$L_i = \frac{\Phi}{A_i \, \Omega_i} = \frac{\Phi}{A_o \, \Omega_o} = L_o \tag{4.71}$$

where Φ is the flux that is passed by the optical system. In practice, there are *insertion losses*, and $L_i < L_o$. In any case, concentrating the light gathered by the optical system into a smaller image also increases the image solid angle, so the radiance of the image is at best equal to that of the source.

Problem 4.27   A lens coaxial with an aperture stop of diameter $5.00\,cm$ is located $25.0\,cm$ from a point source p. Find the solid angle subtended by the aperture stop at p.

Problem 4.28   Find the solid angle subtended by a $f/1$ lens at a point source located at its focus.

Problem 4.29   Find the radiant intensity of a Lambertian emitter that radiates $1.00\,W$ into a hemisphere.

Problem 4.30   Find the maximum solid angle subtended by a $1.00\,cm$-wide detector at a point one meter away.

Problem 4.31

   a) Use Equation 4.21 to show that a single spherical refracting surface can form an image according to

$$\frac{1}{s_o} + \frac{n}{s_i} = \frac{n-1}{R}$$

where R is the radius of curvature of the refracting surface and n is the relative index of refraction. Define a focal length letting the object distance go to infinity:

$$f = \frac{R}{n-1}$$

Use this to show that the image magnification is given by

$$M = -\frac{s_i}{n\,s_o}$$

   b) Discuss how radiance is conserved at this optical interface.

4.8.2   Radiometry of Blackbody Sources

All objects with a non-zero temperature emit and absorb electromagnetic radiation. An object that is in *thermal equilibrium* with its surroundings has equal rates of emission and absorption as a whole, integrated over all wavelengths; thus, *a good emitter is a good absorber*. An *ideal blackbody* can be approximated by a small hole leading to an irregularly shaped cavity within a block of absorbing material such as carbon, as illustrated in Figure

4.35(a). The hole is a good absorber (and thus a good radiator) because a photon that enters the hole will reflect many times before finding a way out of the cavity, so the probability for absorption is very high. If the hole is sufficiently small, energy radiated through it will have a negligible effect on the energy density within the cavity, and thermal equilibrium will be maintained.

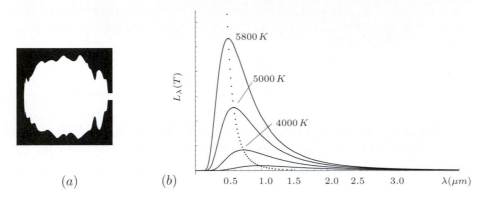

(a) (b)

Figure 4.35 (a) A small hole leading to an irregularly shaped cavity approximates an ideal blackbody. (b) Spectral radiance $L_\lambda(T)$ plotted versus wavelength λ for temperatures of $5800\,K$, $5000\,K$, $4000\,K$, and $3000\,K$. Data points locate Wien's displacement maxima for temperatures that range from $2000\,K$ and $6000\,K$ in steps of $100\,K$.

When light leaving such a cavity is analyzed spectroscopically, the spectrum of Figure 4.35(b) is observed, where spectral radiance $L_\lambda(T)$ is plotted as a function of wavelength. The properties of this distribution were well documented in the years prior to 1900, and in the process, certain empirical facts were noted. As the temperature increases, emission increases at all frequencies, and the wavelength $\lambda_{\max}$ that locates the peak of the wavelength spectrum shifts toward shorter wavelengths according to *Wien's displacement law*[13]

$$\lambda_{\max} = \frac{0.0028978\,m\,K}{T} \tag{4.72}$$

where $\lambda_{\max}$ is the wavelength at the peak of $L_\lambda(T)$, and T is the temperature in Kelvin. Also, the total flux radiated is found to be proportional to the fourth power of the temperature, as summarized by the Stefan-Boltzmann law: [14]

$$\frac{\Phi}{A} = \sigma T^4 \tag{4.73}$$

where Φ is the total radiated flux, A is the blackbody area, T is the absolute temperature, and σ is *Stefan's constant*: $\sigma = 5.6699 \times 10^{-8} W/m^2 K^4$. If the radiator is not a perfect blackbody, then it has an *emissivity* e of less than one:

$$\frac{\Phi}{A} = e\,\sigma T^4 \tag{4.74}$$

with $e = 0$ for a perfect reflector.

[13] Wilhelm Wien: 1864-1928. German physicist and recipient of the 1911 Nobel Prize.
[14] Jozef Stefan: 1835-1893. Ludwig Boltzmann: 1844-1906. Austrian physicists who founded the field of statistical thermodynamics.

4.8.3   Rayleigh-Jeans Theory and the Ultraviolet Catastrophe

Lord Rayleigh[15] and J. H. Jeans[16] independently modeled a blackbody cavity as a collection of linear oscillators with electrons as the oscillating mass. Because they accelerate, the oscillating charges radiate. At thermodynamic equilibrium, the oscillators within the cavity of Figure 4.35(a) will absorb and emit radiation equally, and there will be a steady state *spectral energy density* $u(\lambda, T)$ within the cavity that determines the spectral radiance $L_\lambda(T)$ that exits the outlet hole. According to classical thermodynamics, the average energy of each oscillator is equal to $k_B T$[17], where k_B is *Boltzman's constant*,

$$k_B = 1.38065 \times 10^{-23} \frac{J}{K} \tag{4.75}$$

In the Rayleigh-Jeans model, an oscillator can participate only if it emits radiation in an available cavity *mode*: a standing wave with nodes at each wall of the cavity. The number of such modes per unit frequency interval per unit volume is known as the *Jeans number* (see the Appendix to this chapter)

$$n(f) = \frac{8\pi f^2}{c^3} \tag{4.76}$$

where c is the vacuum speed of light. We may express the Jeans number as a function of wavelength by noting that since $n(f)\,df = n(\lambda)\,d\lambda$ and $f\lambda = c$ (see Problem 4.33),

$$n(\lambda) = n(f) \left| \frac{df}{d\lambda} \right| = n(f) \frac{c}{\lambda^2} \tag{4.77}$$

giving

$$n(\lambda) = \frac{8\pi}{\lambda^4} \tag{4.78}$$

The Rayleigh-Jeans result is given by the product of the Jeans number and the average energy per mode. In terms of wavelength, this gives

$$u_\lambda(T) = n(\lambda)\,k_B T = \frac{8\pi}{\lambda^4} k_B T \tag{4.79}$$

where the spectral energy density $u_\lambda(T)$ is the radiant energy per unit volume between λ and $\lambda + d\lambda$. Expressing this result as a function of frequency gives

$$u_f(T) = n(f)\,k_B T = \frac{8\pi f^2}{c^3} k_B T \tag{4.80}$$

where the spectral energy density $u_f(T)$ is the radiant energy per unit volume between f and $f + df$.

■ EXAMPLE 4.8

Show that the spectral radiance emitted from the cavity hole can be found by multiplying Equations 4.79 and 4.80 by $\frac{c}{4\pi}$.

---

[15] John William Strutt: 1842-1919. English physicist and recipient of the 1904 Nobel Prize.
[16] J. H. Jeans: 1877-1946. English mathematician and astronomer.
[17] See the discussion of the *equipartition theorem* in, for example, Schroeder [18].

Solution

We begin by noting that the hole in the blackbody cavity of Figure 4.35(a) is a perfect Lambertian source. The blackbody cavity is filled with electromagnetic waves that propagate in every possible direction. Flux leaving the hole area δA in a direction θ crosses a projected area of $\delta A \cos \theta$; thus the exiting flux is proportional to $\cos \theta$ giving a radiance that is constant for all angles, as discussed in Section 4.8.1.

Integrating the radiance over the hemisphere of solid angles, we obtain:

$$\delta \Phi = \delta A \int_0^{2\pi} d\varphi \int_0^{\frac{\pi}{2}} L \cos \theta \sin \theta \, d\theta$$

$$= \delta A \, L \, (2\pi) \left. \frac{\sin^2 \theta}{2} \right|_0^{\frac{\pi}{2}}$$

$$= \delta A \, \pi \, L$$

or,

$$L = \frac{1}{\pi} \frac{\delta \Phi}{\delta A}$$

where $\frac{\delta \Phi}{\delta A}$ is the net power passing through the hole, sometimes referred to as the *exitance* M. Note that the multiplicative factor relating radiance and irradiance is *not* 2π, as one might intuitively guess.

We may relate exitance to the energy densities of Equations 4.79 and 4.80 in a manner that is similar to the arguments used in Section 2.3.2. According to Equation 2.32, a collimated beam with energy density u has an intensity of uc. However, the energy passing through the blackbody hole propagates in all possible directions. We first note that only half of the energy within the blackbody cavity propagates in a direction with a forward component. Now apply conservation of radiance to two areas: the area δA of the blackbody hole, and the area of a hemisphere of radius $c\delta t$, as illustrated in Figure 4.36.

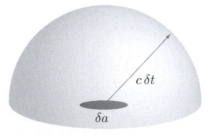

Figure 4.36 The blackbody hole and a hemisphere of radius $c\delta t$.

The radiance at the two areas must be equal. Since the radiation passes through the hemisphere normally, and since the solid angle subtended by the hemisphere is 2π, we obtain

$$\frac{1}{\pi} \frac{\delta \Phi}{\delta A} = \frac{\left(\frac{1}{2} u \right) c}{2\pi}$$

giving

$$\frac{\delta \Phi}{\delta A} = \frac{c}{4} u$$

and a corresponding radiance of

$$L = \frac{c}{4\pi} u$$

According to the result of Example 4.8, the radiance predicted by the Rayleigh-Jeans result is

$$L_\lambda(T) = \frac{2c}{\lambda^4} k_B T \tag{4.81}$$

as a function of wavelength, and

$$L_f(T) = \frac{2f^2}{c^2} kT \tag{4.82}$$

as a function of frequency. Clearly, these results must be wrong. As frequency increases and wavelength decreases, both expressions for radiance increase without limit, in complete contradistinction with observation. Ehrenfest[18] later described this failure as the *ultraviolet catastrophe*. The discrepancy turned out to be very important, for it pointed to an inadequacy of classical physics, paving the way for quantum ideas.

Problem 4.32   The tungsten filament of an incandescent lamp is a wire of diameter $0.10\,mm$, length $1.00\,mm$ and emissivity of 0.9. If the temperature of the filament is at $3200\,K$, find the peak wavelength of blackbody emission and the total power radiated.

Problem 4.33   Show that for an electromagnetic wave of frequency f and wavelength λ that travels in vacuum,

$$df = -\frac{c}{\lambda^2} d\lambda$$

4.8.4   Planck's Quantum Theory of Blackbody Radiation

Max Planck[19] deduced the following expression for blackbody radiation that agrees well with observation:

$$u(f,T) = \frac{8\pi h f^3}{c^3} \frac{1}{\left(e^{\frac{hf}{kT}} - 1 \right)} \tag{4.83}$$

where h is *Planck's constant* and T is the *equilibrium temperature*.

$$h = 6.626196 \times 10^{-34} J \cdot s$$

---

[18] Paul Ehrenfest: 1880-1933. Austrian physicist and mathematician.
[19] Max Karl Ernst Ludwig Planck: 1858-1947. German physicist and winner of the 1918 Nobel Prize.

Planck determined the value of h by using it as a fitting parameter. It is a universal constant that plays a central role in quantum mechanics.

Equation 4.83 can be derived rigorously using the same linear oscillator model of Rayleigh-Jeans along with the following *quantum postulates*:

1. An individual oscillator may not emit or absorb continuously as in the Rayleigh-Jeans model. Each oscillator absorbs or radiates discrete *energy quanta* of magnitude $\Delta\epsilon = hf$.

2. An oscillator cannot have arbitrary energy, but must occupy one of a discrete set of energy states given by $\epsilon_n = nhf$.

The dependence of the Planck distribution on these postulates was first formally established[20] by Einstein[21] in one of three remarkable papers published while a patent clerk in 1905.

We examine the consequences of the above postulates by calculating the average energy per oscillator $\bar{\epsilon}$. According to postulate 2, an individual oscillator can only have discrete values of energy. Since thermal processes are random, the best we can do is to specify a probability that a given available state is occupied. According to classical statistical thermodynamics, this probability is provided by the *Maxwell-Boltzmann distribution*

$$N(n) = N_0 e^{-\frac{\epsilon_n}{kT}} = N_0 e^{-\frac{nhf}{kT}} \tag{4.84}$$

This distribution function was used by Boltzmann with remarkable success to derive the laws of classical thermodynamics from statistical arguments. Because of the exponential dependence, higher energy states are less likely to be populated. Using this distribution function, we calculate the average energy per oscillator as follows

$$
\begin{aligned}
\bar{\epsilon} &= \frac{\sum\limits_{n=0}^{\infty} N(n)\,\epsilon_n}{\sum\limits_{n=0}^{\infty} N(n)} = \frac{N_0 \sum\limits_{n=0}^{\infty} e^{-\frac{nhf}{kT}}(nhf)}{N_0 \sum\limits_{n=0}^{\infty} e^{-\frac{nhf}{kT}}} \\
&= \frac{0 + hf\,e^{-\frac{hf}{kT}} + 2hf\,e^{-\frac{2hf}{kT}} + 3hf\,e^{-\frac{3hf}{kT}} + \cdots}{1 + e^{-\frac{hf}{kT}} + e^{-\frac{2hf}{kT}} + e^{-\frac{3hf}{kT}} + \cdots} \\
&= hf e^{-\frac{hf}{kT}} \left(\frac{1 + 2e^{-\frac{hf}{kT}} + 3e^{-\frac{2hf}{kT}} + \cdots}{1 + e^{-\frac{hf}{kT}} + e^{-\frac{2hf}{kT}} + e^{-\frac{3hf}{kT}} + \cdots} \right) \\
&= hf\,x \left(\frac{1 + 2x + 3x^2 + \cdots}{1 + x + x^2 + x^3 + \cdots} \right)
\end{aligned}
\tag{4.85}
$$

where in the last step, we make the substitution $x = e^{-\frac{hf}{kT}}$. The denominator of the last step is the geometric series (Equation 5.85):

$$\sum_{n=0}^{\infty} x^n = 1 + x + x^2 + \ldots = \frac{1}{1-x} \tag{4.86}$$

The numerator is the derivative of the denominator; thus

$$1 + 2x + 3x^2 + \cdots \sum_{n=0}^{\infty} n\,x^{n-1} = \frac{1}{(1-x)^2} \tag{4.87}$$

---

[20] For an interesting account of this important aspect of the history of physics, see Kuhn [15].
[21] Albert Einstein: 1879-1955. German physicist and recipient of the 1921 Nobel Prize.

Using these results, we obtain

$$\bar{\epsilon} = hf\,x\frac{(1-x)^{-2}}{(1-x)^{-1}} = hf\,x\frac{1}{(1-x)^{-1}} = \frac{hf}{\frac{1}{x}-1}$$

Substituting $x = e^{-\frac{hf}{kT}}$ gives

$$\bar{\epsilon} = \frac{hf}{e^{\frac{hf}{kT}} - 1} \tag{4.88}$$

Multiplying $\bar{\epsilon}$ by the Jeans number (Equation 4.76) gives the Planck distribution of Equation 4.83:

$$u_f(T) = n\,(f)\,\bar{\epsilon} = \left(\frac{8\pi f^2}{c^3}\right)\left(\frac{hf}{e^{\frac{hf}{k_BT}}-1}\right) = \frac{8\pi hf^3}{c^3}\frac{1}{\left(e^{\frac{hf}{k_BT}}-1\right)} \tag{4.89}$$

as a function of frequency, and by Equation 4.78

$$u_\lambda(T) = n(\lambda)\bar{\epsilon} = \frac{8\pi hc}{\lambda^5}\left(\frac{1}{e^{\frac{hc}{\lambda k_BT}}-1}\right) \tag{4.90}$$

when expressed as a function of wavelength. The corresponding spectral radiance in terms of frequency and wavelength is given by

$$L_f(T) = \frac{2hf^3}{c^2}\left(\frac{1}{e^{\frac{hf}{k_BT}}-1}\right) \tag{4.91}$$

$$L_\lambda(T) = \frac{2hc^2}{\lambda^5}\left(\frac{1}{e^{\frac{hc}{\lambda k_BT}}-1}\right) \tag{4.92}$$

■ **EXAMPLE 4.9**

Show that the Planck radiation law expressed as a function of wavelength gives the Wien's displacement result of Equation 4.72 and the Stefan-Boltzman law of Equation 4.73.

Solution

Begin with Equation 4.92 for blackbody irradiance as a function of wavelength.

$$L_\lambda(x) = \frac{2\,(k_BT)^5}{h^4c^3}\left[\frac{x^5}{e^x-1}\right]$$

where we have defined the dimensionless parameter

$$x = \frac{hc}{\lambda k_BT}$$

To obtain a maximum, differentiate and set equal to zero:

$$\frac{2\,(k_BT)^5}{h^4c^3}\left[\frac{5x^4}{e^x-1} - \frac{x^5e^x}{(e^x-1)^2}\right] = 0$$

giving

$$\frac{xe^x}{e^x - 1} = 5$$

or, by rearranging,

$$1 - \frac{x}{5} = e^{-x}$$

This equation can be solved numerically to give $x_{max} = 4.96511$. Thus

$$\lambda_{\max} = \frac{hc}{(4.96511)\,k_B} = \frac{0.0028978\,m\,K}{T}$$

as in Equation 4.72.

To verify the Stefan-Boltzmann law, we integrate the radiance over the half-sphere of solid angle and over all wavelengths. The integration over solid angle gives a factor of π, as in Example 4.8. To integrate over wavelength, we change variables to the dimensionless parameter defined above, with

$$d\lambda = -\frac{hc}{x^2 k_B T}dx$$

The radiated flux per unit blackbody area is

$$\frac{\Phi}{A} = \pi \int_0^\infty L_\lambda d\lambda$$

$$= \pi \frac{2\,(k_B T)^5}{h^4 c^3} \int_\infty^0 \left(\frac{x^5}{e^x - 1}\right) \left(-\frac{hc}{x^2 k_B T}dx\right)$$

$$= \frac{2\pi\,(k_B T)^4}{h^3 c^2} \int_0^\infty \frac{x^3}{e^x - 1}\,dx$$

From integral tables or software:

$$\int_0^\infty \frac{x^3}{e^x - 1}\,dx = \frac{\pi^4}{15}$$

giving

$$\frac{\Phi}{A} = \left(\frac{2\pi^5 k_B^4}{15 h^3 c^2}\right) T^4$$

The quantity inside the parentheses is Stefan's constant.

Since wavelength varies inversely with frequency, uniform wavelength intervals do not give uniform frequency intervals. Thus, the Planck distributions of Equations 4.91 and 4.92 have maxima that occur at different wavelength/frequency values.[22]

[22]For more information, see M. A. Heald, "Where is the 'Wien Peak'?," Am. J. Phys 71, 1322(2003) and references therein.

Problem 4.34   Repeat the calculation of the average energy per oscillator by treating it as an integration over an energy continuum, and show that in this case the average energy per oscillator becomes the classical value of $\bar{\epsilon} = k_B T$. Show that this gives the Rayleigh-Jeans result.

Problem 4.35   Verify the Stefan-Boltzmann law using Equation 4.91 which expresses the Planck radiation law as a function of frequency.

Problem 4.36   Find the peak of the wavelength spectrum of the Sun, assuming that it is a blackbody of temperature 5800 K.

Problem 4.37   Show that the peak Equation 4.91 that expresses the Planck radiation law as a function of frequency occurs when

$$f_{\max} = (2.821)\frac{k_B T}{h}$$

Calculate the corresponding wavelength for a 5800 K blackbody, and compare to the result of Problem 4.36.

Problem 4.38   Show that the flux of photons per unit frequency interval per steradian is given by

$$N_f = \frac{2f^2}{c^2}\frac{1}{e^{\frac{hf}{k_B T}} - 1}$$

Problem 4.39   Referring to Problem 4.38, show that the number of photons emitted by a 5800 K blackbody peaks at 1559 nm when measured as a function of frequency.

Problem 4.40   Show that the flux of photons per unit wavelength interval per steradian is given by

$$N_\lambda = \frac{2c}{\lambda^4}\frac{1}{e^{\frac{hc}{\lambda k_B T}} - 1}$$

Problem 4.41   Referring to Problem 4.40, show that the number of photons emitted by a 5800 K blackbody peaks at 634 nm when measured as a function of wavelength.

Problem 4.42   Assuming a 5800 K blackbody, use numerical integration to find the *median wavelength* of the Planck distribution, defined such that one half of the total radiated energy falls on either side. Show that the same value is obtained with both wavelength and frequency distributions.

4.9   *OPTICAL FIBERS

Optical fibers are thin cylinders of transparent dielectric material that contain and transmit light by total internal reflection. This is useful for a variety of purposes. High-powered laser beams can be manipulated with the precision required for medical applications. Bundles of optical fibers can collect light at the image plane of an optical system, transporting the image information intact to another location. By far, the most important application of fiber optics is in communications. Data transmission over conductors is limited by stray capacitance and inductance, and ultimately by the frequency of the *carrier wave*. Optical frequencies are much higher, and when properly designed, optical fibers can achieve extremely high data transfer rates. Optical fiber technology is the cornerstone of modern high-speed digital networks.

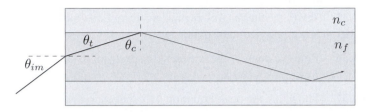

Figure 4.37   A step-index optical fiber with core index n_f and cladding index n_c. A ray incident at θ_{im} on the fiber face is incident at the critical angle at the core-cladding interface.

Figure 4.37 illustrates a *step index* fiber consisting of a central *core* with index of refraction n_f surrounded by a lower-index *cladding* with index n_c. Light will be transmitted without loss along the fiber as long there is total internal reflection at the core-cladding interface. The maximum input angle θ_{im} for a ray that transmits into the fiber may be found from Snell's law. A ray incident in air at this angle is transmitted according to $n_i \sin \theta_{im} = n_f \sin \theta_t$. For this ray to be totally reflected at the fiber-cladding interface, the transmitted angle must be the complement of the critical angle, which from Section 3.5.3 is given by $\theta_c = \sin^{-1}\left(\frac{n_c}{n_f}\right)$. Thus

$$\frac{n_c}{n_f} = \sin\theta_c = \cos\theta_t = \sqrt{1 - \sin^2\theta_t} = \sqrt{1 - \frac{n_i^2 \sin^2\theta_{im}}{n_f^2}}$$

Solving for $n_i \sin\theta_{im}$ gives the *numerical aperture* (NA) of the fiber:

$$NA = n_i \sin\theta_{im} = \sqrt{n_f^2 - n_c^2} \tag{4.93}$$

Note that the core index must be greater than that of the cladding in order for total internal reflection to occur. The cladding must be thick enough to accommodate the *evanescent wave* discussed in Section 3.6.2.

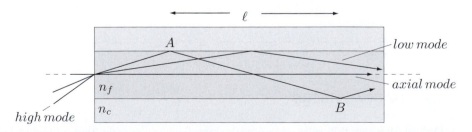

Figure 4.38   A multimode optical fiber. The axial mode propagates a given fiber length in the least time.

Rays with different angles of incidence propagate down a step index fiber as shown in Figure 4.38. Rays with a non-zero angle of incidence travel a longer path than that of an axial path whose angle of incidence is zero. The largest path difference is between the axial path and that of the maximum angle of incidence, the path for which $\theta_i = \theta_{im}$. If the axial distance between points A and B in Figure 4.38 is ℓ, then the longest path between these two points is $\ell_h = \ell / \sin \theta_c = \ell n_f / n_c$. Since the axial path has a transit time A to B of $t_a = \ell / v = \ell n_f / c$, and the maximum path has a transit time between the same points of $t_h = \ell_h n_f / c$, the maximum *time delay* is given by

$$\Delta t = t_h - t_a = \frac{n_f \left(\frac{\ell n_f}{n_c} - \ell \right)}{c}$$

giving a time delay per unit length Δt_ℓ of

$$\Delta t_\ell \equiv \frac{\Delta t}{\ell} = \frac{n_f \left(\frac{n_f}{n_c} - 1 \right)}{c} \qquad (4.94)$$

This time delay places a practical limit on the usable modulation bandwidth for information that can be successfully transmitted across the fiber. For example, digital information is transmitted as a sequence of ones and zeros, represented as discrete changes in the optical intensity between two reference levels. If the time delay of Equation 4.94 exceeds the time interval between adjacent "bits" of digital information, the variation between these reference levels becomes indistinct.

■ **EXAMPLE 4.10**

Estimate the time delay for a $1.00 \, km$ length of fiber with a core index equal to 1.50 that does not have a cladding, and is surrounded by air. Repeat for a fiber with a cladding of index 1.49.

Solution

If the fiber has no cladding, Equation 4.94 gives

$$\Delta t = n_f \left(\frac{n_f}{n_c} - 1 \right) \frac{\ell}{c} = 1.50 \left(\frac{1.50}{1} - 1 \right) \frac{10^3 m}{3.00 \times 10^8 m/s} = 2.50 \times 10^{-6} s$$

If the fiber has a cladding of $n_c = 1.49$,

$$\Delta t = 1.50 \left(\frac{1.50}{1.49} - 1 \right) \frac{10^3 m}{3.00 \times 10^8 m/s} = 3.36 \times 10^{-8} s$$

Thus, the cladded fiber will support over 100 times the bandwidth of the fiber with no cladding.

When the cladding index is close in value to the core index, the critical angle is reduced to the point where the only paths available are very close to axial. In a more detailed analysis that includes a full wave treatment using Maxwell's equations[23], the fiber diameter

[23] Such a treatment is beyond the scope of this discussion. For more information, see Saleh and Teich [17].

becomes very important. Electromagnetic waves can only propagate without loss if the wave amplitude and phase repeats at regular periodic intervals along the fiber, and this restricts the possible non-axial paths that can exist in a fiber with a given cross-sectional diameter. A possible propagation path is called a *mode*, and a larger diameter fiber that can support many modes is called a *multimode fiber*. For this reason, the value of Δt_ℓ given by Equation 4.94 is said to produce *modal distortion* of transmitted information. The modal distortion can be reduced by decreasing the fiber diameter in so called *single-mode fiber*. A single-mode fiber with cladding index that is close in value to the core index supports a single transmission mode, minimizing the modal distortion to support the highest possible bandwidth. Single-mode fibers have diameters that are typically 10 wavelengths or less across.

In addition to the step index model, optical fibers may also be designed with a *graded index* where the index of the core decreases with distance from the core axis. Applications of *graded index optics* (sometimes referred to as *gradient index optics*) (acronym: GRIN) were first investigated by R. K. Luneberg,[24] who designed a spherically symmetric lens that focuses parallel rays perfectly from any incident direction. The difficulty of fabricating the necessary index gradients limited application to all but microwave wavelengths where a layered design could be implemented until modern *ion diffusion* technologies were developed.[25]

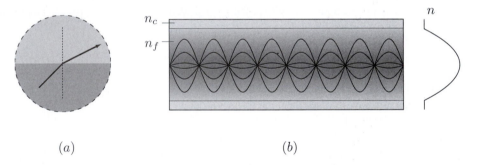

(a) $\qquad\qquad\qquad\qquad\qquad\qquad\qquad$ (b)

Figure 4.39 (a) Refraction across an infinitesimal fiber element for a ray traveling away from the fiber center. (b) A graded index fiber. The fiber index varies as a parabola, decreasing with radial distance from the fiber center. A cladding of index n_c encloses the fiber.

Figure 4.39(a) illustrates refraction when a ray transmits into a region of lower index; in this case refraction causes the ray to refract away from the normal. In a material with graded index, rays refract toward regions of higher index. A familiar example is that of a *mirage*. As the Sun heats a surface such as sand or a roadbed, the density of air is lower at ground level. Rays are refracted *upward* as though they were reflected from a puddle of water, reflecting the sky.

It is possible to vary the index so that all trajectories intersect the axis of symmetry with the same spatial period, as illustrated in Figure 4.39. In this case, all rays periodically pass through identical points of focus, so by Fermat's principle (Section 3.2.1), the transit times must be equal for all trajectories (see Problem 4.46). Intuitively, rays with larger periodic

[24]Rudolf Karl Luneberg: 1903-1949.
[25]Despite these technical difficulties, Nature invented this design hundreds of millions of years ago in, for example, the eye of the box jellyfish. See D. Nilsson et. al., "Advanced optics in a jellyfish eye," Nature Vol 435, p. 201 (2005).

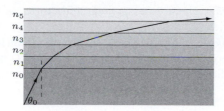

Figure 4.40   Uniform refraction in discrete layers.

amplitude travel longer distances but in regions of lower index, so the travel time is the same. This all but eliminates the modal distortion of step index designs.

Let us investigate the necessary index gradient to achieve this. Let r be the transverse distance from the axis of symmetry. If the periodic trajectories are sinusoidal with period L, then

$$r(z) = r_m \sin(\frac{\pi z}{L}) \tag{4.95}$$

where z measures distance along the fiber axis, and r_m is the maximum value of r for a given trajectory. Each trajectory crosses the axis with a slope that is given by

$$\frac{dr}{dz}\Big|_{z=0} = \frac{r_m \pi}{L} \tag{4.96}$$

This slope is given by the tangent of the angle that the trajectory makes with the axis, and for paraxial rays, it is given by the sine. Since angles of incidence are measured with respect to the normal, this slope is also given by

$$\cos \theta_0 = \frac{r_m \pi}{L} \tag{4.97}$$

where θ_0 is the angle of incidence at $r = 0$, henceforth referred to as the crossing angle. We assume also that the refraction is *uniform*. Consider the discrete layers illustrated in Figure 4.40, and assume that each layer refracts so that $n_0 \sin \theta_0 = n_1 \sin \theta_1 = n_2 \sin \theta_2 = \ldots$. Let r be the distance from the axis of symmetry, and $n(r)$ be the index of refraction at this axial distance. The condition of uniform refraction gives

$$n(r) \sin \theta(r) = const$$

where $\theta(r)$ is the angle of refraction at axial distance r. In particular,

$$n(0) \sin \theta_0 = n(r_m) \tag{4.98}$$

where $n(0)$ is the index on axis, $n(r_m)$ is the index at the maximum value of r for a particular trajectory, and θ_0 is the crossing angle. Since $\sin \theta = \sqrt{1 - \cos^2 \theta}$,

$$n(r_m) = n(0)\sqrt{1 - \left(\frac{\pi r_m}{L}\right)^2} \tag{4.99}$$

Each trajectory has a different value of r_m, so we may drop the subscript. According to the binomial expansion, $\sqrt{1 - x} \approx 1 - \frac{1}{2}x$, giving

$$n(r) = n(0) \left(1 - \left(\frac{\pi^2}{2L^2}\right) r^2\right) \tag{4.100}$$

Thus, a *parabolic index profile* gives periodic trajectories of equal transit times for all paraxial crossing angles.

Material dispersion also limits the usable bandwidth in both GRIN and step index fibers. As discussed in Section 3.7, different frequency components in a communications signal will propagate with different velocities, distorting the shape of a digital pulse that travels along the fiber.

Problem 4.43   A certain multimode fiber has $n_f = 1.46$ and $n_c = 1.45$. Find the numerical aperture and θ_{im} if the input medium is air, the critical angle, the time delay for a $1.00 \, km$ length, and an estimate of the $1 \, km$ bandwidth.

Problem 4.44   Find the trajectory period for a graded index fiber with a 1.00% index variation across a $100 \, \mu m$ diameter. Repeat for a diameter of $50 \, \mu m$.

Problem 4.45   An optical fiber has a core index of 1.499 and a cladding with index 1.489 and is immersed in air. Find the numerical aperture, the maximum input angle, the critical angle and the time delay for a $1.00 \, km$ segment. Estimate the bandwidth for a $5.00 \, km$ segment.

Problem 4.46   Use Fermat's principle to argue that intermodal distortion is eliminated in GRIN fibers that have a parabolic index profile.

Problem 4.47   A short GRIN rod can used as a *GRIN lens*. The *pitch* of a GRIN lens specifies its length as a fraction of the spatial trajectory period L. For example, a 0.25-pitch GRIN lens has a length that is $L/4$. Show that a 0.25-pitch GRIN lens will transmit a point source placed at one end of the rod as a collimated beam.

4.10   *THICK LENSES

A thick lens consists of a combination of two refracting surfaces separated by a distance d. If the paraxial approximation remains valid (and this is not always the case, especially with thick lenses), we can use the ideas and methods just developed for thin-lens combinations. In particular, by using principal points and effective focal lengths we may continue to use the thin-lens equation even with lenses that are thick.

Equation 4.21 describes refraction of light rays at the first lens surface. Divide both sides by n_m to give

$$\frac{1}{s_{o1}} + \frac{n}{s_{i1}} = \frac{n-1}{R_1} \tag{4.101}$$

where $n = n_\ell/n_m$ is the relative index of refraction, and R_1 is the radius of curvature of side 1 of the thick lens. We define a focal length for the first surface by letting $s_{i1} \to \infty$ in Equation 4.101 (see Figure 4.10):

$$f_1 = \frac{R_1}{n-1} \tag{4.102}$$

Combining Equations 4.101 and 4.102 gives

$$\frac{1}{s_{o1}} + \frac{n}{s_{i1}} = \frac{1}{f_1} \qquad (4.103)$$

Equation 4.103 looks like the thin-lens equation with s_{i1} replaced by s_{i1}/n. Thus, surface 1 produces a magnification given by

$$m_1 = -\frac{s_{i1}}{n\, s_{o1}} \qquad (4.104)$$

For surface 2, the incident medium has index n_ℓ:

$$\frac{n}{s_{o2}} + \frac{1}{s_{i2}} = \frac{1-n}{R_2} = -\frac{n-1}{R_2} \qquad (4.105)$$

In Equation 4.105, an infinite object distance produces an image at the focal point of the second surface

$$f_2 = -\frac{R_2}{n-1} \qquad (4.106)$$

Combining Equations 4.105 and 4.106 gives

$$\frac{n}{s_{o2}} + \frac{1}{s_{i2}} = \frac{1}{f_2} \qquad (4.107)$$

With this definition, Equation 4.107 looks like the thin-lens equation with s_{o1} replaced by s_{o1}/n. Thus, surface 2 produces a magnification given by

$$m_2 = -\frac{n\, s_{i2}}{s_{o2}} \qquad (4.108)$$

As a check, note that for a thin lens, we can combine Equations 4.102 and 4.106 using the formula for close contact:

$$\frac{1}{f} = \frac{1}{f_1} + \frac{1}{f_2} = (n-1)\left(\frac{1}{R_1} - \frac{1}{R_2}\right)$$

as in Equation 4.25.

The front and back focal points of a thick lens may be obtained as was done for lens combinations. According to Equation 4.103, surface 1 produces an image located by

$$s_{i1} = \frac{n\, f_1\, s_{o1}}{s_{o1} - f} \qquad (4.109)$$

This image becomes the object for surface 2, with object distance given by $s_{o2} = d - s_{i1}$. According to Equations 4.107 and 4.103, the image distance for surface 2 is given by

$$\begin{aligned} s_{i2} &= \frac{f_2 s_{o2}}{s_{o2} - n f_2} = \frac{f_2\,(d - s_{i1})}{d - s_{i1} - n f_2} \\ &= \frac{f_2 d - f_2\left(\frac{n f_1 s_{o1}}{s_{o1} - f_1}\right)}{d - n f_2 - \frac{n f_1 s_{o1}}{s_{o1} - f_1}} \end{aligned} \qquad (4.110)$$

Taking the limit as $s_{o1} \to \infty$ gives the back focal point:

$$f_b = \frac{f_2\,(d - n f_1)}{d - n\,(f_1 + f_2)} \qquad (4.111)$$

Similarly, letting $s_{i2} \to \infty$ gives the front focal point. This will be the case when

$$d - n f_2 = \frac{n f_1 s_{o1}}{s_{o1} - f_1}$$

giving

$$f_f = \frac{f_1 (d - n f_2)}{d - n (f_1 + f_2)} \tag{4.112}$$

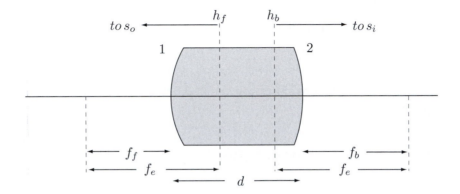

Figure 4.41   Front, back and effective focal lengths and principal points for a thick lens.

4.10.1   *Principal Points and Effective Focal Lengths of Thick Lenses

We again seek principal points with corresponding effective image and object distances and an effective focal length (see Figure 4.41) so that

$$\frac{1}{s_o} + \frac{1}{s_i} = \frac{1}{f_e} \tag{4.113}$$

The lens magnification is the product of the magnification for each surface, and must also be given by the effective quantities:

$$m = \left(-\frac{s_{i1}}{n s_{o1}}\right) \left(-\frac{n s_{i2}}{s_{o2}}\right) = \left(\frac{s_{i1}}{s_{o1}}\right) \left(\frac{s_{i2}}{s_{o2}}\right) = -\frac{s_i}{s_o} \tag{4.114}$$

To find the effective focal length, we let $s_{o1} \to \infty$ and use Equations 4.103 and 4.107. Thus, $s_{i1} \to n f_1$, $s_{o1} \to s_o$ and $s_i \to f_e$. The object distance for lens 2 is then $s_{o2} = d - s_{i1} \to d - n f_1$, giving

$$\frac{n f_1}{d - n f_1} s_{i2} = -f_e$$

The image distance s_{i2} is given by

$$s_{i2} = \frac{f_2 s_{o2}}{s_{o2} - n f_2} = \frac{f_2 (d - n f_1)}{d - n (f_1 + f_2)}$$

Combining the last two results gives the effective focal length:

$$f_e = -\left(\frac{n f_1}{d - n f_1}\right) \frac{f_2 (d - n f_1)}{d - n (f_1 + f_2)} = \frac{n f_1 f_2}{n (f_1 + f_2) - d} \tag{4.115}$$

or, equivalently,

$$\frac{1}{f_e} = \frac{1}{f_1} + \frac{1}{f_2} - \frac{d}{nf_1f_2} \qquad (4.116)$$

The front principal point is determined by (see Figure 4.41)

$$h_f = f_e - f_f = \frac{nf_1f_2}{n(f_1+f_2)-d} - \frac{f_1(d-nf_2)}{d-n(f_1+f_2)}$$

$$= \frac{f_1d}{n(f_1+f_2)-d}\left(\frac{nf_2}{nf_2}\right)$$

giving

$$h_f = \frac{f_ed}{nf_2} \qquad (4.117)$$

Principal points are measured relative to the points where the lens surface intersects the optical axis, as shown in Figure 4.41. Principal points that lie to the left of the corresponding lens surface are negative. According to this sign convention, the value of h_b in Figure 4.41 is

$$h_b = -(f_e - f_b) = -\frac{f_ed}{nf_1} \qquad (4.118)$$

■ **EXAMPLE 4.11**

Find the effective focal lengths, the front and back focal points, and principal points for the following lenses, all made of glass with index 1.50. In each case, find the image location and image size for a 1.00 cm-high object with object distance $s_o = 10.0\,cm$. (a) A biconvex lens with $R_1 = 10.0\,cm$, $R_2 = -10.0\,cm$, and $d = 1.00\,cm$. (b) A convex lens with $R_1 = 10.0\,cm$, $R_2 = -5.00\,cm$, and $d = 1.00\,cm$. (c) A concave lens, with $R_1 = -10.0\,cm$, $R_2 = +5.00\,cm$, and $d = 1.00\,cm$.

Solution

(a) A biconvex lens with $R_1 = 10.0\,cm$, $R_2 = -10.0\,cm$ and $d = 1.00\,cm$. Begin by computing values for f_1 and f_2:

$$f_1 = \frac{R_1}{n-1} = \frac{10.0\,cm}{0.5} = 20.0\,cm$$

$$f_2 = -\frac{R_2}{n-1} = -\frac{-10.0\,cm}{0.5} = 20.0\,cm$$

The effective focal length is given by

$$f_e = \frac{nf_1f_2}{n(f_1+f_2)-d} = \frac{1.50\,(20.0\,cm)^2}{59.0\,cm} = 10.2\,cm$$

The principal points are given by

$$h_f = \frac{f_ed}{nf_2} = \frac{(10.2\,cm)(1.0\,cm)}{30.0\,cm} = 0.339\,cm$$

$$h_b = -\frac{f_ed}{nf_1} = -\frac{(10.2\,cm)(1.00\,cm)}{30.0\,cm} = -0.339\,cm$$

The front focal length is given by

$$f_f = \frac{f_1 \left(d - n f_2 \right)}{d - n \left(f_1 + f_2 \right)} = \frac{20.0 \left(1.00 - 30.0 \right)}{1.00 - 1.50 \left(40.0 \right)} \, cm = 9.83 \, cm$$

Since the lens is biconvex, $f_1 = f_2$ so $f_b = f_f = 9.83 \, cm$. As a check, note that $f_f = f_e - h_f$ and $f_b = f_e + h_b$ (exactly when round-off errors are avoided). A diagram of this lens that shows the principal points is shown in Figure 4.42(a).

To find the image location, use the thin-lens equation:

$$s_i = \frac{f_e s_o}{s_o - f_e}$$

In this case, $f_e = 10.2 \, cm$, which is larger than s_o, so the image is virtual:

$$s_i = \frac{\left(10.2 \, cm \right) \left(10.0 \, cm \right)}{-0.2 \, cm} = -510 \, cm$$

The image is located 510 cm to the left of h_b. The magnification is

$$m = -\frac{s_i}{s_o} = -\frac{-510 \, cm}{10.0 \, cm} = 51.0$$

giving an upright virtual image that is 51.0 cm tall.

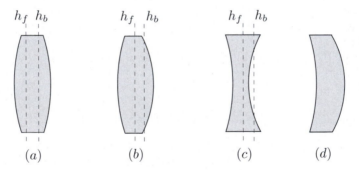

Figure 4.42   (a)-(c) Lenses used in the Example 4.11. (d) Meniscus lens of Problem 4.49. The surface curvatures are exaggerated for clarity.

(b) A convex lens with $R_1 = 10.0 \, cm$, $R_2 = -5.00 \, cm$, and $d = 1.00 \, cm$. The focal lengths for both surfaces are

$$f_1 = \frac{R_1}{n - 1} = 20.0 \, cm$$

$$f_2 = -\frac{R_2}{n - 1} = 10.0 \, cm$$

The effective focal length is given by

$$f_e = \frac{n f_1 f_2}{n \left(f_1 + f_2 \right) - d} = \frac{300}{44.0} = 6.82 \, cm$$

The principal points are located by

$$h_f = \frac{f_e d}{n f_2} = \frac{6.82}{15.0} = 0.455 \, cm$$

$$h_b = -\frac{f_e d}{n f_1} = -\frac{6.82}{30.0} = -0.227 \, cm$$

The front focal length is

$$f_f = \frac{f_1 (d - n f_2)}{d - n (f_1 + f_2)} = \frac{20.0 (1.00 - 15.0)}{1.00 - 1.50 (30.0)} \, cm = 6.36 \, cm$$

and the back focal length is

$$f_b = \frac{f_2 (d - n f_1)}{d - n (f_1 + f_2)} = \frac{10.0 (1.00 - 30.0)}{1.00 - 1.50 (30.0)} \, cm = 6.59 \, cm$$

Check: $f_f = f_e - h_f$ and $f_b = f_e + h_b$. See Figure 4.42(b).

The effective focal length of this lens is $f_e = 6.82 \, cm$, which is smaller than s_o, so the image is real:

$$s_i = \frac{(6.82 \, cm) \, (10.0 \, cm)}{3.18 \, cm} = 21.4 \, cm$$

The image is located 21.4 cm to the right of h_b. The magnification is

$$m = -\frac{s_i}{s_o} = -\frac{21.4 \, cm}{10.0 \, cm} = -2.14$$

giving an inverted real image that is 2.14 cm tall.

(c) A concave lens, with $R_1 = -10.0 \, cm$, $R_2 = +5.00 \, cm$, and $d = 1.00 \, cm$. The focal lengths for both surfaces are

$$f_1 = \frac{R_1}{n - 1} = -20.0 \, cm$$

$$f_2 = -\frac{R_2}{n - 1} = -10.0 \, cm$$

The effective focal length is given by

$$f_e = \frac{n f_1 f_2}{n (f_1 + f_2) - d} = \frac{300}{-46.0} = -6.52 \, cm$$

The principal points are located by

$$h_f = \frac{f_e d}{n f_2} = \frac{-6.52}{-15.0} = 0.435 \, cm$$

$$h_b = -\frac{f_e d}{n f_1} = -\frac{-6.52}{-30.0} = -0.217 \, cm$$

The front focal length is

$$f_f = \frac{f_1 (d - n f_2)}{d - n (f_1 + f_2)} = \frac{-20.0 (1.00 + 15.0)}{1.00 - 1.50 (-30.0)} cm = -6.96 cm$$

and the back focal length is

$$f_b = \frac{f_2 (d - nf_1)}{d - n(f_1 + f_2)} = \frac{-10.0 (1.00 + 30.0)}{1.00 - 1.50 (-30.0)} cm = -6.74 cm$$

Check: $f_f = f_e - h_f$ and $f_b = f_e + h_b$. See Figure 4.42(c).

The effective focal length for this lens is negative ($f_e = -6.52\,cm$) giving a virtual image for all object distances. The effective image distance is

$$s_i = \frac{(-6.52\,cm)\,(10.0\,cm)}{16.5\,cm} = -3.95\,cm$$

The image is located 3.95 cm to the left of h_b. The magnification is

$$m = -\frac{s_i}{s_o} = -\frac{-3.95\,cm}{10.0\,cm} = 0.395$$

giving an upright virtual image that is 0.395 cm tall.

In part (a), the principal points for a biconvex lens were located about $\frac{1}{3}d$ on either side of the lens center. This is a general result when the index of the lens is 1.5.

Problem 4.48 Derive Equation 4.118 for the back principal point of a thick lens.

Problem 4.49 Repeat the calculations of Example 4.11 for the *meniscus* lens shown in Figure 4.42(d), with $R_1 = 10.0\,cm$, $R_2 = 5.00\,cm$, $d = 1.00\,cm$, and $n = 1.50$.

Problem 4.50 A certain thick lens has $R_1 = -10.0\,cm$, $R_2 = +5.0\,cm$, $d = 1.00\,cm$, and $n = 1.50$. Find the effective focal length, the front and back focal points, and principal points. If a $2.00\,cm$-high object is located $10.0\,cm$ in front of this lens, find the image location and image size.

4.11 *INTRODUCTION TO MATRIX METHODS IN PARAXIAL GEOMETRICAL OPTICS

In principle, images formed by any optical system may be located by tracing rays as they refract or reflect at all interfaces in the optical train. If we restrict the analysis to include only paraxial rays, a matrix approach can be used, as described in this section.

We begin by defining matrices for the processes of translation, refraction, and reflection. In addition to the paraxial approximation, we further assume that all optical elements have rotational symmetry and are aligned coaxially along a single optical axis.

4.11.1 The Translation Matrix

We define a ray according to the parameters illustrated in Figure 4.43. The ray is defined according to its intersection with a *reference plane*, as shown in Figures 4.43(a) and (b).

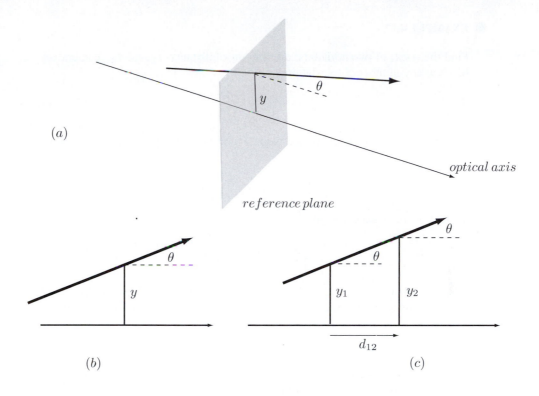

Figure 4.43   (a), (b) A ray passes through a reference plane that is normal to the optical axis. The ray is described by parameters y and θ, where y is the distance between the points of intersections of the reference plane with the optical axis and the ray, and θ is the angle the ray makes with the optical axis. (b) A ray passing through two reference planes separated by distance d_{12}.

The parameter y is the perpendicular distance between the optical axis and the intersection point, and the angle θ is the angle that the ray makes with the horizontal. Ray *translation* between two reference planes is illustrated in Figure 4.43(c). The rays are assumed to be *meridional*, meaning that they lie in the same plane as the optical axis. The ray parameter y varies between the two reference planes according to

$$y_2 = y_1 + d_{12} \tan \theta_1$$

Since we are using the paraxial approximation, $\tan \theta_1$ may be replaced with θ_1, giving

$$\begin{aligned} y_2 &= y_1 + d_{12}\theta_1 \\ \theta_2 &= \theta_1 \end{aligned} \tag{4.119}$$

Equations 4.119 may be rewritten as follows

$$\begin{pmatrix} y_2 \\ \theta_2 \end{pmatrix} = \begin{bmatrix} 1 & d_{12} \\ 0 & 1 \end{bmatrix} \begin{pmatrix} y_1 \\ \theta_1 \end{pmatrix} = M_T \begin{pmatrix} y_1 \\ \theta_1 \end{pmatrix} \tag{4.120}$$

where M_T is called the *translation matrix*, since it translates the ray between two reference planes within the same medium that are separated by horizontal distance d_{12}:

$$M_T = \begin{bmatrix} 1 & d_{12} \\ 0 & 1 \end{bmatrix} \tag{4.121}$$

■ **EXAMPLE 4.12**

Find the result of two horizontal translations of distance T_1 and T_2. Generalize the result to any number of translations.

Solution

Let the ray at an initial reference plane P_0 be given by $\begin{pmatrix} y_0 \\ \theta \end{pmatrix}$. The first translation gives

$$\begin{pmatrix} y_1 \\ \theta \end{pmatrix} = \begin{bmatrix} 1 & T_1 \\ 0 & 1 \end{bmatrix} \begin{pmatrix} y_0 \\ \theta \end{pmatrix}$$

and the second translation gives

$$\begin{pmatrix} y_2 \\ \theta \end{pmatrix} = \begin{bmatrix} 1 & T_2 \\ 0 & 1 \end{bmatrix} \begin{pmatrix} y_1 \\ \theta \end{pmatrix} = \begin{bmatrix} 1 & T_2 \\ 0 & 1 \end{bmatrix} \begin{bmatrix} 1 & T_1 \\ 0 & 1 \end{bmatrix} \begin{pmatrix} y_0 \\ \theta \end{pmatrix}$$

Thus, the net translation is given by the matrix product $M_{T_2} M_{T_1}$. Matrix multiplication is not commutative, so the order is important. Multiply the two matrices as follows[26]:

$$\begin{bmatrix} 1 & T_2 \\ 0 & 1 \end{bmatrix} \begin{bmatrix} 1 & T_1 \\ 0 & 1 \end{bmatrix} = \begin{bmatrix} 1+0 & T_1 + T_2 \\ 0+0 & 0+1 \end{bmatrix}$$

Notice that this could be done any number of times, so translations by the amounts T_1 followed by T_2, T_3, ... T_N would give

$$\begin{pmatrix} y_N \\ \theta \end{pmatrix} = M_N M_{N-1} ... M_1 \begin{pmatrix} y_0 \\ \theta \end{pmatrix} = \begin{bmatrix} 1 & \sum_{i=1}^{N} T_i \\ 0 & 1 \end{bmatrix} \begin{pmatrix} y_0 \\ \theta \end{pmatrix}$$

4.11.2 The Refraction Matrix

Rays refract at dielectric interfaces according to Snell's law

$$n_i \sin \theta_i = n_t \sin \theta_t$$

which for small angles, becomes

$$n_i \theta_i = n_t \theta_t$$

Figure 4.44 illustrates a ray that refracts at a spherical interface of positive radius R (the opposite curvature would have a negative value of R). The refraction occurs at a specific point; hence $y_1 = y_2$. According to the figure, $\theta_1 = \theta_i - \phi$, which in the paraxial approximation becomes

$$\theta_1 = \theta_i - \frac{y_1}{R}$$

Similarly,

$$\theta_2 = \theta_t - \frac{y_1}{R}$$

[26] See any introductory calculus text; e.g. Thomas [13].

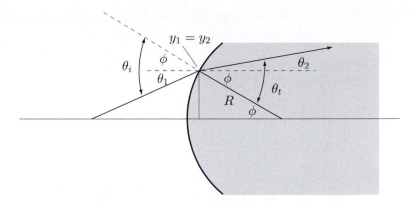

Figure 4.44   Refraction at a spherical interface with positive curvature.

Combining these with the paraxial form of Snell's law gives

$$\theta_2 = \frac{n_i}{n_t}\theta_i - \frac{y_1}{R} = \frac{n_i}{n_t}\left(\theta_1 + \frac{y_1}{R}\right) - \frac{y_1}{R} = -\frac{1}{R}\left(1 - \frac{n_i}{n_t}\right)y_1 + \frac{n_i}{n_t}\theta_1$$

This gives the following system of equations:

$$y_2 = y_1$$
$$\theta_2 = -\frac{1}{R}\left(1 - \frac{n_i}{n_t}\right)y_1 + \frac{n_i}{n_t}\theta_1 \qquad (4.122)$$

which may be rewritten as follows:

$$\begin{pmatrix} y_2 \\ \theta_2 \end{pmatrix} = \begin{bmatrix} 1 & 0 \\ -\frac{(n_t - n_i)}{n_t R} & \frac{n_i}{n_t} \end{bmatrix} \begin{pmatrix} y_1 \\ \theta_1 \end{pmatrix} = M_R \begin{pmatrix} y_1 \\ \theta_1 \end{pmatrix}$$

where M_R is the *refraction matrix*:

$$M_R = \begin{bmatrix} 1 & 0 \\ -\frac{(n_t - n_i)}{n_t R} & \frac{n_i}{n_t} \end{bmatrix} \qquad (4.123)$$

■ **EXAMPLE 4.13**

Use the refraction matrix to determine the matrix for a thin lens of focal length f.

Solution

We represent a thin lens as two refracting spherical surfaces with negligible translation in between. Thus

$$M_\ell = M_2 M_1 = \begin{bmatrix} 1 & 0 \\ -\frac{1 - n_\ell}{R_2} & \frac{n_\ell}{1} \end{bmatrix} \begin{bmatrix} 1 & 0 \\ -\frac{n_\ell - 1}{n_\ell R_1} & \frac{1}{n_\ell} \end{bmatrix} = \begin{bmatrix} 1 & 0 \\ -(n_\ell - 1)\left(\frac{1}{R_1} - \frac{1}{R_2}\right) & 1 \end{bmatrix}$$

By Equation 4.24, this becomes

$$M_\ell = \begin{bmatrix} 1 & 0 \\ -\frac{1}{f} & 1 \end{bmatrix} \qquad (4.124)$$

This is the matrix that represents a thin lens of focal length f.

_____

4.11.3   The Reflection Matrix

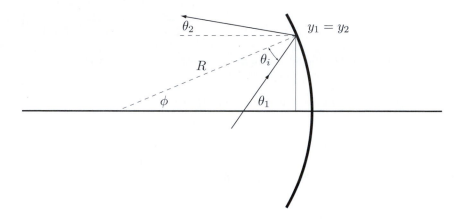

Figure 4.45   Reflection at a spherical interface.

Figure 4.45 illustrates reflection at a spherical surface. Like refraction, this occurs at a single point, so $y_1 = y_2$. From the law of reflection,

$$\theta_2 = \theta_i - \phi$$

and

$$\theta_1 = \theta_i + \phi$$

giving

$$\theta_2 = \theta_1 - 2\varphi$$

The spherical surface in Figure 4.44 is by convention positive, making the spherical surface in Figure 4.45 *negative*. Since this mirror is converging, we must include the minus sign explicitly. Using the small angle approximation, we have

$$\theta_2 = \theta_1 - 2\left(\frac{y_1}{-R}\right)$$

The system of equations that describe reflection at a spherical surface are thus

$$y_2 = y_1$$
$$\theta_2 = \frac{2}{R}y_1 + \theta_1 \qquad (4.125)$$

which may be rewritten as follows

$$\begin{pmatrix} y_2 \\ \theta_2 \end{pmatrix} = \begin{bmatrix} 1 & 0 \\ \frac{2}{R} & 1 \end{bmatrix} \begin{pmatrix} y_1 \\ \theta_1 \end{pmatrix} = M_{Refl} \begin{pmatrix} y_1 \\ \theta_1 \end{pmatrix}$$

where M_{Refl} is the *reflection matrix*

$$M_R = \begin{bmatrix} 1 & 0 \\ \frac{2}{R} & 1 \end{bmatrix} \tag{4.126}$$

Problem 4.51 Find a matrix that represents a thin, symmetric biconvex lens when both surfaces have a radius of curvature of $20.0\,cm$ in magnitude. Assume an index of refraction of 1.5. Repeat when the lens has a thickness of $1.00\,cm$.

4.11.4 The Ray Transfer Matrix

Taken as linear sequence of interfaces and translations, we can use the toolkit of matrices just defined to model optical systems of arbitrary complexity. Figure 4.46 illustrates a ray traversing between two reference planes P_1 and P_2. Each interface and translation yields new ray parameters; thus, the matrix multiplication proceeds in the following order:

$$\begin{pmatrix} y_2 \\ \theta_2 \end{pmatrix} = M_N M_{N-1} ... M_2 M_1 \begin{pmatrix} y_1 \\ \theta_1 \end{pmatrix} = M \begin{pmatrix} y_1 \\ \theta_1 \end{pmatrix} \tag{4.127}$$

where the matrix index increases as illustrated in Figure 4.46.

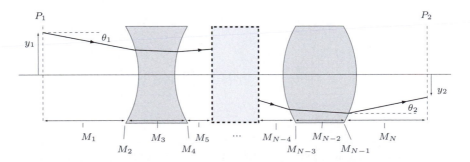

Figure 4.46 A ray travels between reference planes P_1 and P_2 in an optical system. The center box represents an arbitrary number of optical elements. Each interface and translation contributes a matrix with index that increases from left to right, as shown.

The matrix M in Equation 4.127 is called the *ray transfer matrix*. The elements of this 2×2 matrix can be used to determine many properties of the optical system. In order to refer to them, we label the elements of M with the letters A—D as follows:

$$M = \begin{bmatrix} A & B \\ C & D \end{bmatrix} \tag{4.128}$$

Figure 4.47(a) illustrates an optical system with a zero value for element A. In this case, Equation 4.46 gives

$$\begin{pmatrix} y_2 \\ \theta_2 \end{pmatrix} = \begin{bmatrix} A & B \\ C & D \end{bmatrix} \begin{pmatrix} y_1 \\ \theta_1 \end{pmatrix} = \begin{bmatrix} 0 & B \\ C & D \end{bmatrix} \begin{pmatrix} y_1 \\ \theta_1 \end{pmatrix}$$

This yields two equations, one of which is

$$y_2 = B\,\theta_1$$

We interpret this equation as follows: parallel input rays with angle θ_1 all pass through y_2 in reference plane P_2. Thus, in this case P_2 locates the back focal plane of the optical system, as illustrated in Figure 4.47(a). Similarly, if $D = 0$, then

$$\begin{pmatrix} y_2 \\ \theta_2 \end{pmatrix} = \begin{bmatrix} A & B \\ C & 0 \end{bmatrix} \begin{pmatrix} y_1 \\ \theta_1 \end{pmatrix}$$

and

$$\theta_2 = C\,y_1$$

In this case, all rays emanating from y_1 are transmitted with the same angle θ_2; thus the reference plane P_1 is located at the front focal plane of the optical system, as illustrated in Figure 4.47(d).

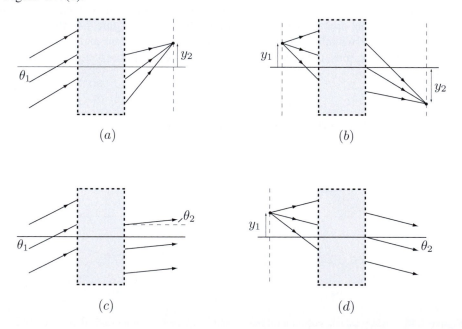

(a)

(b)

(c)

(d)

Figure 4.47   Consequences of zero values for the ray transfer matrix elements. (a) If $A = 0$, then P_2 is located at the back focal plane. (b) If $B = 0$, then P_2 is an image plane of an object located at P_1. (c) If $C = 0$, then the system behaves as a telescope. (d) If $D = 0$, then P_1 is located at the front focal plane.

If $C = 0$, then

$$\begin{pmatrix} y_2 \\ \theta_2 \end{pmatrix} = \begin{bmatrix} A & B \\ 0 & D \end{bmatrix} \begin{pmatrix} y_1 \\ \theta_1 \end{pmatrix} \Rightarrow \theta_2 = D\,\theta_1$$

Thus, entering parallel rays at angle θ_1 are transmitted as parallel rays at angle θ_2. As we have seen in Section 4.7.5, this is the case for a telescope with eyepiece adjusted for the relaxed eye. For this reason, systems with zero values of matrix element C are called *telescopic systems*.

If $B = 0$, we have

$$\begin{pmatrix} y_2 \\ \theta_2 \end{pmatrix} = \begin{bmatrix} A & 0 \\ C & D \end{bmatrix} \begin{pmatrix} y_1 \\ \theta_1 \end{pmatrix} \Rightarrow y_2 = A\,y_1$$

In this case, rays leaving y_1 at P_1 all pass through y_2 in P_2. Thus, reference plane P_2 is an image plane of an object located at P_1 (i.e., P_1 and P_2 are conjugate planes). Since $y_2 = A\,y_1$, the element A gives the magnification.

■ **EXAMPLE 4.14**

A lens combination consists of two thin, symmetric biconvex lenses separated by 5.00 cm. Let the front and back radii of lens 1 be 20.0 cm, and the front and back radii of lens 2 be 10.0 cm. Find the ray transfer matrix, and determine the location of the front and back focal points. Find the image location, size, and orientation when a 3.00 cm-high object is located 8.00 cm in front of lens 1.

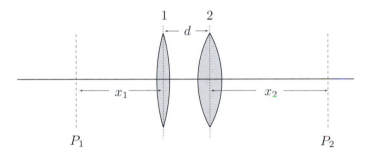

Figure 4.48   (a) A combination of two thin biconvex lenses.

Solution

Let P_1 be located a horizontal distance x_1 to the left of lens 1, and P_2 a horizontal distance x_2 to the right of lens 2, as illustrated in Figure 4.48. The ray transformation between P_1 and P_2 is given by

$$\begin{pmatrix} y_2 \\ \theta_2 \end{pmatrix} = M_7\,(M_6 M_5)\,M_4\,(M_3 M_2)\,M_1 \begin{pmatrix} y_1 \\ \theta_1 \end{pmatrix}$$

The matrices M_2 and M_3, enclosed in parentheses above, represent the back and front surfaces of lens 1. According to the results of Example 4.13, the product is

$$M_3 M_2 = \begin{bmatrix} 1 & 0 \\ -\frac{1}{f_1} & 1 \end{bmatrix}$$

where

$$\frac{1}{f} = (n-1)\left(\frac{1}{R_f} - \frac{1}{R_b}\right)$$

with R_f the radius of curvature of the front surface, and R_b that of the back. Lens 1 has $R_f = +20.0\,cm$, $R_b = -20.0\,cm$, and $n = 1.5$, giving

$$\frac{1}{f_1} = (1.5 - 1)\left(\frac{2}{20.0\,cm}\right)$$

giving $f_1 = 20.0\,cm$ for lens 1. Similarly, lens 2 has a focal length of $f_2 = 10.0\,cm$. Notice that these lenses are identical to those used in Examples 4.5 and 4.3(a). The ray transformation is thus given by

$$\begin{pmatrix} y_2 \\ \theta_2 \end{pmatrix} = M_7 \begin{bmatrix} 1 & 0 \\ -\frac{1}{f_2} & 1 \end{bmatrix} M_4 \begin{bmatrix} 1 & 0 \\ -\frac{1}{f_1} & 1 \end{bmatrix} M_1 \begin{pmatrix} y_1 \\ \theta_1 \end{pmatrix}$$

The matrix M_4 represents a translation of d to go from lens 1 to lens 2:

$$\begin{pmatrix} y_2 \\ \theta_2 \end{pmatrix} = M_7 \begin{bmatrix} 1 & 0 \\ -\frac{1}{f_2} & 1 \end{bmatrix} \begin{bmatrix} 1 & d \\ 0 & 1 \end{bmatrix} \begin{bmatrix} 1 & 0 \\ -\frac{1}{f_1} & 1 \end{bmatrix} M_1 \begin{pmatrix} y_1 \\ \theta_1 \end{pmatrix}$$

$$= M_7 \begin{bmatrix} 1 & d \\ -\frac{1}{f_2} & 1 - \frac{d}{f_2} \end{bmatrix} \begin{bmatrix} 1 & 0 \\ -\frac{1}{f_1} & 1 \end{bmatrix} M_1 \begin{pmatrix} y_1 \\ \theta_1 \end{pmatrix}$$

$$= M_7 \begin{bmatrix} 1 - \frac{d}{f_1} & d \\ -\frac{1}{f_2} - \frac{1}{f_1}\left(1 - \frac{d}{f_2}\right) & 1 - \frac{d}{f_2} \end{bmatrix} M_1 \begin{pmatrix} y_1 \\ \theta_1 \end{pmatrix}$$

$$= M_7 \begin{bmatrix} 1 - \frac{d}{f_1} & d \\ -\frac{1}{f_e} & 1 - \frac{d}{f_2} \end{bmatrix} M_1 \begin{pmatrix} y_1 \\ \theta_1 \end{pmatrix}$$

where

$$\frac{1}{f_e} = \frac{1}{f_1} + \frac{1}{f_2} - \frac{d}{f_1 f_2}$$

gives the effective focal length of the combination (see Equation 4.39). In our case, this is equal to

$$\frac{1}{f_e} = \frac{1}{20.0\,cm} + \frac{1}{10.0\,cm} - \frac{5.00\,cm}{(20.0\,cm)(10.0\,cm)}$$

giving $f_e = 8.00\,cm$, as in Example 4.5. Thus, the transformation from P_1 to P_2 is given by

$$\begin{pmatrix} y_2 \\ \theta_2 \end{pmatrix} = M_7 \begin{bmatrix} 1 - \frac{5}{20} & 5 \\ -\frac{1}{8} & 1 - \frac{5}{10} \end{bmatrix} M_1 \begin{pmatrix} y_1 \\ \theta_1 \end{pmatrix}$$

$$= M_7 \begin{bmatrix} \frac{3}{4} & 5 \\ -\frac{1}{8} & \frac{1}{2} \end{bmatrix} M_1 \begin{pmatrix} y_1 \\ \theta_1 \end{pmatrix}$$

The matrix M_1 represents the initial translation by x_1, and M_7 represents the final translation x_2:

$$\begin{pmatrix} y_2 \\ \theta_2 \end{pmatrix} = \begin{bmatrix} 1 & x_2 \\ 0 & 1 \end{bmatrix} \begin{bmatrix} \frac{3}{4} & 5 \\ -\frac{1}{8} & \frac{1}{2} \end{bmatrix} \begin{bmatrix} 1 & x_1 \\ 0 & 1 \end{bmatrix} \begin{pmatrix} y_1 \\ \theta_1 \end{pmatrix}$$

$$= \begin{bmatrix} \frac{3}{4} - \frac{x_2}{8} & 5 + \frac{x_2}{2} \\ -\frac{1}{8} & \frac{1}{2} \end{bmatrix} \begin{bmatrix} 1 & x_1 \\ 0 & 1 \end{bmatrix} \begin{pmatrix} y_1 \\ \theta_1 \end{pmatrix}$$

$$= \begin{bmatrix} \frac{3}{4} - \frac{x_2}{8} & x_1\left(\frac{3}{4} - \frac{x_2}{8}\right) + 5 + \frac{x_2}{2} \\ -\frac{1}{8} & \frac{1}{2} - \frac{x_2}{8} \end{bmatrix} \begin{pmatrix} y_1 \\ \theta_1 \end{pmatrix}$$

$$= \begin{bmatrix} A & B \\ C & D \end{bmatrix} \begin{pmatrix} y_1 \\ \theta_1 \end{pmatrix}$$

where A—D represent the elements of the ray transfer matrix. The back focal plane is found by setting $A = 0$ and solving for x_2

$$\frac{x_2}{8} = \frac{3}{4}$$

giving $f_b = 6.00\,cm$, as found in Examples 4.5 and 4.3(a). Similarly, setting $D = 0$ and solving for x_1 gives $f_f = 4.00\,cm$ as before. If we let $x_1 = 8.00\,cm$ and set $B = 0$, we obtain

$$8\left(\frac{3}{4} - \frac{x_2}{8}\right) = -5 - \frac{x_2}{2}$$

giving $x_2 = 22.0\,cm$, as in Example 4.5. Evaluating A using $x_2 = 22.0\,cm$ gives a magnification of -2.00, as before. Thus, the $6.00\,cm$ image is inverted. See Figure 4.19.

4.11.5 Location of Cardinal Points for an Optical System

In Section 4.10.1, we found the cardinal points for a two-lens combination. From this analysis, it should seem evident that following a similar approach for all but the simplest optical systems can be algebraically difficult. In this regard, the advantages of the matrix approach become especially evident. A ray transfer matrix contains a complete description of the associated optical system, and in this section we describe how to use the associated matrix elements to easily determine the cardinal points of an optical system of arbitrary complexity.

It will facilitate our analysis to choose the input reference plane P_1 to coincide with the first surface of the optical system, and the output reference plane P_2 to coincide with the final surface. We refer to the matrix that utilizes these reference planes as the *system matrix* M_s.

Figure 4.49 illustrates an optical system bounded by P_1 and P_2 at the initial and final vertex[27]. We assume, for simplicity, that the region preceding P_1 and following P_2 have the same index of refraction[28]. As in Section 4.10.1, the rays in Figures 4.49(a) and (b) define the cardinal points: in Figure 4.49(a) an incoming parallel ray is redirected at back principal plane h_b so that it passes through the back focal point f_b, and in Figure 4.49(b) a ray first passing through front focal point f_f is redirected parallel to the optical axis at front principal plane h_f.

Begin with Figure 4.49(a). Equation 4.127 gives

$$\begin{pmatrix} y_2 \\ \theta_2 \end{pmatrix} = M_s \begin{pmatrix} y_1 \\ 0 \end{pmatrix} = \begin{bmatrix} A & B \\ C & D \end{bmatrix} \begin{pmatrix} y_1 \\ 0 \end{pmatrix} = \begin{pmatrix} A\,y_1 \\ C\,y_1 \end{pmatrix}$$

where M_s is the system matrix, and where the incoming parallel ray is characterized by a zero value of θ_1. This gives the following set of equations:

$$y_2 = A\,y_1$$

---

[27] The *vertex* of an optical element is located at its point of intersection with the optical axis.
[28] For more information regarding cardinal points for optical systems in which these indices are different, see Gerrard and Burch [6].

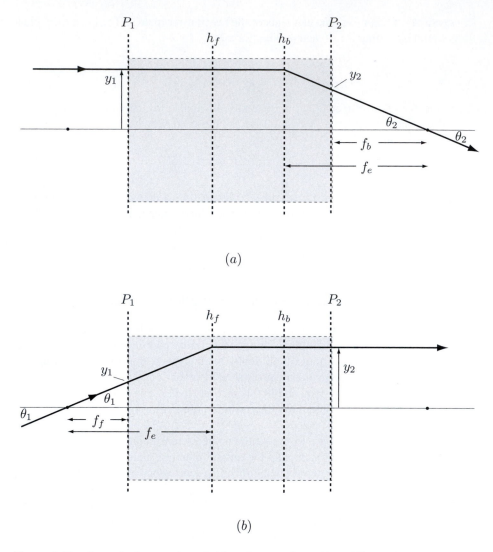

(a)

(b)

Figure 4.49   An optical system bounded by reference planes P_1 and P_2. (a) The incoming ray is parallel to the optical axis. (b) The outgoing ray is parallel to the optical axis.

$$\theta_2 = C\, y_1$$

Note that the angle θ for any ray is derived from the ray slope, so θ_2 in Figure 4.49(a) is *negative*. From the figure, and using the small angle approximation,

$$\theta_2 = -\frac{y_1}{f_e}$$

where f_e is the effective focal length, as defined in Section 4.10.1. Thus,

$$f_e = -\frac{1}{C} \tag{4.129}$$

This was confirmed for a two-lens combination in Example 4.14. From the similar triangles,

$$\frac{f_b}{y_2} = \frac{f_e}{y_1}$$

giving

$$y_2 = -C\,f_b\,y_1 = A\,y_1$$

Thus

$$f_b = -\frac{A}{C} \qquad (4.130)$$

We must maintain the sign convention of Section 4.10.1 which assigns negative values to principal points that lie to the left of the associated reference vertex. Thus, by the same convention that gives Equation 4.118,

$$h_b = -(f_e - f_b)$$

giving

$$h_b = -\frac{A-1}{C} \qquad (4.131)$$

In Figure 4.49(b),

$$\begin{pmatrix} y_2 \\ 0 \end{pmatrix} = M_s \begin{pmatrix} y_1 \\ \theta_1 \end{pmatrix} = \begin{bmatrix} A & B \\ C & D \end{bmatrix} \begin{pmatrix} y_1 \\ \theta_1 \end{pmatrix}$$

giving

$$y_2 = A\,y_1 + B\,\theta_1$$
$$0 = C\,y_1 + D\,\theta_1$$

Since

$$\theta_1 = \frac{y_1}{f_f}$$

$$f_f = -\frac{D}{C} \qquad (4.132)$$

As in Section 4.10.1,

$$h_f = f_e - f_f$$

Thus

$$h_f = \frac{D-1}{C} \qquad (4.133)$$

Equations 4.129 — 4.133 are summarized in Table 4.2..

Table 4.2.   Cardinal points determined by system matrix elements.

	Effective Focal Length	$f_e = -\frac{1}{C}$	
Front Principal Plane	$h_f = \frac{D-1}{C}$	Back Principal Plane	$h_b = -\frac{A-1}{C}$
Front Focal Point	$f_f = -\frac{D}{C}$	Back Focal Point	$f_b = -\frac{A}{C}$

In ray diagrams with single thin lenses, rays that pass through the center of the lens are undeflected. For thick lenses and lens combinations, front and back *nodal points* can be defined. A ray that enters the system and passes through the front nodal point emerges in a parallel direction as though it originated at the back nodal point. For more information, see Gerrard and Burch, *Introduction to Matrix Methods in Optics* [6]).

■ **EXAMPLE 4.15**

Use the system matrix of a two-lens combination to verify the formulas for cardinal points found in Section 4.10.1.

Solution

In Example 4.14, the system matrix for a two-lens combination was found to be

$$
M_s = \begin{bmatrix} 1 - \frac{d}{f_1} & d \\ -\frac{1}{f_e} & 1 - \frac{d}{f_2} \end{bmatrix}
$$

Clearly, $f_e = -\frac{1}{C}$. The front focal point is given by

$$
f_f = -\frac{D}{C} = f_e \left(1 - \frac{d}{f_2}\right) = \left(\frac{f_1 f_2}{f_1 + f_2 - d}\right)\left(\frac{f_2 - d}{f_2}\right) = \frac{f_1 (d - f_2)}{d - (f_1 + f_2)}
$$

as in Equation 4.32. Similarly, the back focal point is located by

$$
f_b = -\frac{A}{C} = f_e \left(1 - \frac{d}{f_1}\right) = \left(\frac{f_1 f_2}{f_1 + f_2 - d}\right)\left(\frac{f_1 - d}{f_1}\right) = \frac{f_2 (d - f_1)}{d - (f_1 + f_2)}
$$

as in Equation 4.31. The front principal plane is located at

$$
h_f = \frac{D - 1}{C} = -f_e \left(1 - \frac{d}{f_2} - 1\right) = \frac{f_e \, d}{f_2}
$$

as in Equation 4.40. Similarly, the back principal plane is located by

$$
h_b = -\frac{A - 1}{C} = f_e \left(1 - \frac{d}{f_1} - 1\right) = -\frac{f_e d}{f_1}
$$

as in Equation 4.41.

■ **EXAMPLE 4.16**

A thick biconvex lens has $R_1 = 10.0 \, cm$, $R_2 = -10.0 \, cm$, thickness $d = 1.00 \, cm$ and is made from a material with index of refraction 1.50. Find the system matrix, and use this to determine the front and back focal points and principal planes. Determine the image location and size when a $1.00 \, cm$-high object is located a distance $20.0 \, cm$ in front of the front principal plane. Compare to the results of Example 4.11(a).

Solution

The system matrix results from the combination of two refracting surfaces and a translation of d. Proceed as in Example 4.13:

$$
M_s = \begin{bmatrix} 1 & 0 \\ -\frac{(1-n)}{R_2} & n \end{bmatrix} \begin{bmatrix} 1 & d \\ 0 & 1 \end{bmatrix} \begin{bmatrix} 1 & 0 \\ -\frac{(n-1)}{n R_1} & \frac{1}{n} \end{bmatrix}
$$

Substitution gives

$$M_s = \begin{bmatrix} 1 & 0 \\ -\frac{1}{20} & 1.5 \end{bmatrix} \begin{bmatrix} 1 & 1 \\ 0 & 1 \end{bmatrix} \begin{bmatrix} 1 & 0 \\ -\frac{1}{30} & \frac{1}{1.5} \end{bmatrix} = \begin{bmatrix} 0.967 & 0.667 \\ -0.0983 & 0.967 \end{bmatrix}$$

Using Table 4.2., we find

$$f_e = -\frac{1}{C} = -\frac{1}{-0.0983} = 10.2\, cm$$

$$h_f = \frac{D-1}{C} = \frac{0.967-1}{-0.0983} = 0.336\, cm$$

$$h_b = -\frac{A-1}{C} = -0.336\, cm$$

$$f_f = -\frac{D}{C} = -\frac{0.967}{-0.0983} = 9.84\, cm$$

$$f_b = -\frac{A}{C} = 9.84\, cm$$

which, to within round-off errors, agrees with the results of Example 4.11(a). To locate the image using matrix methods, we must use the distance from the object to the first optical surface:

$$s_{o1} = s_o - h_f = 19.7\, cm$$

Proceed as in Example 4.14:

$$M = \begin{bmatrix} 1 & s_{o1} \\ 0 & 1 \end{bmatrix} \begin{bmatrix} 0.967 & 0.667 \\ -0.0983 & 0.967 \end{bmatrix} \begin{bmatrix} 1 & s_{i2} \\ 0 & 1 \end{bmatrix}$$

$$= \begin{bmatrix} 1 & 19.7 \\ 0 & 1 \end{bmatrix} \begin{bmatrix} 0.967 & 0.667 \\ -0.0983 & 0.967 \end{bmatrix} \begin{bmatrix} 1 & s_{i2} \\ 0 & 1 \end{bmatrix}$$

$$= \begin{bmatrix} -0.970 & 19.7 \\ -0.0983 & 0.967 \end{bmatrix} \begin{bmatrix} 1 & s_{i2} \\ 0 & 1 \end{bmatrix}$$

$$= \begin{bmatrix} -0.970 & 19.7 - 0.970\, s_{i2} \\ -0.0983 & 0.967 - 0.0983\, s_{i2} \end{bmatrix}$$

To find the image distance, set element B equal to zero and solve, to give

$$s_{i2} = \frac{19.7}{0.970} = 20.3\, cm$$

As a check, use $s_o = 20.0\, cm$ and $f_e = 10.2\, cm$

$$s_i = \frac{f_e s_o}{s_o - f_e} = \frac{(10.2\, cm)\,(20.0\, cm)}{(20.0 - 10.2)\, cm} = 20.8\, cm$$

Thus

$$s_{i2} = s_i + h_b = 20.8\, cm + (-0.336\, cm) = 20.5\, cm$$

which agrees with the above to within round-off error. The magnification is

$$m = -\frac{s_i}{s_o} = -\frac{20.8}{20.0} = -1.04$$

Thus, the $1.04\, cm$-high image is inverted. A ray diagram is given in Figure 4.50.

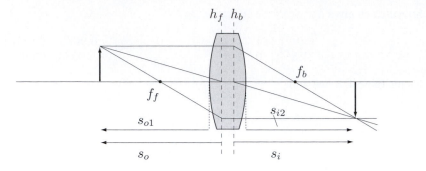

Figure 4.50   An image formed by a thick converging lens in air.

Problem 4.52   It can be shown that the *determinant* of the system matrix is unity, provided that the index of refraction is the same on both sides of the optical system (See, for example, Gerrard and Burch, *Introduction to Matrix Methods in Optics* [6]). The determinant of a 2×2 matrix is defined as

$$\det \begin{bmatrix} A & B \\ C & D \end{bmatrix} = AD - BC$$

Show that this is true for each case below. Assume in each case that the index on both sides of the optical system is 1.00. This can be used as a convenient computational check.

    a) A thin lens

    b) A thick lens of thickness d

    c) A plane slab of thickness d

    d) The optical system of Example 4.15

    e) The optical system of Example 4.16

Problem 4.53   A thick biconvex lens has $R_1 = 10.0\,cm$, $R_2 = -5.0\,cm$, thickness $d = 1.00\,cm$ and is made from a material with index of refraction 1.50. Find the system matrix, and use this to determine the front and back focal points and principal planes. Determine the image location and size when a $1.00\,cm$-high object is located a distance $10.0\,cm$ in front of the front principal plane.

Problem 4.54   A thick biconcave lens has $R_1 = -10.0\,cm$, $R_2 = +5.0\,cm$, thickness $d = 1.00\,cm$ and is made from a material with index of refraction 1.50. Find the system matrix, and use this to determine the front and back focal points and principal planes. Determine the image location and size when a $1.00\,cm$-high object is located a distance $10.0\,cm$ in front of the front principal plane.

Problem 4.55   A thick meniscus lens has $R_1 = 10.0\,cm$, $R_2 = 5.0\,cm$, thickness $d = 1.00\,cm$ and is made from a material with index of refraction 1.50. Find the system matrix, and use this to determine the front and back focal points and principal planes.

Determine the image location and size when a $1.00\,cm$-high object is located a distance $10.0\,cm$ in front of the front principal plane.

Problem 4.56   Show that the system matrix for a vertical parallel slab of index n and thickness L is given by

$$M_s = \begin{bmatrix} 1 & \frac{L}{n} \\ 0 & 1 \end{bmatrix}$$

Show that an incoming ray transmits through the slab in a direction that is parallel to the incoming ray direction, but displaced downward vertically by

$$\Delta y = \left(\frac{n-1}{n} \right) L\theta$$

where θ is the angle of the incoming ray. Include a sketch.

Problem 4.57   Repeat Example 4.14 when each biconvex lens is given a thickness of $1.00\,cm$.

Problem 4.58   Repeat Problem 4.8 on page 135 using matrix methods.

Problem 4.59   A *beam expander* consists of two lenses of focal lengths f_1 and f_2 separated by a positive distance d, with $d = f_1 + f_2$. Show that the system matrix is given by

$$M_s = \begin{bmatrix} -\frac{f_2}{f_1} & f_1 + f_2 \\ 0 & -\frac{f_1}{f_2} \end{bmatrix}$$

Show that an input collimated beam of width w_1 is transformed into an output collimated beam of width w_2, with

$$w_2 = \frac{f_2}{f_1} w_1$$

Draw two designs, one where both lenses are converging, and another where one lens is diverging. In both cases, trace incoming parallel rays as they transmit through the beam expander.

Additional Problems

Problem 4.60   A spherical concave mirror has a focal point located $10.0\,cm$ from the center of the mirror. Find the image distance, image height, and magnification for a $2.00\,cm$-high object whose object distance is (a) $20.0\,cm$ and (b) $5.00\,cm$. Draw a ray diagram for each case.

Problem 4.61   A spherical convex mirror has a focal point located $10.0\,cm$ from the center of the mirror. Find the image distance, image height, and magnification for a $2.00\,cm$-high object whose object distance is (a) $20.0\,cm$ and (b) $5.00\,cm$. Draw a ray diagram for each case.

Problem 4.62   A thin convex lens has a focal point located $10.0\,cm$ from the center of the lens. Find the image distance, image height, and magnification for a $2.00\,cm$-high object whose object distance is (a) $20.0\,cm$ and (b) $5.00\,cm$. Draw a ray diagram for each case.

Problem 4.63   A thin concave lens has a focal point located $10.0\,cm$ from the center of the lens. Find the image distance, image height, and magnification for a $2.00\,cm$-high object whose object distance is (a) $20.0\,cm$ and (b) $5.00\,cm$. Draw a ray diagram for each case.

Problem 4.64   Find the magnifying power of a $10.0\,cm$ focal length converging lens when the virtual image is located at (a) the near point of the eye and (b) infinity.

Problem 4.65   A certain microscope has a tube length of $L = 16.0\,cm$ and a single objective lens. What would the focal length of the objective be if a 20x eyepiece is to give a final magnification of 400?

Problem 4.66   A telescope uses a $25.4\,cm$-diameter $f/10$ objective. What is the focal length of an eyepiece that will give an overall magnification of 200?

Problem 4.67   Find the effective focal lengths, the front and back focal points, and principal points for the following lenses, all made of glass with index 1.50. In each case, find the image location and image size for a $2.00\,cm$-high object with object distance $s_o = 10.0\,cm$. (a) $R_1 = -10.0\,cm$, $R_2 = +10.0\,cm$, and $d = 1.00\,cm$. (b) $R_1 = -5.00\,cm$, $R_2 = +10.0\,cm$, and $d = 1.00\,cm$. (c) $R_1 = +5.0\,cm$, $R_2 = -10.0\,cm$, and $d = 1.00\,cm$. (d) $R_1 = +5.0\,cm$, $R_2 = +10.0\,cm$, and $d = 1.00\,cm$. In each case, draw a diagram of the lens and indicate the location of the principal points.

Appendix: Calculation of the Jeans Number

Begin by considering a cubic box of width L with perfectly reflecting walls. For a standing wave to exist within the cavity, L must be an integral number of half wavelengths:

$$L = n_x \frac{\lambda}{2}$$

with $n_x = 1, 2, 3, ...$ The corresponding frequency is

$$f = \frac{c}{\lambda} = \frac{c}{2L} n_x$$

The waves need not be aligned with the sides of the box. A standing wave with arbitrary orientation within the cavity has frequency

$$f = \frac{c}{\lambda} = \frac{c}{2L} \sqrt{n_x^2 + n_y^2 + n_z^2}$$

The possible values of n_x, n_y, and n_z can be visualized as points in a three-dimensional space. Any particular point lies a distance from the origin given by

$$R = \sqrt{n_x^2 + n_y^2 + n_z^2}$$

Thus,

$$f = \frac{c}{\lambda} = \frac{c}{2L} R$$

or, expressing the radius as a function of frequency, we obtain

$$R = \frac{2L}{c} f$$

The number of available modes of frequency less than f is equal to the number of points within the radius R. If this number is large, we can approximate this with the volume within the first octant: $\frac{1}{8} \frac{4\pi}{3} R^3$. Since there are two polarization states, we multiply by an additional factor of 2:

$$N = 2 \frac{1}{8} \frac{4\pi}{3} \left(\frac{2L}{c} \right)^3 f^3$$

Substituting $V = L^3$ gives

$$N = \frac{8\pi V}{3c^3} f^3$$

Differentiate this expression to obtain the number of modes per unit frequency interval:

$$\frac{dN}{df} = \frac{8\pi V}{c^3} f^2$$

The Jeans number is the number of modes per unit frequency interval per unit volume (Equation 4.76):

$$n(f) = \frac{8\pi f^2}{c^3}$$

CHAPTER 5

SUPERPOSITION AND INTERFERENCE

Twinkle, twinkle, little star.
How I wonder what you are...

—Jane Taylor

Contents

5.1 INTRODUCTION

In superposition, waves are combined, and when waves are combined they interfere. We begin by considering the superposition of harmonic monochromatic waves. The mathematical description of superposition is greatly facilitated by the ideas of *Fourier Analysis*, and we shall take time to develop these ideas fully. In doing so, we will discover that waves must be modeled as *packets* with a finite spread in frequency and wavelength, and an associated group and phase velocity. There are many applications of *interferometry* that are not only technically useful, but that also reveal some of the most subtle aspects of electromagnetic waves and photons. Specific applications of thin film technology are discussed for both single and multiple layers. The effects of interference provide the foundation for nearly all aspects of physical optics, including effects of diffraction and image formation to be discussed in subsequent chapters.

5.2 SUPERPOSITION OF HARMONIC WAVES

A wave is any function that satisfies the differential wave equation. As discussed in Section 1.5, the superposition principle guarantees that we may combine waves algebraically to give other valid wave solutions. Figure 5.1 illustrates the algebraic combination of two harmonic waves with equal amplitude, frequency, and wavelength; in other words, the two wave being combined differ only in phase. As discussed in Section 1.4, we may plot a harmonic traveling wave as a function of either position or time, so the horizontal axes of the plots of Figure 5.1 are not labeled to allow for either possibility. When waves are combined *in phase*, they combine to give a larger amplitude wave, and when they are $\pi \, rad$ out of phase, they tend to cancel. The general term for this is *interference*, with the situation of Figure 5.1(a) referred to as *constructive interference*, and that of Figure 5.1(e) referred to as *destructive interference*.

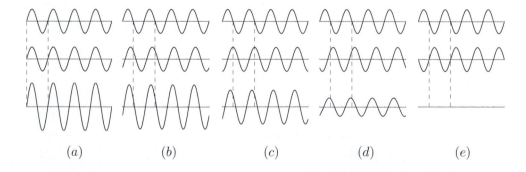

(a) $\quad\quad\quad$ (b) $\quad\quad\quad$ (c) $\quad\quad\quad$ (d) $\quad\quad\quad$ (e)

Figure 5.1 In each column, the top wave (wave 1) sums algebraically with the middle wave (wave 2) to give the bottom wave. In each case, waves 1 and 2 have the same amplitude and wavelength. (a) Waves 1 and 2 are in phase, giving constructive interference. (b)—(d) Wave 2 is shifted in phase by $\pi/4$, $\pi/2$ and $3\pi/4$ relative to wave 1. (e) Wave 2 is shifted in phase by π relative to wave 1, giving destructive interference. Since the amplitudes of waves 1 and 2 are equal, the destructive interference in (e) is complete.

5.3 INTERFERENCE BETWEEN TWO MONOCHROMATIC ELECTROMAGNETIC WAVES

We begin our discussion of interference by examining the combination of two traveling monochromatic electromagnetic waves. As we will see later, no source is perfectly monochromatic; nevertheless, this assumption allows us to utilize the harmonic wave solutions to Maxwell's equations determined in Chapter 2, and will provide an intuitive starting point that can be refined as we proceed.

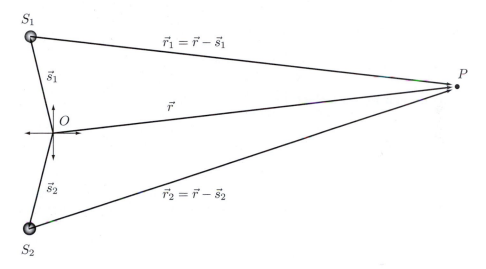

Figure 5.2   Two sources S_1 and S_2 combine at P.

Figure 5.2 shows two sources S_1 and S_2, located by position vectors $\vec{s}_1$ and $\vec{s}_2$ relative to an arbitrary coordinate origin labeled O. Also shown is an observation point P (located by position vector $\vec{r}$ relative to O), where the superposition will be observed. The nature of the two sources S_1 and S_2 are at this point completely arbitrary except that each is monochromatic. In particular, the wavefronts emitted by S_1 and S_2 are arbitrary. So, for example, S_1 and S_2 could be two independent point sources that give spherical wavefronts at P with radius of curvature equal to the separation between P and each source. Or, they could be two independent laser sources, each with a highly-collimated beam that propagates from each source to P.

Let the disturbance at P due to source S_1 be the harmonic traveling electromagnetic wave given by

$$\vec{E}_1 = \vec{E}_{01}e^{i\left(\vec{k}_1\cdot\vec{r}_1 - \omega_1 t + \varphi_1\right)} \tag{5.1}$$

where $\vec{E}_{01}$ determines the polarization, $\vec{k}_1$ is the wavenumber with magnitude $k_1 = 2\pi/\lambda_1$, λ_1 is the wavelength, ω_1 is the angular frequency, and φ_1 is the initial phase. We assume that the polarization vector $\vec{E}_{01}$ is real, and thus that φ represents the entire initial phase of the wave. If $\vec{s}_1$ locates S_1 and $\vec{r}$ locates P, then the vector $\vec{r}_1 = \vec{r} - \vec{s}_1$ represents the path from S_1 to P (see Figure 5.2). Similarly, wave 2 that travels from S_2 to P is given by

$$\vec{E}_2 = \vec{E}_{02}e^{i\left(\vec{k}_2\cdot\vec{r}_2 - \omega_2 t + \varphi_2\right)} \tag{5.2}$$

where we again assume that $\vec{E}_{02}$ is real. It will facilitate notation to make the following substitutions:

$$\vec{E}_1 = \vec{E}_{01}e^{i\left(\vec{k}_1 \cdot \vec{r}_1 - \omega_1 t + \varphi_1\right)} = \vec{E}_{01}e^{i\delta_1}$$

$$\vec{E}_2 = \vec{E}_{02}e^{i\left(\vec{k}_2 \cdot \vec{r}_2 - \omega_2 t + \varphi_2\right)} = \vec{E}_{02}e^{i\delta_2}$$

with

$$\delta_1 = \vec{k}_1 \cdot \vec{r}_1 - \omega_1 t + \varphi_1 \tag{5.3}$$

and

$$\delta_2 = \vec{k}_2 \cdot \vec{r}_2 - \omega_2 t + \varphi_2 \tag{5.4}$$

The combined disturbance at P is then

$$\vec{E} = \vec{E}_1 + \vec{E}_2 = \vec{E}_{01}e^{i\delta_1} + \vec{E}_{02}e^{i\delta_2} \tag{5.5}$$

5.3.1   Linear Power Detection

If the disturbance at P is measured by a *linear power detector* such as a photodetector or photomultiplier, then *irradiance* is detected, which according to Section 2.3.3 is proportional to the square of the total field $\vec{E}$. For this reason, such detectors are commonly called *square-law detectors*. Each field is a vector, so we compute the square as a scalar product. Since we are using the complex representation, we must be careful to compute the *magnitude squared* of the total field $\vec{E}$. The square of the total field is given by

$$
\begin{aligned}
E^2 = \vec{E} \cdot \vec{E}^* &= \left(\vec{E}_1 + \vec{E}_2\right) \cdot \left(\vec{E}_1 + \vec{E}_2\right)^* \\
&= \left(\vec{E}_{01}e^{i\delta_1} + \vec{E}_{02}e^{i\delta_2}\right) \cdot \left(\vec{E}_{01}e^{i\delta_1} + \vec{E}_{02}e^{i\delta_2}\right)^* \\
&= \left(\vec{E}_{01}e^{i\delta_1} + \vec{E}_{02}e^{i\delta_2}\right) \cdot \left(\vec{E}_{01}e^{-i\delta_1} + \vec{E}_{02}e^{-i\delta_2}\right) \\
&= E_{01}^2 + E_{02}^2 + \vec{E}_{01} \cdot \vec{E}_{02}\left(e^{i(\delta_2 - \delta_1)} + e^{-i(\delta_2 - \delta_1)}\right) \\
&= E_{01}^2 + E_{02}^2 + 2\vec{E}_{01} \cdot \vec{E}_{02}\cos\left(\delta_2 - \delta_1\right)
\end{aligned}
\tag{5.6}
$$

According to Section 2.3.3, the irradiance is given by

$$I = \epsilon v \frac{E^2}{2} = \epsilon v \frac{E_{01}^2}{2} + \epsilon v \frac{E_{02}^2}{2} + \epsilon v \vec{E}_{01} \cdot \vec{E}_{02}\cos\left(\delta_2 - \delta_1\right)$$

or

$$I = I_1 + I_2 + \epsilon v \left(\vec{E}_{01} \cdot \vec{E}_{02}\right)\cos\delta \tag{5.7}$$

where $\delta = \delta_2 - \delta_1$ is the *phase difference* between wave 1 and wave 2.

The last term in Equation 5.7 determines the interference effects. Note first that if $\vec{E}_{01}$ and $\vec{E}_{02}$ are perpendicular, then $\vec{E}_{01} \cdot \vec{E}_{02} = 0$. Thus *crossed polarizations do not interfere*. If $\vec{E}_{01}$ and $\vec{E}_{02}$ are parallel, then

$$I = I_1 + I_2 + \epsilon v \left(E_{01}E_{02}\right)\cos\delta$$

Since $I_1 = \epsilon v \frac{E_{01}^2}{2}$,

$$E_{01} = \sqrt{\frac{2I_1}{\epsilon v}}$$

with a similar relation for E_{02}. Thus, parallel polarizations gives

$$I = I_1 + I_2 + 2\sqrt{I_1 I_2} \cos \delta \tag{5.8}$$

The last term in Equation 5.8 is called the *interference term*.

When $\delta = 2m\pi$ with $m = 0, \pm 1, \pm 2, ...$, I in Equation 5.8 is a maximum, and the interference is constructive. Similarly, when δ is an odd-integer multiple of π, the interference is destructive.

5.3.2   Interference Between Beams with the Same Frequency

If $\omega_1 = \omega_2 = \omega$, then Equations 5.3 and 5.4 become

$$\delta_1 = \vec{k}_1 \cdot \vec{r}_1 - \omega t + \varphi_1 \tag{5.9}$$

and

$$\delta_2 = \vec{k}_2 \cdot \vec{r}_2 - \omega t + \varphi_2 \tag{5.10}$$

with the phase difference δ given by

$$\delta = \delta_2 - \delta_1 = \left(\vec{k}_2 \cdot \vec{r}_2 - \vec{k}_1 \cdot \vec{r}_1 \right) + (\varphi_2 - \varphi_1) \tag{5.11}$$

Note that in this case, the phase difference δ is constant in time.

5.3.2.1   *Young's Double-Slit Experiment*   As noted in Section 2.4, this important experiment by Young[1] established the wave properties of light. Figure 5.3 shows the experimental arrangement. In order to observe interference, Young found it necessary to illuminate the double-slit aperture with light that had first passed through a narrow single-slit aperture, as shown in Figure 5.3(b). Intuitively, light that passes through S creates extended wavefronts that illuminate both S_1 and S_2, creating electromagnetic disturbances at both slits that are *correlated* in time. Interference is not observed if the double-slit aperture is illuminated by an extended thermal source; it is a testament to Young's physical and experimental intuition that he knew to include the aperture S in the experimental design.

To analyze Young's experiment, we assume that S is illuminated with plane, monochromatic electromagnetic waves. This creates disturbances at both slits that propagate toward an observation point y. In order to arrive at y, the disturbance from S_1 travels the distance r_1, and the disturbance from S_2 travels a distance r_2. If we assume that each slit is so narrow that we can treat its cross section as a single point (see Figure 5.3), then the dot products in the phase terms of Equations 5.9 and 5.10 become simple products. The phase terms become

$$\delta_1 = k r_1 - \omega t + \varphi_1 \tag{5.12}$$

and

$$\delta_2 = k r_2 - \omega t + \varphi_2 \tag{5.13}$$

where $k = \frac{2\pi}{\lambda}$. The phase difference δ is then

$$\delta = k (r_2 - r_1) + (\varphi_2 - \varphi_1) \tag{5.14}$$

The term $r_2 - r_1$ is the *path difference* between the two beams.

[1]Thomas Young (1773-1829): English physicist and physician.

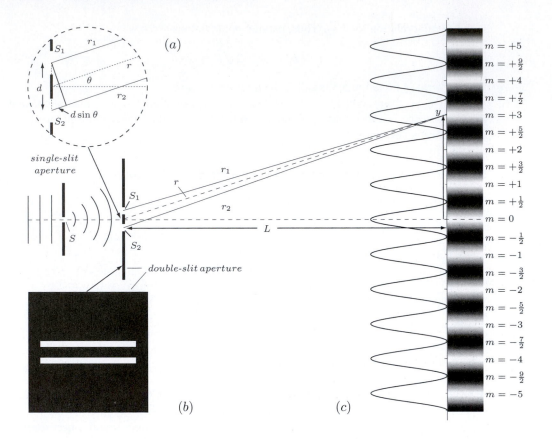

Figure 5.3   Young's double-slit experiment. (a) Geometry detail. Since $L >> d$, the two rays that leave S_1 and S_2 traveling toward the observation point y are nearly parallel. The path difference between beams 1 and 2 is $d \sin \theta$. (b) The double-slit aperture consists of two very narrow rectangular openings separated by d. In Figures (a) and (c), the aperture is shown in cross section with the long slit dimension extending into the page. (c) irradiance profile at the observation screen, along with the corresponding fringe pattern observed as irradiance variations. Irradiance maxima located at positive angles have positive order numbers m with integer values. Minima have half-integer values. Orders below the center line are negative.

In a typical double-slit experiment, the interference pattern due to slits spaced by a fraction of a millimeter are observed on a screen located a few meters from the aperture. Thus, we may assume that $L >> d$, where L is the distance from the aperture to the observing screen, and d is the spacing between the two slits. Under these circumstances, the paths r_1 and r_2 are almost parallel (see the inset in Figure 5.3), with a corresponding path difference of $d \sin \theta$ as illustrated. Thus

$$\delta = k \, (d \sin \theta) + (\varphi_2 - \varphi_1) \qquad (5.15)$$

When S is located symmetrically as in Figure 5.3, then $\varphi_2 - \varphi_1 = 0$.

Constructive interference occurs when $\delta = m(2\pi)$, where $m = 0, \pm 1, \pm 2, ...,$ giving

$$d \sin \theta = m \lambda \qquad (5.16)$$

The term m is called the *order number*. For destructive interference, m has half integer values, as indicated Figure 5.3(c). If θ is small, we may apply the small angle approximation

$$\frac{y}{L} = \tan\theta \approx \sin\theta$$

giving

$$y = m\lambda\frac{L}{d} \tag{5.17}$$

where again it is understood that integer values of m locate bright fringes and half-integer values of m locate dark fringes, as indicated in Figure 5.3.

If the irradiance contribution from each slit is equal, Equation 5.8 gives

$$I_{total} = 2I + 2I\cos\delta = 4I\cos^2\frac{\delta}{2} \tag{5.18}$$

The maximum irradiance in each fringe is four times the irradiance from either slit, but the average irradiance over many fringes is $2I$, in accord with the conservation of energy.

In obtaining such data, Young was not only able to demonstrate unambiguously the wave nature of light, but he also made the first direct measurement of its wavelength.

■ EXAMPLE 5.1

Determine the wavelength of monochromatic light that illuminates a double-slit aperture with $d = 0.100\,mm$ if bright fringes spaced by $2.945\,mm$ are observed on a screen located $5.00\,m$ from the aperture.

Solution

Bright fringes will be spaced by

$$\Delta y = \lambda\frac{L}{d}$$

giving

$$\lambda = \Delta y\frac{d}{L} = \left(2.945 \times 10^{-2}\,m\right)\frac{\left(1.00 \times 10^{-4}\,m\right)}{5.00\,m} = 5.89 \times 10^{-7}m = 589\,nm$$

Despite the substantial conceptual success of this experiment, there are difficulties with the analysis. Observation shows fringes that are *not* evenly spaced, and that diminish in irradiance as m increases, and that some of the fringes are missing. The discrepancy results from our approximations. No aperture is small enough to be taken as a single point, and an exact analysis would integrate contributions over an aperture of finite size. This is the approach of *diffraction* analysis, which we describe fully in Chapter 6.

Problem 5.1   Fringes from a double-slit aperture illuminated by light of wavelength $530\,nm$ are observed on a screen $5.00\,m$ away. If slits are separated by $0.100\,mm$, how far apart are the bright fringes separated?

Problem 5.2   Using the above analysis and the data from Example 5.1, show that the spacing between the $m = +100$ and $m = -100$ bright fringes is $7.29\,m$.

Problem 5.3   Show that the above analysis predicts a maximum value of m. Estimate this for the data of Example 5.1.

Problem 5.4   Describe the interference from two monochromatic beams of the same frequency with $I_1 = 2I_2$. This could be observed, for example, by placing an absorbing filter with 50% transmission over one of the slit openings in Figure 5.3. Assume that the phase of the attenuated beam is unaffected by the absorbing filter.

Problem 5.5   Describe the interference that results when a transparent slab of optical thickness nd is placed over the top slit in Figure 5.3. In particular, what happens to the fringe pattern if the slab produces a phase shift of π? Repeat if the slab is placed over the bottom slit.

5.3.3   Thin-Film Interference

According to the preceding analysis, interference occurs whenever wave are combined. In a Young's double-slit experiment, the phase difference that causes the observed interference fringes is caused by a variation in the optical path for each of the two combined beams. A similar optical path difference can occur in thin films, giving interference that causes, for example, the iridescent colors in an oil slick floating on a puddle of water.

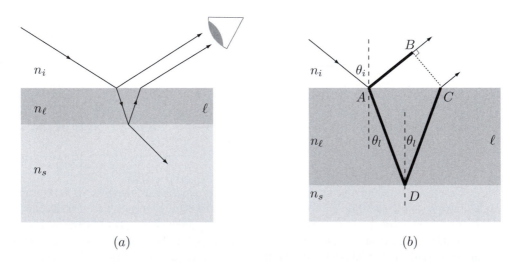

(a) (b)

Figure 5.4   (a) Interference from a thin film of thickness ℓ and index n_ℓ deposited on a substrate of index n_s. The interference results from beams reflected from the upper and lower surfaces of the film, as illustrated. (b) Geometry used to calculate the interference in (a).

Figure 5.4(a) shows a thin layer of index n_ℓ and thickness t deposited on a substrate n_s. *Thin-film interference* is caused by the two reflected rays indicated: one ray reflects from the film externally at the incident interface, and the other internally at the transmitted interface. The resulting interference is determined by two sources of phase shift:

$$\Delta\varphi = \Delta\varphi_{path} + \Delta\varphi_{ref} \qquad (5.19)$$

where $\Delta\varphi_{path}$ is the phase shift due to additional optical path for the internally reflected ray, and $\Delta\varphi_{ref}$ is the reflection phase change that occurs differently for either of the two reflected beams.

The phase shift $\Delta\varphi_{path}$ is determined from the bolded line segments of Figure 5.4(b). The *optical path* (see Section 4.2) difference is

$$\Delta_{path} = n_\ell \frac{2\ell}{\cos\theta_\ell} - n_i \overline{AB} \qquad (5.20)$$

Line $\overline{AB}$ has length $\overline{AC}\sin\theta_i$, with $\overline{AC} = 2t\tan\theta_r$. Thus

$$\overline{AB} = 2\ell \sin\theta_i \tan\theta_\ell \qquad (5.21)$$

The optical difference becomes

$$\Delta_{path} = n_\ell \frac{2\ell}{\cos\theta_\ell} - 2n_i\ell \sin\theta_i \tan\theta_\ell \qquad (5.22)$$

which with Snell's law becomes

$$\Delta_{path} = 2n_\ell\ell \cos\theta_\ell \qquad (5.23)$$

The associated relative phase shift is $k\Delta_{path}$:

$$\Delta\varphi_{path} = k\left(\Delta_{path}\right) = \frac{4\pi n_\ell}{\lambda_0}\ell \cos\theta_\ell \qquad (5.24)$$

where λ_0 is the vacuum wavelength. Note that it is tempting to interpret this phase shift as being due to the wavelength inside the layer ($\lambda_\ell = \lambda_0/n_\ell$), but note also that one of the beams travels in the incident index; fortuitous cancellation gives the simple result of Equation 5.24. Use Snell's law to express $\Delta\varphi_{path}$ in terms of the incident angle θ_i:

$$\Delta\varphi_{path} = \frac{4\pi\ell}{\lambda_0}\sqrt{n_\ell^2 - n_i^2 \sin^2\theta_i} \qquad (5.25)$$

Rules for reflection phase changes were determined in Section 3.5.5:

1. When the index of refraction of the incident medium is less than that of the transmitted medium, the reflected beam is phase-shifted by π radians.

2. If the incident index is greater than the transmitted index, there is no additional phase shift for the reflected ray.

3. There is no additional phase shift in the transmitted beam.

The reader will recall that these rules apply at all angles of incidence.

The total phase shift of Equation 5.19 is given by Equation 5.25 or Equation 5.24, plus perhaps another factor of π, depending on the indices n_i, n_ℓ, and n_s.

■ **EXAMPLE 5.2**

Oil with index of refraction $n_\ell = 1.25$ floats on water with index $n_s = 1.33$. For normal incidence, what minimum film thickness will reflect red light of wavelength $\lambda = 650\,nm$ most strongly? At what incident angle will the same film reflect blue light of wavelength $\lambda = 450\,nm$ most strongly?

Solution

Since the index increases at each interface, each surface gives a phase shift of π, so $\Delta\varphi_{ref} = 0$. Thus,

$$\Delta\varphi_{path} = \frac{4\pi n_\ell}{\lambda_0}\ell\cos\theta_\ell$$

The minimum nonzero film thickness to give constructive interference at normal incidence is given when $\Delta\varphi_{path} = 2\pi$. Thus

$$\frac{4\pi n_\ell}{\lambda_0}\ell = 2\pi$$

giving

$$\ell = \frac{\lambda_0}{2n_\ell} = \frac{650\,nm}{2\,(1.25)} = 260\,nm$$

To find the angle for which the same layer reflects blue strongly, begin with Equation 5.24 and solve for θ_ℓ

$$2\pi = \frac{4\pi\ell n_\ell}{\lambda_0}\cos\theta_\ell \Rightarrow \cos\theta_\ell = \frac{\lambda_0}{2n_\ell\ell} = \frac{450\,nm}{2\,(1.25)\,(260\,nm)} = 0.692$$

giving $\theta_\ell = 46.2°$. Snell's law gives the corresponding incident angle:

$$\sin\theta_i = \frac{n_\ell}{n_i}\sin\theta_\ell = (1.25)\,(0.722) = 0.902$$

and

$$\theta_i = 64.4°$$

─────────────────

■ **EXAMPLE 5.3**

A soap film suspended vertically by a wire loop becomes very thin at the top just prior to breaking. Describe the interference in the reflected beam as the film thickness just goes to zero.

Solution

As illustrated in Figure 5.5(a), a reflection phase shift occurs on the first film surface that does not occur on the back surface. If the film thickness is small enough to ignore the contribution of $\Delta\varphi_{path}$, the net phase shift at all angles is π, which

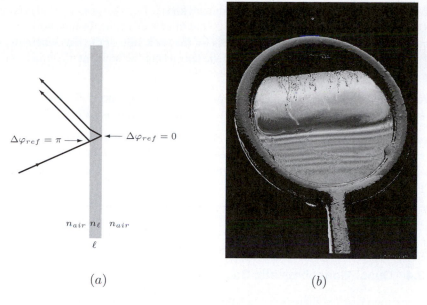

(a)   (b)

Figure 5.5   A thin soap film of negligible thickness ℓ. (a) The front surface gives a reflection phase shift of π, but the back surface does not. (b) Photograph showing reduced reflection at the top of the film, indicating destructive interference. (Richard Megna/Fundamental Photographs)

produces destructive interference. The photograph of Figure 5.5(b) shows a dark region at the top of the film, where gravity reduces the film thickness.

■ **APPLICATION NOTE 5.1**

Antireflection Coatings

Optical surfaces are often coated with thin films to reduce reflection. For camera optics, the film is typically designed to give destructive interference for normal incidence at $550\,nm$, which is at the peak of the solar spectrum. The interference builds back toward constructive for nonzero angles of incidence, which accounts for the violet tint of light reflected from coated camera optics.

    Ideally, optical coatings should have a hardness that allows routine cleaning without fear of scratching the coating. In a more thorough analysis that is given in Section 5.11, we will find that the optimal antireflection coating is achieved when the coating index n_ℓ is equal to the square root of the substrate index. Magnesium Fluoride (MgF_2) with $n_\ell = 1.37$ is commonly used for antireflection coatings on glass substrates.

Problem 5.6 Calculate the minimum film thickness of a MgF_2 ($n_\ell = 1.37$) antireflection coating applied to a crown glass ($n = 1.52$) lens immersed in air. Assume normal incidence. Find the required minimum film thickness for the back side of the lens where the incident medium is glass, and for the front side of the lens where the incident medium is air. Design for a vacuum wavelength of $550\, nm$.

Problem 5.7 Design a normal incidence antireflection coating for zinc selenide ($n = 2.40$) for use at a wavelength of $10.6\, \mu m$. Assume a coating index of 1.55.

Problem 5.8 Design an antireflection coating for a slab of crown glass ($n = 1.52$) that is meant to be used at an angle of incidence of $45°$. Use MgF_2 ($n_\ell = 1.37$) as a coating material, design for a vacuum wavelength of $550\, nm$, and assume that the incident index is that of air.

5.3.4 Quasi-Monochromatic Sources

As mentioned earlier, there is no such thing as a perfectly monochromatic source. Suppose that a source was available whose output consisted of a perfect sine wave; the act of turning the source on at a particular time and turning it off later gives a sinusoidal profile that is zero beyond certain time limits, and thus is not a true sine wave. Even if we leave the source on for a long time so that such effects can be neglected, the photon picture forces us to visualize a monochromatic beam of independent light particles with independent phases, giving a composite beam with random phase fluctuations. Section 5.4 on Fourier analysis provides further insights into why it is impossible to prepare a perfectly monochromatic beam that are quite independent of the photon picture. Even the most monochromatic laser sources produce outputs that contain a range of frequencies. A source with a frequency spread that is a small fraction of the average frequency is called *quasi-monochromatic*.

It may seem paradoxical that stationary interference fringes can be observed with sources that are not precisely monochromatic. We shall refine our understanding of this interesting question over the remainder of this and the next chapter. Briefly, the property of light that enables the formation of stationary interference fringes is called *coherence*. By first passing filtered light through the small hole labeled S in Figure 5.3, Young established a *spatially coherent* illumination across apertures separated by the slit separation d. The extent of the spatial coherence is inversely proportional to the size of S, so interference is only observed when d is sufficiently small. As noted above, spatial coherence results from a property of S (as yet undefined) to illuminate both apertures in a Young's slit experiment with a single wavefront, and for this reason it is called *wavefront splitting interference*. Conversely, in thin film interference a single beam is divided and then recombined to produce *amplitude splitting interference*. In this case, interference is observed only for films that are not to thick. For example, interference colors are observed in thin films of oil and thin layers of soapy water, but usually not in slabs of window glass. For reasons that will be discussed presently, amplitude splitting interference requires a type of coherence referred to as *temporal coherence*. Interestingly, (and perhaps remarkably), lasers can produce beams that are both spatially and temporally coherent. For example, laser light exhibits the odd effect of *speckle* produced by reflection from a rough surface, where interference results from beams reflected from peaks and valleys of spatially distinct structures.

5.3.5   Fringe Geometry

As noted previously, optical beams always contain a range of frequencies. *Stationary fringes* occur whenever coherent beams are combined that have a constant phase difference across a frequency range of both beams.

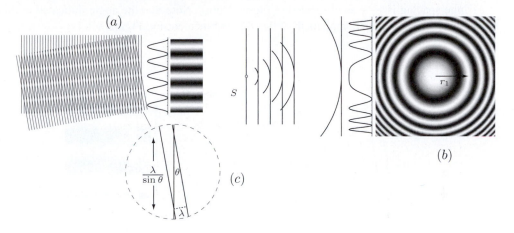

Figure 5.6   (a) Interference of plane waves give a linear fringe pattern. (b) Interference of a plane wave with a spherical wavefront.

The *fringe geometry* is determined by the beam propagation and wavefront characteristics. Figure 5.6(a) illustrates the linear fringe pattern that results when two plane-parallel beams with different propagation directions are combined. In this case, bright fringes are separated by

$$d = \frac{\lambda}{\sin \theta} \tag{5.26}$$

where λ is the beam wavelength and θ is the angle between the two beam directions (see inset).

Figure 5.6(b) shows fringes that result when a spherical wavefront originating at the point S combines with a plane wavefront. The path difference for the disturbances combining at any point on the observation screen is

$$\Delta z = R - \sqrt{R^2 - (x^2 + y^2)} \approx \frac{(x^2 + y^2)}{2R}$$

where x and y are measured on the observing screen. The last expression assumes that bright fringes occur whenever this path difference is equal to an integral number of wavelengths:

$$\frac{(x^2 + y^2)}{2R} \equiv \frac{r_m^2}{2R} = m\lambda$$

where $r_m^2 = x^2 + y^2$. Thus, the fringes have a radius that varies as the square-root of m: $(\sqrt{m}) \, r_1$ where $r_1 = \sqrt{2R\lambda}$ is the radius from the center of the pattern to the maximum of the $m = 1$ fringe.

5.3.5.1   *Lloyd's Mirror*   In a Lloyd's mirror experiment, monochromatic light illuminates a dielectric slab at grazing incidence, as shown in Figure 5.7. Interference occurs

between reflected light and the portion of the incident illumination that is not reflected. Parallel fringes result, as shown in the figure. If the observing screen is placed next to the end of the dielectric slab, a dark fringe is observed at the point of intersection. This result is independent of polarization, and is experimental verification that externally incident rays are reflected with a π phase shift at grazing incidence, as discussed in Section 3.5.5. The fringe spacing is $\frac{\lambda}{\sin\theta}$ as illustrated in Figure 5.6(a). Note that the angle indicated in this figure is the twice the complement of the angle of incidence in Figure 5.7.

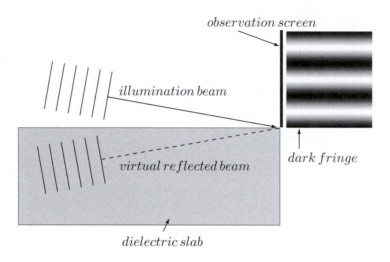

Figure 5.7   Lloyd's mirror. The dark fringe at the intersection of the dielectric slab and the observing screen indicate a π phase change for external grazing incidence.

5.3.5.2 *Newton's Rings*   Newton's[2] rings are observed when a convex lens is placed in contact with a flat dielectric slab, as shown in Figure 5.8. The interference results from reflections from the interfaces on either side of the air gap indicated in the figure. Beam 1 reflects internally from the glass-air interface of the lens. Beam 2 reflects externally from the first slab surface. Since there is no phase shift for the internally reflected beam and the phase shift is π for the externally reflected beam, the center fringe is dark. This result is independent of polarization, and is experimental verification that externally incident rays are reflected with a π phase shift for normal incidence, as discussed in Section 3.5.5.

5.3.6   Interference Between Beams with Different Frequencies

If the frequencies are not equal, the total disturbance at P in Figure 5.2 is given by Equation 5.5 and is now a function of time. It will simplify the analysis to let the field amplitudes be

---

[2]Issac Newton: 1643-1727. English physicist, mathematician and astronomer.

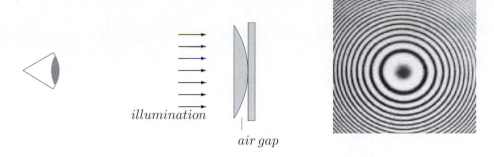

illumination

air gap

Figure 5.8   Newton's rings. The reflected light shows a dark fringe at the center, indicating a π phase change of external normal incidence. (Photo courtesy Bausch & Lomb)

equal, giving $E_{01} = E_{02} = E_0$. Assume also that the polarizations are parallel. Thus

$$
\begin{aligned}
E &= E_0 e^{i\delta_1} + E_0 e^{i\delta_2} \\
&= E_0 \left(e^{i\delta_1} + e^{i\delta_2} \right) \\
&= E_0 \left(e^{i\frac{(\delta_2 - \delta_1)}{2}} + e^{-i\frac{(\delta_2 - \delta_1)}{2}} \right) e^{i\frac{(\delta_1 + \delta_2)}{2}} \\
&= 2E_0 \left(\frac{e^{i\frac{(\delta_2 - \delta_1)}{2}} + e^{-i\frac{(\delta_2 - \delta_1)}{2}}}{2} \right) e^{i\frac{(\delta_1 + \delta_2)}{2}}
\end{aligned}
$$

The observable disturbance is the real part of the last result:

$$
E = 2E_0 \cos \left(\frac{\delta_2 - \delta_1}{2} \right) \cos \left(\frac{(\delta_1 + \delta_2)}{2} \right) \tag{5.27}
$$

Consider the case where the two beams travel in the same direction, as indicated in Figure 5.9(a). Since the frequencies are not equal, the two waves combine in and out of phase as x varies. Adding the two waves results in a disturbance whose amplitude varies between conditions of constructive and destructive interference, as illustrated in Figure 5.9(b). In other words, the disturbance is *amplitude modulated*.

Since both waves travel along positive x, Equations 5.3 and 5.4 become

$$
\delta_1 = k_1 x - \omega_1 t + \varphi_1 \tag{5.28}
$$

and

$$
\delta_2 = k_2 x - \omega_2 t + \varphi_2 \tag{5.29}
$$

If the initial phases are both zero, Equations 5.27, 5.28, and 5.29 give

$$
E = 2E_0 \cos \left(k_{AM} x - \omega_{AM} t \right) \cos \left(\bar{k} x - \bar{\omega} t \right) \tag{5.30}
$$

where

$$
\bar{k} = \tfrac{1}{2} \left(k_1 + k_2 \right)
$$
$$
\bar{\omega} = \tfrac{1}{2} \left(\omega_1 + \omega_2 \right)
$$

are the *average* values of wavenumber and angular frequency. Similarly,

$$
k_{AM} = \tfrac{1}{2} \left(k_2 - k_1 \right)
$$

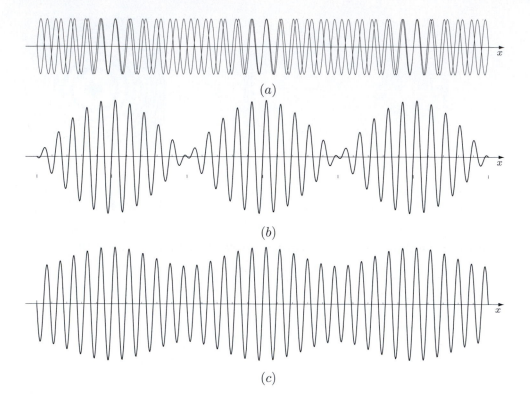

Figure 5.9   (a) Two waves of different wavelength merge in and out of phase. (b) Superposition of the waves in (a). (c) Superposition of waves with different amplitudes.

$$\omega_{AM} = \tfrac{1}{2} \left(\omega_2 - \omega_1 \right)$$

are the wavenumber and angular frequency of the amplitude modulation (AM).

If the two field amplitudes are not equal, there is less amplitude modulation, as illustrated in Figure 5.9(c).

5.3.6.1 *Coherent Detection*   Detection of intensity with a linear power detector, often referred to as *direct detection*, is a time-averaged measurement of wave power that in the process destroys all knowledge of instantaneous wave amplitude and phase. In *coherent detection*, the beating between a signal beam and a reference *local oscillator* is detected as an AC signal, as illustrated in Figure 5.10(a), which as we will see provides a means for detecting both amplitude and phase. The signal beam and the local oscillator (LO) beam are combined along a common direction of propagation using a beam splitter, and the combined beam then illuminates a linear power detector. In practice, the local oscillator beam irradiance is much greater than the irradiance of the signal beam. Let E_1 represent the local oscillator field, E_2 represent the signal field, and let the propagation be along x, giving

$$E_1 = E_{LO}e^{i\delta_{LO}}$$
$$E_2 = E_s e^{i\delta_s}$$

where

$$\delta_{lo} = k_{LO}x - \omega_{LO}t + \varphi_{LO} \tag{5.31}$$

and

$$\delta_s = k_s x - \omega_s t + \varphi_s \tag{5.32}$$

Letting $\bar{\delta} = (\delta_{LO} + \delta_s)/2$ and $\Delta\delta = (\delta_{LO} - \delta_s)$, the sum of both fields is given by

$$E = E_1 + E_2 = E_{LO}e^{i\delta_{LO}} + E_s e^{i\delta_s}$$
$$= \left(E_{LO}e^{i\frac{\Delta\delta}{2}} + E_s e^{-i\frac{\Delta\delta}{2}}\right)e^{i\bar{\delta}}$$

The detected irradiance is proportional to the magnitude-squared of the total field:

$$E^2 = EE^* = \left(E_{LO}e^{i\frac{\Delta\delta}{2}} + E_s e^{-i\frac{\Delta\delta}{2}}\right)\left(E_{LO}e^{i\frac{\Delta\delta}{2}} + E_s e^{-i\frac{\Delta\delta}{2}}\right)^*$$
$$= \left(E_{LO}e^{i\frac{\Delta\delta}{2}} + E_s e^{-i\frac{\Delta\delta}{2}}\right)\left(E_{LO}e^{-i\frac{\Delta\delta}{2}} + E_s e^{+i\frac{\Delta\delta}{2}}\right)$$
$$= E_{LO}^2 + E_s^2 + E_{LO}E_s\left(e^{i\Delta\delta} + e^{-i\Delta\delta}\right)$$
$$= E_{LO}^2 + E_s^2 + 2E_{LO}E_s\cos\Delta\delta$$

The detector combines signals from each point on its surface. Signal components that are 180° out of phase will cancel, so it is important that the signal and local oscillator fields be precisely aligned. This places a limitation on the *field of view* of the coherent detector, as illustrated in Figure 5.10(b). According to this diagram, out-of-phase signals will begin to be generated at the detector edges when the signal and local oscillator beams are misaligned by θ_{max} which is given by

$$\tan\theta_{\max} = \frac{\lambda}{D} \tag{5.33}$$

Notice that smaller detectors allow larger values of misalignment. It is common to use detectors that are only a few wavelengths across.

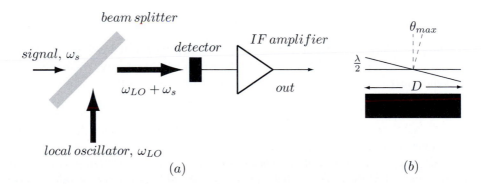

Figure 5.10   (a) A typical schematic for coherent detection. (b) Misalignment of the signal and local oscillator beams results in beat signals that destructively interfere.

For optimized optical alignment over a small detector, we may neglect any variation in position or phase across the detector surface. The detected signal then becomes

$$I = I_{LO} + I_s + 2E_{LO}E_s\cos\omega_{IF}t \tag{5.34}$$

where $\omega_{IF} = \omega_{LO} - \omega_s$ is the *beat frequency*, sometimes referred to as the *intermediate frequency*. This time-dependent signal is typically amplified by a intermediate frequency

(IF) amplifier. These high-bandwidth amplifiers typically do not pass the first two DC terms of Equation 5.34, so the coherent detection signal is given by

$$I_{if} = 2E_{LO}E_s \cos \omega_{IF} \tag{5.35}$$

If the local oscillator is a coherent laser beam with a constant amplitude E_{LO}, then the beat amplitude is linearly proportional to the signal field *amplitude* E_s, even though a square-law detector is used to record the signal. Similarly, a coherent local oscillator allows the signal *phase* to be recorded as well. The beat signal may be thought of as a replica of the signal beam that is *frequency downshifted* to the intermediate frequency ω_{if}. This process is often referred to as *photoelectric mixing*, or by analogy to the similar electrical detection process long used in radio technology, *heterodyne detection*. The entire system diagrammed in Figure 5.10(a) is called a *heterodyne receiver*.

■ **APPLICATION NOTE 5.2**

Coherent Doppler Lidar

The Doppler effect causes electromagnetic radiation originating from moving sources to be frequency shifted. From special relativity[3], the observed frequency is given by

$$f' = f_0 \sqrt{\frac{1 + \frac{v}{c}}{1 - \frac{v}{c}}}$$

where f' is the detected frequency, f_0 is the frequency detected by an observer at rest with respect to the source, and v is the source velocity. The sign convention is that v is negative if the source and observer are receding, and positive otherwise. If $v \ll c$, then approximation gives

$$f' \cong f_0 \left(1 + \frac{v}{c}\right) \tag{5.36}$$

Doppler radar uses radio waves to measure this frequency shift in a frequency regime where antennas measure the amplitude of the radio waves directly. In *Doppler lidar*, coherent detection achieves a similar goal at optical frequencies where antennas cannot be used.

Figure 5.11 illustrates a typical arrangement. Light from the scattering laser is reflected from a moving target, causing the reflected light to be Doppler shifted. The frequency shift is twice the value indicated in Equation 5.36. The first frequency shift occurs in the frame of reference of the target, with the scattering laser moving relative to the target. This radiation is then scattered back toward the lidar receiver from a source with the same relative velocity. The net frequency shift is

$$\Delta f = f_0 \frac{2v}{c} = \frac{2v}{\lambda_0} \tag{5.37}$$

where λ_0 is the wavelength of the unshifted scattering laser.

Many other designs are possible. The arrangement shown in Figure 5.11 uses different lasers for the scattering and local oscillator beams, but a single laser could serve both purposes. Notice that in this design, a portion of the scattering laser

---

[3]See *Spacetime Physics*, by Taylor and Wheeler [21].

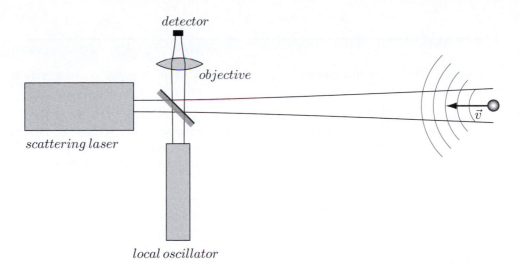

Figure 5.11   Coherent Doppler lidar.

can enter the local oscillator cavity and vice versa, potentially causing the lasers to become unstable.

■ **EXAMPLE 5.4**

Carbon dioxide lasers are commonly used for lidar since they emit a $10.6\,\mu m$-wavelength beam that is within the *atmospheric window* where clean air is relatively transparent. Find the frequency shift that occurs when such a beam is scattered from a target that moves toward the receiver at $1000\,m/s$. Express your answer as a fraction of the scattering laser frequency.

Solution

According to Equation 5.37, the frequency shift is

$$\Delta f = \frac{2v}{\lambda} = \frac{2\,(1000\,m/s)}{10.6 \times 10^{-6}m} = 1.89 \times 10^{8} = 189\,MHz$$

The fastest solid-state optical detectors currently available have bandwidths of more than 100 times this frequency.

The fractional frequency shift is also given by Equation 5.37:

$$\frac{\Delta f}{f_0} = \frac{2v}{c} = \frac{2\,(1000\,m/s)}{3 \times 10^{8}m/s} = 6.67 \times 10^{-6}$$

Such a small fractional shift would be difficult to measure with spectrometers that use prisms or diffraction gratings. Notice that this result is independent of the scattering laser frequency.

Problem 5.9   Show that when $v \ll c$ the Doppler shift is given approximately by $f' \cong f_0 \left(1 + \frac{v}{c}\right)$.

Problem 5.10   Find the Doppler shift of *X-band radar* radiation ($f = 8.4 \times 10^9 \, Hz$) that is reflected from a speeding car traveling at 80 miles per hour. Assume that the car is traveling toward the stationary radar source, then repeat for a car traveling away from the stationary radar source. In each case, express your answer as a signed fraction of the scattering radiation frequency.

5.4   FOURIER ANALYSIS

Monochromatic plane waves have simple mathematical formulas, and we make frequent use of them for purposes of illustration. However, as noted above and in Chapter 2, no wave is truly monochromatic, but instead is a composite wave that consists of many frequency components. The beating wave of Section 5.3.6 is a simple example of a composite wave with two discrete frequency components, but even that is unrealistic. Electromagnetic waves found in Nature always contain an infinite number of frequency components no matter how "monochromatic" they are. This applies not only to thermal sources such as neon signs, but also to coherent beams such as those emitted by helium-neon lasers.

The Principle of Superposition introduced in Section 1.5 allows us to model real electromagnetic waves. Using the appropriate mathematical tools, we can combine the monochromatic harmonic traveling wave solutions found in Chapter 2 to accurately represent any real traveling electromagnetic wave with arbitrary frequency composition. The mathematical tools we require are provided by the *Fourier integrals*[4]. The related topic of *Fourier series* is discussed in the Appendix at the end of this chapter.

5.4.1   Fourier Transforms

Fourier series may be used to represent an arbitrary periodic function, while Fourier transforms can be used to represent functions that are not periodic. It should seem obvious that no physical function is truly periodic, for even a precise oscillator or laser source must at some point be switched on and off. For this and other reasons that will become obvious as we proceed, Fourier transforms are much more applicable to the study of optics than are Fourier series. There is a natural and intuitive conceptual transition from Fourier series to transforms that is outlined in the Appendix. Here, we simply define Fourier transforms, and proceed with the relevant interpretation and application.

We begin by considering two parameters α and β whose product is dimensionless and can be measured in radians. Harmonic waves have a spatial period λ and temporal period T, so the relevant parameter pairs for our purposes will be x, k and t, ω. Suitable choices for α and β are referred to as *conjugate variables*. The Fourier transform "transforms" the dependence of a function $f(\alpha)$ into a completely equivalent representation $F(\beta)$ in the

---

[4]Joseph Fourier (1768-1830): French mathematician who made many contributions to mathematical physics.

following way:

$$F(\beta) = \int_{-\infty}^{\infty} f(\alpha)\, e^{-i\alpha\beta}\, d\alpha \tag{5.38}$$

Note that the quantity in the power of the exponent is unitless. The integration variable is α, so the integration leaves only a dependence on variable β. We refer to $F(\beta)$ as the Fourier transform of the function $f(\alpha)$. The *inverse Fourier transform* makes the opposite transformation:

$$f(\alpha) = \frac{1}{2\pi} \int_{-\infty}^{\infty} F(\beta)\, e^{i\alpha\beta}\, d\beta \tag{5.39}$$

Note the change of sign in the power of the exponent, and the extra factor of $\frac{1}{2\pi}$ in the inverse transform; reasons for these changes are explained in the Appendix.

The functions $f(\alpha)$ and $F(\beta)$ are referred to as *Fourier transform pairs*. Both functions have certain mathematical requirements such as being single-valued, not having an infinite number of maxima, minima, or discontinuities in any finite interval, and so on. In general, we proceed with the confidence that ill-behaved functions are usually not required to describe real physical phenomena. Readers who are interested in the foundations of Fourier analysis and the associated mathematical restrictions should consult Bracewell [4] or a similar text.

5.4.2   Position Space, k-Space Domain

We begin by considering the conjugate variables x and k. In this case, the Fourier transform of a function $f(x)$ is given by

$$F(k) = \int_{-\infty}^{\infty} f(x)e^{-ikx}\, dx \tag{5.40}$$

The inverse Fourier transform of the function $F(k)$ is

$$f(x) = \frac{1}{2\pi} \int_{-\infty}^{\infty} F(k)e^{ikx}\, dk \tag{5.41}$$

When applied to a specific physical situation, we regard $f(x)$ and $F(k)$ as *equivalent* representations of the same physical system. In other words, a knowledge of $f(x)$ completely specifies $F(k)$ and vice versa. The function $f(x)$ is referred to as the *position-space* representation of the system, and $F(k)$ is the representation in *k-space*. This distinction will become more clear in the next section as we begin to apply Fourier transforms to specific physical situations.

■ **EXAMPLE 5.5**

Find the Fourier transform of a single rectangular pulse:

$$f(x) = \begin{cases} +1 & |x| \le a \\ 0 & |x| > a \end{cases}$$

Solution

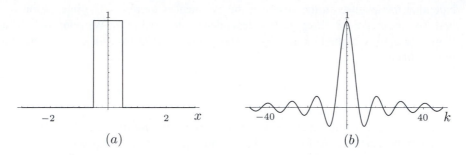

Figure 5.12 (a) The pulse $f(x)$ with $a = \frac{1}{2}$. (b) The Fourier transform $F(k)$.

Begin with Equation 5.40:

$$F(k) = \int_{-\infty}^{\infty} f(x)e^{-ikx}\,dx = \int_{-a}^{a} e^{-ikx}\,dx$$

$$= \frac{-1}{ik}e^{-ikx}\Big|_{-a}^{+a} = \frac{2}{k}\sin(ka) = 2a\frac{\sin(ka)}{ka}$$

A plot of $f(x)$ and its Fourier transform $F(k)$ are shown in Figure 5.12.

The Fourier transform of the rectangular pulse is an example of a function that occurs frequently in the study of optics, referred to as the *sinc function* (pronounced like sink) and defined as

$$\mathrm{sinc}\,x \equiv \frac{\sin \pi x}{\pi x} \tag{5.42}$$

Defined in this way, the sinc function has the following properties (see Problem 5.11):

$$\lim_{x \to 0} \mathrm{sinc}\,x = 1 \tag{5.43}$$

$$\mathrm{sinc}\,N = 0 \qquad N = \pm 1,\ \pm 2,\ \ldots \tag{5.44}$$

$$\int_{-\infty}^{\infty} \mathrm{sinc}\,x\,dx = 1 \tag{5.45}$$

We will also have frequent occasion to utilize the *Gaussian* function, defined as

$$Gauss\left(\frac{x}{\sigma}\right) \equiv e^{-\frac{x^2}{2\sigma^2}} \tag{5.46}$$

where σ is the *spread*. The Gaussian is a bell-shaped function, as illustrated in Figure 5.13. The $1/e$ points of the Gaussian are located at $x = \pm\sqrt{2}\sigma$, separated by $x = 2\sqrt{2}\sigma$ as shown in Figure 5.13(a). In the next example, we explore the Gaussian function and its Fourier transform, which will be especially relevant as we model the properties of real optical sources.

■ **EXAMPLE 5.6**

Show that the Fourier transform of a Gaussian function is another Gaussian with reciprocal spread.

Solution

In position-space, let $f(x)$ be a Gaussian with amplitude A:

$$f(x) = Ae^{-\frac{x^2}{2\sigma^2}}$$

The Fourier transform of $f(x)$ is given by Equation 5.40:

$$F(k) = \int_{-\infty}^{\infty} f(x)e^{-ikx}dx$$

$$= \int_{-\infty}^{\infty} Ae^{-\frac{x^2}{2\sigma^2}}e^{-ikx}dx$$

$$= A\int_{-\infty}^{\infty} \exp[-\frac{x^2}{2\sigma^2} - ikx + \frac{k^2\sigma^2}{2} - \frac{k^2\sigma^2}{2}]dx$$

$$= Ae^{-\frac{k^2\sigma^2}{2}}\int_{-\infty}^{\infty} \exp[-\frac{1}{2\sigma^2}\left(x^2 + i2k\sigma^2x - k^2\sigma^4\right)]dx$$

$$= Ae^{-\frac{k^2\sigma^2}{2}}\int_{-\infty}^{\infty} \exp[-\frac{1}{2\sigma^2}\left(x + ik\sigma^2\right)^2]dx$$

Change variables as follows:

$$u = \frac{1}{\sqrt{2}\sigma}\left(x + ik\sigma^2\right)$$

$$du = \frac{1}{\sqrt{2}\sigma}dx$$

In terms of the new variable u, $F(k)$ is given by

$$F(k) = Ae^{-\frac{k^2\sigma^2}{2}}\int_{-\infty}^{\infty} e^{-u^2}\left(\sqrt{2}\sigma du\right)$$

From integral tables or software, we find that

$$\int_{0}^{\infty} e^{-a^2u^2}du = \frac{\sqrt{\pi}}{2a}$$

Since the Gaussian is even,

$$F(k) = \sqrt{2}\sigma Ae^{-\frac{k^2\sigma^2}{2}}2\int_{0}^{\infty} e^{-u^2}du$$

giving

$$F(k) = \sqrt{2\pi}\sigma A e^{-\frac{k^2\sigma^2}{2}}$$

which is the Gaussian of Figure 5.13(b). Notice that the spread in k-space is the inverse of that in position-space.

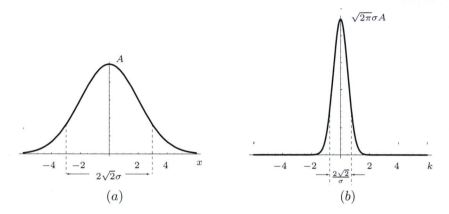

(a)   (b)

Figure 5.13   (a) The Gaussian $f(x)$ in position space. (b) The Fourier transform $F(k)$ is a Gaussian function in k-space .

Continue with the inverse Fourier transform (Equation 5.41):

$$f(x) = \frac{1}{2\pi} \int_{-\infty}^{\infty} F(k)e^{ikx}dk$$

$$= \frac{1}{2\pi} \int_{-\infty}^{\infty} F(k)e^{ikx}dk = \frac{1}{2\pi}\sqrt{2\pi}\sigma A \int_{-\infty}^{\infty} e^{-\frac{k^2\sigma^2}{2}}e^{ikx}dk$$

$$= \frac{\sigma}{\sqrt{2\pi}}A \int_{-\infty}^{\infty} \exp[\frac{-k^2\sigma^2}{2} + ikx + \frac{x^2}{2\sigma} - \frac{x^2}{2\sigma}]dk$$

$$= \frac{\sigma}{\sqrt{2\pi}}Ae^{-\frac{x^2}{2\sigma^2}} \int_{-\infty}^{\infty} \exp[-\frac{\sigma^2}{2}\left(k^2 - i\frac{2kx}{\sigma^2} - \frac{x^2}{\sigma^4}\right)]dk$$

$$= \frac{\sigma}{\sqrt{2\pi}}Ae^{-\frac{x^2}{2\sigma^2}} \int_{-\infty}^{\infty} \exp[-\frac{\sigma^2}{2}\left(k - i\frac{x}{\sigma^2}\right)^2]dk$$

Change variables:

$$u = \frac{\sigma}{\sqrt{2}}\left(k - i\frac{x}{\sigma^2}\right); \quad du = \frac{\sigma}{\sqrt{2}}dk; \quad dk = \frac{\sqrt{2}}{\sigma}du$$

to give

$$f(x) = \frac{\sigma}{\sqrt{2\pi}}\frac{\sqrt{2}}{\sigma}Ae^{-\frac{x^2}{2\sigma^2}} \int_{-\infty}^{\infty} e^{-u^2}du = \frac{1}{\sqrt{\pi}}Ae^{-\frac{x^2}{2\sigma^2}}\sqrt{\pi}$$

or

$$f(x) = Ae^{-\frac{x^2}{2\sigma^2}}$$

which gives the original function again.

5.4.3 Frequency-Time Domain

Fourier transforms may also be expressed with frequency and time. The transform is given by

$$F(\omega) = \int_{-\infty}^{\infty} f(t)e^{-i\omega t} dt \qquad (5.47)$$

with the inverse transform given by

$$f(t) = \frac{1}{2\pi} \int_{-\infty}^{\infty} F(\omega)e^{i\omega t} d\omega \qquad (5.48)$$

where ω is the angular frequency (units: rad/s). While it may seem odd to express the inverse transform in terms of negative frequencies, a real-valued $f(t)$ may be expressed as an integral over positive frequencies (see Problem 5.12).

Problem 5.11 Demonstrate the properties of the sinc function given in Equation 5.43.

Problem 5.12 Show that if $f(t)$ is real, then $F^*(\omega) = F(-\omega)$ and thus $f(t)$ may be expressed as

$$f(t) = \frac{1}{2\pi} \int_{0}^{\infty} \left(F(\omega)e^{i\omega t} + F^*(\omega)e^{-i\omega t} \right) d\omega$$

Problem 5.13 Find the Fourier transform of a Gaussian that is not centered about zero:

$$f(x) = Ae^{-\frac{(x-x_0)^2}{2\sigma^2}}$$

Problem 5.14 Find the Fourier transform of the triangle pulse:

$$f(x) = \begin{cases} (1 - |\frac{x}{a}|) & |x| \leq a \\ 0 & |x| > a \end{cases}$$

where a is a constant. Show that the answer can be expressed as

$$F(k) = a \left(\frac{\sin\left(\frac{ka}{2}\right)}{\frac{ka}{2}} \right)^2$$

Problem 5.15   Find the Fourier transform of an exponential pulse:

$$f(t) = Ae^{-\frac{|t|}{a}}$$

Sketch the function $f(t)$ and its transform $F(\omega)$.

5.5   PROPERTIES OF FOURIER TRANSFORMS

It is instructive to deduce some general properties of Fourier transforms that can be used, for example, to simplify calculation by avoiding unnecessary integration.

5.5.1   Symmetry Properties

It is often useful to note whether a function is *even* or *odd*. A function is even if $f(-x) = f(x)$. In other words, even functions are symmetric about the origin. Similarly, a function is odd if it is antisymmetric about the origin: $f(-x) = -f(x)$. For example, $\cos(x)$ is even and $\sin(x)$ is an odd function. The product of an even and an odd function is odd, but the product of two odd functions or two even functions is even (see Problem 5.16). Sums of even and odd functions have no particular symmetry in that they are neither even nor odd. A function is said to be *hermitian* if the real part is even and the imaginary part is odd. If the real part is odd and the imaginary part even, then the function is *antihermitian*.

We begin with Equation 5.38, and apply the Euler relation:

$$F(\beta) = \int_{-\infty}^{\infty} f(\alpha) e^{-i\alpha\beta} d\alpha$$

$$= \int_{-\infty}^{\infty} f(\alpha) \cos(\alpha\beta) d\alpha - i \int_{-\infty}^{\infty} f(\alpha) \sin(\alpha\beta) d\alpha \tag{5.49}$$

Let $f(\alpha)$ be complex, with real part $f_{\text{Re}}(\alpha)$ and imaginary part $f_{\text{Im}}(\alpha)$. Thus

$$F(\beta) = \int_{-\infty}^{\infty} f_{\text{Re}}(\alpha) \cos(\alpha\beta) d\alpha + \int_{-\infty}^{\infty} f_{\text{Im}}(\alpha) \sin(\alpha\beta) d\alpha$$

$$+ i \left(\int_{-\infty}^{\infty} f_{\text{Im}}(\alpha) \cos(\alpha\beta) d\alpha - \int_{-\infty}^{\infty} f_{\text{Re}}(\alpha) \sin(\alpha\beta) d\alpha \right) \tag{5.50}$$

If $f(\alpha)$ is real, then $f_{\text{Im}}(\alpha) = 0$ and $f(\alpha) = f_{\text{Re}}(\alpha)$, giving

$$F(\beta) = \int_{-\infty}^{\infty} f(\alpha) \cos(\alpha\beta) d\alpha - i \int_{-\infty}^{\infty} f(\alpha) \sin(\alpha\beta) d\alpha \tag{5.51}$$

If in addition to being real, $f(\alpha)$ is also even, then the second integral in Equation 5.51 vanishes, giving a Fourier transform that is real:

$$F(\beta) = \int_{-\infty}^{\infty} f(\alpha) \cos(\alpha\beta) d\alpha \tag{5.52}$$

The result of Equation 5.52 is also even, as can be seen by simple substitution.

Table 5.1. summarizes various symmetry properties of Fourier transforms. Proofs are left for the problems.

Table 5.1. Selected symmetry properties of Fourier transforms

$f(\alpha)$	$F(\beta)$
Real even	Real even
Real odd	Imaginary odd
Real, no symmetry	Hermitian
Imaginary even	Imaginary even
Imaginary odd	Real odd
Imaginary, no symmetry	Antihermitian
Complex even	Complex even
Complex odd	Complex odd
Hermitian	Real, no symmetry
Antihermitian	Imaginary, no symmetry

5.5.2 Linearity

Suppose that $F(\beta) = \int_{-\infty}^{\infty} f(\alpha) e^{-i\alpha\beta} d\alpha$ and $H(\beta) = \int_{-\infty}^{\infty} h(\alpha) e^{-i\alpha\beta} d\alpha$. If A and B are constants, then the Fourier transform of $A f(\alpha) + B h(\alpha)$ is given by $A F(\beta) + B H(\beta)$. To show this, proceed as follows:

$$\int_{-\infty}^{\infty} [A f(\alpha) + B h(\alpha)] e^{-i\alpha\beta} d\alpha = A \int_{-\infty}^{\infty} f(\alpha) e^{-i\alpha\beta} d\alpha + B \int_{-\infty}^{\infty} h(\alpha) e^{-i\alpha\beta} d\alpha$$

$$= A F(\beta) + B H(\beta)$$

$$(5.53)$$

Intuitively, this property accounts for the fact that the spectrum of a sum of signals is given by the sum of the individual spectra.

5.5.3 Transform of a Transform

Let $F(\beta) = \int_{-\infty}^{\infty} f(\alpha) e^{-i\alpha\beta} d\alpha$ and consider the Fourier transform of $F(\alpha)$:

$$\int_{-\infty}^{\infty} F(\alpha) e^{-i\alpha\beta} d\alpha = \int_{-\infty}^{\infty} F(\alpha) e^{i\alpha(-\beta)} d\alpha$$

$$(5.54)$$

$$= f(-\beta)$$

where in the last step we used the definition of the inverse Fourier transform (Equation 5.48). This can provide a quick way to compute Fourier transforms that would otherwise be tedious to integrate. In the Abbe theory of image formation (discussed in Chapter 8), this property also accounts for the fact that real images formed by a single lens are inverted.

Problem 5.16   Show that the product of an even and an odd function results in a function that is odd. Show that the product of two even or two odd functions results in a function that is even.

Problem 5.17   Show that the integral of an odd function over limits symmetric about the origin is zero.

Problem 5.18   Show that the Fourier transform of an odd real function gives a result that is imaginary and odd.

Problem 5.19   Show that
  a) The Fourier transform of an imaginary function that is even gives a result that is imaginary and even.
  b) The Fourier transform of an imaginary function that is odd gives a result that is real and odd.
  c) The Fourier transform of a real function that is odd gives a result that is imaginary and odd.

Problem 5.20   Show that
  a) The Fourier transform of a hermitian function is real with no symmetry.
  b) The Fourier transform of a real function with no symmetry gives a result that is hermitian.

Problem 5.21   Show that
  a) The Fourier transform of an antihermitian function is imaginary with no symmetry.
  b) The Fourier transform of an imaginary function with no symmetry gives a result that is antihermitian.

Problem 5.22   Show that the Fourier transform of the sinc function of Equation 5.42 is given by the pulse function of Example 5.5.

5.6 WAVEPACKETS

A perfectly monochromatic wave cannot exist, for to be perfectly monochromatic the wave must be perfectly harmonic, and so must have existed for all times in the past. Wavetrains that begin and end at finite points can be represented with a Fourier transform.

■ **EXAMPLE 5.7**

Find the Fourier transform of a finite monochromatic wavetrain:

$$f(x) = \begin{cases} E_0 \cos(k_0 x) & -L \leq x \leq +L \\ 0 & otherwise \end{cases}$$

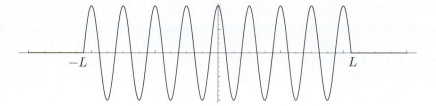

Figure 5.14 A finite wavetrain.

Solution

The wavetrain is given in position space, so we use Equation 5.40:

$$F(k) = \int_{-\infty}^{\infty} f(x)e^{-ikx}dx = \int_{-L}^{L} E_0 \cos(k_0 x)e^{-ikx}dx$$

$$= \frac{E_0}{2} \int_{-L}^{L} \left(e^{ik_0 x} + e^{-ik_0 x}\right)e^{-ikx}dx = \frac{E_0}{2} \int_{-L}^{L} \left(e^{-i(k-k_0)x} + e^{-i(k+k_0)x}\right)dx$$

$$= \frac{E_0}{2} \left[\frac{e^{-i(k-k_0)x}}{-i(k-k_0)} + \frac{e^{-i(k+k_0)x}}{-i(k+k_0)} \right]\Bigg|_{-L}^{L}$$

$$= E_0 L \left[\frac{\sin(k-k_0)L}{(k-k_0)L} + \frac{\sin(k+k_0)L}{(k+k_0)L} \right]$$

giving

$$F(k) = E_0 L \operatorname{sinc}(k-k_0)L + E_0 L \operatorname{sinc}(k+k_0)L$$

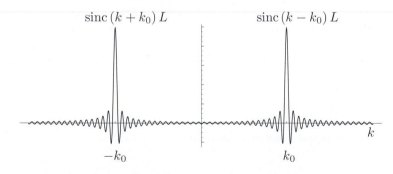

Figure 5.15 The Fourier transform of Figure 5.14.

Figure 5.15 shows a plot of $F(k)$. Two k-space packets are found, one centered about $-k_0$ and another centered around $+k_0$. The negative-k packet represents a backward-traveling wave, and the other a forward-traveling wave. Both packets are possible, and both possibilities are represented in the transform.

Figure 5.16(a) shows the Fourier transform of the finite wavetrain of Example 5.7 over a smaller range of positive k-values. By beginning and ending the wavetrain at finite points in space, we have introduced new wavelength components in the previously monochromatic wave. We may estimate the spread in new k-values by measuring the distance between the first zero-crossing in $F(k)$. From Figure 5.16, the spread Δk is approximated by

$$(k - k_0)\,L = \Delta k\,L = \pi \Rightarrow \Delta k \geq \frac{\pi}{L}$$

Similarly, from Figure 5.14, $\Delta x \geq 2L$. The product $\Delta x\,\Delta k$ is independent of the length of the wavetrain:

$$\Delta x\,\Delta k \approx 2\pi \tag{5.55}$$

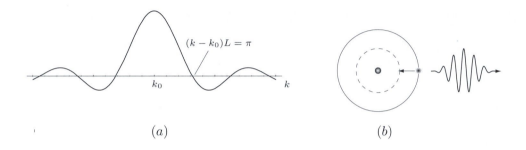

(a) $\qquad\qquad\qquad\qquad\qquad$ (b)

Figure 5.16 (a) Zoom detail of Figure 5.15. (b) An electron bound to an atom makes a transition from a state of higher energy to a lower energy state. In the process, a photon is emitted.

A similar estimation is possible in the frequency-time domain. An atom radiates a photon whenever one of its electrons makes a transition from a higher energy state to a lower energy state, as illustrated in Figure 5.16(b). For conditions of temperature and pressure in the typical discharge lamp, the photon wave packet oscillates over a time interval of about $10^{-8}\,s$. This finite time duration causes the photon to contain a continuous distribution of frequencies.

■ **EXAMPLE 5.8**

Find the Fourier transform of a *Gaussian wavepacket*:

$$f(t) = Ae^{-\frac{t^2}{2\sigma^2}}e^{i\omega_0 t} \tag{5.56}$$

The real part of $f(t)$ is plotted in Figure 5.17. Estimate the frequency spread if the time interval between the $\frac{1}{e}$ points of the Gaussian profile of $f(t)$ is $4.00 \times 10^{-8}s$.

Solution

Take the Fourier transform in the time-frequency domain (Equation 5.47). Proceed

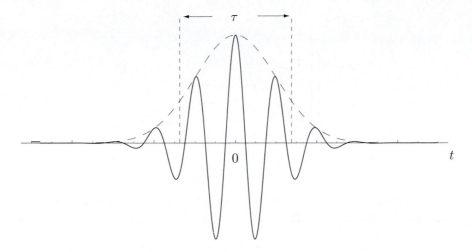

Figure 5.17   Gaussian wavepacket in the time domain.

as in Example 5.6:

$$F(\omega) = \int_{-\infty}^{\infty} f(t)e^{-i\omega t}\,dt$$

$$= A\int_{-\infty}^{\infty} e^{-\frac{t^2}{2\sigma^2}} e^{i\omega_0 t} e^{-i\omega t}\,dt$$

$$= A\int_{-\infty}^{\infty} \exp[-\frac{t^2}{2\sigma^2} - i\,(\omega - \omega_0)\,t + \frac{(\omega - \omega_0)^2\,\sigma^2}{2} - \frac{(\omega - \omega_0)^2\,\sigma^2}{2}]dt$$

$$= Ae^{-\frac{(\omega-\omega_0)^2\sigma^2}{2}} \int_{-\infty}^{\infty} \exp[-\frac{1}{2\sigma^2}\left(t^2 + i2\,(\omega - \omega_0)\,\sigma^2 t - (\omega - \omega_0)^2\,\sigma^4\right)]dt$$

$$= Ae^{-\frac{(\omega-\omega_0)^2\sigma^2}{2}} \int_{-\infty}^{\infty} \exp[-\frac{1}{2\sigma^2}\left(t + i\,(\omega - \omega_0)\,\sigma^2\right)^2]dt$$

Change variables:

$$u = \frac{1}{\sqrt{2}\sigma}\left(t + i\,(\omega - \omega_0)\,\sigma^2\right); \quad dt = \sqrt{2}\sigma\,du$$

to give

$$F(\omega) = \sqrt{2}\sigma A e^{-\frac{(\omega-\omega_0)^2\sigma^2}{2}} \int_{-\infty}^{\infty} e^{-u^2}\,du$$

or

$$F(\omega) = \sqrt{2\pi}\sigma A e^{-\frac{(\omega-\omega_0)^2\sigma^2}{2}}$$

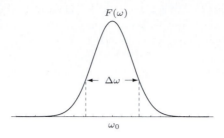

Figure 5.18   Gaussian wavepacket in the frequency domain.

To estimate $\delta\omega$, we again note (see Example 5.6) that the spread of $F(\omega)$ is the reciprocal of that for $f(t)$. The $\frac{1}{e}$ points occur at $t = \pm\sqrt{2}\sigma$, so $\tau = 2\sqrt{2}\sigma$. Thus,

$$\Delta\omega = \frac{2\sqrt{2}}{\sigma} = \frac{\left(2\sqrt{2}\right)^2}{\tau} = \frac{8}{\tau} \tag{5.57}$$

Substituting $\tau = 4.00 \times 10^{-8} s$ gives $\Delta\omega = 2 \times 10^8 \, rad/s$ or $\Delta f = 3.18 \times 10^7 \, Hz$. This is much smaller than the width of a visible spectral line in a typical discharge lamp, which is also broadened by the Doppler effect, and by perturbations caused by collisions with neighboring atoms.

From Equation 5.57 of the last Example, it appears that

$$\Delta\omega \, \Delta t \approx 8 \tag{5.58}$$

for a wavepacket with Gaussian profile, where we have used the $\frac{1}{e}$ packet duration τ as an estimate of Δt. Since Equations 5.55 and 5.58 we both based on estimates provided by specific examples, and since the right-hand sides of both equations are of order unity, it is appropriate to summarize them as follows:

$$\Delta x \, \Delta k > 1 \tag{5.59}$$

$$\Delta\omega \, \Delta t > 1 \tag{5.60}$$

These are sometimes called the *uncertainty relations for conjugate variables*, and they are a property of all *classical* waves. To obtain the corresponding *Heisenberg*[5] *uncertainty*[6] *relations* relations for quantum mechanics, one need only multiply both sides of Equations 5.59 and 5.60 by $\hbar$. Since energy and momentum for photons are given by $p = \hbar k$ and $E = \hbar\omega$ (see Section 2.4), this gives

$$\Delta x \, \Delta p > \hbar \tag{5.61}$$

$$\Delta E \, \Delta t > \hbar \tag{5.62}$$

The width $\Delta\omega$ is called the *linewidth*. The *natural lineshape* is approximately *Lorentzian*, as found in Section 3.7. For radiation emitted by an ensemble of hot atoms, there are two

---

[5] W. K. Heisenberg: 1901-1976. German physicist who contributed much of the foundation of quantum mechanics.
[6] These are also often referred to as the Heisenberg *indeterminacy* relations.

important contributions to $\Delta\omega$. The Doppler effect causes atoms moving toward the detector to emit higher frequencies, and those moving away from the detector to emit lower frequencies. A thermal distribution of random velocities results in a *Gaussian* frequency profile, as in the Example above. This effect is frequently referred to as *Doppler broadening*. There is also *collisional broadening*, caused by collisional perturbations that broaden the Lorentzian profile.

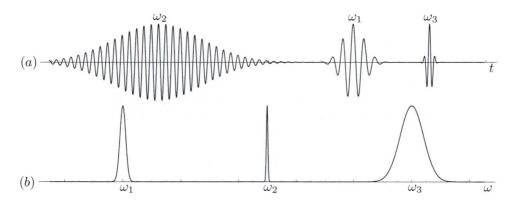

Figure 5.19 (a) A composite wave containing three Gaussian wavepackets. (b) The frequency spectrum of (a).

Figure 5.19(a) shows a composite wave plotted in the time domain that consists of three Gaussian components: a medium-frequency packet with long duration and frequency ω_2, a low-frequency packet of medium duration and frequency ω_1, and a short duration packet with the highest frequency ω_3. Figure 5.19(b) shows the corresponding *frequency spectrum*. The line centered about ω_2 is narrow because the corresponding packet has the longest duration. Similarly, the ω_3 packet has the shortest duration and a correspondingly large frequency spread.

Since $\lambda = c/f$, we note the following useful relations:

$$d\lambda = -\frac{c}{f^2}df = -\lambda\frac{df}{f}$$

with similar relations for ω and f. These are usually summarized as follows:

$$\frac{\Delta\lambda}{\lambda} = -\frac{\Delta f}{f} \tag{5.63}$$

with a similar expression for ω. When λ increases, f and ω decrease; hence, the minus sign.

■ **EXAMPLE 5.9**

Estimate the wavelength spread in a $589\ nm$ line emitted from a discharge tube if the frequency spread Δf is $5.00 \times 10^{10}\ Hz$.

Solution

Ignoring the negative sign in Equation 5.63, we have

$$\Delta\lambda = \lambda\left(\frac{\Delta f}{f}\right) = \frac{\lambda^2}{c}\Delta f = \frac{(589 \times 10^{-9}m)^2}{3.00 \times 10^8 m/s}\left(5.00 \times 10^{10}Hz\right) = 0.0578\,nm$$

Notice that in this case, $\lambda/\Delta\lambda$ is about 10^4.

Problem 5.23   The *natural linewidth* of an atomic transition is a minimum linewidth that does not include effects such as collisional and Doppler broadening. If τ is the lifetime of the excited state, this linewidth can be estimated with the energy-time Heisenberg uncertainty relation by substituting the lifetime for the time interval Δt. Use this approach to estimate the natural linewidth for the transition responsible for the red line of the helium-neon laser, which has a wavelength of $633\,nm$. Assume an average lifetime of $1.00 \times 10^{-8}\,s$. Calculate the spread in both frequency and wavelength. Include an estimate of $\frac{\lambda}{\Delta\lambda}$.

Problem 5.24   Show that

$$\frac{\lambda}{\Delta\lambda} = -\frac{k}{\Delta k}$$

5.7   GROUP AND PHASE VELOCITY

Since a monochromatic wave has no beginning or end, it cannot send a signal. Information can only travel on a wave that is *modulated*, and signals may only be sent using wavepackets. Wavepackets such as those illustrated in Figure 5.19 consist of a range of k-components that are all *in phase* at the center of the packet. In other words, at the peak of the packet, each k-component has a constant phase value:

$$\frac{d\phi}{dk} = \frac{d}{dk}\left(kx - \omega t\right) = 0$$

Since x and t do not depend upon k,

$$x = \left(\frac{d\omega}{dk}\right)t$$

The term within the parentheses is called the *group velocity*

$$v_g = \frac{d\omega}{dk} \tag{5.64}$$

The name derives from the image of a collection of k-components moving as a group. The group velocity v_g will in general have a value that is different from the *phase velocity* found in Chapter 2:

$$v_p = \frac{\omega}{k} = \frac{c}{n} = f\lambda \tag{5.65}$$

The angular frequency ω expressed as a function of k is called a *dispersion relation*. For example,

$$\omega = \frac{c}{n}k \tag{5.66}$$

If there is no dispersion, then n is a constant and $v_g = v_p$. If there is dispersion, then $n = n(k)$, and the phase velocity depends upon frequency:

$$v_g = \frac{d\omega}{dk} = \frac{d}{dk}(k\,v_p) = v_p + k\left(\frac{dv_p}{dk}\right) \tag{5.67}$$

where

$$\frac{dv_p}{dk} = \frac{d}{dk}\left(\frac{c}{n}\right) = -\frac{c}{n^2}\left(\frac{dn}{dk}\right)$$

giving

$$v_g = v_p\left[1 - \frac{k}{n}\left(\frac{dn}{dk}\right)\right] \tag{5.68}$$

In a similar way, one obtains (see Problem 5.25)

$$v_g = v_p\left[1 + \frac{\lambda}{n}\left(\frac{dn}{d\lambda}\right)\right] \tag{5.69}$$

■ EXAMPLE 5.10

Conductors are characterized by a *plasma frequency* ω_p that for a particular conductor is a constant corresponding to wavelengths well into the ultraviolet (see Example 3.13). The conductor becomes *transparent* for frequencies greater than the plasma frequency, with an index of refraction that is *less* than one (Equation 3.101):

$$n = \sqrt{1 - \frac{\omega_p^2}{\omega^2}}$$

As noted in the discussion following Equation 3.101, this implies a phase velocity that is *greater than c*. Find the group velocity for this case, and show that it is less than c.

Solution

Since in this case, n is given as a function of ω, we should solve for a dispersion relation that gives ω as an explicit function of k. Using the above expression for n:

$$\frac{\omega}{k} = \frac{c}{n} = \frac{c}{\sqrt{1 - \frac{\omega_p^2}{\omega^2}}}$$

Thus,

$$\omega^2\left(1 - \frac{\omega_p^2}{\omega^2}\right) = c^2 k^2$$

giving

$$\omega = \sqrt{c^2 k^2 + \omega_p^2}$$

The group velocity is

$$v_g = \frac{d\omega}{dk} = \left(\frac{1}{2\sqrt{c^2k^2 + \omega_p^2}}\right)(c^2 2k) = \frac{c^2 k}{\omega}$$

where in the last step, we substituted from above. Using Equation 5.65, we find

$$v_g = \frac{c^2 k}{\omega} = c\sqrt{1 - \frac{\omega_p^2}{\omega^2}}$$

Since $\omega > \omega_p$, $v_g < c$.

If n is expressed as a function of ω, then we may also compute the group velocity as follows

$$\frac{dk}{d\omega} = \frac{d}{d\omega}\left(\frac{\omega\, n(\omega)}{c}\right) = \frac{n(\omega)}{c} + \frac{\omega}{c}\frac{dn(\omega)}{d\omega} = \frac{1}{v_g}$$

giving (see Problem 5.26)

$$v_g = \frac{c}{n + \omega\frac{dn}{d\omega}} \tag{5.70}$$

For normal dispersion, $n > 1$ and $\frac{dn}{d\omega} > 0$ giving $v_g < c$. However, in the vicinity of an absorption feature there is *anomalous dispersion* (see Figure 3.24) where $\frac{dn}{d\omega} < 0$, and in this case it is possible to have $v_g > c$. In such cases, the group velocity is not a useful measure of signal velocity[7].

The phenomenon of *beats* discussed in Section 5.3.6 provides an intuitive example. According to Equation 5.30, an envelope that travels at velocity $\frac{\Delta\omega}{\Delta k}$ modulates the wave that travels at $\frac{\bar{\omega}}{\bar{k}}$. A situation where n decreases as ω increases would give $\Delta k < 0$ when $\Delta\omega > 0$. In this case, the superposition of two forward-traveling waves would result in a backward-traveling modulation envelope, again illustrating a case where v_g cannot represent a signal velocity in the vicinity of a region of anomalous dispersion.

Most optical applications use transparent dielectric media with normal dispersion, and in most cases the group velocity is frequency dependent. This causes the frequency components of an optical pulse to propagate with different speeds, thus causing the pulse profile to change as the pulse propagates. In optical fibers, this can cause the square profile of digital data pulses to distort, which if uncorrected can lead to data corruption.

Problem 5.25   Show that

$$v_g = v_p\left[1 + \frac{\lambda}{n}\left(\frac{dn}{d\lambda}\right)\right]$$

Problem 5.26   Show how the result of Example 5.10 follows from Equation 5.70.

Problem 5.27   Find the index of refraction, the phase velocity, and the group velocity for $100\,nm$ photons traveling through copper with a plasma frequency of $1.64 \times 10^{16}\,rad/s$.

---

[7]See, for example *Classical Electrodynamics* by J.D. Jackson [12].

5.8   INTERFEROMETRY

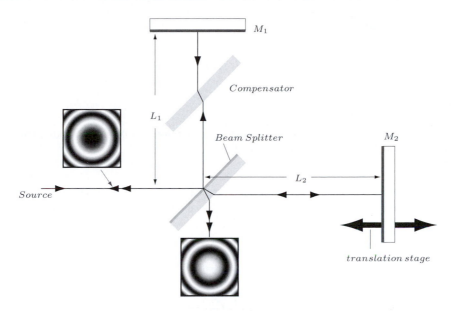

Figure 5.20   A Michelson interferometer.

An *interferometer* is a device designed to facilitate the study of interference. *Optical interferometers* are designed to operate at optical frequencies, and are typically constructed with *mirrors* and partially reflecting mirrors referred to as *beam splitters*. Interferometers constructed with beam splitters are sometimes referred to as *amplitude-splitting* interferometers, as opposed to a *wavefront-splitting* Young's double-slit interferometer. A mirror is designed to specularly reflect 100% of incident light, and can be made of metal, or with a dielectric slab that is coated with a metallic or dielectric film[8]. A beam splitter is made by coating a transparent dielectric slab with a metallic or dielectric coating that transmits a portion of the incident beam. A 50-50 beam splitter transmits 50% along the incident beam direction (neglecting refraction), and reflects 50%. Note that the metallic or dielectric film that does the reflecting is typically only on one side of the dielectric slab, and it is often important to know which side this is[9].

A *Michelson interferometer*[10] is illustrated in Figure 5.20. A beam of light incident from the source is incident on a 50-50 beam splitter oriented at 45°. The reflected and transmitted beams both travel to mirrors aligned to precisely reflect each beam. Before recombining at the beam splitter, the two beams travel different *optical* distances L_1 and L_2. The combined

---

[8]High reflectivity films using multiple dielectric coatings are discussed in Section 5.11.
[9]In the laboratory, this can be checked by carefully observing the *two* reflections of a small piece of paper. If the coating is on the side closest to the paper, the corresponding reflection will be brighter. Note that metallic coatings can be damaged by physical contact with the paper.
[10]Albert A. Michelson (1852-1931): American physicist who made many important contributions, especially in optics.

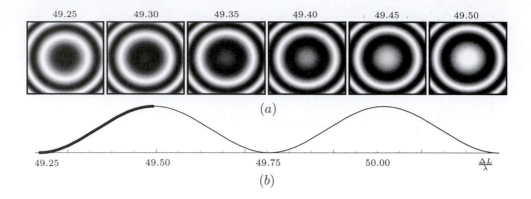

(a)

(b)

Figure 5.21   (a) Fringe variation in a Michelson interferometer as the mirror is moved by $\lambda/4$. (b) Interferogram. The thick segment corresponds to the fringes in (a).

beam that does not travel back into the source can illuminate an observing screen, where interference fringes are observed as indicated in the figure. One of the mirrors is attached to a precision translation stage so that ΔL can be varied.

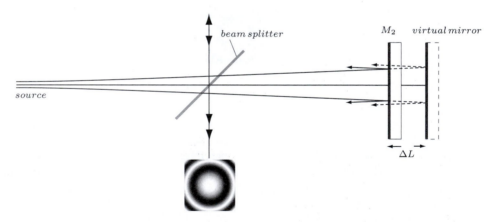

Figure 5.22   Two overlapping diverging beams. The beam from the virtual mirror has a time delay.

Beams 1 and 2 will interfere constructively whenever their optical paths differ by an integer number of wavelengths. Since each beam traverses its arm of the interferometer twice, the path difference is given by

$$\delta = 2L_2 - 2L_1 = 2\Delta L$$

Thus, as the movable mirror translates, the central bright fringe changes from dark to bright and vice versa whenever ΔL changes by $\lambda/4$. Figure 5.23(a) illustrates changes in the fringe pattern as ΔL varies from 49.25λ to 49.5λ. For a mirror separation of 49.25λ the center of the pattern is dark, but as the optical path difference increases toward 49.5λ, the central region of the fringe pattern builds toward constructive interference. As the mirror continues to move, a detector placed behind a pinhole at the center of the fringe pattern would record the curve of Figure 5.23(b). Such a curve is called an *interferogram*.

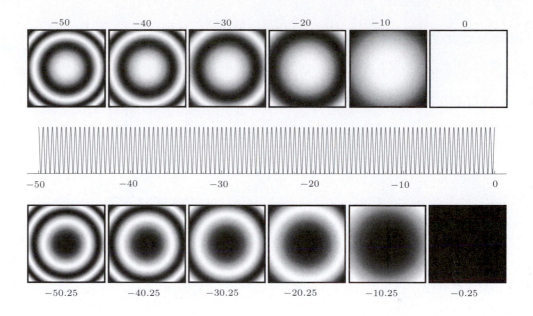

Figure 5.23   Fringe patterns as ΔL approaches zero. The patterns below the interferogram are the complements of those above. At $\Delta L = 0$, the entire field of one interferometer output is bright, and the other field is dark.

Circular fringes are observed when the illumination has a finite angular beam divergence. The interference results from the overlap of two beams, as shown in Figure 5.22. Reflection in the beam splitter results in a *virtual mirror* displaced by ΔL from the real mirror as illustrated. The beam that reflects from the virtual mirror is *time-shifted* relative to the other beam, and thus has a different wavefront curvature if ΔL is nonzero. Figure 5.23 illustrates fringe patterns for various values of ΔL when the beam divergence is $9.5°$. When ΔL is large, conditions for constructive and destructive interference occur for more than one available path through the interferometer. Circular fringes[11] occur because of the cylindrical symmetry about the beam axis. If $\Delta L = 0$, then both beams have the same wavefront curvature, and this results in interference that causes the entire observation field to have uniform irradiance, as indicated in Figure 5.23.

White light fringes may be observed with a Michelson interferometer adjusted so that ΔL is within a few wavelengths of zero, as illustrated in Figure 5.24. For small values of ΔL, different frequencies overlap to give circular fringes with a splendid iridescence, as illustrated in Figure 5.24(c). For larger values of ΔL, the different colors combine in a way that is effectively random, and the distinct colors fade into a uniform white illumination. The corresponding interferograms are illustrated in Figures 5.24(a) and (b).

White light fringes can be observed only if the interferometer is equipped with a *compensator plate*, as illustrated in Figure 5.20. The compensator plate is identical in size and composition to the beam splitter substrate. With this in place, a careful examination of Figure 5.20 verifies an equal thickness of dielectric in each arm of the interferometer. Thus, the *dispersion* in each arm is equal. Without the compensator, the optical path difference

---

[11]Circular fringes only occur with interferometers constructed with perfectly flat optical surfaces. Interferometers constructed with distorted optical surfaces give distorted fringe patterns.

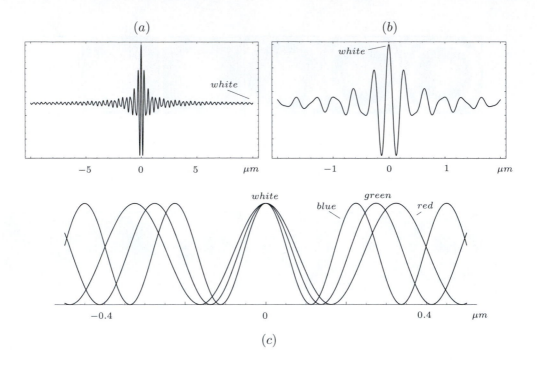

Figure 5.24 (a) Interferogram for white light. (b) Zoom detail of (a). (c) Interferograms for different colors combine to give a white fringe at $\Delta L = 0$, and again for larger values of ΔL.

through the interferometer would be frequency dependent, and could vary by many microns over the optical spectrum, making it impossible to observe white light fringes.

To obtain more monochromatic illumination, one could, for example, illuminate the interferometer with a single line from a discharge lamp. As discussed in Section 5.6, such a line will contain a continuous range of frequencies distributed over a finite linewidth. Since the frequency range is narrow, interference may be observed over a much larger range of ΔL, as illustrated in Figure 5.25. As ΔL increases, the *visibility V* of the fringe pattern begins to decrease:

$$V = \frac{I_{\max} - I_{\min}}{I_{\max} + I_{\min}} \tag{5.71}$$

The *coherence length L_c* is defined as the value of ΔL that gives a visibility of one half. For example, the coherence length of white light is less than a micron, the coherence length of a spectral line can be many tens of microns, and L_c for a frequency-stabilized laser with a long cavity length can be several meters. The *coherence time t_c* is the time required for the composite wave to travel one coherence length:

$$L_c = c\, t_c \tag{5.72}$$

To observe white light fringes in the laboratory, the interferometer should first be aligned with light that has a very long coherence length, such as that from a helium-neon laser. By monitoring the fringe size (see Figure 5.23) ΔL can be adjusted to a value that is close enough to zero to allow fringes to be observed using a white light source that is filtered

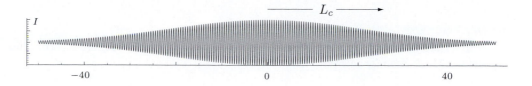

Figure 5.25 The coherence length determined from a Michelson interferogram.

by a narrow bandpass filter[12]. Adjusting ΔL for maximum fringe visibility will usually produce a value small enough to allow white light fringes to be observed when the filter is removed. If not, repeat with a filter that has a wider bandpass.

■ APPLICATION NOTE 5.3

Fourier Transform Spectroscopy

A *Fourier transform spectrometer* is a Michelson interferometer with a movable mirror that is capable of traveling long distances. The interferogram consists of two terms: a bias irradiance level that remains after ΔL exceeds the coherence length, and a normalized modulation term $I_k(x)$

$$I = \frac{I_s}{2}\left(1 + I_k(x)\right)$$

where I_s is the source irradiance, and the subscript on $I_k(x)$ expresses the fact that the range of wavelengths contained in the source affects the shape of the resulting interferogram. Using a Fourier transform, we may re-express the modulation term as a function of k:

$$I(k) = \int_{-\infty}^{\infty} I_k(x)e^{-ikx}dx$$

where x is the position of the movable mirror relative to the $\Delta L = 0$ point. The function $I(k)$ gives irradiance as a function of k, or equivalently, wavelength. In other words, $I(k)$ is the *optical spectrum* of the source.

As found in Example 5.8 and expressed in Equation 5.59, the capability to resolve spectral features separated by small values of Δk is possible only by moving the mirror through large values of Δx

$$\Delta k \sim \frac{1}{\Delta x}$$

The *resolving power* of the spectrometer is defined as $\lambda/\Delta\lambda$, which according to Problem 5.24 is also equal in magnitude to $k/\Delta k$

$$\Re = \frac{\lambda}{\Delta\lambda} = \frac{k}{\Delta k} \sim \frac{2\pi\Delta x}{\lambda}$$

A mirror travel of one meter gives a resolving power in the visible that exceeds 10^6.

---

[12] The filter can be placed in front of the illumination source, or the eye.

Interferometers can be constructed using a variety of geometries. Figure 5.26(b) illustrates a *Mach-Zehnder* interferometer design.

Problem 5.28   The coherence length of a spectral source can be estimated by utilizing the uncertainty relations as an equality. Estimate the coherence length of white light if the interferogram is recorded by a detector that has uniform responsivity between $400\,nm$ and $700\,nm$, and no responsivity outside of this range. Repeat the calculation if the input of the detector is covered by an optical filter centered at $550\,nm$ with a bandpass of $1.00\,nm$.

Problem 5.29   Using the uncertainty relations as equalities, estimate the frequency and wavelength spread of a laser that operates at $530\,nm$ with a coherence length of $30\,cm$.

Problem 5.30   Find the fringe visibility for an interferometer adjusted to give an interferogram (measured with a linear power detector) with minima that are 10% of the maxima.

Problem 5.31   The movable mirror in a Michelson interferometer is displaced to give an interferogram with 1000 local maxima. If the illumination wavelength is $633\,nm$, how far did the mirror move?

5.8.1   *Energy Conservation and Complementary Fringe Patterns

There are two sets of fringes in a Michelson interferometer: one at the observing screen in Figure 5.20, and a set that travels back into the source. When the visibility is high, the peak irradiance in Figure 5.25 is equal to the source irradiance[13]. If there are no losses in the interferometer beam path, then conservation of energy requires that when there is constructive interference at any point on one observing screen, the corresponding point on the other set of fringes is a point of destructive interference. Similar remarks apply for the Mach-Zehnder interferometer, as illustrated in Figure 5.26(b) (see Problem 5.35). With flat optics, either interferometer can be adjusted so that ΔL is zero, giving an entire observation field that is uniformly bright or dark. Here, it seems, is a paradox: if $\Delta L = 0$ and one field is bright, why is the other field dark?

Figure 5.26(a) shows a simplified schematic of a Michelson interferometer illuminated by a source S. The substrate of the beam splitter B and the associated compensator plate are not shown, but contribute to the optical paths L_1 and L_2 to eliminate all effects of dispersion. Only the semi-reflective film of the beam splitter is shown. Mirror M_1 is in the L_1 path, and mirror M_2 is in path L_2. The output beam E_2 consists of two contributions: E_{22}, which travels the path $S \to B \to M_2 \to B \to E_2$, and E_{21}, which travels the path $S \to B \to M_1 \to B \to E_2$. Similarly, $E_1 = E_{11} + E_{12}$, with E_{11} traveling along $S \to B \to M_1 \to B \to E_1$ and E_{12} traveling along $S \to B \to M_2 \to B \to E_1$. The

---

[13]For values of ΔL well beyond L_c, the two beams leaving the interferometer have half the source irradiance.

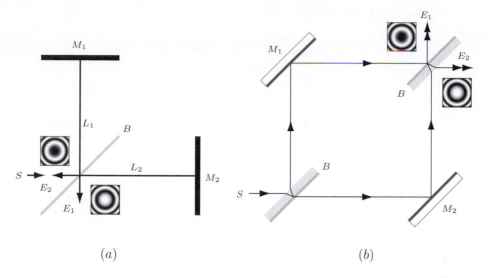

Figure 5.26 Complementary fringe patterns. (a) Michelson interferometer. (b) Mach-Zehnder interferometer.

entire optical path of the interferometer is assumed to be *lossless*, meaning that there are no effects of absorption and/or scattering to remove energy from either of the two beams.

The reflective film of the beam splitter is assumed to reflect and transmit with arbitrary phase shifts that may depend on the direction of propagation. Thus, in Figure 5.26(a), the beam splitter reflects and transmits with complex amplitude coefficient r and t for beams incident on the side of the beam splitter closest to the source, and with r' and t' for beams incident on the opposite side. We may express each complex coefficient in polar form:

$$r = |r|\,e^{i\phi_r} \qquad r' = |r'|\,e^{i\phi'_r}$$
$$t = |t|\,e^{i\phi_t} \qquad t' = |t'|\,e^{i\phi'_t} \tag{5.73}$$

Beam E_1 is the beam in Figure 5.26(a) that does not travel back into the source. Using Equations 5.73, it is given by

$$E_1 = E_{11} + E_{12} = \left(rt\,e^{i\phi_{11}} + tr'e^{i\phi_{12}}\right)E_i \tag{5.74}$$

where E_i is the source amplitude. The phase terms ϕ_{11} and ϕ_{12} account for all other phase changes in the optical path:

$$\phi_{11} = \frac{4\pi}{\lambda}L_1 + \phi_{M_1} \qquad \phi_{12} = \frac{4\pi}{\lambda}L_2 + \phi_{M_2} \tag{5.75}$$

where ϕ_{M1} and ϕ_{M2} account for any phase shifts produced by the mirrors M_1 and M_2. Similarly, the beam that travels back toward the source is given by

$$E_2 = E_{21} + E_{22} = \left(r^2\,e^{i\phi_{21}} + tt'e^{i\phi_{22}}\right)E_i \tag{5.76}$$

with

$$\phi_{21} = \frac{4\pi}{\lambda}L_1 + \phi_{M_1} \qquad \phi_{22} = \frac{4\pi}{\lambda}L_2 + \phi_{M_2} \tag{5.77}$$

Conservation of energy requires that

$$|E_i|^2 = |E_1|^2 + |E_2|^2 \tag{5.78}$$

Substitution and cancellation of $|E_i|^2$ gives

$$\begin{aligned}
1 &= |r|^2 |t|^2 + |t|^2 |r'|^2 + 2 \left|rr't^2\right| \cos\left(\phi_{12} - \phi_{11} + \phi_r' - \phi_r\right) \\
&\quad + |r|^4 + |t|^2 |t'|^2 + 2 \left|r^2 tt'\right| \cos\left(\phi_{22} - \phi_{21} + \phi_t + \phi_t' - 2\phi_r\right)
\end{aligned} \tag{5.79}$$

where

$$\phi_{12} - \phi_{11} = \phi_{22} - \phi_{21} = \frac{4\pi}{\lambda}\Delta L + (\phi_{M_2} - \phi_{M_1}) \equiv \Delta\phi_L$$

Since $\Delta\phi_L$ is arbitrary, the two cosine terms must sum to zero for all values of ΔL. The remaining terms must sum to one:

$$|r|^4 + |t|^2 |t'|^2 + |r|^2 |t|^2 + |t|^2 |r'|^2 = 1$$

Rearranging gives

$$|r|^2 \left(|r|^2 + |t|^2\right) + |t|^2 \left(|r'|^2 + |t'|^2\right) = 1$$

which is true provided

$$|r|^2 + |t|^2 = 1 = |r'|^2 + |t'|^2 \tag{5.80}$$

In order to sum to zero, the coefficients of the cosine terms must be equal:

$$\left|rr't^2\right| = \left|r^2 tt'\right|$$

giving

$$|r'|\,|t| = |r|\,|t'|$$

or

$$\frac{|r|}{|t|} = \frac{|r'|}{|t'|}$$

By Equation 5.80,

$$\frac{|r|}{\sqrt{1 - |r|^2}} = \frac{|r'|}{\sqrt{1 - |r'|^2}} \tag{5.81}$$

which has the solutions (see Problem 5.34)

$$\begin{aligned}
|r| &= |r'| \\
|t| &= |t'|
\end{aligned} \tag{5.82}$$

Substitution back into Equation 5.79 gives

$$1 = 1 + 2\left|r^2\right| \left|t^2\right| \left[\cos\left(\Delta\phi_L + \phi_r' - \phi_r\right) + \cos\left(\Delta\phi_L + \phi_t + \phi_t' - 2\phi_r\right)\right]$$

which is true only if the two cosine terms cancel. Thus, the arguments must differ by π:

$$\phi_r' - \phi_r = \phi_t + \phi_t' - 2\phi_r \pm \pi \tag{5.83}$$

or

$$\phi_t - \phi_r = \phi_t' - \phi_r' \pm \pi \tag{5.84}$$

The apparent paradox referred to in the opening paragraph of this section is resolved by an asymmetry at the beam splitter, as is verified by Equations 5.74 and 5.76. This asymmetry and the conservation of energy results in the factor of π in Equations 5.83 and 5.84 that causes complementary fringe patterns in the two interferometer outputs. Note that it did not matter to which side of Equation 5.83 the factor of $\pm\pi$ was added. Regardless of this uncertainty, an interferometer with flat non-absorbing optics can always be adjusted with a small value of ΔL that gives one output that is uniformly bright, with the other uniformly dark.

Problem 5.32   A *symmetric beam splitter* is one where $r^2 = t^2 = 0.5$. Show that if there are no losses, then $\phi_t - \phi_r = \pm\frac{\pi}{2}$.

Problem 5.33   Show that
   a) $e^{i(\phi'_t - \phi_r)} + e^{i(\phi'_r - \phi_t)} = 0$
   b) $r^*t' + t^*r' = 0$

Problem 5.34   Show that Equation 5.81 has the solutions given by Equation 5.82.

Problem 5.35   Figure 5.26(b) illustrates a Mach-Zehnder interferometer. Let E_1 be the beam that travels toward the top of the page, and E_2 be the beam traveling to the right. Let M_1 be the mirror in the upper left corner, and M_2 be the mirror in the lower right corner. Let both beam splitters be identical, with reflection and transmission characterized by amplitude ratios r and t when illuminated on the side with the semi-reflecting film, and r' and t' when illuminated from the side with the dielectric substrate.
   a) Show that

$$E_1 = \left(r're^{i\phi_{11}} + t'^2 e^{i\phi_{12}}\right) E_i$$

$$E_2 = \left(r'te^{i\phi_{21}} + t'r'e^{i\phi_{22}}\right) E_i$$

   b) Show that

$$|r'|^2 \left(|r|^2 + |t|^2\right) + |t'|^2 \left(|r'|^2 + |t'|^2\right) = 1$$

   and hence Equation 5.80.
   c) Verify Equations 5.82 and 5.84 and show that the two interferometer outputs give complementary fringe patterns.

5.9   SINGLE-PHOTON INTERFERENCE

Consider a Mach-Zehnder interferometer adjusted with $\Delta L = 0$ so that one interferometer output is uniformly bright, and the other output uniformly dark, as discussed in the previous section. Now use a source that is so dim that one is certain that only one photon is traversing the interferometer at any point in time. Assume for simplicity that each beam splitter transmits and reflects equally and that the horizontal output of Figure 5.26(b) is the one that is bright. Thus, a photon reaching the first beam splitter transmits or reflects with 50% probability, but when it reaches the second identical beam splitter the probability for reflection is zero! There is no easy resolution to this fascinating result. For more information, see *The Quantum Challenge* by Greenstein and Zajonc [8]. In particular, see their Figure 2-7 for data, and the subsequent discussion of the "Delayed Choice Experiment." As in Section 2.4.1, it appears as if the photon takes *both* paths. See Section 6.5 for more on the "sum over paths" interpretation.

5.10 MULTIPLE-BEAM INTERFERENCE

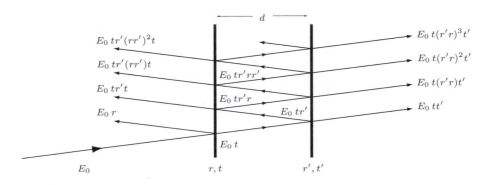

Figure 5.27   Multiple beam interference between two parallel partially reflecting surfaces.

Consider two parallel, flat reflective surfaces separated by distance d, as shown in Figure 5.27. The first surface has amplitude reflection and transmission coefficients r and t, and the second surface has amplitude reflection and transmission coefficients r' and t'. Normal incidence is assumed; the angle used in the figure is for purposes of illustration. Multiple reflections between the two surfaces result in two series of reflected and transmitted terms, as shown. Let the phase difference between successive transmitted terms be δ_0 due to the round-trip optical travel

$$\delta_0 = k(2nd)$$

where n is the index of refraction in the region between the two reflecting surfaces. The transmitted series is

$$
\begin{aligned}
E_t &= E_0 tt' + E_0 t \left(r'r\right) t' e^{i\delta_0} + E_0 t \left(r'r\right)^2 t' e^{i2\delta_0} + E_0 t \left(r'r\right)^3 t' e^{i3\delta_0} + \dots \\
&= E_0 tt' \left[1 + \left(r'r\right) e^{i\delta_0} + \left(r'r\right)^2 e^{i2\delta_0} + \left(r'r\right)^3 e^{i3\delta_0} + \dots\right] \\
&= E_0 tt' \left[1 + \left(r're^{i\delta_0}\right) + \left(r're^{i\delta_0}\right)^2 + \left(r're^{i\delta_0}\right)^3 + \dots\right]
\end{aligned}
$$

The last result is in the form of a *geometric series*

$$\sum_{n=0}^{\infty} x^n = 1 + x + x^2 + \dots = \frac{1}{1-x} \tag{5.85}$$

provided $|x| < 1$, which is obviously valid in our case. Thus, the infinite series of transmitted terms is given by

$$E_t = \frac{E_0 tt'}{1 - rr'e^{i\delta_0}}$$

If we let $r = r'$ and $t = t'$, we obtain

$$E_t = \frac{E_0 t^2}{1 - r^2 e^{i\delta_0}}$$

Recall that for normal incidence, the reflectance R is given by rr^*, and the transmittance is given by $T = tt^*$. Since irradiance is proportional to EE^*, the transmitted irradiance is given by

$$I_t = \frac{I_0 T^2}{|1 - r^2 e^{i\delta_0}|^2}$$

The denominator of the last result is given by

$$\begin{aligned}
|1 - r^2 e^{i\delta_0}|^2 &= \left(1 - r^2 e^{i\delta_0}\right)\left(1 - r^{*2} e^{-i\delta_0}\right) \\
&= 1 - \left(r^2 e^{i\delta_0} + r^{*2} e^{-i\delta_0}\right) + R^2
\end{aligned}$$

The amplitude reflectance r may in general be complex to allow for reflection phase shifts. If we express r^2 in polar form and incorporate any extra reflection phase shifts into a new phase shift term δ, we obtain

$$\delta = \frac{4\pi n d}{\lambda} + 2\phi_R \tag{5.86}$$

where ϕ_R is the reflection phase change. This gives

$$\begin{aligned}
|1 - r^2 e^{i\delta_0}|^2 &= 1 - \left(r^2 e^{i\delta_0} + r^{*2} e^{-i\delta_0}\right) + R^2 \\
&= 1 - \left(R e^{i\delta} + R e^{-i\delta}\right) + R^2 \\
&= 1 - 2R\cos\delta + R^2 \\
&= (1 - R)^2 + 2R\left(1 - \cos\delta\right)
\end{aligned}$$

From trigonometry,

$$(1 - \cos\delta) = 2\sin^2\left(\frac{\delta}{2}\right)$$

so that

$$\begin{aligned}
|1 - r^2 e^{i\delta_0}|^2 &= (1 - R)^2 + 4R\sin^2\left(\frac{\delta}{2}\right) \\
&= (1 - R)^2 \left[1 + \frac{4R}{(1 - R)^2}\sin^2\left(\frac{\delta}{2}\right)\right]
\end{aligned}$$

We define the *coefficient of finesse* F:

$$F = \frac{4R}{(1 - R)^2} \tag{5.87}$$

to give

$$I_t = \frac{I_0 T^2}{(1 - R)^2}\left(\frac{1}{1 + F\sin^2\left(\frac{\delta}{2}\right)}\right) \tag{5.88}$$

If there is no absorption, then $T = 1 - R$, giving

$$I_t = \frac{I_0}{1 + F\sin^2\left(\frac{\delta}{2}\right)} \tag{5.89}$$

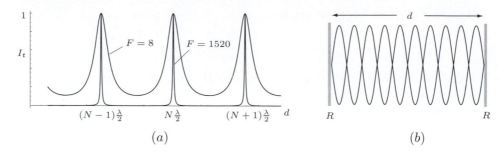

Figure 5.28   (a) Transmittance of two identical flat parallel mirrors separated by distance d. A mirror reflectance of 0.5 gives $F = 8$, and $R = 0.95$ gives $F = 1520$. (b)Standing waves when the mirror spacing d is equal to an integer number half wavelengths.

The coefficient of finesse F can become large for mirror reflectances R that approach unity. Consider two flat, parallel mirrors separated by a distance d. A large value of F causes the transmission to be low except within a region where $sin[\delta/2]$ is close to zero. Neglecting reflection phase shifts that are independent of d, this occurs when $k(nd) = N\pi$. Thus, positions of maximum transmission are separated by $\lambda/2$, as indicated in Figure 5.28(a). At each of these discrete points, there is 100% transmission[14] even with individual mirrors that are very reflective. In such a case, the two mirrors form a *resonant cavity*. At resonance, energy that transmits into the cavity multiply-reflects between the cavity mirrors. Furthermore, all reflections interfere constructively to produce a relatively large irradiance. Intuitively, one expects an irradiance between the mirrors to be inversely proportional to mirror transmittance. For example, a mirror that reflects 95% and transmits 5% needs an incident irradiance of $20 I_0$ to transmit the irradiance I_0.

At resonance, the radiation field between the mirrors satisfies a *standing wave condition*, as indicated in Figure 5.28(b). Standing waves are produced when waves that travel in opposite directions constructively interfere. For example, consider the combination of the following equal amplitude monochromatic waves:

$$E = E_0 \left[\sin\left(kx - \omega t\right) + \sin\left(kx + \omega t\right)\right]$$

Use

$$\sin A + \sin B = 2\sin \tfrac{1}{2}\left(A + B\right)\cos \tfrac{1}{2}\left(A - B\right)$$

to give

$$E = \left(2E_0 \sin kx\right)\cos \omega t$$

The portion within the parentheses is a position-dependent standing wave amplitude such as that in Figure 5.28(b). There are *nodes* where the field is zero, and *antinodes* with maximum field fluctuations. Nodes and antinodes are spaced by $\lambda/2$. Thus, moving one mirror by $\lambda/2$ transitions the cavity from one resonance to the next.

---

[14]Most reflective coatings have some absorption. The number of reflections and associated absorption losses increase as F increases.

Problem 5.36 Given Equation 5.85 for an infinite geometric series, show that the finite series is given by

$$\sum_{n=0}^{N} x^n = \frac{1 - x^{N+1}}{1 - x}$$

Assume that x is a real number with $|x| < 1$.

Problem 5.37 Find the coefficient of finesse for Fabry-Perot mirrors with a reflectivity of 50%. Repeat when the reflectivity is 99.9%.

Problem 5.38 A set of Fabry-Perot mirrors transmits a one-watt beam. If each mirror has a reflectivity of 90%, what is the incident beam power in the region between the two mirrors?

5.10.1 The Scanning Fabry-Perot Interferometer

A *scanning Fabry-Perot interferometer*[15] consists of two parallel mirrors, one of which is attached to a movable base. As the mirror is scanned, wavelengths that resonate in the cavity transmit through the interferometer at distinct values of d.

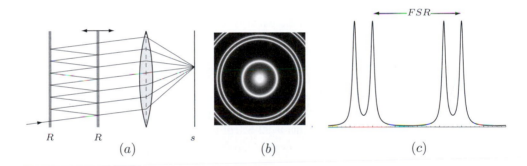

R R (a) s (b) (c)

Figure 5.29 (a) Scanning Fabry-Perot interferometer. (b) Fringe pattern produced by two closely-spaced spectral lines. (c) Interferogram of (b). The pattern repeats as the interferometer scans over the free spectral range.

Higher values of F give a sharper transmission bandpass and greater spectral resolution. To find the half-width of the bandpass, solve

$$\frac{1}{1 + F \sin^2\left(\frac{\delta_{1/2}}{2}\right)} = \frac{1}{2}$$

giving

$$\sin\left(\frac{\delta_{1/2}}{2}\right) = \frac{1}{\sqrt{F}}$$

[15]Charles Fabry: 1867-1945. French physicist who discovered the ozone layer and made important contributions to interferometry. Alfred Perot: 1863-1925. French physicist and colleague of Charles Fabry. See J. F. Mulligan, "Who Were Fabry and Perot?" American Journal of Physics 64, 1109 (1996).

For large F we may use the small angle approximation. The full width at half maximum (FWHM) is then

$$FWHM = 2\delta_{1/2} = \frac{4}{\sqrt{F}} \tag{5.90}$$

The *finesse*[16] is defined as the ratio of a full 2π phase change to the FWHM:

$$\mathcal{F} = \frac{2\pi}{FWHM} = \frac{\pi\sqrt{F}}{2} \tag{5.91}$$

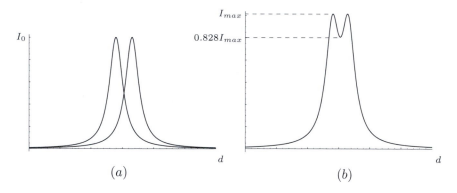

Figure 5.30   (a) Two closely spaced spectral lines separated by their FWHM. (b) The sum of (a), showing the saddle point.

The capability to distinguish closely spaced spectral lines is called the *resolving power*. Consider two equal-irradiance spectral lines that are much narrower than the profile measured by the Fabry-Perot interferometer. Let the two profiles intersect at the half-power points, as shown in Figure 5.30(a). The measured profile is the sum illustrated in Figure 5.30(b) which has a *saddle point* that is 82.8% of the maximum irradiance. Noting that in Equation 5.86, $\phi_R << \frac{2\pi(2nd)}{\lambda}$, we neglect the reflection phase changes to give

$$\Delta\delta = -\frac{2\pi(2nd)}{\lambda^2}\Delta\lambda$$

Rearranging gives

$$\frac{\lambda}{\Delta\lambda} = -\frac{\delta}{\Delta\delta}$$

The resolving power is defined as $\lambda/\Delta\lambda$. For δ we substitute $N(2\pi)$, where the *order number* N is the number of half wavelengths in d when there is resonance (see Figure 5.28):

$$N = \frac{nd}{\lambda/2} = \frac{2nd}{\lambda} = \frac{\delta}{2\pi} \tag{5.92}$$

For $\Delta\delta$, we use the $FWHM$ and Equation 5.90 to give

$$RP = \left|\frac{\lambda}{\Delta\lambda}\right| = \frac{\delta}{FWHM} = \frac{N(2\pi)}{\frac{2\pi}{\mathcal{F}}} = N\mathcal{F} = \frac{N\pi\sqrt{F}}{2} \tag{5.93}$$

---

[16]Be careful not to confuse the coefficient of finesse of Equation 5.87 with the finesse given by Equation 5.91. The notation is standard.

Notice that the resolving power increases with mirror spacing.

A given spectral line transmits each time d satisfies a standing wave condition, as illustrated in Figure 5.29(c). *Overlap* occurs when

$$(N + 1)\lambda = N(\lambda + \Delta\lambda)$$

or[17]

$$\Delta\lambda = \frac{\lambda}{N}$$

To avoid overlapping, the wavelength spread of the analysis beam must be limited to the *free spectral range* $\Delta\lambda_{fsr}$

$$\Delta\lambda_{fsr} = \frac{\lambda}{N} = \frac{\lambda^2}{2nd} \tag{5.94}$$

where λ is the average wavelength in the analysis beam. Since

$$\left|\frac{\Delta\lambda_{fsr}}{\lambda}\right| = \frac{\Delta f_{fsr}}{f} = \frac{\lambda}{2nd}$$

the free spectral range in terms of frequency is given by

$$\Delta f_{fsr} = \frac{f\lambda}{2nd} = \frac{c}{2nd} \tag{5.95}$$

Thus, wideband illumination results in a transmitted beam that consists of a number of narrow frequency components separated by $c/2nd$. To obtain a transmitted beam with a narrow frequency range, the illumination must be narrower than the free spectral range.

■ **EXAMPLE 5.11**

The isotopes hydrogen and deuterium have slightly different emission spectra. For the red Hα line of the Balmer series, the wavelength is $656.28\,nm$ for hydrogen and $656.10\,nm$ for deuterium ($\Delta\lambda = 0.18\,nm$). Describe how this splitting could be measured using Fabry-Perot mirrors with a reflectance of 90%.

Solution

The coefficient of finesse is

$$F = \frac{4R}{(1 - R)^2} = \frac{3.6}{(.1)^2} = 360$$

giving a finesse of

$$\mathcal{F} = \frac{\pi\sqrt{F}}{2} = \frac{\pi\sqrt{360}}{2} = 29.8$$

The splitting will be bracketed nicely with a free spectral range of $0.5\,nm$. Assuming that air fills the space between the mirrors, this requires a mirror spacing of

$$d = \frac{\lambda^2}{2\Delta\lambda_{fsr}} = \frac{(656.2\,nm)^2}{2\,(0.5\,nm)} = 4.31 \times 10^5 nm = 0.431\,mm$$

---

[17]This should not be confused with the wavelength spread associated with resolving power.

Such a small mirror separation leaves little room for alignment error! Clearly, the Fabry-Perot interferometer is best suited for very high resolution measurements. Under these conditions, N is equal to

$$N = \frac{4.31 \times 10^5\,nm}{656.2\,nm/2} = 1312$$

This gives a resolving-power of

$$RP = NF = (1312)\,(29.8) = 3.91 \times 10^4$$

and a minimum wavelength resolution of

$$\Delta\lambda = \frac{\lambda}{RP} = \frac{656.2\,nm}{3.91 \times 10^4} = 0.0168\,nm$$

which is more than a factor of 10 smaller than the splitting.

■ **APPLICATION NOTE 5.4**

The Laser Cavity

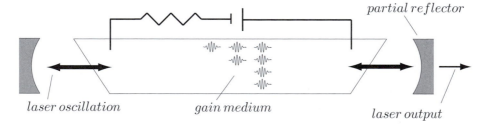

Figure 5.31   A laser cavity.

The word *LASER* is an acronym for Light Amplification by the Stimulated Emission of Radiation. In *stimulated emission*, photons interact with excited atoms or molecules of the laser medium to stimulate the emission of additional photons. Photons emitted by stimulated emission have identical phase and propagation direction as the stimulating photon, as opposed to the competing process of *spontaneous emission* where emission proceeds randomly. For example, the radiation from a neon sign results from spontaneous emission, and the beam from a neon laser results from stimulated emission. Stimulating photons must reach the excited states before these decay spontaneously, and this is usually achieved by placing the laser medium within a Fabry-Perot cavity.

Figure 5.31 shows a laser cavity formed by two mirrors that enclose a *gain medium*; in the case illustrated, this medium consists of a gas that is excited by collisions in an electrical discharge. Each mirror typically has a small amount of curvature (large radius of curvature) that serves to confine the multiple reflections. A partially reflecting mirror on the output end of the laser transmits the output beam.

The reflectivity of the output mirror is typically close to 100%, so the irradiance within the laser cavity must be substantially higher than that of the output. For example, if the output mirror transmits 10%, the irradiance in the cavity must be about ten times greater than that of the output. The high cavity irradiance increases the probability of stimulated emission.

The Fabry-Perot cavity also makes the laser output very monochromatic. For example, consider a helium-neon laser with a cavity formed by mirrors of 99% reflectivity. If the cavity length is $10.0\,cm$, then the resolving-power of the cavity is nearly 10^9 by Equation 5.93. Theoretically, this gives a $\Delta\lambda$ that is a very small fraction of a nm, and a corresponding Δf of less than a MHz. Longer cavity lengths can be even more monochromatic.

Changes in cavity length cause changes in the wavelength and frequency of the output. Thermal variations cause frequency changes with time scales on the order of minutes. Acoustic changes cause more rapid fluctuations. In either case, the frequency can vary over the entire free spectral range of the cavity.

Problem 5.39   Find the resolving-power and free spectral range of a scanning Fabry-Perot interferometer that uses identical mirrors of 80% reflectivity and a mirror separation of $20\,cm$. Assume an operating wavelength of $540\,nm$.

Problem 5.40   A $633\,nm$ helium-neon laser has 99% reflecting mirrors spaced $10.0\,cm$ apart. Find the free spectral range. Use the Fabry-Perot resolving power to estimate the spectral width (in both wavelength and frequency) of the output. Repeat for a cavity length of $1.00\,m$.

Problem 5.41   A Fabry-Perot cavity is made from a $1.00\,cm$-thick slab of germanium, with index of refraction 4.02. Find the finesse, the coefficient of finesse, the resolving power and the free spectral range when the wavelength is $3.00\,\mu m$.

5.11   INTERFERENCE IN MULTILAYER FILMS

As discussed in Section 5.3.3, single layers of thin dielectric films can be used to create antireflection coatings. It is also possible to design coatings for high reflectivity using multiple thin-film layers.

Figure 5.32 shows a thin film of thickness ℓ and index of refraction n_ℓ. It is surrounded on the left by an incident region with index of refraction n_0, and on the right by a transmitted region with index of refraction n_t. Incident light that transmits into the film will multiply reflect between the two dielectric interfaces, and we could certainly solve for the net transmittance and reflectance as in the last section, by detailed analysis of the infinite series of transmitted and reflected terms. Instead, we shall solve the *boundary value problem*. In this approach, Maxwell's equations are solved in each of the three regions of Figure 5.32, and these solutions are matched at the two boundaries using the *boundary conditions* determined in Section 3.3.1. Furthermore, we have already found the necessary solutions to Maxwell's equations; they are the electromagnetic wave solutions found in Section 2.3.1.

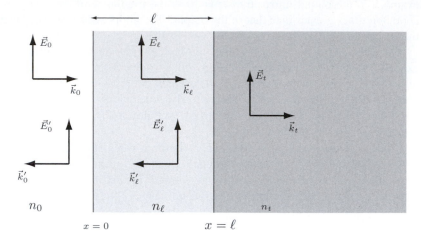

Figure 5.32   A thin film of thickness ℓ and index n_ℓ deposited on a substrate of index n_t.

Figure 5.32 illustrates the transverse electromagnetic wave solutions in each of the three regions. The incident region contains a forward-traveling wave with field amplitude $\vec{E}_0$ and wave vector $\vec{k}_0$ that represents the incident illumination. This region also contains a backward-traveling solution with $\vec{E}_0'$ and $\vec{k}_0'$ that represents the net reflected radiation. Both solutions must be used as we proceed, because they are *linearly independent*; a forward-traveling wave can not be represented as a linear combination of backward-traveling waves, and vice versa. The film region also contains a forward-traveling solution with $\vec{E}_\ell$ and $\vec{k}_\ell$ and a backward-traveling solution with $\vec{E}_\ell'$ and $\vec{k}_\ell'$ that account for all multiple reflections within this region. Finally, the transmitted region only contains a forward-traveling wave with $\vec{E}_0$ and $\vec{k}_0$, since we assume a semi-infinite transmitted region that cannot physically produce a backward component.

According to the boundary conditions found in Section 3.3.1 for non-magnetic dielectrics, the tangential components of $\vec{E}$ and $\vec{B}$ are continuous at each boundary. According to Section 2.3.1, E and B are related by

$$B = \frac{n}{c}E \qquad (5.96)$$

If we assume normal incidence, the vectors $\vec{E}$ and $\vec{B}$ are both tangential. According to the first boundary condition, the tangential components of electric field must be continuous at $x = 0$:

$$E_0 + E_0' = E_\ell + E_\ell' \qquad (5.97)$$

The tangential components of $\vec{B}$ must also be continuous at $x = 0$. Here, we must be careful to note that for the backward-traveling wave, if $\vec{E}$ points up as shown in Figure 5.32, then $\vec{B}$ must point *into* the page. Taking out of the page as positive gives

$$B_0 - B_0' = B_\ell - B_\ell'$$

which by Equation 5.96 gives

$$n_0 \left(E_0 - E_0'\right) = n_\ell \left(E_\ell - E_\ell'\right) \qquad (5.98)$$

Table 5.2. Selected materials used for optical coatings.

Material	Useful Wavelength Range[a] (μm)	n_r [a]
Aluminum Oxide (Al_2O_3)	$0.2 - 5.0$	1.62
Barium Fluoride (BaF_2)	$0.4 - 12.0$	1.48
Calcium Fluoride (CaF_2)	$0.13 - 10.0$	1.43
Cerium Fluoride (CeF_3)	$0.3 - 5.0$	1.63
Cerium Oxide (CeO_2)	$0.4 - 16.0$	2.35
Cryolite (Na_3AIF_6)	$0.25 - 14.0$	1.33
Hafnium Oxide(HfO_2)	$0.23 - 7.0$	1.95
Indium-Tin Oxide(ITO)	$0.4 - 1.0$	2.00
Lanthanum Fluoride (LaF_3)	$0.25 - 14.0$	1.58
Magnesium Fluoride (MgF_2)	$0.12 - 8.0$	1.38
Scandium Oxide(Sc_2O_3)	$0.35 - 7.0$	1.89
Silicon Monoxide (SiO)	$0.8 - 8.0$	1.55
Silicon Oxide (SiO_2)	$0.2 - 9.0$	1.46
Tantalum Oxide (Ta_2O_5)	$0.35 - 8.0$	2.10
Titanium Monoxide (TiO)	$0.4 - 10.0$	2.35
Titanium(III) Oxide (Ti_2O_3)	$0.4 - 12.0$	2.4
Yttrium Oxide (Y_2O_3)	$0.4 - 8.0$	1.87
Zinc Selenide ($ZnSe$)	$0.4 - 14.0$	2.4
Zinc Sulfide (ZnS)	$0.4 - 14.0$	2.39
Zirconium Dioxide (ZrO_2)	$0.3 - 8.0$	2.40

[a]High power applications may require a more restricted range. [b]Nominal, optical range, if available. See Section 3.7.

At the second interface where $x = \ell$, we must include the effects of propagation. It is important to note here that we are computing a *steady-state* solution for transmission and reflection. The reflected and transmitted beams will be measured by a power detector that records an average over many cycles. The result will be time-independent, and we thus ignore the time term in each electromagnetic wave solution. Accordingly, we represent a forward-traveling wave using e^{+ikx}, and a backward-traveling wave using e^{-ikx}. Using this convention, the boundary condition for $\vec{E}$ at $x = \ell$ gives

$$E_\ell e^{ik\ell} + E_\ell' e^{-ik\ell} = E_t \tag{5.99}$$

where we have included the $e^{ik\ell}$ in the term E_t by allowing it to become complex, since in the transmitted region we have no need to distinguish between forward- and backward-traveling solutions. Similarly, the boundary condition at $x = \ell$ for magnetic fields gives

$$n_\ell \left(E_\ell e^{ik\ell} - E_\ell' e^{-ik\ell} \right) = n_t E_t \tag{5.100}$$

Our goal is to solve Equations 5.97-5.100 for the reflected and transmitted beams E_0' and E_t when the incident illumination E_0 is known. To do this, we eliminate the fields internal to the layer: E_ℓ and E_ℓ'. From Equations 5.99 and 5.100 we find

$$E_\ell = \frac{1}{2} \left(1 + \frac{n_t}{n_\ell} \right) E_t e^{-ik\ell} \tag{5.101}$$

$$E'_\ell = \frac{1}{2}\left(1 - \frac{n_t}{n_\ell}\right)E_t e^{ik\ell} \tag{5.102}$$

from which we find

$$E_\ell + E'_\ell = \left(\cos k\ell - i\frac{n_t}{n_\ell}\sin k\ell\right)E_t \tag{5.103}$$

$$E_\ell - E'_\ell = \left(-i\sin k\ell + \frac{n_t}{n_\ell}\cos k\ell\right)E_t \tag{5.104}$$

From Equation 5.97,

$$1 + \frac{E'_0}{E_0} = \left(\cos k\ell - i\frac{n_t}{n_\ell}\sin k\ell\right)\frac{E_t}{E_0} \tag{5.105}$$

and from Equation 5.98

$$n_0 - n_0\frac{E'_0}{E_0} = (-in_\ell\sin k\ell + n_t\cos k\ell)\frac{E_t}{E_0} \tag{5.106}$$

To simplify the notation, we let $r = \frac{E'_0}{E_0}$ and $t = \frac{E_t}{E_0}$. Equations 5.105 and 5.106 can be written in matrix form:

$$\begin{pmatrix} 1 \\ n_0 \end{pmatrix} + \begin{pmatrix} 1 \\ -n_0 \end{pmatrix}r = \begin{pmatrix} \cos k\ell & -\frac{i}{n_\ell}\sin k\ell \\ -in_\ell\sin k\ell & \cos k\ell \end{pmatrix}\begin{pmatrix} 1 \\ n_t \end{pmatrix}t = M\begin{pmatrix} 1 \\ n_t \end{pmatrix}t \tag{5.107}$$

where M is called the *transfer matrix*:

$$M = \begin{pmatrix} \cos k\ell & -\frac{i}{n_\ell}\sin k\ell \\ -in_\ell\sin k\ell & \cos k\ell \end{pmatrix} \tag{5.108}$$

Given the incident and reflected fields at $x = 0$, the transfer matrix determines the fields at $x = \ell$. If there are N layers with indices of refraction $n_{\ell_1}, n_{\ell_2}, ..., n_{\ell_N}$ with thicknesses $\ell_1, \ell_2, ..., \ell_N$, Equation 5.107 becomes[18]

$$\begin{pmatrix} 1 \\ n_0 \end{pmatrix} + \begin{pmatrix} 1 \\ -n_0 \end{pmatrix}r = M_1 M_2 ... M_N\begin{pmatrix} 1 \\ n_t \end{pmatrix}t \tag{5.109}$$

The order of matrix multiplication is important; in the above, M_1 is the layer that the incident light illuminates first. Let the result of N layers result in a final transfer matrix with the matrix elements indicated below:

$$\begin{pmatrix} 1 \\ n_0 \end{pmatrix} + \begin{pmatrix} 1 \\ -n_0 \end{pmatrix}r = \begin{pmatrix} m_{11} & m_{12} \\ m_{21} & m_{22} \end{pmatrix}\begin{pmatrix} 1 \\ n_t \end{pmatrix}t$$

giving the following set of equations:

$$1 + r = (m_{11} + n_t m_{12})t \tag{5.110}$$

$$n_0 - n_0 r = (m_{21} + n_t m_{22})t \tag{5.111}$$

These can be solved for r and t to give (see Problem 5.42)

$$r = \frac{n_0 m_{11} + n_0 n_t m_{12} - m_{21} - n_t m_{22}}{n_0 m_{11} + n_0 n_t m_{12} + m_{21} + n_t m_{22}} \tag{5.112}$$

$$t = \frac{2n_0}{n_0 m_{11} + n_0 n_t m_{12} + m_{21} + n_t m_{22}} \tag{5.113}$$

The reflectance and transmittance are then given by $R = |r|^2$ and $T = |t|^2$.

---

[18]This result is not obvious. For more information, see *Principles of Optics* by Born and Wolf [2]. Note that they refer to the transfer matrix as the *characteristic matrix*.

5.11.1   Antireflection Films

As introduced in Application Note 5.1, it is possible to use a single dielectric film as an antireflection coating. The transfer matrix for a single layer is given by Equation 5.108. Letting $n_0 = 1$, Equation 5.112 becomes

$$r = \frac{n_\ell (1 - n_t) \cos k\ell - i \left(n_t - n_\ell^2\right) \sin k\ell}{n_\ell (1 + n_t) \cos k\ell - i \left(n_t + n_\ell^2\right) \sin k\ell} \tag{5.114}$$

If we let the optical thickness of the layer be $\frac{\lambda}{4}$, then $k\ell = \pi/2$, giving[19]

$$r = \frac{n_t - n_\ell^2}{n_t + n_\ell^2} \tag{5.115}$$

At the design wavelength, $R = 0$ when $n_\ell = \sqrt{n_t}$. For glass with index $n_t = 1.5$, the ideal index would be 1.22. Magnesium fluoride (MgF_2) with index $n_\ell = 1.38$ is commonly used since it is durable and amenable to existing deposition techniques. Note that the index increases across each interface, giving a π phase shift for each reflected component (see Section 3.5.5); thus the half wavelength round-trip travel for the wave that reflects from the substrate combines with the front-surface reflection in destructive interference, giving complete cancellation if $n_\ell = \sqrt{n_t}$. The film reflectance increases for wavelengths other than the design wavelength, as illustrated in Figure 5.33. Incident angles that depart from normal will also increase the film reflectance. Optics with antireflection coatings for the visible typically use $550\,nm$ as the design wavelength, since this is at the peak of the solar spectrum. The glare from such coatings typically shows a mixture of red and violet.

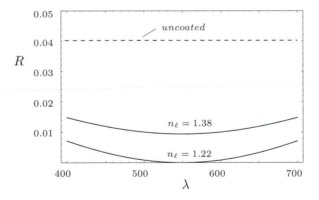

Figure 5.33   Antireflection coating consisting of a single quarter-wavelength layer applied to glass of index 1.5. The ideal index of $\sqrt{1.5}$ gives zero reflection at the design wavelength of $550\,nm$. Magnesium fluoride with index 1.38 gives a reflection of only about 1% at $550\,nm$. The uncoated reflectance is about 4%.

As Figure 5.33 indicates, a single film of MgF_2 on a $n_t = 1.5$ substrate reduces reflection to only about 1% at the design wavelength. Further reduction can be achieved by using more than one layer. A *quarter-quarter* coating consists of two quarter-wave layers of alternating high and low index, as illustrated in Figure 5.34(a). According to Equations

---

[19]In other words, the physical thickness of the layer is one-quarter wavelength as measured *within the film*.

5.108 and 5.109, the transfer matrix for the quarter-quarter coating is

$$M = M_1 M_2 = \begin{pmatrix} 0 & -\frac{i}{n_1} \\ -in_1 & 0 \end{pmatrix} \begin{pmatrix} 0 & -\frac{i}{n_2} \\ -in_2 & 0 \end{pmatrix} = \begin{pmatrix} -\frac{n_2}{n_1} & 0 \\ 0 & -\frac{n_1}{n_2} \end{pmatrix}$$

Equation 5.112 gives

$$r = \frac{-\frac{n_2}{n_1} + n_t \frac{n_1}{n_2}}{-\frac{n_2}{n_1} - n_t \frac{n_1}{n_2}} = \frac{n_2^2 - n_t n_1^2}{n_2^2 - n_t n_1^2}$$

The reflectance will be zero when

$$\frac{n_2}{n_1} = \sqrt{n_t} \qquad\qquad (5.116)$$

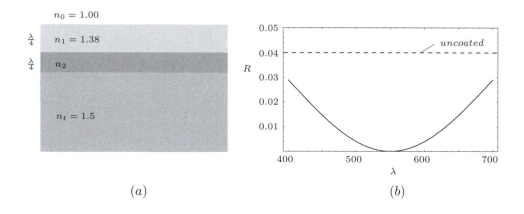

$$(a) \qquad\qquad\qquad\qquad\qquad\qquad (b)$$

Figure 5.34   Quarter-quarter coating on a substrate with index 1.50. The top layer is MgF_2. (b) Reflectance when the n_2 layer has the ideal index of 1.69.

For the situation illustrated in Figure 5.34(a), if MgF_2 is used for the n_1, then zero reflectance is achieved if a material with $n = 1.69$ for the n_2 layer can be found. Table 5.2. lists some candidates. The index is not the only criterion; the film material must be transparent over the intended range of use, it must be durable, and it must be amenable to existing deposition technologies. Figure 5.34(b) shows the reflectance R vs. wavelength for the optimal value of $n_2 = 1.69$. Although the reflectance does indeed reach zero at the design wavelength, it increases sharply as the wavelength departs from the design value. *Broadband antireflection coatings* use larger numbers of film layers to achieve low reflectivity over a larger wavelength range. The analysis can be algebraically complicated and is usually better done with a computer.

Problem 5.42   Derive Equations 5.112 and 5.113.

Problem 5.43   Find the reflectance of a quarter-wave layer of cryolite ($n = 1.33$) deposited on glass of index 1.5.

Problem 5.44   Show that when the incident medium has index of refraction n_0, the reflectance and best performance of the following coatings are as follows:

a) Single quarter-wave coating:

$$R = \left(\frac{n_0 n_t - n_\ell^2}{n_0 n_t + n_\ell^2} \right)^2$$

Best performance when

$$n_\ell = \sqrt{n_0 n_t}$$

b) Quarter-quarter coating:

$$R = \left(\frac{n_0 n_2^2 - n_t n_1^2}{n_0 n_2^2 + n_t n_1^2} \right)^2$$

Best performance when

$$\frac{n_2}{n_1} = \sqrt{\frac{n_t}{n_0}}$$

5.11.2   High-Reflectance Films

A dielectric coating for high-reflectance must use multiple layers. Figure 5.35 shows a 4-layer stack of alternating high and low index layers deposited on a substrate of index n_t. Adjacent layers have a transfer matrix given by

$$M_{HL} = M_H M_L = \begin{pmatrix} 0 & -\frac{i}{n_H} \\ -i n_H & 0 \end{pmatrix} \begin{pmatrix} 0 & -\frac{i}{n_L} \\ -i n_L & 0 \end{pmatrix} = \begin{pmatrix} -\frac{n_L}{n_H} & 0 \\ 0 & -\frac{n_H}{n_L} \end{pmatrix}$$

If the coating consists of $2N$ high-low pairs, the transfer matrix is

$$M = M_{HL}^N = \begin{pmatrix} -\frac{n_L}{n_H} & 0 \\ 0 & -\frac{n_H}{n_L} \end{pmatrix}^N = \begin{pmatrix} \left[-\frac{n_L}{n_H}\right]^N & 0 \\ 0 & \left[-\frac{n_H}{n_L}\right]^N \end{pmatrix}$$

Equation 5.112 gives the reflectance:

$$R = |r|^2 = \left| \frac{\left[-\frac{n_L}{n_H}\right]^N - n_t \left[-\frac{n_H}{n_L}\right]^N}{\left[-\frac{n_L}{n_H}\right]^N + n_t \left[-\frac{n_H}{n_L}\right]^N} \right|^2 = \left| \frac{n_t \left[\frac{n_H}{n_L}\right]^{2N} - 1}{n_t \left[\frac{n_H}{n_L}\right]^{2N} + 1} \right|^2 \tag{5.117}$$

Since the ratio $\frac{n_H}{n_L} > 1$, the reflectance approaches one for even a few pairs of layers.

■ **EXAMPLE 5.12**

Using MgF_2 with $n_L = 1.38$ and ZnS with $n_H = 2.39$, find the reflectance of a coating that utilizes two high-low pairs, as illustrated in Figure 5.35(a). Repeat for a coating that uses four high-low pairs. Assume a substrate with $n_t = 1.5$.

Solution

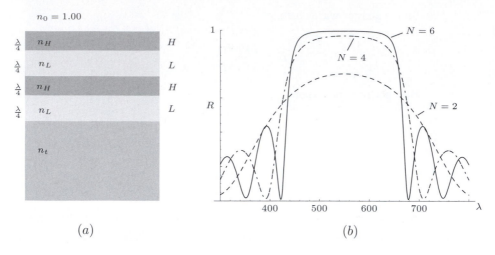

Figure 5.35 (a) A two-pair high-reflectance coating. (b) Reflectance vs. wavelength for 2, 4 and 6 pair coatings. See Example 5.12.

From Equation 5.117, a two pair stack has a reflectance of

$$R = \left| \frac{n_t \left[\frac{n_H}{n_L}\right]^{2N} - 1}{n_t \left[\frac{n_H}{n_L}\right]^{2N} + 1} \right|^2 = \left[\frac{(1.5) * \left(\frac{2.39}{1.38}\right)^4 - 1}{(1.5) * \left(\frac{2.39}{1.38}\right)^4 + 1} \right]^2 = 0.743$$

A four pair stack has reflectance

$$R = \left| \frac{n_t \left[\frac{n_H}{n_L}\right]^{2N} - 1}{n_t \left[\frac{n_H}{n_L}\right]^{2N} + 1} \right|^2 = \left[\frac{(1.5) * \left(\frac{2.39}{1.38}\right)^8 - 1}{(1.5) * \left(\frac{2.39}{1.38}\right)^8 + 1} \right]^2 = 0.968$$

5.11.2.1 *Fabry-Perot Interference Filters*
High-reflectance dielectric coatings can be used to make very high finesse Fabry-Perot cavities. An *interference filter*, sometimes also called a Fabry-Perot *etalon*, uses a fixed cavity thickness that results in a band-pass filter for use at a specific wavelength. Figure 5.36 shows high-reflectance films consisting of three pairs of quarter-wave high-low pairs deposited on either side of a half-wave spacer. Since the cavity has the minimum thickness of half-wavelength, the free spectral range is maximum:

$$\Delta f = \frac{c}{2d} = \frac{c}{\lambda} = f$$

Thus, higher order transmission peaks occur at shorter wavelengths: $\frac{\lambda}{2}, \frac{\lambda}{4}, \dots$. The higher frequencies are usually blocked by absorption in the layers. Out of band transmission can also occur due to changes in reflectivity of the dielectric coatings away from the design wavelength (see Figure 5.35(b)).

The bandpass will change for angles of incidence other than normal. As the angle is changed, the effective layer thickness increases, so the transmission shifts toward longer

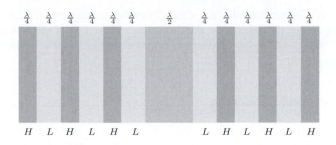

Figure 5.36 An interference filter consisting of quarter-wave high-low pairs deposited on either side of a half-wave spacer.

wavelengths. For sufficiently large angles, the design bandpass can shift into the infrared, and a higher frequency bandpass can shift into the visible from the blue.

Problem 5.45 Show that the reflectance of a $2N$-layer coating where the top layer has index n_L is given by

$$R = \left| \frac{\left[\frac{n_H}{n_L}\right]^{2N} - n_t}{\left[\frac{n_H}{n_L}\right]^{2N} + n_t} \right|^2$$

Problem 5.46 If the incident medium in Figure 5.35(a) has index greater than one, show that the reflectance is given by

$$R = \left| \frac{n_t \left[\frac{n_H}{n_L}\right]^{2N} - n_0}{n_t \left[\frac{n_H}{n_L}\right]^{2N} + n_0} \right|^2$$

Problem 5.47 Find the reflectance at the design wavelength of a six-pair high-reflectance coating using the data from Example 5.12. Compare your result with the plot of Figure 5.35(b).

Problem 5.48 Find the coefficient of finesse for the interference filter shown in Figure 5.36 using the layer and substrate data of Example 5.12.

5.12 COHERENCE

Two fields are *coherent* if they are *correlated*. For example, beams collected from a *thermal source* contain emissions that result from individual accelerating electrons that radiate in a statistically random way. The combination of these random contributions within a beam produce *fluctuations* in irradiance that are also statistically random. Two beams that exhibit coherence have fluctuations that are statically correlated in time. Coherence between two beams that arrive at the *same point* after traveling different optical paths is called *temporal coherence*. *Spatial coherence* exists when fluctuations measured at *different points* in space are correlated.

5.12.1 Temporal Coherence

Temporal coherence is best defined in terms of a Michelson interferometer. In Section 5.8, *coherence length* was defined in terms of the mirror separation ΔL that gives a fringe visibility of one-half, and the *coherence time* is the corresponding light travel time (see Equation 5.72). As discussed in Section 5.8, the coherence length and coherence time are determined by frequency spread in the beam that enters the interferometer. Monochromatic beams have a long coherence time, and broadband sources such as sunlight have very small coherence times.

According to Figure 5.22 and the discussion of single-photon interference in Section 5.9, a Michelson interferometer creates two identical copies of the same beam, then recombines them after producing a *time delay* in one of the copies. The two beams combine at the observation point to give a combined irradiance proportional to

$$|E|^2 = \left\langle \vec{E} \cdot \vec{E}^* \right\rangle = \left\langle \left(\vec{E}_1 + \vec{E}_2 \right) \cdot \left(\vec{E}_1^* + \vec{E}_2^* \right) \right\rangle$$
$$= \left\langle |E_1|^2 + |E_2|^2 + 2\mathrm{Re}[\vec{E}_1 \cdot \vec{E}_2^*] \right\rangle$$

where the angle-brackets denote a *time average* (see Example 2.4):

$$\langle f \rangle = \lim_{T \to \infty} \frac{1}{T} \int_0^T f(t)dt \tag{5.118}$$

We assume that the correlations are *stationary*, meaning that the time average is independent of the origin of time. We also assume that both beams have identical polarizations. Thus

$$I = I_1 + I_2 + 2\epsilon_0 c \mathrm{Re}[\langle E_1 E_2^* \rangle] \tag{5.119}$$

where $I_1 = \epsilon_0 c \left\langle |E_1|^2 \right\rangle$ and $I_2 = \epsilon_0 c \left\langle |E_2|^2 \right\rangle$. Let the time for beam 1 to traverse the interferometer be t, and the corresponding time for beam 2 be $t + \tau$. We define the *mutual coherence function*

$$\Gamma_{12}(\tau) = \langle E_1(t) E_2^*(t + \tau) \rangle = \lim_{T \to \infty} \int_0^T E_1(t) E_2^*(t + \tau) \, dt \tag{5.120}$$

If, after time delay τ beams 1 and 2 are statistically uncorrelated, the real and imaginary parts of $E_1(t)E_2^*(t + \tau)$ should be randomly positive and negative, and in this case the integral of Equation 5.120 gives zero. However, if there is correlation, positive values are more common, and the integral is nonzero. We define the *normalized degree of partial coherence* as

$$\gamma_{12} = \frac{\varepsilon_0 c \Gamma_{12}(\tau)}{\sqrt{I_1 I_2}}$$

Using this definition, Equation 5.119 for the irradiance measured in a Michelson interferometer becomes

$$I = I_1 + I_2 + 2\sqrt{I_1 I_2}\,\mathrm{Re}[\gamma_{12}] \qquad (5.121)$$

The function γ_{12} is a complex-valued function whose value determines the coherence. If $|\gamma_{12}| = 1$, there is complete coherence. If $0 < |\gamma_{12}| < 1$ there is partial coherence, and if $|\gamma_{12}| = 0$ there is complete incoherence. Equation 5.71 defined the *fringe visibility*:

$$V = \frac{I_{\max} - I_{\min}}{I_{\max} + I_{\min}}$$

Since $I_{\max} = I_1 + I_2 + 2\sqrt{I_1 I_2}\,|\gamma_{12}|$ and $I_{\min} = I_1 + I_2 - 2\sqrt{I_1 I_2}\,|\gamma_{12}|$, the visibility becomes

$$V = \frac{2\sqrt{I_1 I_2}\,|\gamma_{12}|}{I_1 + I_2} \qquad (5.122)$$

If $I_1 = I_2$,

$$V = |\gamma_{12}| \qquad (5.123)$$

In other words, the fringe visibility is a measure of correlation, which in turn is a measure of coherence. Thus, *two fields are coherent if they produce interference fringes when combined.*

5.12.2 Spatial Coherence

In Section 5.3.2.1, we described Young's double-slit experiment, and in particular noted the necessity to illuminate the double-slit aperture with light that had first passed through a small single slit (see Figure 5.3). If the single slit is small enough, incoherent light passing through it illuminates the double-slit aperture coherently. Since the two slits are located at different points in space, we refer to this as *spatial coherence*[20]. Stationary fringes will be visible at the observation screen only if the light exiting each double-slit aperture is correlated in *phase*. While it may seem odd that light emanating from independent random radiators in a thermal source can be made coherent simply by passing it through a sufficiently small hole, this is in fact the case, as we now discuss.

Figure 5.37(a) shows a quasi-monochromatic thermal source illuminating an observation screen a distance L away. The two point sources labeled 1 are positioned at the very edge of the thermal source, separated by the source width d. Together, they form a Young's double-aperture interference pattern on the observing screen, with first fringe minimum located by Equation 5.17 with $m = 1/2$:

$$\theta = \pm\frac{\lambda}{2d}$$

The pairs of points labeled by 2 and 3 are closer together, and thus give fringe patterns with wider spacing, as illustrated in Figure 5.37(b). As all points on the thermal source are included, the corresponding fringe patterns overlap to give uniform illumination everywhere except within the region of zero path difference, where the illumination is brighter. A *quasi-monochromatic* source is characterized by random phase shifts that occur, for visible light, on the order of 10^{-8} s, as discussed in Section 5.6, and illustrated in Figure 5.38. As

---

[20] Strictly, this is transverse spatial coherence, as opposed to *longitudinal spatial coherence* where the two points located along a radius line. Longitudinal spatial coherence is similar to temporal coherence.

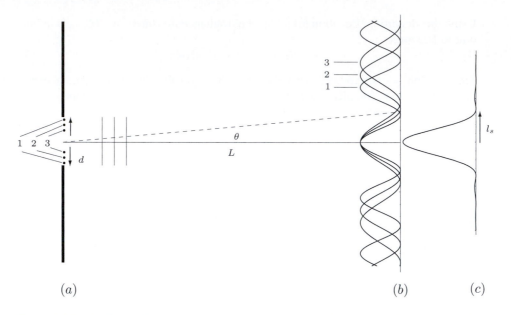

(a) (b) (c)

Figure 5.37   (a) Pairs of independent quasi-monochromatic point radiators placed symmetrically across a single aperture. (b) Each pair of radiators in (a) produces a Young's interference pattern with fringe spacing inversely proportional to the pair separation. (c) The result of adding the fringe patterns due to all points in the aperture.

these phase shifts occur, the fringe pattern for each point pair shifts to either side of the symmetric point of zero path difference, which widens the central bright area somewhat.

A tiny quasi-monochromatic source located a distance L from an observing screen shows enhanced brightness at the zero path difference location. Because the central region is bright, it is an area of constructive interference, and thus is a region of coherence. Furthermore, the bright area remains as more frequencies are added to the source, since all frequencies interfere constructively when the path difference is zero. Thus, there is an area of spatial coherence even for thermal sources that emit with a broad spectrum, such as planets and stars.

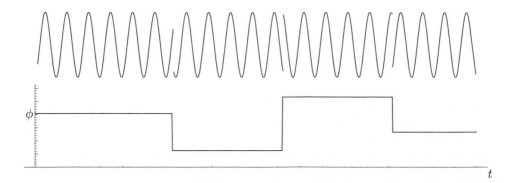

Figure 5.38   A quasi-monochromatic source. For visible light, the phase changes abruptly on a time scale of 10^{-8} s. The corresponding wave train uses a long wavelength for the sake of illustration.

The above intuitive arguments are summarized rigorously in the *Van-Cittert-Zernike theorem*[21]. Here, the area of coherence is determined to lie within the central maximum of the *Fraunhofer diffraction pattern* produced by the source. For example, a thermal source such as a star with angular diameter θ produces a coherence region of *radius*

$$l_s = \frac{1.22\,\lambda}{\theta} \tag{5.124}$$

where l_s is the *spatial coherence width* and θ is measured in radians. The factor of 1.22 comes from mathematical details that will be described fully in Section 6.3.3.

■ **EXAMPLE 5.13**

Find the spatial coherence width for the following sources: (a) the Sun ($\theta \approx 0.5°$), (b) Venus ($\theta \approx 1/60°$), (c) Betelgeuse ($\theta \approx 0.0000131°$). Assume an average wavelength of $550\,nm$, and that each object has circular cross section.

Solution

For the Sun,

$$l_s = \frac{(1.22)\left(550 \times 10^{-9}m\right)}{(0.5°)\left(\frac{\pi\,rad}{180°}\right)} = 7.69 \times 10^{-5}m = 0.0768\,mm$$

For Venus,

$$l_s = \frac{(1.22)\left(550 \times 10^{-9}m\right)}{\left(\frac{1}{60}\right)\left(\frac{\pi\,rad}{180°}\right)} = 2.31 \times 10^{-3}m = 2.31\,mm$$

For Betelgeuse,

$$l_s = \frac{(1.22)\left(550 \times 10^{-9}m\right)}{(0.0000131°)\left(\frac{\pi\,rad}{180°}\right)} = 2.93\,m$$

The Sun is an extended thermal source, and it illuminates a properly oriented observing screen *uniformly*. However, two small apertures more closely spaced than the distance found in Example 5.13 will show Young's interference fringes. To see interference fringes with more widely spaced apertures, one must modify the source as Young did, by first passing it through a single aperture of sufficiently small angular size so as to create an area of coherence at the aperture large enough to contain both openings. According to the Figure 5.37, θ will only be small if L is large, and waves can only travel with such a small angular spread if they have effectively *plane* wavefronts over the dimensions of the source. Wavefronts that reach us from Betelgeuse or even a more distant star can be propagated back to the source as a single large wavefront correlated over the entire surface of the star.

Figure 5.39 shows intuitive representations of photons that exhibit different degrees of coherence. Figure 5.39(a) illustrates light with low temporal and spatial coherence, such as sunlight. In Figure 5.39(b), the spatial coherence has been increased, indicated by the larger wavefront; an example here would be starlight. In Figure 5.39(c), the spatial coherence is

[21] F. Zernike: 1888-1966, received the Nobel prize for inventing the phase contrast microscope.

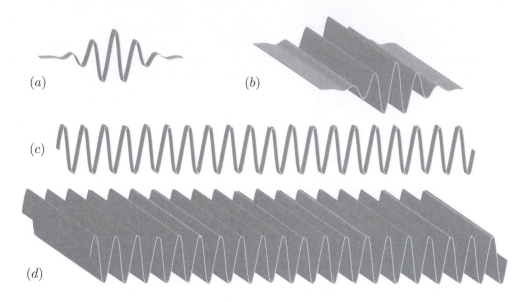

Figure 5.39 Intuitive representations of spatial and temporal coherence: (a) low temporal and low spatial coherence; (b) low temporal, high spatial; (c) high temporal, low spatial; (d) high temporal and high spatial coherence.

low but the temporal coherence is higher, indicated by a more monochromatic wave without as many Fourier frequency components. Light of this kind could be obtained, for example, by passing sunlight through a narrow-band interference filter. Figure 5.39(d) illustrates the case where both the spatial and temporal coherence is high. Examples here include filtered sunlight that is passed through a tiny aperture, and light emitted by many types of lasers.

Coherent light beams sometimes exhibit interesting effects. *Star twinkle*, for example, is largely affected by coherence. Before it encounters the atmosphere, starlight is effectively plane wave, but atmospheric density fluctuations along the optical path causes distortion and folding of the wavefront. Because the light is coherent, interference occurs when portions of the wavefront overlap, which increases the observed irradiance fluctuations. For the most part, planets do not twinkle because they have a large angular size and a correspondingly small spatial coherence length. Another example of the effects of coherence is given by *laser speckle* that occurs when, for example, a laser beam is diffusely reflected from a painted wall. Interestingly, laser light usually has both temporal and spatial coherence, and the speckle results from reflections from various levels of the rough surface that combine on an observing screen to give constructive and destructive interference. Usually, this observing screen is the retina of the eye. Natural light does not exhibit speckle because (1) it has a coherence length that is less than the altitude variations of the rough surface and (2) it has a spatial coherence that is less than the diameter of the illuminated region.

5.12.3 Michelson's Stellar Interferometer

Michelson used interferometry to obtain the first direct measurement of the angular size of a star. The experimental arrangement is shown in Figure 5.40. Light from the same star is collected by mirrors with transverse spacing l_s, and then combined on an observation screen. The collection mirrors were mounted on a framework that maintained the geometry

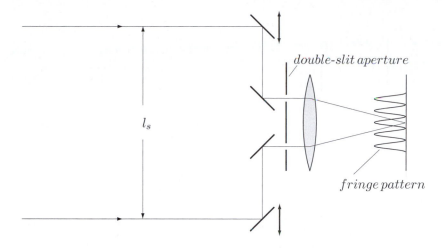

Figure 5.40   The Michelson stellar interferometer.

of Figure 5.40 as the telescope was rotated, and which also allowed the mirror separation l_s to be varied. This extended framework had to be extremely stable to allow observation of a stable fringe pattern. Fringes with nonzero visibility are observed as long as the inequality of Equation 5.124 is valid. The value of l_s that just gives zero fringe visibility determines the angular size θ of the star. Michelson's first observations were on Betelgeuse (see Example 5.13).

5.12.4 Irradiance Interferometry

A Michelson interferometer is sometimes called a *phase interferometer*. Here, two fields are combined *prior to* measurement of irradiance, and correlation is established by the observation of stationary fringes with nonzero visibility (see Equation 5.122). In *irradiance interferometry* intensities are measured first and then tested for correlation. This technique, first demonstrated by R. Hanbury Brown and R. Q. Twiss[22] and thus called the Hanbury Brown-Twiss (HBT) effect, generated much controversy within the physics community. Figure 5.41(a) illustrates their experimental arrangement. Light from a mercury discharge lamp is collected by a lens and passed through a narrow interference filter to isolate one of the available spectral lines. This monochromatic beam was then passed through a small aperture to give an output beam with both temporal and spatial coherence. This beam illuminated a beam splitter, and the two component beams were detected by separate square-law photomultiplier tubes. The irradiance signal from each detector was then tested for correlation. One of the detectors could be translated transversely to the beam path, and the measured correlation was accurately described by Equation 5.124.

The controversy concerned the effect of the beam splitter. In the photon picture, a single-photon is indivisible, and must travel *either* to detector 1 or detector 2; from this perspective, one would not expect to see correlations at all. As noted in Section 2.4, versions of this experiment conducted with single photons at very low light levels demonstrate the expected *anticorrelation*, and thus demonstrate the existence of photons. At higher light

[22]Robert Hanbury Brown (1916-2002): British astronomer and physicist. Richard Q. Twiss (?-2005): British mathematician and engineer.

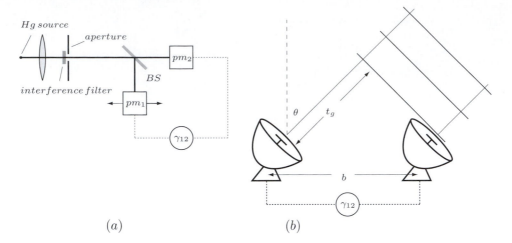

Figure 5.41 (a)The HBT experiment. (b) Two receivers separated by baseline distance b. A single wavefront arrives at either detector with a *geometric time delay* $t_g = b \sin \theta / c$.

levels, "quantum clumping" causes groups of photons to divide at the beam splitter, leading to the HBT observation of correlation.[23]

5.12.5 Telescope Arrays

Figure 5.41 shows two telescopes with the detectors connected to an electronic correlator. When operated as a phase interferometer, detection usually includes a *mixing* step where signals are combined with a *local oscillator LO*, producing beats as in Section 5.3.6. The difference frequency, usually referred to as the *intermediate frequency IF*, is low enough to enable analysis with available analog and digital circuit elements. In the radio spectrum, the signal frequencies are sufficiently low to allow detectors such as antennas to measure the electric field directly, allowing the mixing to occur electronically after detection. For higher frequencies, *optical mixing* typically uses a laser as the local oscillator, with the LO and signal beams combined on the surface of an optical detector for coherent detection, as described in Section 5.3.6.1. In either case, the field amplitude of the signal is detected directly, and the correlation step is equivalent to measuring fringe visibility. In an HBT approach, the telescope detectors measure the signal irradiance, and the irradiance signals are combined at the correlator as described above. Either approach can be used to measure stellar diameters; however, the phase interferometer is far more sensitive.

To obtain correlation results, one of the two signals must usually be time-shifted to account for the *geometric time delay*, which is equal to the light travel time difference along the optical path for light reaching each detector (see Figure 5.41(b))

$$\tau_g = \frac{b \sin \theta}{c}$$

---

[23] See the discussion of the HBT experiment and the associated single-photon of Grangier et. al., in *The Quantum Challenge*, by Greenstein and Zajonc [8]. To obtain a truly single-photon measurement, the Grangier experiment used a two-photon process where one photon triggered the detection of the other.

where b is the *baseline* that separates the two telescopes[24]. Recall that a single optical reflector satisfies Fermat's principle that every ray reaching the focus travels in minimum and therefore equal time. Thus, by including the time shift, the two telescopes behave as a single telescope aperture with collection area equal to the sum and collection diameter equal to b. Any number of telescopes can be connected in this way with any practicable baselines. For example, the Very Large Array (VLA) of the National Radio Observatory (NRAO)[25] consists of 27 telescopes that can be combined to give the resolution of a single telescope with objective diameter of $D = 22$ miles. The Very Long Baseline Array[26] consists of ten $25\,m$-wide radio telescopes distributed across North America with baselines that extend from Hawaii to the Virgin Islands.

Problem 5.49 Sodium light of wavelength $589\,nm$ is focused through a pinhole before illuminating a Young's double aperture with an aperture spacing of $1.00\,mm$. If the spacing between the pinhole and the aperture is $10.0\,cm$, what is the maximum pinhole diameter that will illuminate the aperture coherently?

Problem 5.50 Find the spatial coherence length for the sources of Example 5.13 using radio waves of frequency $3.00\,GHz$.

---

[24]There are typically also electronic time delays associated with the extended electric circuit that connects the two telescopes, which must be included in the time shift.
[25]See http://www.vla.nrao.edu/
[26]See http://www.vlba.nrao.edu/

Additional Problems

Problem 5.51   Determine the slit spacing in a Young's experiment when a fringe spacing of $5.50\,mm$ is observed on an observing screen that is $5.00\,m$ from the aperture and the wavelength is $632.8\,nm$.

Problem 5.52   Describe the Newton's rings pattern that results when the transmitted irradiance is studied. Is the central fringe light or dark?

Problem 5.53   A patrolman bounces microwaves off a vehicle moving at $65\,mph$. If the base wavelength of the microwaves used is $3.00\,cm$, find the frequency shift, and estimate the minimum resolving power of the receiving unit if it is just able to measure such a frequency shift.

Problem 5.54   Find the change in wavelength that occurs when acoustic effects cause the frequency of a helium-neon laser to "jitter" rapidly across a frequency span of $300\,MHz$. Assume an average laser wavelength of $633\,nm$.

Problem 5.55   In Problem 5.54, calculate the coherence length using the uncertainty relations as an equality. Find the coherence length if the jitter is controlled and the laser is stabilized to limit frequency fluctuations to $10.0\,kHz$.

Problem 5.56   Estimate the coherence length of Hα emission from a low-pressure mixture of hydrogen if the frequency width is $3.00 \times 10^9\,Hz$. Describe the interferogram that results when Hα light from an $50:50$ mixture of hydrogen and deuterium is studied with a Michelson interferometer. See Example 5.11 for spectral data.

Problem 5.57   What mirror spacing is necessary to achieve a resolving power of 10^5 at $632.8\,nm$ using Fabry-Perot mirrors with 95% reflectivity? In this case, what is the free spectral range?

Problem 5.58   Find the reflectance at the design wavelength of a five-pair high-reflectance coating using the data from Example 5.12. Compare your result with the plot of Figure 5.35(b).

Problem 5.59   Find the coefficient of finesse for an interference filter that uses a six-pair high-reflectance coating. Use the layer and substrate data of Example 5.12.

Problem 5.60   Estimate the spatial coherence length for sunlight focused through a $1.00\,mm$-diameter hole when it is observed $2.00\,m$ from the aperture.

Problem 5.61   Find the geometric time delay for radio signals that reach two telescopes with a baseline of $10.0\,km$ from a source located $30°$ from zenith.

Appendix: Fourier Series

Fourier series provide a way to represent an arbitrary *periodic* function as a *linear combination* of harmonic terms:

$$f(x) = \frac{A_0}{2} + \sum_{m=1}^{\infty} A_m \cos(mkx) + \sum_{m=1}^{\infty} B_m \sin(mkx) \tag{5.A.1}$$

where $k = 2\pi/\lambda$ and m is a positive integer. The terms A_m and B_m are collectively called the *Fourier coefficients* and are given by

$$A_m = \frac{2}{\lambda} \int_{-\frac{\lambda}{2}}^{\frac{\lambda}{2}} f(x) \cos(mkx) \, dx \tag{5.A.2}$$

$$B_m = \frac{2}{\lambda} \int_{-\frac{\lambda}{2}}^{\frac{\lambda}{2}} f(x) \sin(mkx) \, dx \tag{5.A.3}$$

To demonstrate Equations 5.A.1-5.A.3, we begin with the *orthogonality relations*[27] for the harmonic functions $\cos(m\pi x)$ and $\sin(m\pi x)$:

$$\int_{-\frac{\lambda}{2}}^{\frac{\lambda}{2}} \sin(pkx) \cos(qkx) \, dx = 0 \tag{5.A.4}$$

$$\int_{-\frac{\lambda}{2}}^{\frac{\lambda}{2}} \cos(pkx) \cos(qkx) \, dx = \frac{\lambda}{2} \delta_{p,q} \tag{5.A.5}$$

$$\int_{-\frac{\lambda}{2}}^{\frac{\lambda}{2}} \sin(pkx) \sin(qkx) \, dx = \frac{\lambda}{2} \delta_{p,q} \tag{5.A.6}$$

where p and q are arbitrary integers, and the term δ_{pq} is the *Kronecker delta*:

$$\delta_{pq} = \begin{cases} 1 & \text{if } p = q \\ 0 & \text{if } p \neq q \end{cases} \tag{5.A.7}$$

[27] The term orthogonality derives from a geometrical interpretation where each harmonic function is proportional to a unit vector in an infinite-dimensional vector space called a *Hilbert space*. Equations 5.A.4—5.A.6 are defined as scalar products in this space, and scalar products of orthogonal vectors are zero. For more information, see Boas [1].

■ **EXAMPLE 5.14**

Show that Equation 5.A.6 is true.

Solution

Use the Euler relation of Section 1.6 to represent each harmonic function:

$$
\int_{-\frac{\lambda}{2}}^{\frac{\lambda}{2}} \sin(pkx)\sin(qkx)\,dx = \int_{-\frac{\lambda}{2}}^{\frac{\lambda}{2}} \left(\frac{e^{ipkx} - e^{-ipkx}}{2i}\right)\left(\frac{e^{iqkx} - e^{-iqkx}}{2i}\right)dx
$$

$$
= -\frac{1}{4}\int_{-\frac{\lambda}{2}}^{\frac{\lambda}{2}} \left[\left(e^{i(p+q)kx} + e^{-i(p+q)kx}\right) - \left(e^{i(p-q)kx} + e^{-i(p-q)kx}\right)\right]dx
$$

$$
= -\frac{1}{2}\int_{-\frac{\lambda}{2}}^{\frac{\lambda}{2}} \cos\left[(p+q)\,kx\right]dx + \frac{1}{2}\int_{-\frac{\lambda}{2}}^{\frac{\lambda}{2}} \cos\left[(p-q)\,kx\right]dx
$$

Since p and q are integers, the integrand of the first integral oscillates an integer number of times over the integration interval. Thus, the first integral is zero. For the same reason, the second integral is zero unless $p = q$. In this case the integration gives

$$
\frac{1}{2}\int_{-\frac{\lambda}{2}}^{\frac{\lambda}{2}} dx = \frac{\lambda}{2}
$$

which verifies Equation 5.A.6.

Equations 5.A.4-5.A.6 may be used to find expressions for the Fourier coefficients. Multiply both sides of Equations 5.A.1 by $\sin(pkx)\,dx$ and integrate to give

$$
\int_{-\frac{\lambda}{2}}^{\frac{\lambda}{2}} f(x)\sin(pkx)\,dx = \int_{-\frac{\lambda}{2}}^{\frac{\lambda}{2}} \left[\frac{A_0}{2} + \sum_{m=1}^{\infty} A_m \cos(mkx) + \sum_{m=1}^{\infty} B_m \sin(mkx)\right]\sin(pkx)\,dx
$$

$$
= \int_{-\frac{\lambda}{2}}^{\frac{\lambda}{2}} \frac{A_0}{2}\sin(pkx)\,dx + \int_{-\frac{\lambda}{2}}^{\frac{\lambda}{2}} \left[\sum_{m=1}^{\infty} A_m \cos(mkx) + \sum_{m=1}^{\infty} B_m \sin(mkx)\right]\sin(pkx)\,dx
$$

$$
= 0 + \int_{-\frac{\lambda}{2}}^{\frac{\lambda}{2}} \left[\sum_{m=1}^{\infty} A_m \cos(mkx)\right]\sin(pkx)\,dx + \int_{-\frac{\lambda}{2}}^{\frac{\lambda}{2}} \left[\sum_{m=1}^{\infty} B_m \sin(mkx)\right]\sin(pkx)\,dx
$$

where the first integral is zero since the harmonic term oscillates an integer number of times over the interval from 0 to λ. Interchanging the order of integration and summation in the last two integrals gives

$$\int_{-\frac{\lambda}{2}}^{\frac{\lambda}{2}} f(x) \sin(pkx) \, dx$$

$$= \sum_{m=1}^{\infty} A_m \int_{-\frac{\lambda}{2}}^{\frac{\lambda}{2}} \cos(mkx) \sin(pkx) \, dx + \sum_{m=1}^{\infty} B_m \int_{-\frac{\lambda}{2}}^{\frac{\lambda}{2}} \sin(mkx) \sin(pkx) \, dx$$

According to Equations 5.A.4 and 5.A.6, this gives

$$\int_{-\frac{\lambda}{2}}^{\frac{\lambda}{2}} f(x) \sin(pkx) \, dx = 0 + \sum_{m=1}^{\infty} B_m \left(\frac{\lambda}{2} \delta_{m,p} \right) = B_p$$

or

$$B_p = \frac{2}{\lambda} \int_{-\frac{\lambda}{2}}^{\frac{\lambda}{2}} f(x) \sin(pkx) \, dx$$

We note that p is just a symbol that stands for any integer, so we change it to a new symbol m to give Equation 5.A.3. Equation 5.A.2 may be obtained in a similar way (see Problem 5.63).

It is often useful to note whether a function is *even* or *odd*. A function is even if $f(-x) = f(x)$. In other words, even functions are symmetric about the origin. Similarly, a function is odd if it is antisymmetric about the origin: $f(-x) = -f(x)$. For example, $\cos(x)$ is even and $\sin(x)$ is an odd function. The product of an even and an odd function is odd, but the product of two odd functions or two even functions is even (see Problem 5.16). Since the integral of an odd function between limits symmetric about zero gives zero, odd functions must have $A_m = 0$ for all m, and even functions must have all $B_m = 0$.

■ **EXAMPLE 5.15**

Find the Fourier coefficients for a *step function*:

$$f(x) = \begin{cases} +1 & 0 < x \le \frac{\lambda}{2} \\ -1 & -\frac{\lambda}{2} \le x < 0 \end{cases}$$

Plot partial Fourier series that use the first term only, the first three nonzero terms, and the first 25 nonzero terms.

Solution

From Figure 5.A.1, it is clear that $f(x)$ is odd, and thus $A_m = 0$ for all m. To find the B_m, we must integrate

$$B_m = \frac{2}{\lambda} \left[\int_{-\frac{\lambda}{2}}^{0} (-1) \sin\left(\frac{2m\pi}{\lambda} x \right) dx + \int_{0}^{\frac{\lambda}{2}} (+1) \sin\left(\frac{2m\pi}{\lambda} x \right) dx \right]$$

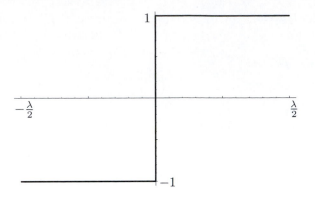

Figure 5.A.1   Step function.

Change variables in the first integral by letting $\bar{x} = -x$. Then $d\bar{x} = -dx$, and $\sin(\bar{x}) = -\sin(x)$. We must also change the limits of integration: when $x = -\frac{\lambda}{2}$, then $\bar{x} = +\frac{\lambda}{2}$. The above becomes

$$B_m = \frac{2}{\lambda} \left[\int_{\frac{\lambda}{2}}^{0} (-1)(-)\sin\left(\frac{2m\pi}{\lambda}x\right)(-d\bar{x}) + \int_{0}^{\frac{\lambda}{2}} (+1)\sin\left(\frac{2m\pi}{\lambda}x\right) dx \right]$$

$$B_m = \frac{2}{\lambda} \left[-\int_{\frac{\lambda}{2}}^{0} \sin\left(\frac{2m\pi}{\lambda}\bar{x}\right) d\bar{x} + \int_{0}^{\frac{\lambda}{2}} \sin\left(\frac{2m\pi}{\lambda}x\right) dx \right]$$

$$= \frac{2}{\lambda} \left[\int_{0}^{\frac{\lambda}{2}} \sin\left(\frac{2m\pi}{\lambda}\bar{x}\right) d\bar{x} + \int_{0}^{\frac{\lambda}{2}} \sin\left(\frac{2m\pi}{\lambda}x\right) dx \right]$$

Finally, we note that $\bar{x}$ is, like x, simply a parameter that denotes position. The two integrals in the last result are equivalent and can be combined to give

$$B_m = \frac{4}{\lambda} \left[\int_{0}^{\frac{\lambda}{2}} \sin\left(\frac{2m\pi}{\lambda}x\right) dx \right] = \frac{4}{\lambda} \left(\frac{-\lambda}{2m\pi}\right) \cos\left(\frac{2m\pi}{\lambda}x\right) \Big|_{0}^{\frac{\lambda}{2}}$$

$$= -\frac{2}{m\pi} [\cos(m\pi) - 1] = \frac{2}{m\pi} [1 - (-1)^m]$$

or, equivalently,

$$B_m = \begin{cases} \frac{4}{m\pi} & m \text{ odd} \\ 0 & m \text{ even} \end{cases}$$

According to Equation 5.A.1, the Fourier series for this step function is

$$f(x) = \frac{4}{\pi} \left[\sin\left(\frac{2\pi x}{\lambda}\right) + \frac{1}{3} \sin\left(\frac{6\pi x}{\lambda}\right) + \frac{1}{5} \sin\left(\frac{10\pi x}{\lambda}\right) + \ldots \right]$$

Figure 5.A.2 shows a plot of this series. Since the functions used to expand the step function are periodic, the expansion is periodic as well. When plotted outside the range $-\frac{\lambda}{2} < x < \frac{\lambda}{2}$, the expansion becomes that of a *square wave*.

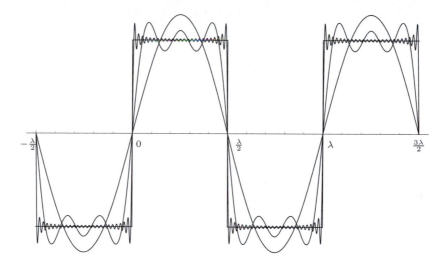

Figure 5.A.2 Fourier expansions of Example 5.15 plotted over two full cycles of the fundamental. In the three expansions, one uses the first nonzero term, another uses three nonzero terms, and the best approximation uses the first 25 nonzero terms in the series.

The terms in a Fourier series are called *harmonics*. The lowest frequency oscillating term is called the *fundamental*. A Fourier expansion containing terms with $m > 0$ is always periodic, and the period is that of the fundamental.

Figure 5.A.2 illustrates the *Gibbs phenomenon*[28] , sometimes also called the *Gibbs overshoot*. This occurs at *discontinuities* such as those when the square wave of Figure 5.A.2 transitions between positive and negative values. As more terms are added, the amount of overshoot converges to a constant value (about 14% for the square wave), and it moves closer to the point of discontinuity. A partial Fourier series always passes through a point that is the average of the values of $f(x)$ on either side of the discontinuity. Although functions that represent real physical quantities are never truly discontinuous, the Gibbs phenomenon is often observed, for example, in electrical signals when circuit elements attenuate the higher frequency components of a rapidly changing signal.

It is not straightforward to utilize evenness and oddness for functions that are not defined symmetrically about zero. However, one may always expand about symmetric limits using Equations 5.A.1—5.A.3, and then modify the result with a shift in origin. For example, we may modify the series of Example 5.15 to give the series for a square wave that is even:

$$f(x) = \frac{4}{\pi}\left[\sin\left(\frac{2\pi\,(x+0.5)}{\lambda}\right) + \frac{1}{3}\sin\left(\frac{6\pi\,(x+0.5)}{\lambda}\right) + \frac{1}{5}\sin\left(\frac{10\pi\,(x+0.5)}{\lambda}\right) + ...\right]$$

$$(5.A.8)$$

---

[28] Josiah Willard Gibbs (1839-1903): an American physicist who made important contributions to mathematical physics and thermodynamics.

■ **EXAMPLE 5.16**

Find the Fourier coefficients for a rectangular pulse-train of variable duty-cycle:

$$f(x) = \begin{cases} +1 & |x - n\lambda| < a \\ 0 & \text{otherwise} \end{cases}$$

where n is any integer. Plot a partial Fourier series that consists of the first 5 nonzero terms when $a = 0.5$ and $\lambda = 2$.

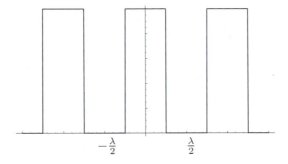

Figure 5.A.3   Rectangular pulse train.

Solution

We will represent this periodic function by expanding one cycle in a Fourier series. This function is even, so the B_m are zero. Begin by finding A_0:

$$A_0 = \frac{2}{\lambda} \int_{-\frac{\lambda}{2}}^{\frac{\lambda}{2}} f(x)\, dx = \frac{2}{\lambda} \int_{-a}^{a} dx = \frac{4a}{\lambda}$$

Now find the A_m:

$$A_m = \frac{2}{\lambda} \int_{-\frac{\lambda}{2}}^{\frac{\lambda}{2}} f(x) \cos(mkx)\, dx = \frac{2}{\lambda} \int_{-a}^{a} \cos(mkx)\, dx$$

$$= \frac{2}{\lambda} \left(\frac{1}{mk} \right) \sin(mkx)\big|_{-a}^{a}$$

giving

$$A_m = \frac{2}{m\pi} \sin\left(2\pi m \frac{a}{\lambda} \right)$$

The term $\frac{a}{\lambda}$ is called the *duty-cycle*. It is the fraction of a period that the pulse is nonzero. The pulse-train of Figure 5.A.3 has a duty-cycle of $1/2$.

The Fourier expansion of this function is given by

$$f(x) = \frac{A_0}{2} + \sum_{m=1}^{\infty} A_m \cos(mkx) = \frac{2a}{\lambda} + \sum_{m=1}^{\infty} \left[\left(\frac{2}{\pi m} \right) \sin\left(2\pi m \frac{a}{\lambda} \right) \right] \cos mkx$$

The first 5 nonzero terms when $a = 0.5$ and $\lambda = 2$ are

$$f(x) = \frac{1}{2} + \sum_{m=1}^{\infty} \left[\left(\frac{2}{\pi m} \right) \sin \left(\frac{\pi m}{2} \right) \right] \cos mkx$$

$$= \frac{1}{2} + \frac{2}{\pi} \left[\cos (\pi x) - \frac{1}{3} \cos (3\pi x) + \frac{1}{5} \cos (5\pi x) - \frac{1}{7} \cos (7\pi x) \right]$$

The five-term series in shown in Figure 5.A.4.

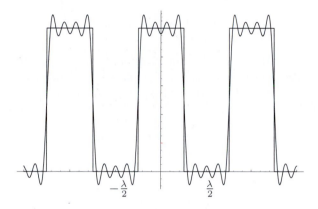

Figure 5.A.4   Fourier expansion of Example 5.16 plotted over three full cycles of the fundamental.

Complex Fourier Series

We may use the Euler relation of Section 1.6 to re-express the formulas for Fourier series as follows:

$$f(x) = \sum_{m=-\infty}^{+\infty} c_n e^{imkx} \tag{5.A.9}$$

with

$$c_m = \frac{1}{\lambda} \int_{-\frac{\lambda}{2}}^{\frac{\lambda}{2}} f(x) e^{-imkx} dx \tag{5.A.10}$$

It should be noted that the c_n given by Equation 5.A.10 are in general complex.

Equations 5.A.9 and 5.A.10 may be demonstrated by re-expressing the orthogonality relations of Equations 5.A.4—5.A.6 in complex form:

$$\int_{-\frac{\lambda}{2}}^{\frac{\lambda}{2}} e^{ipkx} e^{-iqkx} dx = \lambda \delta_{p,q} \tag{5.A.11}$$

where p and q are arbitrary integers and $\delta_{p,q}$ is the Kronecker delta. As in Example 5.14, the integral of Equation 5.A.11 is zero unless $p = q$ where the integral then gives λ (see

Problem 5.62). To demonstrate Equation 5.A.10, multiply both sides of Equation 5.A.9 by e^{-ipkx} and integrate to give

$$\int_{-\frac{\lambda}{2}}^{\frac{\lambda}{2}} f(x)e^{-ipkx}dx = \int_{-\frac{\lambda}{2}}^{\frac{\lambda}{2}} \left[\sum_{m=-\infty}^{+\infty} c_m e^{imkx} \right] e^{-ipkx}dx$$

$$= \sum_{m=-\infty}^{+\infty} c_m \left(\int_{-\frac{\lambda}{2}}^{\frac{\lambda}{2}} e^{imkx}e^{-ipkx}dx \right) = \sum_{m=-\infty}^{+\infty} c_m \left(\lambda \delta_{m,p} \right)$$

$$= \lambda c_p$$

It should be noted that the index m in Equations 5.A.9 and 5.A.10 ranges over positive and negative integer values, whereas in Equations 5.A.1—5.A.3, $m \geq 0$.

■ **EXAMPLE 5.17**

Repeat Example 5.16 using a complex Fourier series.

Solution

Use Equation 5.A.10 to give

$$c_0 = \frac{1}{\lambda} \int_{-\frac{\lambda}{2}}^{\frac{\lambda}{2}} f(x)dx = \frac{1}{\lambda} \int_{-a}^{a} dx = \frac{2a}{\lambda}$$

$$c_m = \frac{1}{\lambda} \int_{-\frac{\lambda}{2}}^{\frac{\lambda}{2}} f(x)e^{-imkx}dx = \frac{1}{\lambda} \int_{-a}^{a} e^{-imkx}dx = \frac{1}{\lambda} \left(-\frac{1}{imk} \right) e^{-imkx} \Big|_{-a}^{a}$$

$$= \left(-\frac{1}{2\pi im} \right) \left[e^{-imka} - e^{imka} \right]$$

giving

$$c_m = \left(\frac{1}{\pi m} \right) \sin \left(2\pi m \frac{a}{\lambda} \right)$$

Notice that in this case $c_m = c_{-m}$, so that

$$f(x) = \sum_{m=-\infty}^{\infty} c_m e^{imkx} = c_0 + \sum_{m=1}^{\infty} \left(c_m e^{imkx} + c_{-m}e^{-imkx} \right)$$

$$= \frac{2a}{\lambda} + \sum_{m=1}^{\infty} 2c_m \left(\frac{e^{imkx} + e^{-imkx}}{2} \right)$$

giving

$$f(x) = \frac{2a}{\lambda} + \sum_{m=1}^{\infty} \left[\left(\frac{2}{\pi m} \right) \sin \left(2\pi m \frac{a}{\lambda} \right) \right] \cos mkx$$

as in Example 5.16

Non-periodic Functions and Fourier Transforms

A pulsetrain such as that illustrated in Figure 5.A.5 is periodic, and can be represented with a Fourier series. However, a single pulse is not periodic, and has no Fourier series representation. Arbitrary *non-periodic* functions can be represented by a *Fourier transform*.

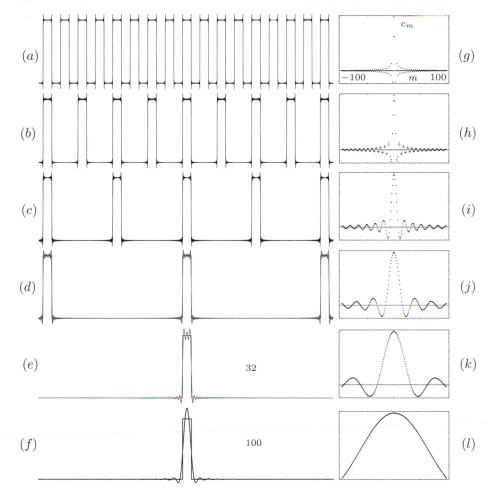

Figure 5.A.5 A pulse train plotted with increasing values of wavelength. In each case, the pulsewidth is constant and equal to 1. (a) $\lambda = 2$. (b) $\lambda = 4$. (c) $\lambda = 8$. (d) $\lambda = 16$. (e) $\lambda = 32$. (f) $\lambda = 100$. (g)—(l): A plot of the c_m vs. index m for the series to the immediate left. In each case, m varies from -100 to 100. As the wavelength increases, the c_m values merge toward a continuous function.

Consider Figure 5.A.5 which illustrates pulsetrains with increasing values of wavelength. The series expansions use the complex expansion coefficients of Example 5.17 with 201 values of m that range over $-100 \le m \le 100$. The pulsewidth for each case is the same, and as the wavelength increases, neighboring pulses move farther away. Note that edge detail disappears for the larger wavelengths, indicating a need for more terms in the series. Also shown for each case is a plot of the c_m versus index m. *As the wavelength tends to infinity, values of c_m begin to merge into a continuous function.* The plots of Figures 5.A.5(g)-(l) are merging toward the Fourier transform of a single rectangular pulse.

As long as the wavelength is finite, we may expand in a Fourier series:

$$f(x) = \sum_{m=-\infty}^{\infty} c_m e^{ik_m x} \Delta m = \frac{\lambda}{2\pi} \sum_{m=-\infty}^{\infty} c_m e^{ik_m x} \Delta k$$

where $\Delta m = 1$, and where we have defined $k_m = m\frac{2\pi}{\lambda}$ and $\Delta k = \frac{2\pi}{\lambda}$. As λ becomes large, $\Delta k \rightarrow dk$ and k_m becomes a continuous variable k. For infinite λ, we replace the sum sign with an integral sign to give

$$f(x) = \frac{\lambda}{2\pi} \sum_{m=-\infty}^{\infty} c_m e^{ik_m x} \Delta k$$

$$= \frac{\lambda}{2\pi} \sum_{m=-\infty}^{\infty} \left(\frac{1}{\lambda} \int_{-\frac{\lambda}{2}}^{\frac{\lambda}{2}} f(x) e^{-ik'_m x} dx \right) e^{ik_m x} \Delta k$$

where we have substituted the integral formula for the Fourier coefficient with a primed-k to indicate that it is independent of the unprimed-k. To make the transition to the Fourier transform, we let $\lambda \rightarrow \infty$, so that $\Delta k \rightarrow dk$. We replace the discrete sum with an integral to give

$$f(x) \rightarrow \frac{1}{2\pi} \int_{\infty}^{-\infty} \left(\int_{\infty}^{-\infty} f(x) e^{-ik' x} dx \right) e^{ikx} dk$$

where $c(k')$ has become a function of the continuous variable k':

$$c(k') = \int_{-\infty}^{\infty} f(x) e^{-ik' x} dx \tag{5.A.12}$$

Note that the primed-k is just notation, which can now be dropped. The function $f(x)$ given by

$$f(x) = \frac{1}{2\pi} \int_{-\infty}^{\infty} F(k) e^{ikx} dk \tag{5.A.13}$$

The integrals of Equations 5.A.12 and 5.A.15 are known as *Fourier integrals*.

The *Fourier transform* is defined as the integral of Equation 5.A.12. In standard notation:

$$F(k) \equiv \int_{-\infty}^{\infty} f(x) e^{-ikx} dx \tag{5.A.14}$$

The *inverse Fourier transform* from Equation 5.A.15, again in standard notation:

$$f(x) = \frac{1}{2\pi} \int_{-\infty}^{\infty} F(k) e^{ikx} dk \tag{5.A.15}$$

Examples and applications of Fourier transforms are explored in Section 5.4.1.

Problem 5.62   Show that Equations 5.A.5, 5.A.4, and 5.A.11 are true.

Problem 5.63   Derive Equation 5.A.2 for the Fourier coefficients of the cosine terms. Verify that A_0 is divided by 2 in Equation 5.A.1.

Problem 5.64   Show by explicit integration that all A_m Fourier coefficients are zero for the function of Example 5.15.

Problem 5.65   Plot the first 5 nonzero terms of Equation 5.A.8.

Problem 5.66   Find the Fourier coefficients of a sawtooth wave:

$$f(x) = \begin{cases} 2x & |x| \leq 0.5 \\ 0 & |x| > 0.5 \end{cases}$$

Plot the series that contains the first five nonzero terms for $-1.5 \leq x \leq +1.5$.

Problem 5.67   Find the Fourier series for a triangle wave using the function of Problem 5.14. Plot the series that includes the first five nonzero terms over three periods of the fundamental.

CHAPTER 6

DIFFRACTION

Intuition is the undoubting conception of a pure and attentive mind, which arises from the light of reason alone, and is more certain than deduction.

—Descartes

I want to emphasize that light comes in this form — particles. It is very important to know that light behaves like particles, especially for those of you who have gone to school, where you were probably told something about light behaving like waves. I'm telling you the way it *does* behave — like particles.

—Feynman

Contents

6.1   INTRODUCTION

In a sense, diffraction is interference done correctly. In Young's double-slit experiment from Chapter 5, light from two distinct sources combined to form a fringe pattern. Given this, how do we make sense of the observation that light from a *single* aperture behaves in much the same way? The answer lies in Huygens's marvelous intuitional leap that described wave propagation in terms of secondary wavelets. Refinements followed with more rigorous analysis, but Huygens' principle still forms the conceptual and analytical bedrock of diffraction phenomena.

In this chapter, perhaps more so than others, we see the power of informed approximation. The general diffraction problem is as yet intractable, yet we make progress by neglecting effects that are too small to be observed. Each approximation has its own domain of validity. Thus, the same aperture and illumination can produce Fraunhofer diffraction patterns in one case and Fresnel diffraction in another. Approximation is an art, and in this chapter we will study some fine examples of this craftsmanship.

It can perhaps be said that the theory of diffraction is the crowning achievement of the wave theory. We have seen examples in preceding chapters of instances where the wave approach seems insufficient at the level of single photons. We conclude this chapter with an introduction to *quantum electrodynamics* within the context of Fresnel diffraction, and a description of FeynmanŠs picture where photons completely replace the model of electromagnetic waves.

6.2   HUYGENS' PRINCIPLE

We begin our discussion of diffraction with *Huygens' principle*[1]. In effect, Huygens guessed the answer to a problem that can be analytically very difficult to solve. According to this *phenomenological* principle, each point on the wavefront of an electromagnetic wave acts as a source of *secondary wavelets*. When summed over an extended, unobstructed wavefront, the secondary wavelets recreate the next wavefront (see Figure 6.1(a)). The secondary sources radiate in phase with the same frequency as the original wavefront. The principle is phenomenological since Huygens could not explain *why* electromagnetic waves behave in this way. Furthermore, Huygens did not attempt to explain why the secondary wavelet sources radiate only in the *forward* direction, as indicated in Figure 6.1(a). The correctness of the principle lies mainly in its predictive power. For example, in addition to providing an explanation for wave propagation, Huygens' principle can be used to explain the laws of reflection and refraction[2], indicating that it is consistent with Maxwell's equations. More importantly, this principle can be used to predict and describe the results of diffraction experiments, as we shall presently see.

Figure 6.1(b) shows a point source s and receive point p on either side of an aperture. An arbitrary point on the aperture opening is located at distance r' relative to s and at r relative to p. According to Section 1.7.2, the spherical wavefront due to s at the aperture point is given by

$$E_A(r', t) = \frac{E_0}{r'} e^{i(kr' - \omega t)} \tag{6.1}$$

---

[1] Christiaan Huygens: 1629-1695. Dutch mathematician, astronomer and physicist.
[2] See, for example, Guenther [10].

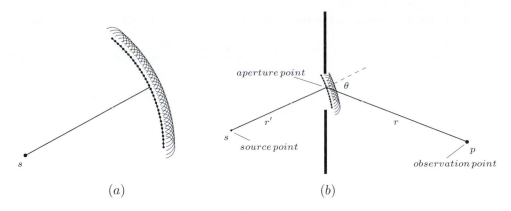

Figure 6.1   (a) Each point on an unobstructed wave as a source of Huygens wavelets. (b) When the wave is obstructed, wavelets at the aperture edges combine to form a diffracted wave.

Since s is a point source, the disturbance arrives at a sphere of radius r' centered on s at a time that we designate as $t = 0$. Over this sphere, the disturbance has amplitude given by

$$E_A = \frac{E_0}{r'}e^{i(kr')} \tag{6.2}$$

The aperture point acts as a source of secondary wavelets that propagate onward to an observation point p as a spherical wave. The disturbance at p is thus

$$dE_p = \frac{E_A}{r}e^{i(kr-\omega t)} = \frac{E_0}{r'r}e^{ik(r+r')}e^{-i\omega t} \tag{6.3}$$

The net disturbance at p is obtained by summing the contributions from all points on the aperture:

$$E_p = E_0 e^{-i\omega t}\int_A \frac{e^{ik(r+r')}}{r'r}dA \tag{6.4}$$

This is the expression for the diffracted wave obtained via Huygens' principle.

The phenomenological arguments of Huygens were subsequently justified mathematically and physically by the *scalar theory* of Fresnel[3] and Kirchhoff[4]. By scalar theory, we mean that effects due to polarization are ignored. The details of the analysis are omitted here, but can be found elsewhere[5]. In the more rigorous theory, the absence of backward traveling secondary wavelets arises naturally from the formalism as an *obliquity factor*. Along the portion of the sphere illustrated in Figure 6.1(b), the obliquity factor is equal to

$$\frac{1 + \cos\theta}{2}$$

where θ is the angle between the source ray and the observation ray at the aperture point, as shown in the figure. The obliquity factor has value $+1$ along the incident ray direction, and is zero in the opposite direction. The scalar theory also includes a few additional

[3] Augustin-Jean Fresnel: 1788-1827. French physicist who established the wave theory of light.
[4] Gustav Kirchhoff: 1824-1887. German physicist.
[5] See, for example, Born and Wolf [2].

multiplicative constants. The Fresnel-Kirchhoff result may be summarized as follows:

$$E_p = i\frac{A}{\lambda}e^{-i\omega t} \int_A \left[\frac{1 + \cos\theta}{2}\right] \frac{e^{ik(r+r')}}{r'r} dA \tag{6.5}$$

where A is a constant and λ is the wavelength. The factor of i indicates a 90° phase shift for the diffracted wave relative to the incident wave.

Despite the apparent simplicity, Equation 6.5 can be difficult to solve without the aid of simplifying assumptions. For example, if θ remains small for all source and receive directions, the obliquity factor does not depart significantly from 1 over the entire diffracting aperture, and we may take the Huygens result of Equation 6.4 as a starting point. If both source and receive points are located far away, then both incident and diffracted fields become *plane wave*. This is commonly referred to as *far-field*, or *Fraunhofer*[6] diffraction. If either the source or receive point are close enough to the diffracting aperture so that the curvature of the illuminating or diffracted wavefronts cannot be neglected, the diffraction approaches the *near-field* or *Fresnel* diffraction regime. One measure of wavefront curvature is given by the quantity δ in Figure 6.2(a):

$$\delta = r - \sqrt{r^2 - h^2} \tag{6.6}$$

The second term of this expression may be expanded to give

$$\delta = r - r\sqrt{1 - \left(\frac{h}{r}\right)^2} \cong r - r\left[1 - \frac{1}{2}\left(\frac{h}{r}\right)^2\right] = \frac{h^2}{2r} \tag{6.7}$$

The usual criterion used to determine the validity of the Fraunhofer approximation is that the total path variation due to curvature in the source and diffracted waves be limited to much less than one wavelength. Using parameters defined in Figure 6.1, this gives

$$\delta_{total} = \frac{h^2}{2}\left(\frac{1}{r'} + \frac{1}{r}\right) \ll \lambda \tag{6.8}$$

When expressed in this way, h may be regarded as the size of the diffracting aperture. As Equation 6.8 approaches equality, the diffraction approaches the Fresnel case. Note that in the Fresnel regime, h is typically large enough for the obliquity factor to be important, so Equation 6.5 must be used to determine the diffracted field.

Problem 6.1   Improve the approximation of Equation 6.7 by including the next term in the expansion used.

6.2.1 Babinet's Principle

Figure 6.2(b) and (c) shows two *complementary apertures* that when taken together form a single opaque screen. Let E_{p1} and E_{p2} be the field at point p measured when the individual

[6]Joseph von Fraunhofer: 1787-1826. German physicist.

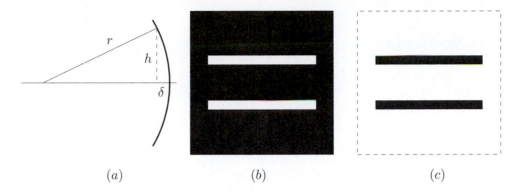

Figure 6.2 (a) Path difference along a spherical wavefront. (b) and (c): Complementary apertures.

apertures are illuminated. The combination of these fields must give the unobstructed wave E_0, giving *Babinet's principle*:

$$E_{p1} + E_{p2} = E_0 \tag{6.9}$$

Babinet's principle is particularly useful for Fraunhofer diffraction, where the illumination is plane wave with zero incident angle. For all nonzero diffracted angles, the unobstructed wave is zero, and the diffracted fields for the complementary apertures have the property:

$$E_{p1} + E_{p2} = 0 \tag{6.10}$$

Thus, with Fraunhofer diffraction, the diffracted fields from complementary apertures differ only by a constant phase difference of π, and have identical *irradiance* patterns for all nonzero diffraction angles. For example, when illuminated by a collimated beam, a small spherical obstacle will produce the same diffraction pattern as a circular hole in an opaque screen.

6.3 FRAUNHOFER DIFFRACTION

As noted in Section 6.2, Fraunhofer diffraction occurs when the curvature of both source and diffracted waves can be neglected. If the illumination is plane wave, then the source s must be located infinitely far away. The diffracted waves will also be plane when the observing screen is located infinitely far away, as shown in Figure 6.3(a). Here, the diffracted field consists of plane waves inclined at angles that locate bright regions in the diffraction pattern. In practice, of course, such distances can only be approximated, but the conditions of Fraunhofer diffraction can be obtained exactly by using focusing lenses, as illustrated in Figure 6.3(b). A point source s located one focal length from a collection lens gives plane wave illumination. A second focusing lens placed on the other side of the aperture collects the diffracted waves. The plane wave components that leave the aperture traveling toward the Fraunhofer diffraction pattern at infinity are all brought to a focus at the focal plane of the second lens. Thus, the second focusing lens creates an exact Fraunhofer diffraction pattern on an observing screen coincident its focal plane, otherwise known as the *Fraunhofer plane*.

To calculate the amplitude and irradiance of Fraunhofer diffraction patterns, we begin with Equation 6.5, and make the following assumptions:

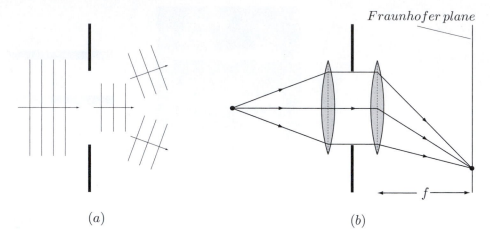

Figure 6.3   (a) In Fraunhofer diffraction, diffracted waves must be plane in order to travel to an observation screen at Infinity. (b) Infinite source and observation distances obtained with lenses.

1. The angular spread of the diffracted wave is small enough so that the obliquity factor can be ignored.

2. The quantity $\frac{e^{ikr'}}{r'}$ is constant over the aperture, and may be moved outside the integral.

3. The variation of $\frac{e^{ikr}}{r}$ is due mainly to the exponential, so r may be regarded as constant and moved outside the integral.

With these simplifications, Equation 6.5 becomes

$$E_p = C \int_A e^{ikr} dA \qquad (6.11)$$

where C is a constant. This is the *Fraunhofer diffraction integral*. The Fraunhofer diffraction pattern for a particular aperture is found by integrating this equation over the aperture opening.

6.3.1   Single Slit

The single-slit aperture consists of an opaque screen with a long, narrow rectangular opening. The long dimension L is so large as to be effectively infinite, and thus produces no diffraction effects. Figure 6.4 shows this aperture in cross section, with the long dimension of the slit extending out of the page. We locate any point on the aperture with the coordinates denoted by capital X and Y, a convention that we maintain throughout this chapter. The slit width b causes diffraction at angles θ defined in the Y-Z plane, as indicated in the figure. The ray r_0 lies in the Y-Z plane, and measures the distance from the geometric center of the aperture to a point on the observing screen. Similarly, r measures this distance from a point on the aperture with a nonzero value of Y. The *path difference* relative to r_0 is $Y \sin \theta$, as illustrated in Figure 6.4:

$$r = r_0 + Y \sin \theta \qquad (6.12)$$

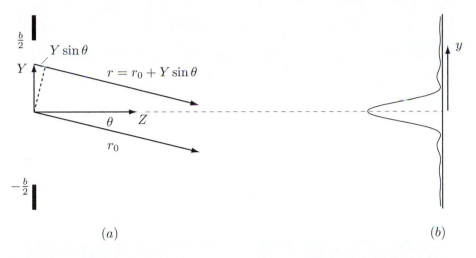

Figure 6.4 (a) Schematic and parameter definitions for determining the Fraunhofer diffraction pattern of a single long slit. The long dimension of the slit (X-direction) extends out of the page, and an arbitrary point on the cross section is located by the coordinate Y. Rays leaving the aperture in the Y-Z plane travel to the observing screen with path difference $Y \sin \theta$ relative to the ray that leaves from the center. (b) The observation screen. Points in the observation plane are located with coordinates x and y.

For Fraunhofer diffraction, r must be large enough to satisfy the criterion of Equation 6.8. Letting $dA = L\,dY$, Equation 6.11 becomes

$$E_p = C \int_{-\frac{b}{2}}^{\frac{b}{2}} e^{ik(r_0 + Y \sin \theta)} L\,dY = CLe^{ikr_0} \int_{-\frac{b}{2}}^{\frac{b}{2}} e^{ikY \sin \theta}\,dY = CLe^{ikr_0} \left(\frac{e^{ikY \sin \theta}}{ik \sin \theta} \bigg|_{-\frac{b}{2}}^{\frac{b}{2}} \right)$$

$$= \frac{2bCLe^{ikr_0}}{kb \sin \theta} \left(\frac{e^{i\left(\frac{1}{2}kb \sin \theta\right)} - e^{i\left(\frac{1}{2}kb \sin \theta\right)}}{2i} \right) = \frac{bCLe^{ikr_0}}{\frac{1}{2}kb \sin \theta} \sin \left(\tfrac{1}{2}kb \sin \theta\right)$$

$$(6.13)$$

Define the new parameter β:

$$\beta = \tfrac{1}{2} kb \sin \theta \qquad (6.14)$$

With this, Equation 6.13 becomes

$$E_p = C' \frac{\sin \beta}{\beta} \qquad (6.15)$$

with C' another constant. From L'Hospital's rule,

$$\lim_{\beta \to 0} \frac{\sin \beta}{\beta} = \lim_{\beta \to 0} \frac{\cos \beta}{1} = 1$$

so the diffraction pattern has electric field given by

$$E_p = E_0 \frac{\sin \beta}{\beta} \qquad (6.16)$$

where E_0 is the electric field at $\beta = 0$, and irradiance given by

$$I = I_0 \left(\frac{\sin \beta}{\beta} \right)^2 \tag{6.17}$$

where I_0 is the irradiance at $\beta = 0$.

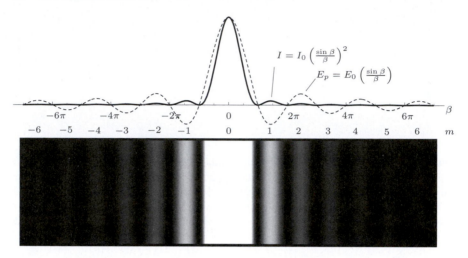

Figure 6.5   Fraunhofer diffraction profile and pattern of a single slit.

Figure 6.5 shows the electric field E_p and irradiance I of a single-slit diffraction pattern plotted versus β. E_p and I are both zero at points where β equals an integer multiple of π. As the analysis in Equation 6.13 shows, E_p is real-valued at all values of β. Thus, the diffracted field is *coherent*. Note in particular that the field within the central bright region located between $-\pi < \beta < +\pi$ is coherent and *in phase*. As with interference, we refer to each phase region with an *order number m*. The central order is referred to as the zeroth order ($m = 0$). As indicated in Figure 6.5, even orders have fields that are in phase, and odd orders are shifted in phase by π relative to the zeroth-order field.

According to the Van-Cittert-Zernike Theorem discussed in Section 5.12.2, the field within the central zeroth-order region is *always* coherent, even if the illumination is incoherent. This region of coherence extends roughly between the first zero crossing points of E_p:

$$\beta = \tfrac{1}{2} \left(\frac{2\pi}{\lambda} \right) b \sin \theta = \pm \pi$$

giving,

$$\sin \theta = \pm \frac{\lambda}{b}$$

Thus, large slit widths give correspondingly small regions of coherence. In Section 5.3.2.1 we discussed Young's double-slit experiment, and in particular noted that Young illuminated his double-slit aperture with light that had previously passed through a single slit (see Figure 5.3). To obtain coherent illumination, the single slit needed a width b small enough to give a zeroth-order coherent beam that extended across each of the double-slit apertures. Had this not been done, the two individual beams would not have been mutually coherent, and no interference fringes would have been observed. As discussed in Section 5.12.2, the coherence decreases rapidly outside of the zeroth-order region, especially when the illumination has a large frequency spread. An extended pattern of definite bright and null fringes, as illustrated in Figure 6.5, is only observed when the single-slit aperture is illuminated with light that is spatially and temporally coherent.

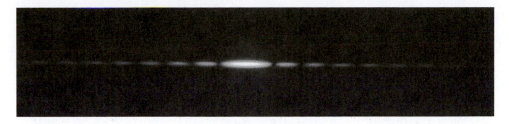

Figure 6.6   Diffraction of a laser beam by a single-slit aperture. (Edward Kinsman /Photo Researchers, Inc.)

The irradiance maxima of the diffraction pattern occur at the local extrema of E_p:

$$\frac{d}{d\beta}\left(\frac{\sin\beta}{\beta}\right) = \frac{\cos\beta}{\beta} - \frac{\sin\beta}{\beta^2} = 0$$

which has the solution $\tan\beta = \beta$. The values of β for which this is true occur at the intersection of the curve $y = \beta$ and $y = \tan\beta$. The first two solutions are at $\beta = 1.4303\pi$ and $\beta = 2.459\pi$, and the 100th solution occurs at $\beta = 100.499\pi$. Thus, as the order number increases, the point of maximum irradiance occurs closer to odd-integer multiples of $\pi/2$.

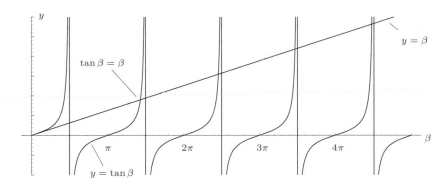

Figure 6.7   Graphical solution of the trancendental equation that locates irradiance maxima in a single-slit diffraction pattern.

■ EXAMPLE 6.1

A $0.100\,mm$-wide slit is illuminated with plane waves from a helium-neon laser of wavelength $632.8\,nm$. Find the distance from the zeroth-order maximum to the first two irradiance nulls, and to the first- and second-order irradiance maxima when the pattern is observed on a wall $Z = 1.0\,m$ away. Repeat for a slit width of $2.00\,\mu m$.

Solution

Let the slit width be $b = 0.100\,mm$. Using Equation 6.8 and letting $h = b$, $r_0 = Z$, and $r' = \infty$, we find

$$\delta_{total} = \frac{h^2}{2r} = \frac{b^2}{2r} = \frac{\left(10^{-4}\,m\right)^2}{2\,(1\,m)} = 5 \times 10^{-9}\,m = 5\,nm$$

which is in fact much less than λ, so we are in the Fraunhofer range. The first irradiance null is located at

$$\beta = \frac{1}{2}\left(\frac{2\pi}{\lambda}\right)b\sin\theta = \pi$$

or

$$\sin\theta = \frac{\lambda}{b} = \frac{632.8 \times 10^{-9}\,m}{10^{-4}\,m} = 6.328 \times 10^{-3}$$

which is small enough to use the small angle approximation. Let y measure a point on the observation screen relative to its intersection with the Z axis. Then

$$\sin\theta \cong \tan\theta = \frac{y}{Z}$$

giving

$$y = Z\theta = 6.328 \times 10^{-3}\,m = 6.328\,mm$$

The second null is at $\beta = 2\pi$ which is still well within the range of small angles, and thus is located at twice this distance from the central maximum. The first-order maximum is located at $\beta = 1.43\pi$, giving

$$y = \pm 1.43\frac{\lambda Z}{b} = \pm 1.43\left(6.328 \times 10^{-3}\,m\right) = \pm 9.05\,mm$$

and the second-order maximum is located at $\beta = 2.459\pi$, or $y = \pm 15.6\,mm$.

For a slit width of $2.00\,\mu m$, The first null is located at $\beta = \pi$, or

$$\sin\theta = \frac{\lambda}{b} = \frac{632.8 \times 10^{-9}\,m}{2.00 \times 10^{-6}\,m} = 0.3164$$

which is too large for the small angle approximation. Thus

$$\theta = \sin^{-1}(0.3164) = 0.3219\,rad = 18.4°$$

giving

$$y = Z\tan\theta = 0.334\,m$$

The second-order null is located at $\beta = 2\pi$:

$$\theta = \sin^{-1}(0.6328) = 0.6852\,rad = 39.3°$$

$$y = Z\tan\theta = 0.817\,m$$

The first-order maximum is located at $\beta = 1.43\pi$:

$$\theta = \sin^{-1}(1.43 * 0.3164) = \sin^{-1}(0.452) = 0.470\,rad = 26.9°$$

$$y = Z\tan\theta = 0.507\,m$$

and the second-order maximum is located at $\beta = 2.459\pi$:

$$\theta = \sin^{-1}(2.459 * 0.3164) = \sin^{-1}(0.778) = 0.892 \, rad = 51.1°$$

$$y = Z \tan \theta = 1.24 \, m$$

We are starting to need a large wall! Clearly, β cannot increase beyond the point where the corresponding value of $\sin \theta$ exceeds one. Thus, there are no maxima beyond the second order, and no nulls beyond third order.

Problem 6.2   The central maximum of a single-slit diffraction pattern has a width of $2.00 \, cm$ when measured on an observing screen $1.00 \, m$ away. If the illumination wavelength is $632.8 \, nm$, what is the slit width?

Problem 6.3   Repeat Example 6.1 when the illumination wavelength is $400 \, nm$.

Problem 6.4   Using the parameters of Example 6.1, find the maximum order number for irradiance nulls when the slit width is $0.100 \, mm$. Repeat when the wavelength is $400 \, nm$.

6.3.2   Rectangular Aperture

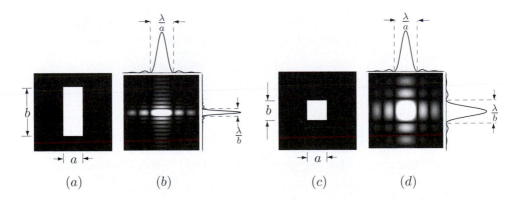

(a) $\qquad\qquad$ (b) $\qquad\qquad\qquad\qquad$ (c) $\qquad\qquad$ (d)

Figure 6.8   (a) A rectangular aperture with height b that is three times the width a. (b) The Fraunhofer diffraction pattern produced by the the aperture in (a) is wider than it is high by about 3 to 1. (c) A square aperture with each side equal to a. (d) The Fraunhofer diffraction pattern of (a) has equal width and height.

Consider an opaque screen with a rectangular opening of height b and width a, as illustrated in Figure 6.8(a). We begin with the Fraunhofer diffraction integral, Equation 6.11. Since both dimensions of the aperture are now finite, both produce diffraction. Dimension b produces diffraction at angles θ as defined in Figure 6.4, and dimension a produces diffraction at a new angle ϕ, defined in a similar way in the X-Z plane. The

Fraunhofer diffraction integral becomes

$$E_p = C \int_A e^{ikr} dA = C \int_{-\frac{b}{2}}^{\frac{b}{2}} \int_{-\frac{a}{2}}^{\frac{a}{2}} e^{ik(r_0 + x\sin\phi + y\sin\theta)} dx\,dy \qquad (6.18)$$

Rearranging gives

$$E_p = Ce^{ikr_0} \left(\int_{-\frac{a}{2}}^{\frac{a}{2}} e^{ikx\sin\phi} dx \right) \left(\int_{-\frac{b}{2}}^{\frac{b}{2}} e^{iky\sin\theta} dy \right) \qquad (6.19)$$

Both integrations proceed as in Section 6.3.1 above, to give

$$E_p = E_0 \left(\frac{\sin\alpha}{\alpha} \right) \left(\frac{\sin\beta}{\beta} \right) \qquad (6.20)$$

$$I = I_0 \left(\frac{\sin\alpha}{\alpha} \right)^2 \left(\frac{\sin\beta}{\beta} \right)^2 \qquad (6.21)$$

where

$$\alpha = \frac{1}{2}ka\sin\phi \qquad (6.22)$$

$$\beta = \frac{1}{2}kb\sin\theta \qquad (6.23)$$

Figure 6.8(b) shows the diffraction pattern for an aperture with a height of three times the width ($a = 3b$). Notice that the shorter dimension produces wider diffraction. Figure 6.8(c) illustrates a square aperture, and Figure 6.8(d) shows the corresponding symmetric diffraction pattern.

■ **EXAMPLE 6.2**

A rectangular aperture that is $a = 0.100\,mm$ wide by $b = 0.300\,mm$ high is illuminated with plane waves from a helium-neon laser of wavelength $632.8\,nm$. Find the dimensions of the zeroth-order diffraction region as measured between the first irradiance nulls when the pattern is observed on a wall $Z = 1.0\,m$ away.

Solution

Fist, let's double check that we are still in the Fraunhofer region. Use b since it is the larger dimension:

$$\delta_{total} = \frac{h^2}{2r} = \frac{b^2}{2r} = \frac{\left(3.00 \times 10^{-4}m\right)^2}{2\,(1.00\,m)} = 45.0 \times 10^{-9}m = 45.0\,nm$$

Depending on the accuracy of our measurements, we might expect to seen some effects of Fresnel diffraction, especially in the vertical direction. Nevertheless, we shall proceed with the Fraunhofer calculations.

The first diffraction nulls along the horizontal direction are located at angles determined by

$$\sin\phi = \frac{\lambda}{a} = \frac{632.8 \times 10^{-9}m}{10^{-4}\,m} = 6.328 \times 10^{-3}$$

Since this angle is measured from the center of the pattern, the horizontal width of the central maxium is

$$\Delta x = Z\,(2\phi) = 12.7\,mm$$

In the vertical direction,

$$\sin\theta = \frac{\lambda}{b} = \frac{632.8 \times 10^{-9}\,m}{3.00 \times 10^{-4}\,m} = 2.11 \times 10^{-3}$$

giving

$$\Delta y = Z\,(2\theta) = 4.22\,mm$$

The aperture and resulting diffraction pattern are illustrated in Figure 6.8(a).

Problem 6.5 Repeat Example 6.2 for a rectangular aperture of width $a = 2.00\,\mu m$ and height $b = 6.00\,\mu m$.

6.3.3 Circular Aperture

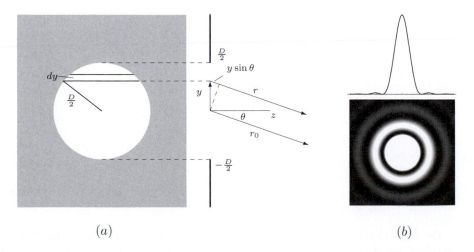

(a) $\qquad\qquad\qquad\qquad\qquad\qquad\qquad\qquad\qquad\qquad\qquad$ (b)

Figure 6.9 (a) Circular aperture. (b) Fraunhofer diffraction pattern and profile for a circular aperture.

The Fraunhofer diffraction pattern of a circular aperture will have the same circular symmetry as the aperture, so we need only calculate the diffraction profile through a cross section that passes through its center. Figure 6.9(a) shows a diagram of a circular aperture of diameter D with an elemental strip of width dy, length $\Delta x = 2\sqrt{\left(\frac{D}{2}\right)^2 - y^2}$, and area $dA = 2\sqrt{\left(\frac{D}{2}\right)^2 - y^2}\,dy$. The field E_p along a central cross section parallel to y on the observing screen is, by the Fraunhofer diffraction integral,

$$E_p = C \int_A e^{ikr}\,dA = Ce^{ikr_0} \int_{\frac{D}{2}}^{\frac{D}{2}} e^{iky\sin\theta}\left(\sqrt{\left(\tfrac{D}{2}\right)^2 - y^2}\right)\,dy$$

We will resort to integration tables to solve this integral. It will help to change variables as follows: define u and σ such that

$$u = \frac{2y}{D} \tag{6.24}$$

$$\sigma = k\frac{D}{2}\sin\theta \tag{6.25}$$

Express the above integration in terms of these variables, noting that $dy = \frac{D}{2}du$ and that when $y = \pm\frac{D}{2}$, $u = \pm 1$:

$$E_p(\sigma) = C' \int_{-1}^{1} e^{i\sigma u}\sqrt{1 - u^2}\,du$$

This evaluates to[7]

$$\int_{-1}^{1} e^{i\sigma u}\sqrt{1 - u^2}\,du = \frac{\pi J_1(\sigma)}{\sigma}$$

where $J_1(\sigma)$ is the *Bessel function of the 1st kind*, and where σ is defined by Equation 6.25.

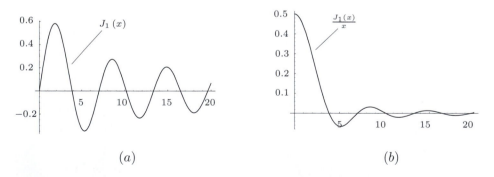

$$(a) \qquad\qquad\qquad (b)$$

Figure 6.10   (a) $J_1(x)$ is an oscillating function that is zero at the origin. (b) $J_1(x)/x$ has value 0.500 as x approaches zero.

The Bessel function $J_1(x)$ is an oscillating function that, like $\sin(x)$, has value zero when the argument is zero, as illustrated in Figure 6.10(a). The limit of $\frac{J_1(x)}{x}$ is finite at zero:

$$\lim_{x\to 0} \frac{J_1(x)}{x} = \frac{1}{2}$$

---

[7] This answer may also be verified using symbolic programming languages such as Mathematica.

as illustrated in Figure 6.10(b). Thus the diffracted field and irradiance are given by

$$E_p = E_0 \left(\frac{2J_1\left(\sigma\right)}{\sigma} \right) \tag{6.26}$$

$$I = I_0 \left(\frac{2J_1\left(\sigma\right)}{\sigma} \right)^2 \tag{6.27}$$

where E_0 and I_0 are the field amplitude and irradiance at $\sigma = 0$. Figure 6.9(b) shows a cross section irradiance pattern I. The central bright fringe of the diffraction pattern is called the *Airy disk*. As indicated in Figure 6.10, $J_1(\sigma)$ has its first zero at $\sigma = 3.832$, so by Equation 6.25,

$$\sin\theta = \frac{2\left(3.892\right)}{kD}$$

For small angles, the angular diameter of the Airy disk is given by

$$\Delta\theta \cong \frac{2.44\lambda}{D} \tag{6.28}$$

■ **EXAMPLE 6.3**

A $D = 0.100\,mm$ wide circular aperture is illuminated with plane waves from a helium-neon laser of wavelength $632.8\,nm$. Find the width of the Airy disk when the pattern is observed on a wall $Z = 1.0\,m$ away.

Solution

The first diffraction null is located an angle given by

$$\sin\theta = \frac{2\left(3.892\right)}{kD} = \frac{\lambda\left(3.892\right)}{\pi D} = \frac{\left(632.8 \times 10^{-9}\,m\right)\left(3.892\right)}{\pi\left(10^{-4}\,m\right)} = 7.84 \times 10^{-3}$$

This is small enough to allow the small angle approximation. The width of the Airy disk is twice this angle. Alternatively, Equation 6.28 may be used:

$$\Delta\theta = \frac{2.44\lambda}{D} = \frac{2.44\left(632.8 \times 10^{-9}\,m\right)}{10^{-4}\,m} = 1.54 \times 10^{-2}$$

which agrees to within round-off error. The width of the central disk is

$$\Delta x = Z\left(\Delta\theta\right) = 15.4\,mm$$

The aperture and resulting diffraction pattern are illustrated in Figure 6.10.

6.3.4 Optical Resolution

Figure 6.11(a) illustrates a telescope collecting plane waves from a distant star. The image formed at the telescope focal plane is the Fraunhofer diffraction pattern formed by the telescope aperture D, and has a size and structure that is completely unrelated to the actual size and structure of the star. Two stars lying within the field of view of the telescope create Airy disks that overlap on the focal plane. According to the *Rayleigh criterion*, two point sources are just *resolved* if they have an angular separation equal to the angular radius of the Airy disk:

$$\Delta\theta \cong \frac{1.22\lambda}{D} \tag{6.29}$$

Such a situation is illustrated in Figure 6.11(b), and Figure 6.11(c) illustrates the case where two point sources are well resolved. For a given wavelength, the resolution may only be improved by increasing the diameter of the objective lens.

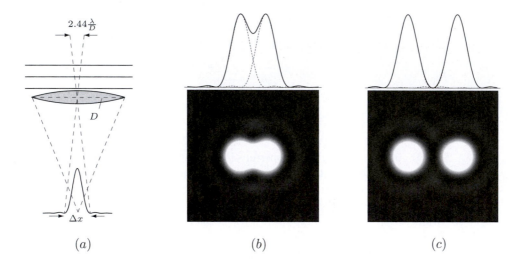

$$(a) \qquad\qquad (b) \qquad\qquad (c)$$

Figure 6.11 (a) Plane waves diffract at the telescope aperture to give an Airy disk at the focal plane. (b) Diffraction-limited images of two point sources at the minimum angular separation specified by the Rayleigh criterion. (c) Same as (b), but with an angular separation of twice the minimum amount.

According to Equation 6.28, the physical size of the Airy disk at the focal plane is determined by the wavelength and the *f-number* of the objective:

$$\Delta x = f \,\Delta\theta = 2.44\frac{f\lambda}{D} = 2.44 \,\lambda \,(f/\#) \tag{6.30}$$

Digital cameras typically use optics in the neighborhood of $f/4$, so for visible light of roughly half-micron, the corresponding Airy disk is about $5\,\mu m$ wide. The detector elements in a CCD[8] array are usually around $10\,\mu m$ wide spaced by about $5\,\mu m$. Lenses capable of focusing a distant point source to the minimum size given by Equation 6.30 are called *diffraction-limited*; however, most lenses have aberrations that cause significant distortion to the diffraction pattern, making it substantially larger than the Fraunhofer result. The net response of an optical system to a distant point source that includes all aberrations is called the *point spread function*, which is discussed more fully in Section 8.3.

---

[8]Charge-Coupled Device: the detectors used in most digital cameras.

6.3.5 More on Stellar Interferometry

As noted in Section 6.3.1, the field within the central diffraction maximum is spatially coherent. Thus, light diffracts from a distant star, arriving at Earth with a spatial coherence length determined by the radius of its Airy disk, which in turn determines the fringe visibility in a Michelson stellar interferometer with a given mirror spacing (see Section 5.12.3 and Figure 5.40.). As in Equation 5.124,

$$\frac{\ell_s}{r} = 1.22\frac{\lambda}{D} \tag{6.31}$$

where ℓ_s is the spatial coherence length, and r is the stellar distance. By Equation 5.124, if ℓ_s can be determined from the fringe visibility, then the star's angular diameter can be determined. If r can also be determined by parallax measurements, then the actual physical diameter D of the star can be determined.

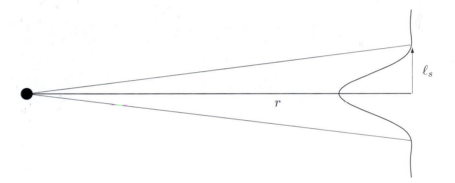

Figure 6.12   Fringe visibility for light from a distant star is determined by the spatial coherence length.

Problem 6.6   Repeat Example 6.3 if the aperture diameter is $2.00\,\mu m$.

Problem 6.7   The Airy disk of a uniformly illuminated circular hole is measured to be $2.00\,cm$ on an observing screen located $2.00\,m$ from the aperture. If the illumination wavelength is $632.8\,nm$, what is the hole diameter?

Problem 6.8   What is the minimum separation of two stars in a binary system 10.0 lightyears away if they are to be just resolved by the Hubble telescope whose diameter is $2.40\,m$. Assume a wavelength of $550\,nm$.

Problem 6.9   Find the diameter of the Airy disk formed by a distant star at the image plane of a telescope that uses a $25\,cm$-wide objective mirror of focal length $2.5\,m$. Assume an average wavelength of $550\,nm$.

6.3.6 Double Slit

Figure 6.13(a) shows a cross section of two long parallel slits of width b and separation d with the long dimensions of the slits extending out of the page. As with the single slit, we assume that the long dimension L is sufficient to allow a one-dimensional analysis with $dA = Ldy$. Begin with the Fraunhofer diffraction integral and proceed as in Section 6.3.1:

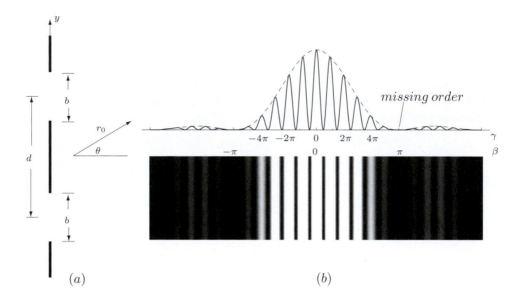

Figure 6.13   (a) Double-slit aperture geometry. (b) Irradiance profile and fringe pattern for the case where $d = 6b$.

$$E_p = C \int_0^b e^{ik(r_0 + y\sin\theta)} L \, dy \; + \; C \int_d^{d+b} e^{ik(r_0 + y\sin\theta)} L \, dy$$

$$= C L e^{ikr_0} \left[\left(\frac{e^{iky\sin\theta}}{ik\sin\theta} \Big|_0^b \right) + \left(\frac{e^{iky\sin\theta}}{ik\sin\theta} \Big|_d^{d+b} \right) \right]$$

$$= \frac{C L e^{ikr_0}}{ik\sin\theta} \left[e^{ikb\sin\theta} - 1 + e^{ik(d+b)\sin\theta} - e^{ikd\sin\theta} \right]$$

$$= \frac{C L e^{ikr_0}}{ik\sin\theta} \left(e^{ikb\sin\theta} - 1 \right) \left(1 + e^{ikd\sin\theta} \right)$$

$$= \frac{C L e^{ikr_0} b}{i\frac{1}{2}kb\sin\theta} (2) \, e^{i\frac{1}{2}kb\sin\theta} e^{i\frac{1}{2}kd\sin\theta} \left(\frac{e^{i\frac{1}{2}kb\sin\theta} - e^{-i\frac{1}{2}kb\sin\theta}}{2i} \right) \left(\frac{e^{i\frac{1}{2}kd\sin\theta} + e^{-i\frac{1}{2}kd\sin\theta}}{2} \right)$$

$$= C' \left(\frac{\sin\beta}{\beta} \right) (\cos\gamma)$$

where $\beta = \frac{1}{2}kb\sin\theta$ and $\gamma = \frac{1}{2}kd\sin\theta$. The irradiance of the diffraction pattern is thus

$$I = I_0 \left(\frac{\sin\beta}{\beta}\right)^2 (\cos\gamma)^2 \tag{6.32}$$

This irradiance pattern is illustrated in Figure 6.13(b).

It is interesting to compare the result just obtained with that found in Section 5.3.2.1 for Young's double slit. Both give a fringe spacing determined by $\frac{1}{2}kd\sin\theta$ (see Equation 5.18), but the more rigorous diffraction analysis includes the $\left(\frac{\sin\beta}{\beta}\right)^2$ envelope. In the Young's analysis, we assumed an infinitely narrow line source at each slit, but in the Fraunhofer analysis, we assume a distribution of Huygens wavelets over an aperture of finite width. Notice that the envelope has points of zero amplitude which, if aligned with a maximum of $(\cos\gamma)^2$ results in a *missing order* (see Figure 6.13(b)). Bright fringes occur when $\gamma = m\pi$ with m any integer, and envelope zeros occur when $\beta = m\pi$, with m a nonzero integer.

A simple interpretation of the double-slit diffraction pattern is that it results from the interference of two single-slit patterns that overlap on the observing screen. As in Young's analysis, it is necessary that both slits be illuminated coherently; otherwise, the interference can not occur.

Problem 6.10   Plot the Fraunhofer irradiance distribution of a double-slit aperture with $d = 2b$. Repeat for $d = 40b$.

Problem 6.11   The missing orders in a double-slit diffraction pattern occur when condition for interference maximum coincides with a zero of the $\left(\frac{\sin\beta}{\beta}\right)^2$ envelope. Show that missing orders occur when the ratio of slit spacing to slit width is an integer, and that this integer determines the order number of the first missing maximum.

Problem 6.12   If the spacing between the missing orders of Figure 6.13 is $6.00\,cm$ and the observation screen is $2.00\,m$ from the aperture, determine the slit spacing and slit width.

6.3.7 N Slits: The Diffraction Grating

We now consider an aperture that consists of N long parallel openings of width b and separation d in an opaque screen. Figure 6.14(a) illustrates the geometry. Proceed as in the double-slit analysis:

$$E_p = Ce^{ikr_0}L\left[\int_0^b e^{iky\sin\theta}dy + \int_d^{d+b} + \int_{2d}^{2d+b} + \dots + \int_{(N-1)d}^{(N-1)d+b} e^{iky\sin\theta}dy\right]$$

$$= Ce^{ikr_0}L\left[\frac{e^{iky\sin\theta}}{ik\sin\theta}\Big|_0^b + \frac{e^{iky\sin\theta}}{ik\sin\theta}\Big|_d^{d+b} + \frac{e^{iky\sin\theta}}{ik\sin\theta}\Big|_{2d}^{2d+b} + \dots + \frac{e^{iky\sin\theta}}{ik\sin\theta}\Big|_{(N-1)d}^{(N-1)d+b}\right]$$

$$= Ce^{ikr_0}L\frac{(e^{ikb\sin\theta}-1)}{ik\sin\theta}\left[1 + e^{ikd\sin\theta} + e^{ik2d\sin\theta} + \dots + e^{ik(N-1)d\sin\theta}\right]$$

$$= Ce^{ikr_0}L\frac{(e^{ikb\sin\theta}-1)}{ik\sin\theta}\left[\frac{1-e^{ikNd\sin\theta}}{1-e^{ikd\sin\theta}}\right]$$

where the last result follows from Equation 5.85 for the sum of a geometric series. Further simplification gives

$$E_p = Ce^{ikr_0}L2be^{i\beta}e^{i(N-1)\gamma}\frac{\sin\beta}{\beta}\left[\frac{\sin N\gamma}{\sin\gamma}\right]$$

where β and γ are given by

$$\beta = \tfrac{1}{2}kb\sin\theta \qquad \gamma = \tfrac{1}{2}kd\sin\theta \qquad\qquad (6.33)$$

Since

$$\lim_{\gamma\to m\pi}\frac{\sin N\gamma}{\sin\gamma} = \lim_{\gamma\to m\pi}\frac{N\cos N\gamma}{\cos\gamma} = \pm N \qquad\qquad (6.34)$$

the ratio of sines may be normalized by including an extra factor of N in the denominator:

$$E_p = E_0\frac{\sin\beta}{\beta}\left[\frac{\sin N\gamma}{N\sin\gamma}\right] \qquad\qquad (6.35)$$

giving the irradiance pattern

$$I = I_0\left(\frac{\sin\beta}{\beta}\right)^2\left(\frac{\sin N\gamma}{N\sin\gamma}\right)^2 \qquad\qquad (6.36)$$

Figure 6.14(b) illustrates the irradiance pattern of Equation 6.36 when $N = 5$. Strong constructive interference occurs at the *principle maxima*, with weaker *secondary maxima* in between. As with the double-slit aperture, the N-slit pattern includes a $\left(\frac{\sin\beta}{\beta}\right)^2$ envelope.

The locations of the irradiance maxima are determined by the *interference term* $\left(\frac{\sin N\gamma}{N\sin\gamma}\right)^2$. By Equation 6.33, the envelope is determined by the slit width b, while the interference term is determined by the slit spacing d. Depending on the slit width, the envelope may be zero at a point where the interference term is maximum, leading to a missing order, as illustrated in Figure 6.14(b).

According to Equations 6.34 and 6.36, the principle maxima occur when $\gamma = m\pi$, with the integer m referred to as the order number. Secondary local maxima occur when $N\gamma$ is an odd-integer multiple of $\pi/2$. In this case, γ is the same odd-integer multiple of $\gamma = \frac{\pi}{2N}$. Replacing $\sin\gamma$ with unity gives:

$$\left(\frac{\sin N\gamma}{N\sin\gamma}\right)^2 \approx \frac{1}{N^2}$$

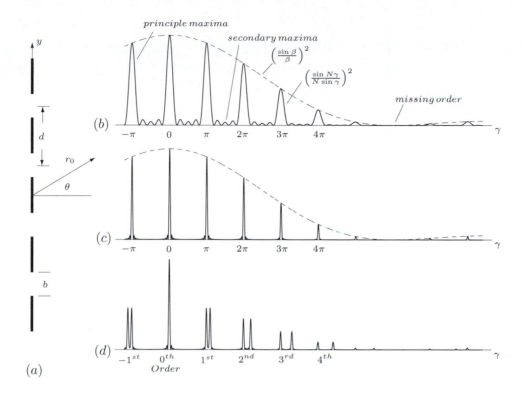

Figure 6.14   (a) N-slit aperture geometry. (b) irradiance profile for the case where $N = 5$ and $d = 6b$. (c) irradiance profile when $d = 6b$ and $N = 25$. (d) A second wavelength is added that is 10% longer than the one in (b) and (c).

Thus, when N becomes large, the secondary maxima are much smaller than the principle maxima and are much more localized, as illustrated in Figure 6.14(c) for $N = 25$.

Principal maxima occur when $\gamma = \frac{1}{2}kd \sin\theta = m\pi$, or

$$d \sin\theta = m\lambda \tag{6.37}$$

Thus, distinct wavelengths give distinctly located principle maxima, as illustrated in Figure 6.14(d).

6.3.8   The Diffraction Grating

A *diffraction grating* is an aperture designed to produce a diffraction profile with distinct principle maxima that can be used for spectroscopy. In the simplest case, it consists of a N-slit aperture which, when illuminated normally, produces the diffraction profile given by Equation 6.36, and principle maxima located by Equation 6.37. When the incidence is not normal, there is path difference associated with both incident and diffracted rays, as illustrated in Figure 6.15. In this case, Equation 6.37 becomes

$$d \left(\sin\theta_i + \sin\theta\right) = m\lambda \tag{6.38}$$

where θ_i is the incident angle, and θ is the diffracted ray. We refer to Equation 6.38 as the *grating equation*.

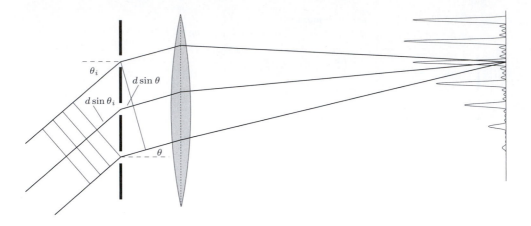

Figure 6.15   A transmission diffraction grating illuminated at angle of incidence θ_i. The diffraction pattern at the Fraunhofer plane is due to path differences in both the incident and diffracted beams.

■ **EXAMPLE 6.4**

Two monochromatic spectral lines of wavelength $400\,nm$ and $700\,nm$ illuminate a $1.00\,cm$-wide diffraction grating that has $5,000$ equally spaced rulings with $d = 2b$. Find the angles that locate the principle maxima when this grating is illuminated at an angle of incidence of $45.0°$.

Solution

The slit spacing is

$$d = \frac{10^{-2}m}{5.00 \times 10^3} = 2.00 \times 10^{-6}m = 2.00\,\mu m$$

Since $d = 2b$, the slit width is $1.00\,\mu m$.

To find the principle maxima, begin with the grating equation 6.38. The zeroth-order maxima occurs at

$$\sin \theta_0 = -\sin \theta_i$$

or $\theta_0 = -45.0°$. Note the sign convention for angles calculated according to the grating equation: θ does not become positive until it "crosses over" the grating normal.

When $\lambda = 400\,nm$, the $m = 1$ order is located at

$$\sin \theta_1 = \frac{(1)\lambda}{d} - \sin \theta_i = (1)\left(\frac{400 \times 10^{-9}m}{2.00 \times 10^{-6}m}\right) - \sin 45° = -0.507 \Rightarrow \theta_1 = -30.5°$$

and the $m = -1$ order is located at

$$\sin \theta_{-1} = \frac{(-1)\lambda}{d} - \sin \theta_i = (-1)(0.2) - \sin 45° = -0.907 \Rightarrow \theta_{-1} = -65.1°$$

There will be no further negative orders, since the grating equation gives values for the $\sin \theta$ whose absolute values exceed unity. Continuing with the positive orders, we find

$$\sin \theta_2 = \frac{(+2)\lambda}{d} - \sin \theta_i = (+1)(0.2) - \sin 45° = -0.307 \Rightarrow \theta_2 = -17.9°$$

$$\sin \theta_3 = \frac{(+3)\lambda}{d} - \sin \theta_i = (+3)(0.2) - \sin 45° = -0.107 \Rightarrow \theta_3 = -6.15°$$

Continue in this fashion to find $\theta_4 = +5.33°$, $\theta_5 = +17.03°$, $\theta_6 = +29.5°$, , $\theta_7 = +43.9°$, and $\theta_8 = +63.2°$ with no further positive orders.

The envelope $\left(\frac{\sin \beta}{\beta}\right)^2$ will be zero when $\sin \beta$ is zero for nonzero values of β. Since $b = \frac{d}{2}$, $\beta = \frac{\gamma}{2}$, so β will be an integer multiple of π whenever m is even. Thus all even orders are missing.

When $\lambda = 700\,nm$, there are no negative orders, since the grating equation gives

$$\sin \theta_{-1} = \frac{(-1)(700\,nm)}{d} - \sin \theta_i = (-1)(0.35) - \sin 45° = -1.06$$

Again, the even orders will be missing, so we need only check the odd positive orders. The first-order principal maximum occurs at

$$\sin \theta_{+1} = \frac{(+1)(700\,nm)}{d} - \sin \theta_i = (+1)(0.35) - \sin 45° = -0.357 \Rightarrow \theta_1 = -20.9°$$

Continue to find that $\theta_3 = 20.1°$ with no further odd positive orders.

Note that the angular spacing between the two $m = +1$ principle maxima is $9.60°$, with the *red* line diffracted through the largest angle. This angular spacing increases to $20.7°$ in third order.

Using broadband illumination can result in spectral *overlap*. For example, using the grating of Example 6.4 with white light containing all wavelengths between $400\,nm$ and $700\,nm$ gives a third-order spectrum that ranges from $-6.15°$ to $20.1°$. However, a $400\,nm$ line will have a fifth-order principal maximum that also falls within this angular range, creating a false data feature. Gratings with imperfect rulings can also have spectral *ghosts*. Typically, these result from periodic ruling errors that result from imperfect screw mechanisms that control the mechanical ruling apparatus. Such errors can be avoided with *holographic* diffraction gratings constructed by exposing a photographic emulsion to an interference pattern produced by carefully controlled laser illumination.

There are several options available to construct a diffracting aperture that behaves as a diffraction grating. Figure 6.16(a) illustrates a *transmission* grating that consists of parallel openings in an opaque screen. In Figure 6.16(b), there are parallel absorbing strips deposited on a reflecting surface to give a *reflection* grating. In practice, this is approximated with irregularly shaped rulings scratched in an optically flat reflecting surface. The rough rulings scatter light away; however, some of this scattered light can also combine with the diffracted beam. Figure 6.16(c) illustrates a *phase* grating. If the material is transparent, it is a *transmission phase grating*, where each stepped region differs in optical thickness from the adjacent regions by $\lambda/2$. In principle, the *diffraction efficiency* of phase gratings is larger since none of the incident light is absorbed or scattered away. Alternatively, a reflective material can produce a *relection phase grating* if each stepped region differs in physical distance by $\lambda/2$ from its neighbors. All gratings in Figures 6.16(a)—(c) give an

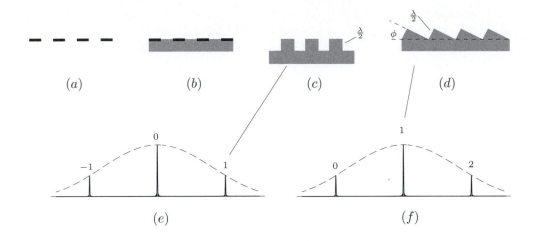

Figure 6.16   (a) A transmission grating. (b) A reflection grating with reflective regions separated by perfectly absorbing regions. (c) A reflective phase grating with reflective planes separated by $\lambda/2$. (d) A blazed diffraction grating with blaze angle ϕ. (e) Diffraction profile where the maximum irradiance is in the zeroth order. (f) A blazed diffraction grating shifts the irradiance maximum to a nonzero order.

envelope term that is centered about the $m = 0$ principle maxima, as illustrated in Figure 6.16(e). The grating of Figure 6.16(d) is *blazed* with reflective surfaces that are inclined at the *blaze angle* ϕ, as shown. Adjusting the blaze angle to correspond with the angle of a particular principal maximum shifts the envelope term so that it is centered about that angle, giving the nonzero maximum the largest irradiance, as illustrated in Figure 6.16(f). This represents a further increase in the efficiency of the grating to diffract light into a particular nonzero diffraction order.

■ **EXAMPLE 6.5**

A diffraction grating is blazed to give maximum diffraction efficiency into first order at the design wavelength λ_0. If the slit width b is equal to the slit spacing d, show all other orders are missing.

Solution

Figure 6.16(d) illustrates a blazed grating. The principle maxima occur when the interference term is maximum; $\gamma = \frac{1}{2}kd\sin\theta = m\pi$ where m is the order number. Because the grating is blazed for first order, the maximum irradiance occurs for the $m = 1$ order. Since $b = d$, $\beta = \gamma$, so all angles that give the above condition for γ also give a similar condition for β:

$$\beta = \frac{1}{2}kd\sin\theta = m\pi$$

This is precisely the condition that gives $\frac{\sin\beta}{\beta} = 0$. Thus all other orders ($m \neq 1$) are missing.

Notice from Figure 6.16 that it is indeed possible to construct a blazed grating with $d \approx b$. The surface steps must be $\lambda_0/2$, and the angle must coincide precisely with the diffraction angle for the $m = 1$ diffraction order. In this case, most of the diffracted energy couples into the $m = 1$ diffraction order.

■ **APPLICATION NOTE 6.1**

Tuned Laser Cavity

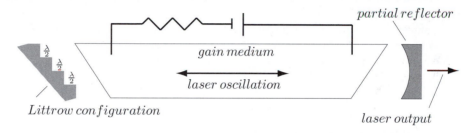

Figure 6.17 A laser cavity that is tuned with a blazed diffraction grating in the Littrow configuration.

As noted in Application Note 5.4, a *laser cavity* is a Fabry Perot interferometer where the region between the plates contains a *gain medium* capable of manufacturing photons. This will be discussed in detail in Chapter 7, but we suffice it here to say the gain medium typically consists of a material that is elevated to an excited state by a mechanism such as an electric current. Often, there is more than one quantum state capable of producing laser output at a discrete set of wavelengths or possibly even over a continuous range of wavelengths. One way of selecting such wavelengths is illustrated in Figure 6.17, where one of the mirrors in a Fabry Perot cavity is replaced by a blazed diffraction grating mounted in the *Littrow configuration*. The grating is designed to have high reflectance in the first order principal maximum. In this mounting arrangement, the reflective surfaces of the grating are adjusted to be exactly perpendicular to the cavity axis, and because of the half-wavelength spacing, they act as a single reflective surface at the design wavelength. The diffraction efficiency of the grating along with the reflectivity of a partial reflector on the other end of the cavity determine the coefficient of finesse for the Fabry Perot cavity. This finesse is destroyed for the other wavelengths since they are diffracted at angles that are not precisely parallel with the cavity axis. By adjusting the angle of the grating, the cavity can be made optimal for the desired output wavelength.

There are two ways to "tune" the laser cavity of Figure 6.17 for a specific wavelength output. One is to vary the spacing between the grating and the mirror, as discussed in Section 5.10. A laser of length L one meter will have a Fabry Perot free spectral range of $C/2L \approx 150\,MHz$. For lasers where the gain medium is an excited gas, this is a small fraction of the Doppler and collisionally broadened linewidth for a single transition. The blazed diffraction grating allows the laser to be tuned for operation on a number of different available transitions. Usually, these available wavelengths are localized enough so that the diffraction grating performs more or less equivalently over all possible output wavelengths.

Lasers can only operate if the gain (e.g., photon production), exceeds all sources of losses in the cavity. Any light that is coupled into orders other than that of the blaze design represents a loss. The high diffraction efficiency of a blazed grating enables their use for this important application.

6.3.8.1 *Chromatic Resolving Power*

The angular location of the principle maxima is determined by the interference term of Equation 6.36. The *chromatic resolving power* of a grating is a measure of its ability to separate two closely spaced spectral lines into distinct principle maxima, which in turn is determined by the number of slits in the grating, as illustrated in Figure 6.14.

At any particular principal maximum, both numerator and denominator of the interference term are zero. By Equation 6.36, the interference term will be zero again when the argument of the numerator changes by π. Using the symbol δ to represent a finite change in this quantity,

$$\delta\left(N\gamma\right) = \delta\left(N\frac{1}{2}kd\sin\theta\right) = N\frac{1}{2}kd\,\delta\left(\sin\theta\right)$$

We can estimate this finite change $\sin\theta$ by using a differential: $\delta\left(\sin\theta\right) = \cos\theta\,\Delta\theta$. This along with the previous result must give a change of π:

$$N\frac{1}{2}\frac{2\pi}{\lambda}d\cos\theta\,\Delta\theta = \pi$$

Solving for $\Delta\theta$ gives

$$\Delta\theta = \frac{\lambda}{Nd\cos\theta} \tag{6.39}$$

If we assume normal incidence, the grating equation gives (see Equation 6.37)

$$\lambda = \frac{d}{m}\sin\theta$$

We calculate the change in wavelength $\Delta\lambda$ as above:

$$\Delta\lambda = \frac{d}{m}\cos\theta\,\Delta\theta \tag{6.40}$$

According to the Rayleigh criterion (see Figure 6.11), two principle maxima will just be *resolved* if the maximum for one coincides with a zero for the other. Thus, we combine the value of $\Delta\theta$ found above with the last result to give

$$\Delta\lambda = \frac{\lambda}{Nm}$$

The chromatic resolving power is defined as

$$\frac{\lambda}{\Delta\lambda} = Nm \tag{6.41}$$

where N is the number of slits in the grating that are illuminated coherently by the beam to be analyzed, and m is the order number. Note that the resolution increases with increasing order number, as indicated in Figure 6.14(d).

■ **APPLICATION NOTE 6.2**

Grating Spectrometer

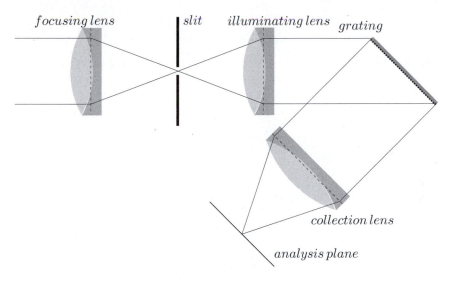

Figure 6.18   A grating spectrometer.

A *grating spectrometer* uses a diffraction grating to separate an analysis beam into its constituent wavelengths. Figure 6.18 shows one possible design. Here, the light to be analyzed has been previously collimated, and it is focused onto an *entrance slit* by a focusing lens. The entrance slit has two purposes. It is one focal length away from an illuminating lens. If the cone of light formed by the focusing lens matches the cone that optimally fills the illuminating lens, then the grating will be fully illuminated. This will be the case when the $f/\#$ of the focusing lens is the same as that of the illuminating lens. More importantly, the slit is oriented so that light that passes through it is spatially coherent across all ruling of the grating. This will be the case when the slit and grating rulings are parallel. Illuminating the grating so that all available rulings are coherently illuminated gives the maximum resolving power indicated by Equation 6.41.

Light diffracted by the grating is collected by the collection lens. In a spectrometer, film or a CCD array is placed at the focal plane of the collection lens, labeled as the analysis plane in the figure. In a *monochromator*, the grating is rotated and an *exit slit* is placed at the analysis plane.

Problem 6.13   A 5.00 cm-wide diffraction grating with 6000 *lines/cm* has a slit width that is one-third the slit spacing. Find the missing orders, and angles of all observable diffraction orders for 400 nm illumination. Repeat for light with wavelength 700 nm. Assume an incident illumination angle of $-45°$.

Problem 6.14   Repeat Problem 6.13 for normal illumination.

Problem 6.15   A diffraction grating for use at $10.6\,\mu m$ has $450\,lines/cm$. If the slit width is one-half the slit spacing, find the missing orders and all observable diffraction orders for normal illumination.

Problem 6.16   How many lines must be ruled on a diffraction grating that can just resolve the sodium doublet ($588.995\,nm$ and $589.592\,nm$) in first order?

Problem 6.17   A HeNe laser has a cavity length of $25\,cm$ and a wavelength of $633\,nm$. How wide must a grating with $8000\,lines/cm$ be in order to just resolve in first order a wavelength change corresponding to the free spectral range of this laser?

6.3.9   Fraunhofer Diffraction as a Fourier Transform

In this section we demonstrate that the Fraunhofer diffraction integral may be approximated as a Fourier transform integral. In this and subsequent chapters, we will find many opportunities to exploit this association. As we proceed, we will use the geometry of Figure 6.19, with coordinates X and Y that locate points in the aperture plane, and x and y that locate points on the observation plane.

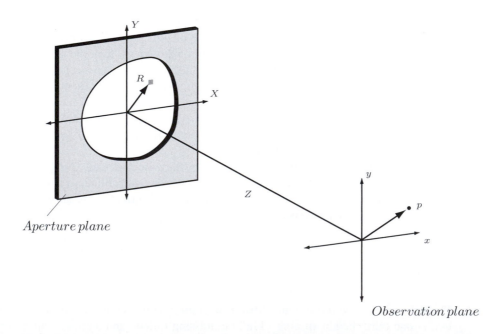

Figure 6.19   Geometry for Fraunhofer diffraction. The distance Z is assumed to be large enough for Fraunhofer diffraction.

The Fraunhofer diffraction integral for the single-slit aperture was found in Section 6.3.1:

$$E_p = C \int_{-\frac{b}{2}}^{\frac{b}{2}} e^{ikY \sin \theta} dY$$

where the constant C is redefined to include all constant terms in Equation 6.13. Let Z be the distance from the aperture to the observing plane. If the angle θ is small, we may write

$$\sin \theta \approx \tan \theta = \frac{y}{Z}$$

Combining this with the previous result gives

$$E_p = C \int_{-\frac{b}{2}}^{\frac{b}{2}} e^{ikY \frac{y}{Z}} dY$$

We may represent the aperture with an *aperture function* $g(Y)$. In the case of a single slit, $g(Y) = 1$ when $|Y| < \frac{b}{2}$ and is zero otherwise. By including the aperture function, we may use infinite integration limits

$$E_p = C \int_{-\infty}^{\infty} g(Y) e^{i\left(k\frac{y}{Z}\right)Y} dY \tag{6.42}$$

We define the *spatial frequency* μ:

$$\mu = k\frac{y}{Z} \tag{6.43}$$

to give

$$E_p(\mu) = C \int_{-\infty}^{\infty} g(Y) e^{i\mu Y} dY \tag{6.44}$$

Apart from the sign of the exponential (which is a matter of convention and can be adjusted with a change of variables), a comparison with Equation 5.40 shows this to be mathematically equivalent to the one-dimensional Fourier transform discussed in Section 5.4.1.

Diffracting apertures can have arbitrary shape with values of transmission and phase retardation that vary over the diffracting screen. The interpretation of Equation 6.44 is that the Fraunhofer diffraction pattern is the Fourier transform of the aperture function $g(Y)$. Thus, observation of the Fraunhofer diffraction pattern from an aperture gives an intuitive way to visualize the Fourier transform of the associated aperture function.

■ **EXAMPLE 6.6**

Analyze the single slit of Section 6.3.1 as a Fourier transform.

Solution

The aperture function for this slit is the rectangular pulse of Example 5.5:

$$g(Y) = \begin{cases} +1 & |Y| \leq \frac{b}{2} \\ 0 & |Y| > \frac{b}{2} \end{cases}$$

In this example, the Fourier transform was found to be a *sinc* function:

$$E_p = C \left(b \frac{\sin\left(\frac{\mu b}{2}\right)}{\frac{\mu b}{2}} \right) = E_0 \frac{\sin\left(\frac{\mu b}{2}\right)}{\frac{\mu b}{2}}$$

According to Equation 6.48,

$$\frac{\mu b}{2} = \frac{ky}{Z}\frac{b}{2} = \frac{1}{2}kb\frac{y}{Z} = \frac{1}{2}kb\sin\theta = \beta$$

where, in the last step we reversed the small angle approximation. Thus, the Fraunhofer diffraction pattern is given by

$$E_p = E_0 \frac{\sin\beta}{\beta}$$

as in Section 6.3.1

The rectangular aperture of Section 6.3.2 can be represented in a similar way as a *two-dimensional Fourier transform*. Begin with the Fraunhofer diffraction integral, and note that the path to p from any point on the aperture is given by

$$r = r_0 + X\frac{x}{Z} + Y\frac{y}{Z} \tag{6.45}$$

where again we have used the small angle approximation. The Fraunhofer diffraction integral becomes

$$E_p = C \int\!\!\int_A e^{ikr} dA = C \int\!\!\int_A e^{ik\left(r_0 + \frac{x}{Z} + \frac{y}{Z}\right)} dA = Ce^{ikr_0} \int\!\!\int_A e^{i\frac{k}{Z}(xX+yY)} dX dY \tag{6.46}$$

where the above integral is over the aperture opening. We define the *aperture function* $g(X, Y)$ so that the Fraunhofer diffraction integral may be rewritten as follows:

$$E_p = Ce^{ikr_0} \int_{-\infty}^{\infty}\int_{-\infty}^{\infty} g[X, Y] e^{i\frac{k}{Z}(xX+yY)} dX dY \tag{6.47}$$

As above, define the *spatial frequencies* ν and μ:

$$\nu = k\frac{x}{Z} \qquad \mu = k\frac{y}{Z} \tag{6.48}$$

Using these definitions, the Fraunhofer diffraction integral becomes

$$E_p(\nu, \mu) = Ce^{ikr_0} \int_{-\infty}^{\infty}\int_{-\infty}^{\infty} g[X, Y] e^{i(\nu X+\mu Y)} dX dY \tag{6.49}$$

The Fourier interpretation of Fraunhofer diffraction is appealing, but we must always remember that it is an approximation. In addition to the usual Fraunhofer approximations, we have also assumed a diffraction pattern that is small enough to allow the small angle approximation.

6.3.10   Apodization

As we saw in Example 5.6, the Fourier transform of a Gaussian function is also a Gaussian. This means that the Fraunhofer diffraction pattern of a beam with a Gaussian profile will have no nonzero diffraction orders. This can be used, for example, to enhance the resolution of a telescope. Normally, a telescope objective collects light equally from all points across the lens aperture. This creates a step irradiance profile at the lens, and an Airy profile at the focal plane. The rings in the Airy profile contain about 16% of the pattern irradiance, and these can completely obscure a dimmer star with nearby angular separation. By coating the lens so that its transmission decreases toward the edges, the Airy rings are reduced in the image, enhancing the image of the dimmer star. An optical element coated in this way is said to be *apodized*.

■ **APPLICATION NOTE 6.3**

Spatial Filters

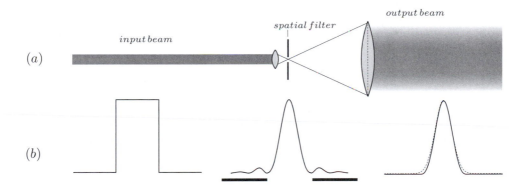

Figure 6.20   (a) An input beam with non-Gaussian structure is spatially filtered by a pinhole at the focus of an objective lens. The filtered beam is collimated by a second lens which is also one focal length away from the pinhole. (b) Profiles of the non-filtered beam, the non-filtered beam just prior to spatial filtering, and the filtered output with the nonzero spatial frequencies removed. In the filtered profile, the dotted line has a Gaussian profile, and the solid line has an Airy profile.

Ideal laser beams typically have Gaussian output beam profiles. However, dust and other optical imperfections in the optical train can create undesirable irradiance variations across the beam. Because the laser beam is coherent, these variations can be quite large. Such beam imperfections can be removed with a *spatial filter*, as illustrated in Figure 6.20. Here, a lens focuses a non-Gaussian beam to its Fraunhofer diffraction pattern where a pinhole of just the right diameter scrapes away the nonzero diffraction orders, passing only the central Airy disk, which is very nearly Gaussian in profile, almost free of non-Gaussian imperfections. The output beam may be optionally collimated by a clean second lens as illustrated.

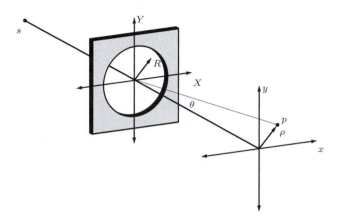

Figure 6.21 Geometry for an aperture with circular symmetry.

6.3.10.1 *Apertures with Circular Symmetry*   The two-dimensional Fourier transform of an aperture function that has circular symmetry is also given by the one-dimensional *zeroth-order Hankel Transform*[9]. Figure 6.21 illustrates the geometry. Let $\rho = \sqrt{x^2 + y^2}$ locate a point in the observation plane, and $R = \sqrt{X^2 + Y^2}$ locates a point in the aperture plane. Then

$$E_p(\rho) = C \int\limits_0^\infty g(R) J_0 \left(\frac{kR\rho}{r_0} \right) 2\pi R\, dR \tag{6.50}$$

where J_0 is the *zeroth-order Bessel function of the first kind*. In many cases, using the Hankel rather than a Fourier transform can provide computational and/or analytical advantages.

■ **EXAMPLE 6.7**

Use the Hankle transform of Equation 6.50 to find the Fraunhofer diffraction pattern of a uniform circular aperture.

Solution

$$E_p(\rho) = C \int\limits_0^\infty g(R) J_0 \left(\frac{kR\rho}{r_0} \right) 2\pi R\, dR$$

$$= 2\pi C \int\limits_0^{D/2} J_0 \left(\frac{kR\rho}{r_0} \right) R\, dR$$

[9]See, for example, Bracewell [4].

Changing variables will simplify the integral. Let $\bar{R} = \frac{2R}{D}$ with $dR = \frac{D}{2}d\bar{R}$ to give

$$E_p = 2\pi C \frac{D^2}{4} \int_0^1 J_0\left(k\frac{D}{2}\frac{\rho}{r_0}\bar{R}\right)\bar{R}d\bar{R}$$

Since $\frac{\rho}{r_0} = \sin\theta$, Equation 6.25 gives

$$E_p = 2\pi C \frac{D^2}{4} \int_0^1 J_0\left(\sigma\bar{R}\right)\bar{R}d\bar{R} \qquad (6.51)$$

Use integral tables or software to verify:

$$\int_0^1 J_0\left(\sigma\bar{R}\right)\bar{R}d\bar{R} = \frac{J_1(\sigma)}{\sigma}$$

Thus

$$E_p = E_0\frac{J_1(\sigma)}{\sigma}$$

as obtained in Section 6.3.3.

6.4 FRESNEL DIFFRACTION

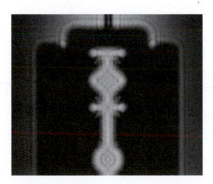

Figure 6.22 Fresnel diffraction from a razor blade. (Ken Kay/Fundamental Photographs)

We now consider the case where both source and diffracted waves have spherical wavefronts. We begin with the Fresnel Kirchhoff integral theorem (see Equation 6.5)

$$E_p = C\int_A \left[\frac{1 + \cos\theta}{2}\right] \frac{e^{ik(r+r')}}{r'r}dA \qquad (6.52)$$

As previously noted in Section 6.2, this equation can be interpreted in terms of Huygens secondary wavelets modified by the obliquity factor $K(\theta)$:

$$K(\theta) = \frac{1}{2}(1 + \cos\theta) \qquad (6.53)$$

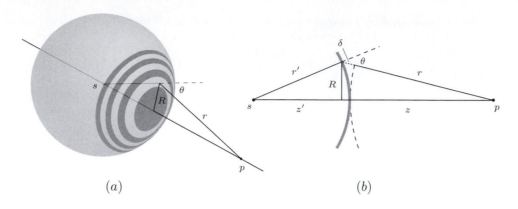

(a) (b)

Figure 6.23   (a) A spherical wavefront from source s with Fresnel zones. (b) Diagram that illustrates the path difference δ.

where theta is the angle between r' and r, as illustrated in Figure 6.23.

Figure 6.23 shows a wavefront that emanates from s located a distance r' away, and a receive point p separated from s by $r' + r$. According to Equation 6.52, each point on this wavefront becomes a source of secondary Huygens wavelets that combine to give the disturbance at p. For wavelets generated away from the line that joins s and p, there is additional path that leads to additional phase shift, as illustrated in Figure 6.23(b). Using the horizontal distances z and z' as defined in the figure, we may represent the total path from s to p as

$$r' + r = \sqrt{z'^2 + R^2} + \sqrt{z^2 + R^2}$$

$$= z'\sqrt{1 + \frac{R^2}{z'^2}} \quad + \quad z\sqrt{1 + \frac{R^2}{z^2}}$$

$$= z'\left(1 + \frac{R^2}{2z'^2} + \dots\right) \quad + \quad z\left(1 + \frac{R^2}{2z^2} + \dots\right)$$

In the Fresnel approximation, the path from s to p given by

$$r' + r = z' + z + \frac{R^2}{2}\left(\frac{1}{z'} + \frac{1}{z}\right) \tag{6.54}$$

We note that Equation 6.54 will only be valid when R is small compared to z' and z. Given this, we may further simplify Equation 6.52 by assuming that r' and r may be replaced by z' and z in the denominator of the integrand, reasoning that the main influence of the last term in Equation 6.54 comes mostly from the exponential term. Finally, if R is small compared to the source and receive distances, we may safely ignore the effects of the obliquity factor. Equation 6.52 then becomes

$$E_p = \frac{C'}{z'z}\int_A e^{ik(r'+r)}dA = C\int_A e^{ik(r'+r)}dA \tag{6.55}$$

where C' and C are both constants. The quantity $r' + r$ is given by Equation 6.54.

The Fresnel curvature term of Equation 6.54 prevents analytical solution of Equation 6.55, so we must proceed numerically. Before doing so, it will be useful to develop an intuitional tool that will allow us to anticipate many of the features of Fresnel diffraction without the need of detailed numerical analysis.

6.4.1 Fresnel Zones

Fresnel zones, also called *half-period zones*, can simplify the interpretation and analysis of Equation 6.55. Fresnel zones are regions bounded by concentric circles of radius R_m such that δ differs by $\frac{\lambda}{2}$ from one boundary to the next. The mth boundary has radius R_m determined by

$$m\frac{\lambda}{2} = \frac{1}{2}R_m^2 \left(\frac{1}{z'} + \frac{1}{z}\right) = \frac{1}{2}\frac{R_m^2}{L} \tag{6.56}$$

where the distance L is given by

$$\frac{1}{L} = \frac{1}{z'} + \frac{1}{z} \tag{6.57}$$

Solving Equation 6.56 for R_m gives

$$R_m = \sqrt{m\lambda L} \tag{6.58}$$

The area of the mth Fresnel zone is given by

$$\pi R_{m+1}^2 - \pi R_m^2 = \pi (m+1)\lambda L - \pi m\lambda L$$
$$= \pi \lambda L$$
$$= \pi R_1^2$$

Thus, the area of all Fresnel zones are equal as long as the Fresnel approximation of Equation 6.54 is valid.

■ **EXAMPLE 6.8**

Find the number of Fresnel zones that fit within a circular aperture of radius $1.00\,cm$ if the source and receive points are $10.0\,cm$ away. Assume HeNe illumination with $\lambda = 632.8\,nm$. What is the radius of the central Fresnel zone?

Solution

By Equation 6.58, $L = \frac{1}{20}\,m$. Solving for m in Equation 6.56 gives

$$m = \frac{R^2}{\lambda L} = \frac{\left(10^{-4}\,m^2\right)\left(20\,m^{-1}\right)}{632.8 \times 10^{-9}\,m} = 3161$$

The central zone has radius

$$R_1 = \sqrt{\lambda L} = \sqrt{\frac{632.8 \times 10^{-9}m}{20\,m^{-1}}} = 1.78 \times 10^{-4}\,m = 0.174\,mm$$

A circular aperture of radius a centered on the line that connects s and p passes a number of zones given by the dimensionless *Fresnel number* N_F:

$$N_F = \frac{a^2}{\lambda L} \tag{6.59}$$

For non-circular apertures, a is a characteristic size that can be used to approximate the number of zones passed by the aperture. When $N_F << 1$, Fraunhofer diffraction suffices, but when $N_F > 1$, Fresnel diffraction must be used.

The Fresnel zone that contains the line passing from s and p is called the *central Fresnel zone*, and all Huygens wavelets that originate within this zone arrive at p with a path difference of less than $\frac{\lambda}{2}$, and a corresponding phase shift of less than π. In the next Fresnel zone, this path difference is $\frac{\lambda}{2} < \Delta r < \lambda$, with phase shift $\pi < \Delta\phi < 2\pi$. The zones continue in this fashion to cover the entire spherical wavefront; however, the Fresnel approximation will cease to be valid when the zone radius R_m becomes comparable to r' or r. The first few of these zones are illustrated in Figure 6.23(a) using a wavelength that is much closer to L than in Example 6.8.

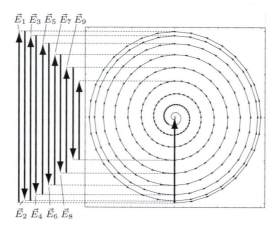

Figure 6.24 Phasors for Huygens wavelets in an unobstructed wave. The field is one half that from the central Fresnel zone.

Within a given Fresnel zone, integration of Equation 6.52 consists of summing the contribution of all secondary wavelets within the zone area. Note that each wavelet arrives at p as a complex number (see Equations 6.1-6.3). Recall that complex numbers are added by summing real and imaginary parts separately, much like adding vector components. We complete this analogy by defining *phasors*, which we visualize as vectors with amplitude equal to the amplitude of the secondary wavelet, and direction given by the amount of additional phase shift caused by the increase in propagation path to p. Figure 6.24 illustrates a *vibration curve* that results when these phasors are added. For purposes of illustration, only 16 phasors for each zone are shown. The vibration curve begins with the phasor that represents secondary wavelets that emerge from the center of the central Fresnel zone. The next phasor represents secondary wavelets emitted a bit farther from the center of the zone; the amplitude of this phasor is almost the same as the first, but the direction is changed according to the additional phase shift that occurs for these wavelets. As we continue through the central zone, this phase shift ranges from zero at the center to π at the outer edge. The net contribution to E_p from the central Fresnel zone is determined the vector sum of all phasors originating from this zone, illustrated as $\vec{E}_1$ in Figure 6.24. The phasors that originate from the second Fresnel zone have phase shifts that range from π to 2π, and sum together to produce the downward pointing phasor $\vec{E}_2$, as illustrated. Zone 3 contributes $\vec{E}_3$ that points up, zone 4 produces $\vec{E}_4$ that points down, and so on. As we progress through higher zones, the phasor amplitude begins to decrease. Part of this decrease comes from the obliquity factor, since zones away from the central zone have larger values of θ. Higher zones also have larger values of r, which also causes the phasor amplitude to decrease. As the phasor amplitude decreases, the vibration curve spirals into the center, and the net sum of the phasors from all of the zones approaches $\frac{\vec{E_1}}{2}$, as illustrated in the figure. Thus, Fresnel's beautiful construct has allowed us to intuitively determine a fascinating result of Equation 6.52: the effect at p of all secondary wavelets emerging from the unobstructed wavefront is about equal to half the contribution from the central Fresnel zone.

We can reach the same conclusion in an alternate way. According to the vibration curve of Figure 6.24, odd Fresnel zones contribute positively and even Fresnel zones decrease the disturbance at p:

$$|E_p| = |E_1| - |E_2| + |E_3| - |E_4| + |E_5| - |E_6| + ... \tag{6.60}$$

We may rearrange this series as follows:

$$|E_p| = \frac{|E_1|}{2} + \left(\frac{|E_1|}{2} - |E_2| + \frac{|E_3|}{2} \right) + \left(\frac{|E_3|}{2} - |E_4| + \frac{|E_5|}{2} \right) + ... \tag{6.61}$$

The terms in the parentheses are close to zero, especially when the decrease in phasor amplitude with zone number is slow. Thus, $E_p = \frac{\vec{E_1}}{2}$ for an unobstructed wave, as before.

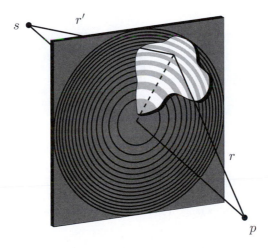

Figure 6.25   The field measured at p is determined by the Fresnel zones passed by the aperture.

When obstructed by an aperture, the disturbance at p is determined by the Fresnel zones that the aperture passes, as illustrated in Figure 6.25. When the aperture shape is symmetric, we can determine many features of the Fresnel diffraction pattern by examining the zones that are passed.

6.4.1.1 Circular Apertures
Fresnel zone analysis is particularly useful for circular apertures and obstacles since in these cases, the symmetry of the aperture matches that of the Fresnel zones. Figure 6.28(a)—(e) shows Fresnel diffraction patterns for cases where the aperture passes an integer number of zones when the source and receive points s and p are located on the axis of symmetry. According to Equation 6.60, when an even number of zones add, the result is almost zero, and when an odd number of zones add the result gives about twice the amplitude of the unobstructed wave, or four times the irradiance. Thus,

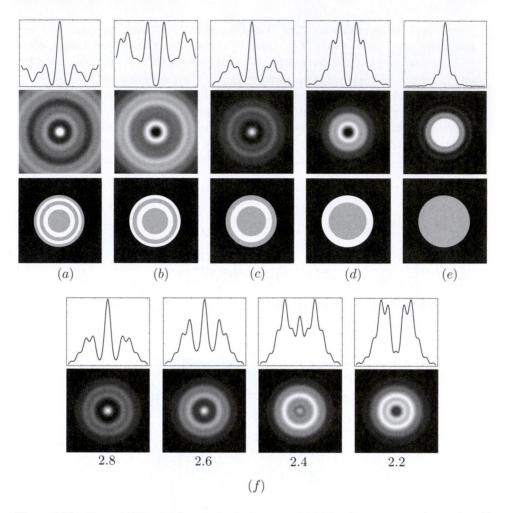

Figure 6.26 Fresnel diffraction from a circular aperture. (a) Diffraction patterns and central profile when the aperture passes 5, 4, 3, 2, and 1 Fresnel zones. (b) Evolution of the Fresnel patterns and profiles as the aperture passes a range from 2.8 to 2.2 Fresnel zones.

in Figures 6.28(a) where five zones are passed, there is a bright fringe at the center of the diffraction pattern. The same is true for Figure 6.28(c) where three zones are passed and again for Figure 6.28(d) where only the central zone is passed by the aperture. In Figures 6.28(b) and (d) even numbers of zones are passed, and the center of the Fresnel diffraction pattern is dark. Figure 6.28(f) shows the pattern transition as the number of zones passed by the aperture ranges from three to two. The diffraction patterns in Figure 6.28 are simulated according to numerical methods to be discussed in Section 6.4.3; Figure 6.27 shows actual photographs of diffraction from a circular aperture arranged to pass integer numbers of Fresnel zones that range from 10-2 (see also Problem 6.42).

For points on the observation plane that lie off the axis of symmetry, incomplete Fresnel zones combine to produce the observed irradiance, as illustrated in Figure 6.28. In this figure, the central point on the observing screen is illuminated by the complete central Fresnel zone. For points somewhat off center, there is less illumination from the central zone, and increasing illumination from the second zone; both effects tend to decrease the

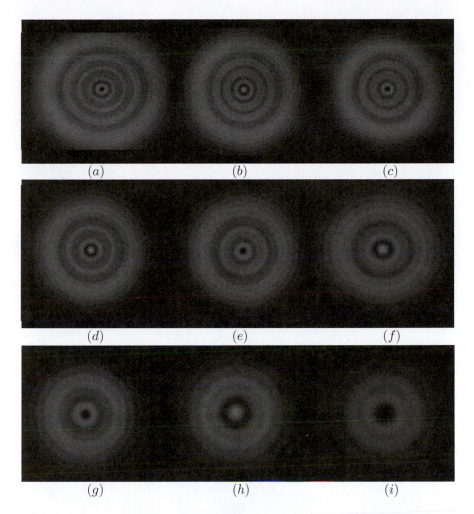

Figure 6.27   Fresnel diffraction from a circular aperture. To record these images, HeNe laser light ($\lambda = 632.8\,nm$) emerging from a spatial filter illuminated a circular aperture with diameter $0.865\,mm$. The diffraction pattern was recorded by allowing it to illuminate the detector array in a digital camera directly. The distances r' (spatial filter to aperture) was maintained at a constant value of $50.5\,cm$; the value of r (aperture to detector array) was then adjusted to pass the number of Fresnel zones indicated, according to Equation 6.59. The Fresnel number for each photograph is (a) 10; (b) 9; (c) 8; (d) 7; (e) 6; (f) 5; (g) 4; ((h) 3; (i) 2. There is some minor interference structure that results from an infrared filter in the cameral optics that should be ignored. Photo credits: Matt Brown, Christian Kaltreider, Casey Peters, and Alex Sell.

pattern irradiance. Points further off axis receive illumination from the third zone which increases the pattern irradiance, and produce the shoulder on the central lobe. Finally, points well off the axis are illuminated more or less equally by alternating zones which sum to give zero irradiance.

6.4.1.2   Circular Obstacles
There is bright spot at the center of the shadow cast by a spherical obstacle or normally illuminated circular disk, called the *spot of Arago*. This spot has an interesting history. In 1818, the French Academy sponsored a competition, to which

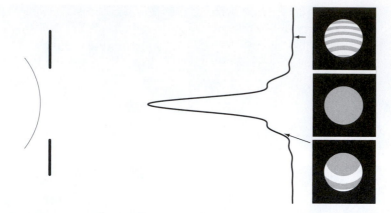

Figure 6.28   Off-axis regions of the diffraction pattern result from the contribution of partial zones.

Fresnel submitted his new wave theory of light. At this point, Newton's corpuscular theory still had many proponents, including Poisson, who was one of the judges in the competition. Poisson used Fresnel's theory to predict the bright spot at the center of a circular shadow, which he intended as a counter-argument, *reductio ad absurdum*. Subsequently, Arago (another of the judges) was able to experimentally observe the spot, and Fresnel won the competition.

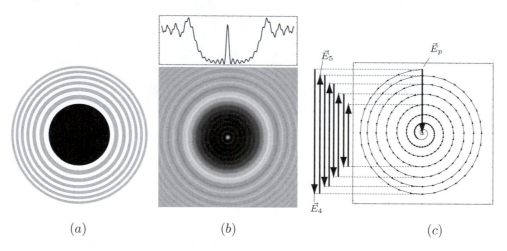

(a) (b) (c)

Figure 6.29   (a) An opaque circular aperture that passes all Fresnel zones > 3 for an observation point on the axis of symmetry. (b) The Fresnel diffraction pattern over a plane that contains the point in (a) The central bright fringe is the spot of Arago. (c) The vibration curve. The maximum field and irradiance in the spot of Arago are about equal to those of the unobstructed wave.

The disturbance at the center of the shadow is due to the Fresnel zones that are passed by the obstacle. For example, a circular obstacle that blocks all but the first three zones produces a disturbance at the center of the shadow represented by the downward vibration curve arrow of Figure 6.29(b) as will any circular obstacle that blocks an odd number of zones. If the effect of the obliquity term is negligible, this results in a central spot that is about equal in irradiance to the unobstructed wave. If an even number of zones are blocked, the vibration curve arrow points up, again indicating a spot irradiance equal to the unobstructed wave. For all other obstacle diameters, the resultant vibration curve arrow is inclined at an arbitrary angle with about the same length, so the bright spot occurs for all points on the axis that pass at least the central Fresnel zone. Presumably, Arago had little trouble observing his spot!.

■ EXAMPLE 6.9

Estimate the maximum irradiance from the annular aperture of Figure 6.30(a). Assume that this aperture passes the five zones: 4—8.

Solution

Figure 6.30(b) shows the vibration curve. The resultant $\vec{E}_p$ has a magnitude that is about twice that of the unobstructed wave. Thus, the irradiance of the spot is about four times that of the unobstructed wave.

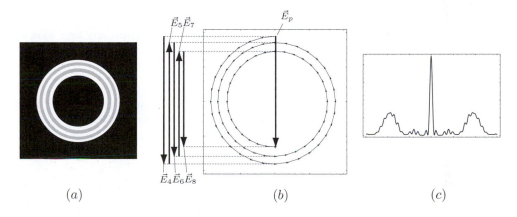

(a) $\qquad\qquad\qquad\qquad$ (b) $\qquad\qquad\qquad$ (c)

Figure 6.30   (a) An annular aperture that passes all Fresnel zones from 3 through 8 for an observation point on the axis of symmetry. (b) The vibration curve. The maximum field and irradiance in the spot of Arago is about four times that of the unobstructed wave.

6.4.1.3 *Fresnel Zone Plate*

According to Equation 6.31, an obstacle that blocks all even Fresnel zones gives a disturbance at p that is represented by a series of positive field terms that sum to a large value, and thus represents a very large observed irradiance at p. If the odd terms are blocked, the resultant field sums to a negative value with a large absolute value, thus also represents a very large irradiance. Such an obstacle is called a *Fresnel zone plate*, illustrated in Figure 6.31. The zone plate of Figure 6.31(b) blocks the even Fresnel zones, and the zone plate of Figure 6.31(c) blocks odd zones. Figure 6.31(a) shows a vibration curve when the even zones are blocked, where all phasors add and cumulatively produce a very large final amplitude.

According to Equation 6.56,

$$\frac{1}{r'} + \frac{1}{r} = \frac{m\lambda}{R_m^2} \qquad\qquad (6.62)$$

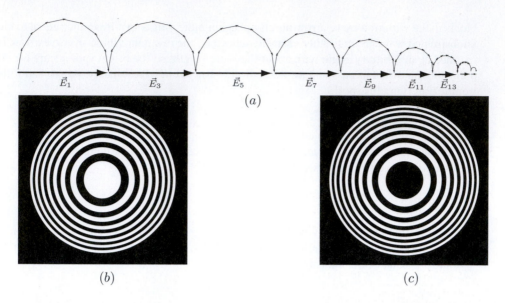

Figure 6.31 (a) The phasors from a Fresnel zone plate combine to give a very large wave at the focus. (b) A Fresnel zone plate that passes the central zone. (c) The complementary aperture to (b) has the same focus.

where R_m is the radius of the mth Fresnel zone. Thus, a point source at s located a distance r' behind the zone plate produces a irradiance maximum at p located a distance r in front of the plate. The zone plate clearly behaves as a lens, with focal length given by

$$f = \frac{R_m^2}{m\lambda} \tag{6.63}$$

By Equation 6.56, $R_m = \sqrt{m}R_1$, so Equation 6.63 gives the same result for all values of m:

$$f = \frac{R_1^2}{\lambda} \tag{6.64}$$

where R_1 is the radius of the central Fresnel zone. Note the dependence on wavelength; Fresnel zone plates behave as a lens with large chromatic aberration.

■ EXAMPLE 6.10

Design a Fresnel zone plate that gives a focused spot $10.0\,cm$ away when illuminated by plane waves of wavelength $632.8\,nm$. Estimate the focused irradiance when the zone plate has 20 annular openings. What is the focal length of a zone plate with $R_1 = 0.500\,cm$?

Solution

By Equation 6.64, the central zone has a radius given by

$$R_1 = \sqrt{\lambda f} = \sqrt{(632.8 \times 10^{-9}\,m)\,(0.100m)} = 2.52 \times 10^{-4}\,m = 0.252\,mm$$

All other zones are bounded by circles of radius $R_m = \sqrt{m}R_1$.

If there are 20 annular openings, then E_p is twenty times the unobstructed wave if we ignore the effects of the obliquity factor. The focused irradiance is then 400 times that of the unobstructed wave.

A zone plate with $R_1 = 0.500\,cm$ has a very long focal length:

$$f = \frac{R_1^2}{\lambda} = \frac{\left(5.00 \times 10^{-3}\,m\right)^2}{632.8 \times 10^{-9}\,m} = 39.5\,m$$

There are focal points of weaker irradiance located at $f/3$, $f/5$, $f/7$, ...; in other words, at odd-integer divisions of the focal length given by Equation 6.64. Consider a set of Fresnel zones where the central zone is $\frac{1}{\sqrt{3}}$ as large as the central disk in the zone plate. In this case, each zone plate opening has three times the area of illuminating zones, and so each opening contains exactly three zones, as illustrated in Figure 6.32(a). By Equation 6.64, this zone pattern gives an irradiance maximum at $f/3$. The irradiance maxima become weaker as the odd integer becomes large, since even numbers of zones combine to give a zero contribution, and the area of the remaining zone gets progressively smaller. The same argument can be used to predict that no irradiance maxima occur for points located at even-integer divisions of f; however, observation shows bright spots at these locations. This anomalous effect can be attributed to the outer, closely spaced zones acting as a circular diffraction grating[10].

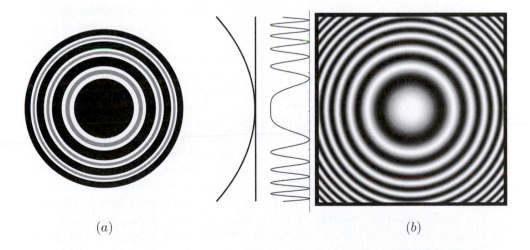

(a)   (b)

Figure 6.32   (a) A Fresnel zone plate illuminated with a source that fills each aperture opening with exactly three Fresnel zones. (b) The interference pattern produced when a plane wave interferes with a spherical wavefront. Exposing this pattern to photographic film produces a zone plate that is the negative of the pattern shown.

Gabor zone plates[11] can be produced optically by photographing the Newton's ring interference pattern that occurs when two mutually coherent spherical wavefronts with

[10] See J. Higbie, "Fresnel zone plate: anomalous foci," Am. J. Phys. 44, 929(1976).
[11] Dennis Gabor: 1900-1979; Hungarian born physicist and inventor of holography.

different radii of curvature combine, as illustrated in Figure 6.32(b). This produces a zone plate where the transition from each transparent to opaque region is continously smooth.

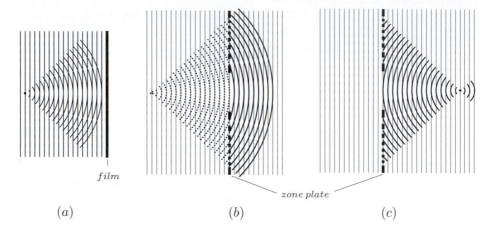

(a) (b) (c)

Figure 6.33   (a) A point source and plane wave interfere on a film plane. (b) The zone plate recorded on the film in (a) is illuminated with a plane wave. Wavefronts from a virtual source behind the plate are reconstructed as the plane wave diffracts at the plate aperture. (c) A real image is also reconstructed by the zone plate.

Fresnel zone plates such as those of Figure 6.31 with step transitions between transparent and opaque regions are called *binary amplitude zone plates*. The multiple foci that occur at odd-integer divisions of the focus given by Equation 6.64 may be thought of as diffraction orders with $m > 1$, similar to those that occur with diffraction gratings. Because of the continuously varying transmission of Gabor zone plates, the higher diffraction orders do not occur, so Gabor zone plates only have the one focus of Equation 6.64. There are, however, both virtual and real images, as illustrated in Figures 6.33(b) and (c). Both images occur because the Gabor zone plate can be constructed by a diverging spherical wavefront as illustrated in Figure 6.33(a), or by a converging wavefront with the same radius of curvature. If the zone plate is illuminated from behind, an observer in front of the plate will see a virtual point source behind the plate, and if the observation distance is great enough, a real image of the point source will be observed at a point that is in front of the plate. Single real and virtual point images will be observed only if the film is exposed and developed within its linear recording range. Over-developing photographic film that has been exposed to a Newton's ring interference pattern steepens the transitions from transparent to opaque, and can cause the multiple foci and corresponding real and virtual images to reappear.

In a *Binary phase zone plates*, each zone area is transparent, with the optical thickness of each alternate zone adjusted to give a relative phase difference of π. Since the two sets of zones are complimentary, Babinet's principle predicts another phase shift of π, so the optical disturbances from each set of zone areas add together in phase. Because every zone is transparent, the net field increases by 2, so the irradiance from a binary phase zone plate is greater than that of the corresponding binary amplitude zone plate by a factor of four. *Bleaching* a Gabor amplitude zone plate creates a *Gabor phase zone plate* that in principle can produce brighter real and virtual images.

According to Figure 6.33, an illuminated Gabor zone plate recreates the wavefronts that could be used to produce it. This process is called *wavefront reconstruction*.

6.4.2 Holography

A schematic for recording a *transmission hologram* is illustrated in Figure 6.34(a). The hologram is recorded on film that is exposed to the interference pattern that results from combining a *reference beam* and an *object beam*, both beams formed from a single light source that is both temporally and spatially coherent.

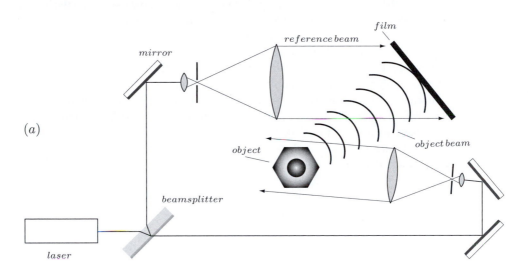

(a)

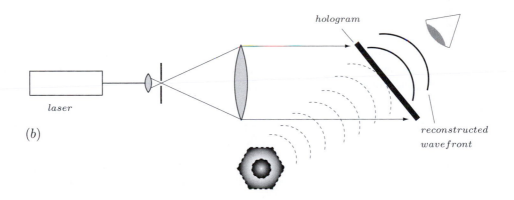

(b)

Figure 6.34   (a) Typical arrangement for recording a hologram. (b) Typical arrangement for viewing a hologram.

In principle, a hologram is very simple. Each point on the object scatters a point source wavefront that combines with the reference beam to produce a Gabor zone plate on the recording film. When the film is re-illuminated with the original reference beam as shown in Figure 6.34(b), wavefront reconstruction recreates the wavefronts of the original object beam, producing a true three-dimensional reconstruction of the original object. Both real and virtual images can be observed.

The interference detail of a hologram represents a recording of both *amplitude and phase* of the original object beam. An ordinary photograph records amplitude only. The extra phase information that is recorded by the film accounts for the three-dimensional image.

Holograms must be recorded with light whose spatial coherence length is greater than total path difference between all portions of the reference and object beams. The reference beam must have greater irradiance than the object beam over all portions of the film plane if the hologram is to reconstruct the object wavefronts. It is not necessary to view the hologram with spatially coherent light, but the viewing illumination must be monochromatic enough to avoid the chromatic aberration of Equation 6.64.

The word hologram derives from the Greek word for "whole." The Gabor zone plate for each source point extends over the entire film plane. Thus, the film may be cut into pieces, and each piece will reconstruct the entire hologram, although at decreased resolution.

The interference pattern produced by an extended object has intricate detail down to the scale of the wavelength of light used to produce the hologram. The film used to record this detail must have adequate *resolution*. Holograms are typically recorded on *spectrographic plates*; these are glass plates coated with a photographic emulsion with a very small *grain size* (typically less than a micron). High-resolution film is very *slow*, in that it requires more light energy to achieve a given level of opacity than is required for film with less resolution. The entire optical arrangement must be very stable. Even the slightest vibration can cause the intricate fringe pattern to shift and average away during the time interval required for exposure.

Problem 6.18   Find the Fresnel number for a $2.00\,mm$-diameter circular aperture illuminated by collimated light of wavelength $632.8\,nm$ for the following observation distances: (a) $15.0\,cm$, (b) $1.50\,m$, and (c) $15.0\,m$.

Problem 6.19   Determine the appropriate observation distances r required to observe the Fresnel diffraction patterns of Figure 6.28(a)-(e) if the aperture diameter is $2.00\,mm$ and the illumination is plane wave with $\lambda = 633\,nm$.

Problem 6.20   Determine the appropriate observation distance r required to observe the Fresnel diffraction patterns of Figure 6.29 if the obstacle diameter is $2.00\,mm$ and the illumination is plane wave with $\lambda = 633\,nm$.

Problem 6.21   Using a ruler, estimate the focal length of the Fresnel zone plates of Figures 6.31(b) and (c) as determined by their actual size. Assume a wavelength of $632.8\,nm$.

Problem 6.22   A Fresnel zone plate is prepared with a drawing program and printed on a laser printer. The printed zone plate has a central zone diameter of $5.00\,cm$. A photograph of this zone plate is to be reduced on high-resolution film to give a working zone plate with a $1.00\,m$ focal length. If this zone plate is to be used with light of $633\,nm$, what factor of reduction is required?

Problem 6.23   Light from a helium-neon laser ($\lambda = 633\,nm$) is focused through a pinhole prior to illuminating a $1.00\,mm$ diameter aperture. If the pinhole is located $50.0\,cm$ behind the diffracting aperture, how far in front of the aperture should an observing screen be located to give a Fresnel number of ten?

6.4.3   Numerical Analysis of Fresnel Diffraction with Circular Symmetry

We return now to the Fresnel diffraction integral of Equation 6.55. When s, p, and the aperture point (X, Y) are collinear, Equation 6.54 gives the total optical path, but when this is not the case, we must also add the path increment used for Fraunhofer diffraction, which for small angles is given by Equation 6.45. Combining both path increments gives

$$E_p = C \int\!\!\int_A e^{ik(r'+r)} dA = C \int\!\!\int_A e^{ik\left(z'+z+X\frac{x}{z}+\frac{y}{z}+\frac{R^2}{2}\left(\frac{1}{z'}+\frac{1}{z}\right)\right)} dA$$

$$= Ce^{ik(z'+z)} \int\!\!\int_A e^{ik\frac{R^2}{2}\left(\frac{1}{z'}+\frac{1}{z}\right)} e^{ik\left(X\frac{x}{z}+Y\frac{y}{z}\right)} dA$$

where $R^2 = X^2 + Y^2$. If $g(A)$ is the aperture function, the integration limits can be infinite. Redefining the constant term to absorb the exponential term gives

$$E_p = C \int_\infty\!\!\int_\infty \left(g(A)e^{ik\frac{R^2}{2}\left(\frac{1}{z'}+\frac{1}{z}\right)}\right) e^{ik\left(\frac{Xx}{z}+Y\frac{y}{z}\right)} dA \qquad (6.65)$$

According to Section 6.3.9, we recognize this as the Fourier transform of the function $g(A)\, e^{ik\frac{R^2}{2}\left(\frac{z+z'}{z'z}\right)}$.

If the aperture is circularly symmetric, we may use a Hankel transform, as described in Section 6.3.10.1:

$$E_p = C \int_\infty g(R)e^{ik\frac{R^2}{2}\left(\frac{1}{z'}+\frac{1}{z}\right)} J_0\left(\frac{kR\rho}{r_0}\right) 2\pi R\, dR \qquad (6.66)$$

If the aperture is circular with radius a, then by Equation 6.59,

$$\frac{kR^2}{2}\left(\frac{1}{z'}+\frac{1}{z}\right) = \frac{R^2}{a^2}\left(\frac{\pi a^2}{\lambda L}\right) = \frac{R^2}{a^2}(\pi N_F)$$

giving

$$E_p = 2\pi C \int_\infty g(R)e^{i\pi N_F\left(\frac{R^2}{a^2}\right)} J_0\left(\frac{kR\rho}{r_0}\right) R\, dR$$

where N_F is the number of Fresnel zones passed by the aperture opening. Change variables by letting $\bar{R} = \frac{R}{a}$ with $dR = a\, d\bar{R}$.

$$E_p = 2\pi C \int_\infty g(\bar{R})e^{i\pi N_F \bar{R}^2} J_0\left(\frac{ka\bar{R}\rho}{r_0}\right) a\bar{R}\, a\, d\bar{R}$$

$$= 2\pi C (N_F\lambda L) \int_\infty g(\bar{R})e^{i\pi N_F \bar{R}^2} J_0\left(\sigma\bar{R}\right) \bar{R}\, d\bar{R}$$

where σ is defined in Equation 6.25. Finally, redefine the constant C but retain the factor N_F since each zone contributes to E_p. For a circular opening in an opaque screen, $g(R) = 1$

within the opening, and is zero outside:

$$E_p = C\,N_F \int_0^1 e^{i\pi N_F \bar{R}^2} J_0\left(\sigma\bar{R}\right) \bar{R}\,d\bar{R} \tag{6.67}$$

Note that if $N_F \ll 1$, then the above result approaches that of Fraunhofer diffraction (see Equation 6.51).

There are limitations to Equation 6.67 that result from the approximations used to derive it. First, we have ignored the obliquity factor and all other effects that would otherwise cause the vibration curve of Figure 6.24 to "spiral in" for large values of N_F. Of course, the Fresnel approximation must also be valid. Despite these limitations, this equation can be very useful. All circularly symmetric Fresnel profile plots and diffraction patterns used in the illustrations for this chapter were obtained by numerical integration of Equation 6.67[12].

■ EXAMPLE 6.11

Find expressions for the Fresnel diffraction pattern of a circular aperture that contains 8 Fresnel zones, and the annular opening of Example 6.9.

Solution

For a circular opening that contains 8 Fresnel zones, Equation 6.67 becomes

$$E_p = (8) \int_0^1 e^{i\pi 8 \bar{R}^2} J_0\left(\sigma\bar{R}\right) \bar{R}\,d\bar{R}$$

where we have set the constant C to one.

The annular opening of Example 6.9 consists of an obstacle with $N_F = 3$ and an opening with $N_F = 8$:

$$E_p = (8) \int_{\sqrt{\frac{3}{8}}}^1 e^{i\pi 8 \bar{R}^2} J_0\left(\sigma\bar{R}\right) \bar{R}\,d\bar{R}$$

The results of numerical integration of these expressions are illustrated in Figure 6.35

Problem 6.24   Determine the integrals to use calculate the Fresnel diffraction pattern due to (a) a circular aperture that passes 9 Fresnel zones, (b) an annular aperture that passes zones 5—9, and (c) an annular aperture that passes zones 6—9.

Problem 6.25   Write a computer program that plots the central profile of the diffraction patterns of Problem 6.24.

---

[12]It shortens the computation time to first use Equation 6.67 to generate a radial profile, which can then be used as a look-up table to generate the pattern plot.

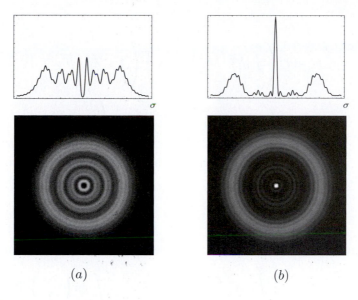

Figure 6.35   Fresnel profiles and diffraction patterns for (a) a circular aperture that contains 8 zones, and (b) an annular opening that contains the five zones 4—8. The scales used for (a) and (b) are identical.

6.4.4   Fresnel Diffraction from Apertures with Cartesian Symmetry

As in the previous section, we begin with Equation 6.65 that expresses the Fresnel diffraction integral as a Fourier transform

$$E_p = C \int\limits_{-\infty}^{\infty} \int\limits_{-\infty}^{\infty} \left(g\left(X, Y\right) e^{ik\left(\frac{1}{z'}+\frac{1}{z}\right)\frac{\left(X^2+Y^2\right)}{2}} \right) e^{i\frac{k}{z}\left(Xx+Yy\right)} dX dY \qquad (6.68)$$

In principle, this expression can be evaluated numerically; however in practice, it simplifies the analysis to maintain the observation point at the origin of the observation plane ($x = 0$ and $y = 0$), and then to "move the aperture" by varying the limits of integration. Assuming that $g(X, Y)$ is unity within the aperture opening and zero otherwise, Equation 6.68 then becomes

$$E_p = C \int\limits_{Y_1}^{Y_2} \int\limits_{X_1}^{X_2} e^{ik\left(\frac{1}{z'}+\frac{1}{z}\right)\frac{\left(X^2+Y^2\right)}{2}} dX dY \qquad (6.69)$$

where the integration limits are defined by each position of the aperture. Change variables to U and V as follows:

$$U = X\sqrt{\frac{2\left(z'+z\right)}{\lambda z' z}} \qquad V = Y\sqrt{\frac{2\left(z'+z\right)}{\lambda z' z}} \qquad (6.70)$$

to give

$$E_p = C \left(\int_{X_1}^{X_2} e^{i \frac{2\pi}{\lambda} \left(\frac{z'+z}{z'z} \right) \frac{X^2}{2}} \, dX \right) \left(\int_{Y_1}^{Y_2} e^{i \frac{2\pi}{\lambda} \left(\frac{z'+z}{z'z} \right) \frac{Y^2}{2}} \, dY \right)$$

$$= C \left(\frac{\lambda z' z}{2 \left(z' + z \right)} \right) \left(\int_{U_1}^{U_2} e^{i\pi \frac{U^2}{2}} \, dU \right) \left(\int_{V_1}^{V_2} e^{i\pi \frac{V^2}{2}} \, dV \right)$$

(6.71)

where U_1 and U_2, V_1 and V_2 are obtained using Equation 6.70. Redefining the constant term gives

$$E_p = C \left(\int_{U_1}^{U_2} e^{i\pi \frac{U^2}{2}} \, dU \right) \left(\int_{V_1}^{V_2} e^{i\pi \frac{V^2}{2}} \, dV \right)$$

(6.72)

The integrals in Equation 6.72 may be expressed in terms of the *Fresnel integrals*

$$C(u) = \int_0^u \cos \left(\frac{\pi t^2}{2} \right) \, dt \qquad S(u) = \int_0^u \sin \left(\frac{\pi t^2}{2} \right) \, dt$$

(6.73)

Using the Euler relation, we express Equation 6.72 as

$$E_p = C \left(\int_0^{U_2} e^{i\pi \frac{U^2}{2}} \, dU - \int_0^{U_1} e^{i\pi \frac{U^2}{2}} \, dU \right) \left(\int_0^{V_2} e^{i\pi \frac{V^2}{2}} \, dV - \int_0^{V_1} e^{i\pi \frac{V^2}{2}} \, dV \right)$$

$$= C \left(C(U_2) - C(U_1) + i \left[S(U_2) - S(U_1) \right] \right) \left(C(V_2) - C(V_1) + i \left[S(V_2) - S(V_1) \right] \right)$$

$$= C \left(D_X \right) \left(D_Y \right)$$

(6.74)

where D_X and D_Y denote the results of X and Y integration.

Numeric values of the Fresnel integrals can be found in tables and with software, and can be found graphically using the *Cornu spiral*[13] shown in Figure 6.36.

■ **EXAMPLE 6.12**

Use the Cornu spiral to find the numeric value and magnitude of

$$\int_{-1}^{\infty} e^{i\pi \frac{u^2}{2}} \, du$$

Solution

Express in terms of the Fresnel integrals:

$$\int_{-1}^{\infty} e^{i\pi \frac{u^2}{2}} \, du = \int_{-1}^{\infty} \left[\cos \left(i\pi \frac{u^2}{2} \right) + i \sin \left(i\pi \frac{u^2}{2} \right) \right] \, du$$

$$= C(\infty) - C(-1) + i \left[S(\infty) - S(-1) \right]$$

---

[13]M. A. Cornu (1841-1902): French physicist. Despite the name, Cornu's spiral was investigated earlier by both Bernoulli and Euler.

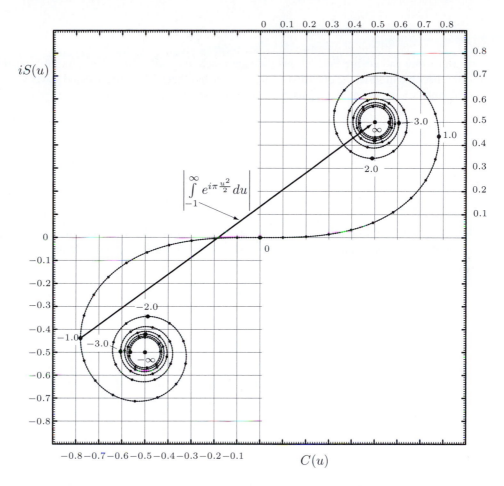

Figure 6.36   The Cornu spiral. Any point on the spiral has vertical coordinate equal to $iS(u)$ and horizontal coordinate equal to $C(u)$ with values of u given along the spiral as shown. The directed line segment has magnitude equal to the value of the integral indicated in the figure.

Using Figure 6.36, verify that

$$C(\infty) = S(\infty) = \tfrac{1}{2}$$

$$C(-1) = -0.780 \quad S(-1) = -0.438$$

Thus,

$$\int_{-1}^{\infty} e^{i\pi \frac{u^2}{2}} du = 0.500 - (-0.780) + i[0.500 - (-0.438)] = 1.28 + 0.938i$$

The magnitude is equal to the length of the line that connects the integration limits (see Figure 6.36):

$$\left| \int_{-1}^{\infty} e^{i\pi \frac{u^2}{2}} du \right| = \sqrt{(1.28)^2 + (0.938)^2} = 1.59$$

Both sine and cosine Fresnel integrals approach one half as $u \to \infty$. As the aperture limits become infinite, the disturbance at p approaches that of the unobstructed wave, whose irradiance we denote as I_∞. Thus, the diffracted irradiance is

$$I_p = \frac{I_\infty}{4} |D_x|^2 |D_x|^2 \tag{6.75}$$

Problem 6.26   Using the method of Example 6.12 find values for the relevant Fresnel integrals and the final numeric result for both complex value and magnitude of

$$\int_{-1}^{2} e^{i\pi \frac{u^2}{2}} \, du$$

6.4.4.1 *Semi-infinite Straightedge*   Figure 6.37 shows an obstacle in the shape of a semi-infinite plane with one edge located at V_1. Setting $U_1 = +\infty$, $U_2 = -\infty$ and $V_2 + \infty$ in Equation 6.74 gives

$$I_p = \frac{I_\infty}{4} |(1+i)|^2 \left| \tfrac{1}{2}(1+i) - C(V_1) - iS(V_1) \right|^2 \tag{6.76}$$

The approach used in Equation 6.69 removes explicit dependence on the observation coordinates (x, y). These can be restored by using their geometric relationship with the aperture coordinates (U, V).

■ **EXAMPLE 6.13**

Find the distance from the geometric shadow to the first irradiance maximum when the source s is $10\,cm$ behind the straightedge, and the observation point p is $100\,cm$ in front of the straightedge. Use $\lambda = 632.8\,nm$. Repeat when the illumination is plane wave.

Solution

Use Equation 6.70 with $z' = 10\,cm$ and $z = 100\,cm$ to give

$$V = Y \sqrt{\frac{2\,(1.10\,m)}{(632.8 \times 10^{-9}\,m)\,(0.100\,m^2)}} = (5896)\,Y$$

or equivalently,

$$Y = \left(1.70 \times 10^{-4}\right) V$$

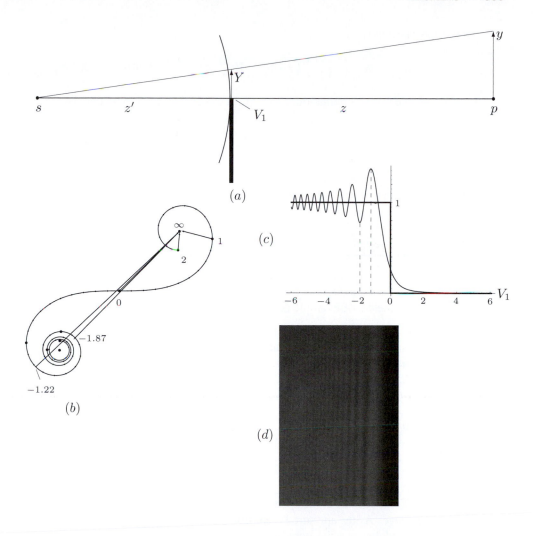

Figure 6.37 (a) Geometry for calculating Fresnel diffraction by a straightedge. The obstacle extends infinitely into and out of the plane of the paper (U direction) and extends infinitely in the $-V$ direction. The value of V_1 shown is on the line that connects s and p, so in this case, p is located on the geometric shadow of the obstacle. Points on the observation plane map to points on the aperture plane by a ray that begins at the source s. (b) The Cornu spiral, with line segments extending between the limits of integration: V_1 and $V_2 = \infty$. (c) The irradiance measured at p. When $V_1 = 0$, p is on the geometric shadow. (d) Photograph of irradiance pattern. (Ken Kay/Fundamental Photographs)

Using similar triangles (see Figure 6.37),

$$y = \left(\frac{z' + z}{z'} \right) Y$$

The geometric shadow is located at $y = 0$. From Figure 6.37, the first fringe maximum is located at $V = -1.22$, or $Y = -0.485\,mm$. In the observation plane, the fringe spacing is

$$y = \left(\frac{1.10}{0.1}\right)(-4.85 \times 10^{-4}\,m) = -5.34\,mm$$

With plane wave illumination, $z' \rightarrow \infty$. In this case, Equation 6.70 gives

$$V = \sqrt{\frac{2}{\lambda z}} = \sqrt{\frac{2}{(632.8 \times 10^{-9}\,m)\,(1.00\,m)}} = (1778)\,Y$$

with the first bright fringe located at $Y = -0.686\,mm$. With $z' \rightarrow \infty$, this is also the value of y in the observation plane.

Problem 6.27   Find the distance from the geometric shadow to the first irradiance minimum for the parameters of Example 6.13.

Problem 6.28   Estimate the spacing between the two irradiance maxima closest to the geometric shadow of a plain wave illuminated straightedge if the source distance is $10.0\,cm$ and the observation distance is $20.0\,cm$. Assume $\lambda = 632.8\,nm$.

6.4.4.2   *Single Slit*   Figure 6.37(a) illustrates a long, narrow slit of width b. The value of V_1 is now finite, and Equation 6.74 gives

$$I_p = \frac{I_\infty}{2}\,|C\,(V_2) + iS\,(V_2) - C(V_1) + iS\,(V_1)|^2 \tag{6.77}$$

where V_1 and V_2 are located at the aperture edges, as shown. When $V_1 = 0$, V_2 locates the other aperture edge. As the aperture is repositioned, the ΔV spacing is determined by the aperture width and thus remains constant:

$$\Delta V = V_2 - V_1 = b\sqrt{\frac{2\,(z' + z)}{\lambda z' z}} \tag{6.78}$$

As the aperture is moved, this constant ΔV spacing generates line segments on the Cornu spiral of Figure 6.37(b), which in turn determine the irradiance at p for each aperture location. Using ΔV, we may re-express Equation 6.77 so that it only depends upon one variable. It is convenient to use the aperture center, which is located at

$$V_0 = \frac{1}{2}\,(V_1 + V_2) \tag{6.79}$$

Using V_0 as a variable, each integration will have limits $V_0 - \frac{\Delta V}{2}$ and $V_0 + \frac{\Delta V}{2}$. Equation 6.77 becomes

$$I_p\,(V_0) = \frac{I_\infty}{2}\,|C\,(V_0 + \tfrac{\Delta V}{2}) + iS\,(V_0 + \tfrac{\Delta V}{2}) - C(V_0 - \tfrac{\Delta V}{2}) - iS\,(V_0 - \tfrac{\Delta V}{2})|$$

$$\tag{6.80}$$

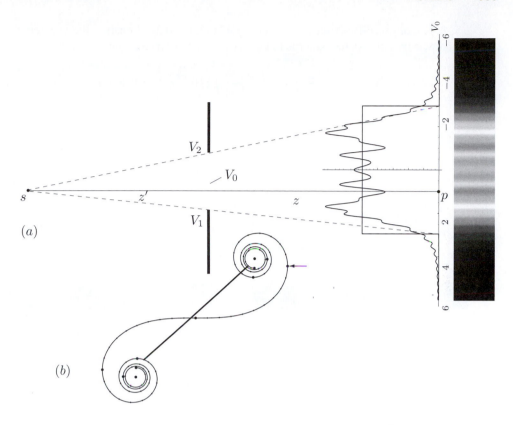

Figure 6.38 (a) Geometry for calculating Fresnel diffraction by a long narrow slit. The aperture extends infinitely into and out of the plane of the paper (U direction). The diffraction pattern is plotted for the parameters in Example 6.14. Note that $V_0 = +1$ for the aperture location illustrated. (b) Cornu spiral with line segment for p illustrated in a.

■ **EXAMPLE 6.14**

A $1.00\,mm$ wide slit is located $10.0\,cm$ in front of a source s. Describe the Fresnel diffraction pattern that is observed on a screen located $100\,cm$ from the slit. Assume $\lambda = 632.8\,nm$.

Solution

From Equation 6.78, the slit width gives a ΔV of

$$\Delta V = b\sqrt{\frac{2\,(z' + z)}{\lambda z' z}} = \left(10^{-3}\,m\right)\sqrt{\frac{2.2\,m}{632.8 \times 10^{-10}\,m^3}} = 5.90$$

This is straightforward to evaluate numerically, either by using the Cornu spiral (see Figure 6.38(b)), or by software[14].

---

[14] For example, in Mathematica the syntax is:
deltaV = 5.90
irradiance[v_] := 0.5* Abs[FresnelC[v + deltaV/2] - FresnelC[v - deltaV/2] + I*(FresnelS[v + deltaV/2] - Fres-

A plot of this diffraction pattern is shown in Figure 6.38. Locating the geometric shadow edges allows the patterns to be scaled to the observing plane.

$$\Delta y = \left(\frac{z' + z}{z'}\right) b = \left(\frac{1.1\,m}{0.1\,m}\right) \left(10^{-3}\,m\right) = 1.00\,cm$$

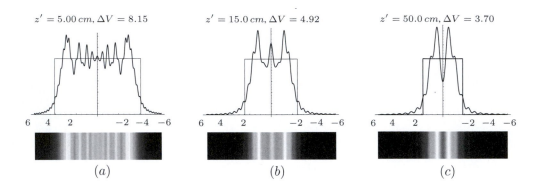

Figure 6.39   Fresnel diffraction from a single slit when $z = 100\,cm$ and $\lambda = 632.8\,nm$, plotted for different values of the source distance z'. (a) $z' = 5.00\,cm$. (b) $z' = 15.0\,cm$. (c) $z' = 30.0\,cm$.

6.4.4.3 *Rectangular Aperture*

We consider here a rectangular aperture of height that is three times the width, such as the one used for the Fraunhofer diffraction patterns of Figure 6.8. The two-dimensional diffraction pattern is the product of two single-slit patterns, as indicated by Equations 6.74 and 6.75. Proceeding as for the single slit, we find

$$D_x = \sqrt{2}\left[C\left(U_0 + \tfrac{\Delta U}{2}\right) + iS\left(U_0 + \tfrac{\Delta U}{2}\right) - C\left(U_0 - \tfrac{\Delta U}{2}\right) - iS\left(U_0 - \tfrac{\Delta U}{2}\right) \right] \quad (6.81)$$

$$D_y = \sqrt{2}\left[C\left(V_0 + \tfrac{\Delta V}{2}\right) + iS\left(V_0 + \tfrac{\Delta V}{2}\right) - C\left(V_0 - \tfrac{\Delta V}{2}\right) - iS\left(V_0 - \tfrac{\Delta V}{2}\right) \right] \quad (6.82)$$

with

$$I_p = \frac{1}{4}\left|D_x\right|^2 \left|D_y\right|^2 \quad (6.83)$$

Figure 6.40 shows patterns and profiles from a $1.00\,mm$-high, $0.300\,mm$-wide rectangular aperture for a variety of source and observation distances. When the source and observation distances are small, many Fresnel zones are passed by the aperture, and the Fresnel diffraction pattern is higher than it is wide. As the source and observation distances become large, less Fresnel zones are passed by the aperture, and the diffraction pattern evolves toward the Fraunhofer case, with the width becoming larger than the height. Compare Figure 6.40(g)—(i) with the Fraunhofer patterns of Figure 6.8.

```
nelS[v - deltaV/2])]^2
gshadow[x_] := UnitStep[-(x - deltaV/2)*(x + deltaV/2)]
Plot[{irradiance[y], gshadow[y]}, {y, -6, 6}, PlotRange -> All, PlotPoints -> 1000]
```

Problem 6.29 Find a relationship between the width ΔU of a rectangular aperture, and the number of Fresnel zones passed horizontally. Then show that the number of zones passes is equal to

$$N_{F,x} = \frac{(\Delta U)^2}{8}$$

Find a similar expression the number of zones passed vertically: $N_{F,y}$. Determine $N_{F,x}$ and $N_{F,y}$ in Figures 6.40(a)—(i).

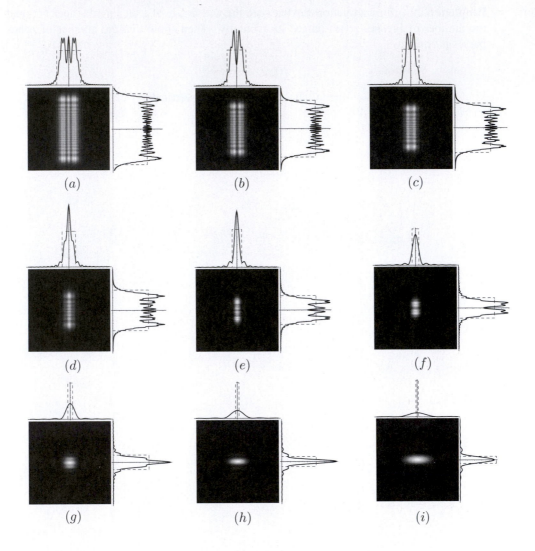

Figure 6.40   Fresnel diffraction from a rectangular aperture with vertical dimension $b = 1.00\,mm$ and horizontal dimension $a = 0.300\,mm$.   The wavelength is $632.8\,nm$.   As the source and observation distances increase, the diffraction pattern evolves toward Fraunhofer diffraction (see Figure 6.8(a)-(b)).   (a) $z' = 2.50\,cm$ and $z = 5.00\,cm$ giving $\Delta V = 13.8$ and $\Delta U = 13.8/3 = 4.60$.   (b) $z' = 5.00\,cm$, $z = 5.00\,cm$ and $\Delta V = 11.2$.   (c) $z' = 5.00\,cm$, $z = 10.0\,cm$ and $\Delta V = 9.74$.   (d) $z' = 10.0\,cm$, $z = 10.0\,cm$ and $\Delta V = 7.95$.   (e) $z' = 20.0\,cm$, $z = 20.0\,cm$ and $\Delta V = 5.62$.   (f) $z' = 40.0\,cm$, $z = 40.0\,cm$ and $\Delta V = 3.98$.   (g) $z' = 80.0\,cm$, $z = 80.0\,cm$ and $\Delta V = 2.81$.   (h) $z' = 160\,cm$, $z = 160\,cm$ and $\Delta V = 5.62$.   (i) $z' = 320\,cm$, $z = 320\,cm$ and $\Delta V = 1.41$.

6.5   INTRODUCTION TO QUANTUM ELECTRODYNAMICS

In our progress to this point, we have seen examples of the intricacies associated with resolving the wave and photon pictures of light, especially in cases where the photon flux is very small. In Section 2.4.1, we discussed the Young's double-slit experiment conducted with illumination so dim that only one photon could interact with the slit at any one point in time. In Section 5.9, we discussed a "delayed-choice" experiment using a Mach-Zehnder interferometer with similar illumination. In both experiments, the data appear to suggest that photons take all possible paths. In Quantum Electrodynamics (QED), this is in fact the interpretation.

QED is a theory of the interaction of light with electrons. It improves upon Schrodinger's equation which does not include the concept of a photon. Largely developed between 1929—1948, QED is now a mature theory that has been experimentally verified to unprecedented accuracy. The 1965 Nobel Prize was shared by Feynman[15], Schwinger[16], and Tomonaga[17] for their contributions to the development of this theory.

Before examining some of the details of QED as it relates to photons, it is useful to re-examine Fresnel diffraction in the following intuitive way. Consider Figure 6.41(a) that shows an aperture illuminated by a point source s. According to Equation 6.55, the irradiance measured at the recieve point p is given by

$$E_p = C \int_A e^{ik(r+r')} dA \qquad (6.84)$$

where as usual, dA is an element of area on the aperture opening, r' is the distance from s to dA, and r is the distance from dA to p. In Figure 6.41(b), we approximate the integral of Equation 6.84 as a sum of complex numbers, each the result of integrating over a small patch ΔA. As we sum these individual contributions, the resulting complex numbers add like vectors to approximate the Cornu spiral, as shown. The irradiance measured at p is determined by magnitude-squared of the arrow that represents the sum of the individual contributions.

Intuitively, we have associated the model of Huygens wavelets with Equation 6.84. In this model, a spherical wave propagates from s to the aperture, where the resulting wavefront becomes a source of Huygens wavelets that combine at p to give the diffracted field E_p. Individual wavelets arise from each of the paths indicatated in Figure 6.41(a): the path from s to A along r' determines the initial wavelet amplitude and phase, and the path from A along r determines the final wavelet amplitude and phase at p. Each of the "little" arrows of Figure 6.41(b) approximates the contribution to Equation 6.84 from the corresponding path in Figure 6.41(a). Each little arrow has a direction that is determined by the phase of the exponential in Equation 6.84: according to this equation, the little arrow makes a complete revolution each time the path includes a complete wavelength, so the final orientation of the little arrow is determined by the number of wavelenths in the total path $r' + r$.

The straight line path from s to p contributes the little arrow in Figure 6.41(b) that is bolded for emphasis. This path and those close by contribute arrows that are nearly collinear, and thus provide most of the amplitude of the final result. From Figure 6.36, this main contribution lies between values on the Cornu spiral determined by $-1 < u < +1$. According to the definitions for U and V in Equation 6.70, these points correspond to points on the aperture with Fresnel number N_F of one-half (see Equation 6.59)

$$-\frac{1}{2} < \frac{a^2}{\lambda L} < \frac{1}{2} \qquad (6.85)$$

---

[15] R. P. Feynman: 1918-1988. Extraordinary American physisict, author and teacher.
[16] Julian Schwinger: 1918-1994. American theoretical physicist.
[17] Sin-Itiro Tomonaga: 1906-1979. Japanese theoretical physicist.

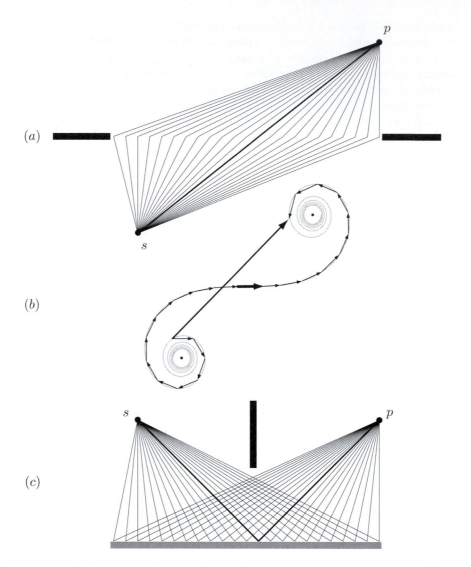

Figure 6.41   (a) An aperture illuminated by source point s, and analyzed at receive point p. (b) The Fresnel diffraction integral is approximated by the contributions of many paths that intersect the aperture opening at different points. The contribution from the straight line path from s to p is bolded for emphasis. (c) A reflective surface. The path corresponding to the law of reflection contributes the bold arrow in (b).

where

$$\frac{1}{L} = \frac{1}{r'} + \frac{1}{r}$$

Paths that intersect the aperture outside of this range contribute less to the final diffracted amplitude because the corresponding little arrow lies on the spiral portion of the Cornu curve.

   If the aperture were gradually widened to include more paths in Figure 6.41(a), then the field measured at p would show only small variations as the little arrows twisted their way around each end of the Cornu spiral. In the limit, the field amplitude of the unobstructed wave connects the two points determined by $u = \pm\infty$; in this case, *all possible paths* are included in the final field amplitude. The bold straight line path from s to p and corresponding bolded little arrow on the Cornu spiral are in the middle of the range that contibutes most to the final amplitude, however nearby paths that do not take the straight line route contribute nearly as much provided they intersect the aperture plane at points within a circle of radius a, where according to Equation 6.86,

$$a = \sqrt{L\frac{\lambda}{2}} \tag{6.86}$$

Thus, the Huygens model describes the unobstructed wave propagation between two points s and p by including effects due to all possible non-straight line paths that connect these points. This interesting consequence of Equation 6.84 comes to us from Maxwell's equations via the Fresnel-Kirchhoff integral theorem. According to classical electromagnetism, this is the way electromagnetic waves propagate.

6.5.1 Feynman's Interpretation

In Feynman's view, there are no electromagnetic waves. As summarized in the prologe to this chapter, there are only photons. The multiple paths indicated in Figure 6.41(a) that generate secondary Huygens wavelets correspond to real physical photon paths in Feynman's interpretation.

QED emphasizes *events* and *probabilities*. For example, suppose a source of photons is placed at s that emits photons isotropically at a random rate. A photon counter placed at p records an event whenever a photon is detected. If s is the only available source, then a detection event means that a photon left s and arrived at p. According to Feynman, *any* path that begins at s and ends at p is possible. The possibility that the photon travels along a particular path is assigned a probabililty with an equation that is mathematically similar to the Fresnel diffraction Equation 6.84. In analyzing this equation via the approach of Figure 6.41(b), we again have a little arrow that rotates once per wavelenth to eventually lie along the Cornu spiral. In Feynman's picture, the little arrow is determined by the complex number that results from the following expression:

$$\tilde{a}_i = a_0 e^{i\frac{S_{path}}{\hbar}} \tag{6.87}$$

where a_0 is a constant, and S_{path} is given by

$$S_{path} = \int_{t_s}^{t_p} \hbar\omega \, dt = \hbar\omega \left(t_p - t_s\right) = \hbar\omega \, \Delta t \tag{6.88}$$

In the last equation, $\hbar\omega$ is photon energy and Δt is the light travel time along a particular path. The net probability includes contributions from all possible paths

$$P = N_p \left| \sum_{all\ paths} e^{i\frac{S_{path}}{\hbar}} \right|^2 \tag{6.89}$$

where N_p is a normalization factor. When a large number of photon events are recorded the photon flux predicted by Equation 6.89 agrees with the Fresnel diffraction analysis, as discussed in Section 2.4.1 and illustrated in Figure 2.9. In QED, the classical electromagnetic wave irradiance represents the probability of detecting a photon at a particular point. When the photon flux is high, the wave picture gives an intuitive model. When the photon flux is low, we must work with probabilities.

Since Feynman's approach agrees with Fresnel diffraction results obtained with Equaton 6.84, we can use QED to obtain all results of classical optics. However, QED has greater predictive power, since it can also be used to describe the motion of *electrons*. For electrons, S_{path} is given by the *action* integral:

$$S_{path} = \int\limits_{t_s}^{t_p} L dt \qquad (6.90)$$

where L is the *Lagrangian*, and is given by the kinetic energy minus potential energy

$$L = KE - PE$$

Since a photon has no mass, there is no PE term in its action integral. Identifying the photon energy with KE gives Equation 6.88. In this way, QED offers an integrated explanation for both photons and matter. In particular, Fermat's principle of least time, illustrated by the bold paths of Figure 6.41(a) and (c) is seen to be a special case of the principle of *least action*, which can be used to deduce Newton's laws of motion[18].

For more detail on the QED of both photons and electrons, see Feynman's introductory text: *QED, The Strange Theory of Light and Matter* (Princeton, 1985) [5].

---

[18] See "Quantum physics explains Newton's laws of motion" Jon Ogborn and Edwin F. Taylor, Physics Education, Vol. 40, No. 1, p26—34 (2005). This and related papers are currently archived at http://www.eftaylor.com/leastaction.html.

Additional Problems

Problem 6.30 The central maximum of a single-slit diffraction pattern has a width of $30.0\,cm$ when measured between adjacent nulls on an observing screen $1.00\,m$ away. If the illumination wavelength is $633\,nm$, what is the slit width?

Problem 6.31 A single-slit aperture has a slit width of $2.00\,\mu m$. Find the number of irradiance maxima when this slit is illuminated by light with wavelength $400\,nm$. Repeat when the wavelength is $700\,nm$.

Problem 6.32 The central order for a diffraction pattern from a rectangular aperture is measured to be $15.0\,mm$ high and $5.00\,mm$ wide on an observing screen $1.00\,m$ away. If the illumination wavelength is $633\,nm$ what is the aperture's (a) height and (b) width?

Problem 6.33 The central order for a diffraction pattern from a rectangular aperture is measured to be $10.0\,cm$ high and $30.0\,cm$ wide on an observing screen $1.00\,m$ away. If the illumination wavelength is $633\,nm$ what is the aperture's (a) height and (b) width?

Problem 6.34 A $1.00\,mm$-wide single-slit aperture is illuminated by plane waves of wavelength $633\,nm$. How far away should the observing screen be placed if the path error of Equation 6.8 is to be less than one percent of the wavelength?

Problem 6.35 A carbon dioxide laser with wavelength $10.6\,\mu m$ is designed to output a beam with diameter $10.0\,cm$. Estimate the size of this beam after it has traveled $3.80 \times 10^5\,km$ to the moon. Repeat for a HeNe laser of the same beam diameter.

Problem 6.36 How far away can a truck be if its headlights are $2.00\,m$ apart and they are to be just resolved by the night adjusted eye with a pupil diameter of $7.00\,mm$? Assume a wavelength of $550\,nm$, and that the headlights are bright enough to be seen at this distance. Comment on whether or not you believe that your eyesight is diffraction-limited.

Problem 6.37 Estimate the diameter of the diffraction-limited Airy disk produced at the focal plane of the Hubble telescope if it has a objective diameter of $2.4\,m$ and a focal length of $57.6\,m$. Assume a wavelength of $500\,nm$.

Problem 6.38 A spectral line of wavelength $656\,nm$ is studied in fourth order using a diffraction grating illuminated normally. If it is suspected that a fifth order is overlapping, what would the overlapping wavelength be?

Problem 6.39 The diffraction pattern of a uniformly illuminated double-slit aperture consists of bright fringes spaced $5.00\,cm$ apart on an observation screen located $2.00\,m$ from the aperture. If the illumination wavelength is $633\,nm$ and the fifth-order fringes are missing, what is the slit width and slit separation?

Problem 6.40 By direct integration, find a formula for the Fraunhofer diffraction pattern of a triple-slit aperture, and plot the result when the slit spacing is six times the slit width.

Problem 6.41 Use Equation 6.50 to find the Fraunhofer diffraction pattern of an annular slit with inner radius equal to half the outer radius. Interpret your result with Babinet's principle. How would this diffraction pattern be affected if the illumination were not perfectly collimated?

Problem 6.42 Using information provided in the caption of Figure 6.27, calculate the value of r used to record each subfigure.

Problem 6.43   Determine the integral to use to calculate the Fresnel diffraction pattern due to an annular disk that passes all zones greater than 3. What artifacts would you expect from numerical analysis that did not include an infinite number of zones? Write a computer program that calculates a diffraction profile.

Problem 6.44   Find the distance from the geometric shadow to the first irradiance maximum when a $\lambda = 400\,nm$ point source s is $10.0\,cm$ behind the straightedge, and the observation point p is $50\,cm$ in front of the straightedge. Find the distance from the geometric shadow to the first irradiance minimum.

Problem 6.45   Using the method of Example 6.12 find values for the relevant Fresnel integrals and the final numeric result for both complex value and magnitude of

$$\int\limits_{-1}^{4.9} e^{i\pi \frac{u^2}{2}}\,du$$

Problem 6.46   Write a computer program or use the Cornu spiral to reproduce the Fresnel single-slit diffraction profiles of Figure 6.39. Make additional plots that show similar profiles for observation distances of $10.0\,cm$ and $20.0\,cm$.

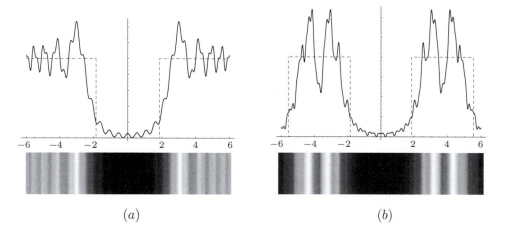

(a)   (b)

Figure 6.42   (a) Fresnel diffraction from a narrow obstacle of width $\Delta V = 3.70$. (b) Fresnel diffraction from a double slit with slit width $\Delta V = 3.70$ and slit separation that is twice the slit width.

Problem 6.47   Find an expression that gives the Fresnel diffraction pattern of a narrow obstacle of width ΔU. Plot the profile, and compare your result to Figure 6.42(a).

Problem 6.48   If the illumination in Figure 6.42(a) is plane parallel with wavelength $632.8\,nm$, what is the observation distance?

Problem 6.49   Find an expression that gives the Fresnel diffraction pattern of a double-slit of width ΔU and slit spacing twice the slit width. Plot the profile, and compare your result to Figure 6.42(b).

CHAPTER 7

LASERS

A splendid light has dawned upon me about the absorption and emission of radiation.
—Einstein, in a letter to Michael Besso, 1916.

Contents

7.1 INTRODUCTION

The word LASER is an acronym that stands for Light Amplification by the Stimulated Emission of Radiation. Before discussing stimulated emission, we give an overview in Section 7.2 of the quantum mechanics of materials that determine energy levels where population inversions can occur. In Section 7.3, we discuss stimulated emission and how this can enable amplification. Elements of laser design are discussed in Section 7.4, illustrated by examples that use some common gain mediums. Most laser designs use a Fabry-Perot cavity that usually gives output beams with a Gaussian profile. An overview of Gaussian beam propagation is discussed in Section 7.7, followed by a more rigorous treatment in Section 7.8. The propagation of Gaussian beams through optical systems is discussed in Sections 7.8.3 and 7.8.4. In Section 7.9.1, we take a closer look at how laser cavity design affects the output beam characteristics.

7.2 ENERGY LEVELS IN ATOMS, MOLECULES, AND SOLIDS

A laser generates photons, and photons are produced when a material makes transitions between available *energy levels*. The exact nature of these energy levels can be determined by the methods of quantum mechanics. In this section, we give a brief overview of some specific cases, noting at the outset that a full quantum mechanical treatment is beyond the scope of our discussion.

7.2.1 Atomic Energy Levels

The simplest atomic system is hydrogen, which consists of a single proton in the nucleus, and a single electron bound by the electric Coulomb force. When an electric current is passed through a container of hydrogen atoms, the hydrogen *spectrum* is observed. This spectrum consists of photons with *discrete* values of energy, as opposed to the continuous spectrum emitted by a blackbody radiator. The discrete spectrum of hydrogen was first analyzed by Bohr[1]. According to the *Bohr model of hydrogen*, electron orbits are restricted to discrete values that correspond to *quantized energy levels* given by[2]

$$E_n = -\frac{m_e e^4}{8h^2 \epsilon_0^2} \frac{1}{n^2} = -(13.6\,eV)\frac{1}{n^2} \tag{7.1}$$

where m_e is the electron mass, e is the element of charge, h is Planck's constant, ϵ_0 is the permitivitty of free space, and the integer n is the *principle quantum number*. Substitution gives the *hydrogen ionization energy* of $13.6\,eV$, which is the energy required to completely remove the bound electron. The negative sign in Equation 7.1 is due to the fact that more tightly bound states must have lower energy. Electrons excited to a higher energy state with $n = n_i$ decay to a lower energy state ($n = n_f$) and in the process emit a photon with energy equal to the energy difference between the initial and final states:

$$E_{photon} = E_{n_i} - E_{n_f} = -(13.6\,eV)\left[\frac{1}{n_i^2} - \frac{1}{n_f^2}\right] \tag{7.2}$$

■ **EXAMPLE 7.1**

Find the energy required to excite a hydrogen atom from its ground state to the first excited state. Compare this to the three lowest photon energies of the *Balmer series* that have a final state of $n_f = 2$.

---

[1] Neils Bohr: 1885-1962. Danish physicist who received the Nobel Prize in 1922 for his contributions to the development of quantum mechanics.
[2] See any text on Modern Physics, e.g., Krane [14].

Solution

Using Equation 7.1, we calculate the energy of the ground state to be $-13.6\,eV$, and that of the first excited state to be $-3.4\,eV$. Thus, it takes $10.2\,eV$ to excite a hydrogen atom from the ground state to the first excited state. When this excited state decays back to the ground state, an ultraviolet photon of energy $10.2\,eV$ is emitted.

In the Balmer series, $n_f = 2$. When $n_i = 3$, Equation 7.2 gives

$$E_{photon} = E_3 - E_2 = -\,(13.6\,eV)\left[\frac{1}{9} - \frac{1}{4}\right] = 1.89\,eV$$

To find the corresponding photon wavelength, use $E_{photon} = hf$ and $f\lambda = c$:

$$\lambda = \frac{hc}{E_{photon}} = \frac{\left(6.63 \times 10^{-34}\,J\,s\right)\left(3.00 \times 10^{8}\,\frac{m}{s}\right)}{(1.89\,eV)\left(1.60 \times 10^{-19}\,\frac{J}{eV}\right)} = 658\,nm$$

This is the red line of the hydrogen spectrum, which for historical reasons is referred to as the Hα line. Similarly, when $n_i = 4$, $E_{photon} = 2.55\,eV$ and $\lambda = 488nm$ (Hβ, blue-green), and when $n_i = 5$, $E_{photon} = 2.86\,eV$ and $\lambda = 435nm$ (Hγ, violet).

The integer n that determines the discreteness of Equation 7.1 is referred to as a *quantum number* In a more complete quantum mechanical treatment that takes as a starting point[3], the hydrogen atom solution is found to depend on two additional quantum numbers that result from quantization of the orbital angular momentum of the bound electron. The quantum number ℓ determines the magnitude of the orbital angular momentum vector, and the quantum number m_ℓ determines its spatial orientation. If there are no external electric and magnetic fields, the energy does not depend on values of ℓ and m_ℓ, and in this case the energy is said to be *degenerate* in ℓ and m_ℓ. When the only interaction considered is that of the Coulomb attraction between the electron and the proton, the energy levels given by Schrodinger's equation are precisely those obtained by Bohr in Equation 7.1.

Only certain values of ℓ and m_ℓ are allowed, meaning that the details of the solution of Schrodinger's equation result in *selection rules* that limit the possible values that quantum numbers can attain. In particular, ℓ must be less than n, so for example, when $n = 2$, ℓ can only have values of zero and one. For each value of ℓ, there are $2\ell + 1$ values of m_ℓ that range from $-\ell$ to ℓ in steps of one. The electron is also found to have an *anomalous magnetic moment*, historically referred to as *spin*. The quantum number associated with the spin magnitude is referred to as s, and can only have a value of one half, so the allowed values m_s can only be $\pm\frac{1}{2}$. Thus, the full set of quantum numbers for a single electron atom are $(n, \ell, m_\ell, s, m_s)$. *Schrodinger's equation!hydrogen atom solution*

The quantum mechanics of atoms with more than one electron are subject to the *Pauli exclusion principle*[4] which requires that each electron within an atom have different sets of quantum numbers. So for example in helium, there are two electrons and if $n = 1$, ℓ and m_ℓ must both be zero. These two electrons can have different sets of quantum numbers if their spin orientations are different; in other words, provided one electron has $m_s = \frac{1}{2}$ and

---

[3]Erwin Schrodinger: 1887-1961. Austrian physicist who received the Nobel Prize of 1933 for his important contributions to quantum mechanics.
[4]Wolfgang Pauli: 1900-1958. Austrian-Swiss physicist who made important contributions to quantum mechanics.

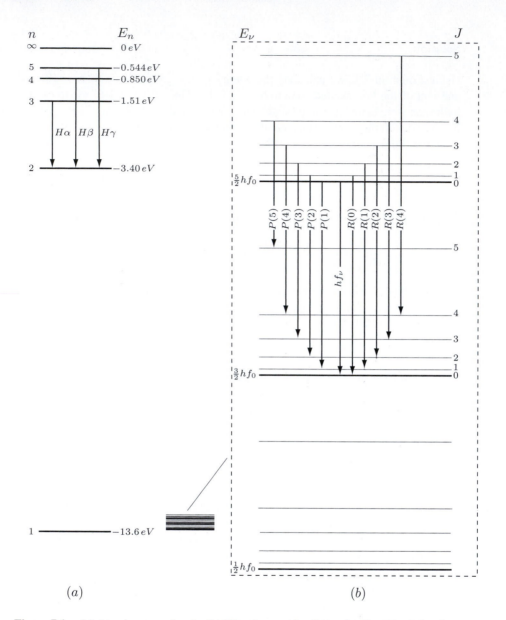

Figure 7.1 (a) Atomic energy levels. (b) Vibration-rotation lines of a diatomic molecule.

the other has $m_s = -\frac{1}{2}$. This completely fills the $n = 1$ level, which for historical reasons is called the *1s subshell*.

Table 7.1. gives the ground state *electron configurations* for hydrogen through krypton. From this table, we see that hydrogen has a single $1s$ electron, and helium has two $1s$ electrons in a completely filled subshell. The third electron in a lithium atom must reside in the $n = 2$ level, where possible values of ℓ are zero and one. It is found that the third electron in lithium resides in the $2s$ subshell, and the fourth electron in beryllium fills the $2s$ subshell. Subshells for which $\ell = 1$ are referred to as p subshells, and those with $\ell = 2$ are

Table 7.1. Ground state electron configurations for $Z = 1$ through 36.

Z	Element	$1s$	$2s$	$2p$	$3s$	$3p$	$3d$	$4s$	$4p$
1	H	1							
2	He	2							
3	Li	2	1						
4	Be	2	2						
5	B	2	2	1					
6	C	2	2	2					
7	N	2	2	3					
8	0	2	2	4					
9	F	2	2	5					
10	Ne	2	2	6					
11	Na	2	2	6	1				
12	Mg	2	2	6	2				
13	Al	2	2	6	2	1			
14	Si	2	2	6	2	2			
15	P	2	2	6	2	3			
16	S	2	2	6	2	4			
17	Cl	2	2	6	2	5			
18	Ar	2	2	6	2	6			
19	K	2	2	6	2	6		1	
20	Ca	2	2	6	2	6		2	
21	Sc	2	2	6	2	6	1	2	
22	Ti	2	2	6	2	6	2	2	
23	V	2	2	6	2	6	3	2	
24	Cr	2	2	6	2	6	5	1	
25	Mn	2	2	6	2	6	5	2	
26	Fe	2	2	6	2	6	6	2	
27	Co	2	2	6	2	6	7	2	
28	Ni	2	2	6	2	6	8	2	
29	Cu	2	2	6	2	6	10	1	
30	Zn	2	2	6	2	6	10	2	
31	Ga	2	2	6	2	6	10	2	1
32	Ge	2	2	6	2	6	10	2	2
33	As	2	2	6	2	6	10	2	3
34	Se	2	2	6	2	6	10	2	4
35	Br	2	2	6	2	6	10	2	5
36	Kr	2	2	6	2	6	10	2	6

referred to as d subshells, again for historical reasons[5]. From Table 7.1., we find that boron has 2 $1s$ electrons, 2 $2s$ electrons, and a single $2p$ electron. A total of six electrons can reside in a p subshell, since there are three possible values of m_ℓ, each with two possible values of m_s. As can be seen from Table 7.1., the $2p$ subshell is filled sequentially by

---

[5]The notation originated from spectroscopy prior to the development of quantum mechanics. The letters derive from the words: *sharp*, *principle* and *diffuse*.

the elements aluminum through argon. States do not always fill sequentially, as can be determined from the listed electron configurations of potassium and calcium.

Electrons in lower subshells tend to "shield" the outer electrons from the nuclear charge. For example, *noble gases* such as helium, neon, argon and krypton all have completely filled outer subshells. The *alkali* atoms such as lithium and sodium all have outer electrons that are very effectively shielded by completely filled lower subshells, and are thus much less tightly bound. For example, the excited states of helium and neon are over twenty electron volts above the ground state, compared to only a few electron volts for lithium and sodium.

The electron configurations of the outer electrons determine an atom's chemical reactivity. For example, noble gases tend not to react with other atoms because the outer electrons all reside in subshells that are completely filled. Alkali atoms are much more reactive due to the weakly bound outer electron. Many properties of the *periodic table* can be accounted for from the electron configuration of an atom's outer electrons.

Problem 7.1   Verify that the ionization energy of hydrogen is $13.6\,eV$.

Problem 7.2   Find the first six wavelengths of the Paschen series of hydrogen that have the $n = 3$ state as the final state.

Problem 7.3   Two examples of alkali atoms were given in the discussion above: lithium and sodium. Is there another one listed in Table 7.1.? If so, give the element and explain your reasoning.

7.2.2   Molecular Energy Levels

Molecules are composed of atoms held together by intramolecular forces that we may model as a Hooke's law spring, as illustrated for the diatomic molecule of Figure 7.2. Such a molecule can vibrate in simple harmonic motion, and it can also rotate about its center of mass. When this system is analyzed by the methods of quantum mechanics, both types of energy are quantized. The vibrational energy levels are determined by the quantum number ν which takes on positive integer values:

$$E_\nu = h f_0 \left(\nu + \tfrac{1}{2} \right) \tag{7.3}$$

where f_0 is the natural frequency of vibration. This gives a ladder of equally spaced energy levels separated by the vibrational energy $h f_v$. The rotational motion is also quantized, with energy levels determined by an angular momentum quantum number J, which again take positive integer values:

$$E_J = \frac{h^2}{8\pi I} J \left(J + 1 \right) \tag{7.4}$$

where I is the moment of inertia of the molecule. In general, a molecule can both vibrate and rotate, and thus has *vibration-rotation* energy levels given by the sum of the last two results:

$$E_{\nu,J} = h f_0 \left(\nu + \tfrac{1}{2} \right) + \frac{h^2}{8\pi I} J \left(J + 1 \right) \tag{7.5}$$

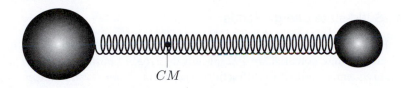

Figure 7.2   A molecule modeled as two atoms connected by a Hooke's law spring. The molecule can vibrate, and it can rotate about the center of mass (CM).

In a typical molecule, the rotational energy levels are about 500 times more closely spaced than the vibrational energy level spacing. Figure 7.1(c) illustrates vibration-rotation energy levels with the rotational spacing increased for the purposes of illustration. Transitions between vibration-rotation states are subject to the selection rule that J can only change by plus one or minus one. When the upper level has the lower value of J, the transition energy is less than hf_0:

$$E_{\nu+1,J-1} - E_{\nu,J} = hf_0 - \frac{h^2}{4\pi I}J \tag{7.6}$$

the set of all such transitions is called the *P-branch*, which consists of transition energies that are equally spaced in both energy and frequency. Similarly, the set of transitions where J for the upper level is higher is the *R-branch*:

$$E_{\nu+1,J-1} - E_{\nu,J} = hf_0 + \frac{h^2}{4\pi I}(J+1) \tag{7.7}$$

P-branch and R-branch transitions are illustrated in Figure 7.1(c).

The simple model just presented does a fairly good job of accounting for many features of molecular spectroscopy; however, there are important corrections. For example, for many molecules, the vibrational levels are *anharmonic*, meaning that the spacing between vibrational levels is not quite constant. Also, centripetal forces can change the moment of inertia for higher rotational states, causing unequal spacing in the P and R branches.

Problem 7.4   The vibration-rotation lines of the HCl molecule are separated in energy by $2.62 \times 10^{-3}\,eV$. What is the moment of inertia of the molecule?

Problem 7.5   The natural frequency of a classical harmonic oscillator of mass m and spring constant k is given by

$$f_0 = \frac{1}{2\pi}\sqrt{\frac{k}{m}}$$

For the HBr molecule, this frequency is $7.68 \times 10^{13}\,Hz$.

    a) Calculate the effective spring constant for this intermolecular force. Since the bromine atom is much heavier than hydrogen, you may assume that the bromine remains stationary during the vibration.

    b) Estimate the frequency of vibration for a DBr molecule where the hydrogen is replaced by deuterium, which has roughly twice the mass of hydrogen.

7.2.3 Solid-state Energy Bands

In the Bohr model of hydrogen just discussed, the electron is bound to a nucleus by a central electrostatic potential. In 1923, Louis de Broglie[6] postulated *matter waves* which were subsequently verified in diffraction experiments from crystal lattices. In the Bohr model, only certain discrete orbits are allowed, and these can be interpreted as those orbits within which the electron waves constructively interfere as a standing wave. In a crystalline solid, atoms are arranged in a periodic array. The atomic nuclei are located at fixed *lattice sites*, which creates an electric potential that varies periodically across the entire crystal. If such a system is analyzed rigorously with Schrodinger's equation, it is found that there are allowed electron energies that result in constructive interference of the electron matter waves, and forbidden electron energies that cannot exist within the crystal. Because the number of lattice sites is so large, the allowable electron energies lie within continuous *energy bands*. Electrons in the so-called *valence band* participate in the bonding of the solid. In metals, there are also electrons in a *conduction band* that do not participate in the bonding forces, and are thus free to move about freely within the conductor. If an electric field is established within a metal, conduction electrons experience an electrostatic force that causes a collective movement and a corresponding electric current.

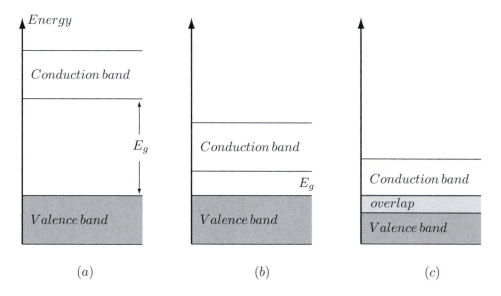

Figure 7.3   (a) Band structure of an insulator. The band gap E_g is large, and thermal fluctuations only rarely transition electrons from the valence band to the conduction band. (b) Band structure of a semiconductor, with a much smaller band gap. (c) In a metal, there are conduction electrons that are free to move about within the solid.

Figure 7.3 illustrates the band structure of insulators, semiconductors and conductors. The top of the valence band and the bottom of the conduction band are called the *band edges*, and the spacing between the band edges is called the *band gap E_g*. *Insulators* have large band gaps and empty conduction bands, and thus are poor conductors of electric current. *Semiconductors* have smaller band gaps. At absolute zero they are insulating, but

---

[6]Louis de Broglie: 1892-1987. Recipient of the 1929 Nobel Prize for postulating matter waves in his 1923 Ph.D dissertation.

thermal fluctuations at nonzero temperatures can eject electrons from the valence band, leaving a positive ion in the valence band that is referred to as a *hole*. The electron in the conduction band is then free to move within the solid, and can provide electric current. The hole vacancy within the valence band can also move in response to an electric force; the hole changes location when an electron from a neighboring lattice site hops over to fill the original vacancy, creating a new vacancy at the neighboring site. The electron and hole move in opposite directions in response to an applied electric field, but contribute equally to the resulting current as illustrated in Figure 7.4(a). Metallic *conductors* have overlapping valence and conduction bands as illustrated in Figure 7.3(c).

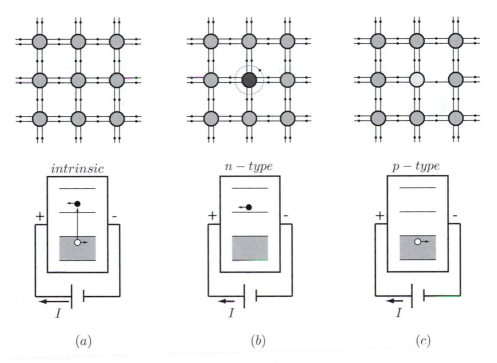

Figure 7.4   (a) Intrinsic (pure) silicon. Current can only be conducted if an electron is ejected from the valence band into the conduction band. The ejected electron leaves a hole that can move within the valence band as a positive charge carrier. Each silicon atom has two valence electrons in the p-subshell. (b) A donor impurity atom with three valence electrons in the p-subshell. After covalent bonding, there is an extra electron that is weakly bound to the impurity site. Thermal fluctuations are sufficient to eject this electron into the conduction band, leaving a fixed positive impurity ion. The mobile conduction electron can provide electric current. (c) An acceptor impurity atom with one valence electron in the p-subshell. After covalent bonding, the neutral impurity site has outer shell location available. An electron from a neighboring site can fill this site, leaving a fixed negative impurity site and a mobile hole that can move freely within the valence band to produce electric current.

Crystals can be *doped* with impurity atoms to control the conducting properties of the semiconductor. For example, atoms in a silicon lattice form *covalent bonds* by sharing valence electrons with neighbors to form closed valence subshells. According to Table 7.1., Si needs four additional electrons to completely fill the p-subshell. In the Si lattice, each Si atom has four nearest neighbors, and it can "share" one electron with each neighbor to completely fill the p-subshell, resulting in the covalent bond illustrated in Figure 7.4(b).

Phosphorus has three electrons in the p-subshell, so if impurity atoms of P are added to Si, the lattice sites that contain a P atom contain one weakly bound electron that does not participate in the covalent bond. Thermal fluctuations are sufficient to promote this electron to the conduction band, leaving a positive P ion at the impurity site. The crystal of course remains uncharged, and the now-free electron can conduct electrical current as illustrated in Figure 7.4(b). The P atom is called a *donor* because it can donate an electron to the conduction band. Undoped Si is referred to as *intrinsic*. Silicon doped with donor impurities is referred to as *n-type*, since the impurity sites donate mobile electrons that can conduct current, as illustrated in Figure 7.4(b).

Similarly, boron only has one electron in the p-subshell, so lattice sites that contain a boron atom have an incompletely filled subshell as illustrated in Figure 7.4(c). A valence electron from a neighboring site can *hop* into this subshell, leaving the fixed boron atom negatively charged, and leaving a electron vacancy (*hole*) in the neighboring location. Once the hole is created, valence electrons from other neighboring sites can hop into it, creating another hole in another location. In this way, the hole appears move freely within the valence band. Because a hole is an absent electron, it behaves as a positive charge. Impurity sites such as boron in a silicon lattice are called *acceptors* since they can accept valence electrons to create a mobile hole. Silicon doped with boron is referred to as *p-type* since current is primarily conducted by mobile holes as illustrated in Figure 7.4(c).

A *pn junction* is formed when a p-type material is brought into contact with an n-type material, as illustrated in Figure 7.5. Donor electrons diffuse across the boundary, leaving positive ions on the n-side and creating negative ions on the p-side. This creates an electric field within the boundary region that quickly builds up to a value that stops further diffusion. At equilibrium, this boundary layer has no unfilled donor or acceptor sites, and thus it is called the *depletion layer*. Within the depletion layer, there is an electric field $\vec{E}_d$ that points from the n-side to the p-side that causes the material to behave as a *diode*. If a voltage is applied with the positive terminal connected to the p-side (*forward bias*, Figure 7.5(d)), the applied field $\vec{E}_0$ tends to cancel the depletion field $\vec{E}_d$, and current begins to flow across the junction. However, if a *reverse bias* voltage is applied as illustrated in Figure 7.5(c), the applied electric field adds to $\vec{E}_d$, expanding the depletion region and further retarding current flow. Thus, the diode conducts current in one direction but not the other, as illustrated by the plot of current *vs* voltage in Figure 7.5(b). Pn junctions do not obey Ohm's law, and so are referred to as *non-ohmic*.

7.2.4 Semiconductor Devices

A *photoconductor* is made from an intrinsic semiconducting material with a band gap E_g that is smaller than the photon energy to be detected. Photons that illuminate the photoconductor can be absorbed, causing an electron to be transitioned from the valence band to the conduction band. Both the electron and the corresponding hole can move in response to an applied electric field to produce a *photo-current*. Thermally induced electron-hole pairs give *dark current* which can be kept small by cooling the photoconductor.

In a *photodiode*, electron hole pairs are generated within the depletion layer, and are subsequently swept away by the depletion layer field $\vec{E}_d$. Current is generated even with no applied voltage; this can function as a *solar cell* that converts light energy to electrical energy. In practice, the photodiode works a bit better when a reverse bias is applied since this expands the depletion layer; in such case it is referred to as a *reverse bias photodiode*. Electrons and holes generated within an *avalanche photodiode* are accelerated

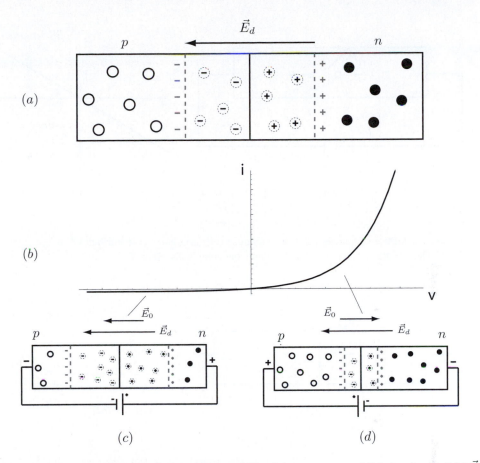

Figure 7.5 (a) P-n junction. Donor electrons diffuse from n to p to establish the electric field $\vec{E}_d$ across the depletion layer. (b) Plot of current versus voltage that illustrates the non-ohmic behavior of the junction. (c) Reverse bias, where the applied field $\vec{E}_0$ points in the same direction as $\vec{E}_d$. (d) Forward bias: the applied field opposes $\vec{E}_d$.

by the electric field within the depletion layer and acquire sufficient energy to liberate more electron-hole pairs, leading to amplification.

When a forward bias is applied to photodiodes made of certain materials, electrons and holes recombine *radiatively* in the depletion layer to produce photons in what is essentially the reversal of the process shown in Figure 7.7(a). Such a device is referred to as a *light-emitting diode* (LED).

■ APPLICATION NOTE 7.1

Photonic Bandgap Materials

In a *photonic bandgap material*, also referred to as a *photonic bandgap crystal*, the index of refraction of the material is varied periodically throughout the material. In a semiconductor, the crystal lattice creates a periodic potential for electrons that leads to allowed and forbidden energy bands when analyzed with Schrodinger's equation. In a photonic bandgap material, periodic variation of the index of refraction

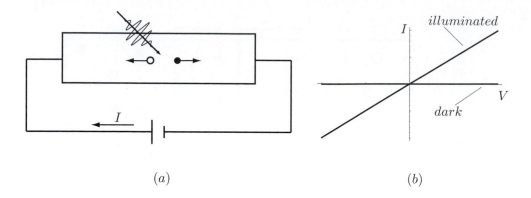

(a)   (b)

Figure 7.6 Photoconductor (a) A photo-induced electron-hole pair contributes to the photocurrent I. (b) The resistance is lower when the photoconductor is illuminated.

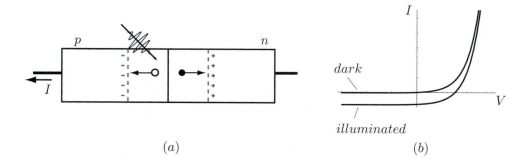

(a)   (b)

Figure 7.7 Photodiode. (a) A photo-induced electron-hole pair is swept away from the depletion layer by depletion layer field to give a photocurrent. (b) When illuminated, the photo-current adds to that of the diode current. The photodiode is typically operated with a reverse bias voltage. When the bias voltage V is zero, the device is a solar cell.

creates a similar band structure for *photons* when analyzed with Maxwell's equations. Intuitively, photon wavelengths that interfere constructively with the index structure propagate within the material, and those that interfere destructively are rejected.

Figure 7.8 illustrates possible geometries. Figure 7.8(a) illustrates a one-dimensional layered structure similar to the multilayer films discussed in Section 5.11. Figure 7.8(b) illustrates a two-dimensional structure and Figure 7.8(c) illustrates a three-dimensional structure. The index variations are typically a fraction of the photon wavelength, and in this case they are called *subwavelength structures*.

Techniques for manufacturing the very small structures necessary for application at optical wavelengths are only now becoming feasible. Subwavelenth structures can be integrated into optical fibers, and applications of photonic bandgap materials are becoming common in communications technologies.

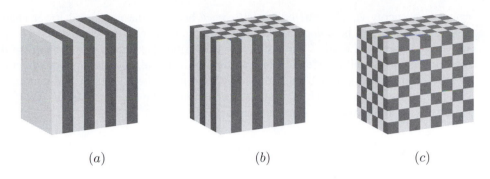

(a) (b) (c)

Figure 7.8   Step index variations in a (a) 1D, (b) 2D, and (c) 3D photonic bandgap structure.

Problem 7.6   A phosphorous doped silicon crystal may be regarded as a collection of hydrogen-like atoms at each phosphorous site. The effect of the periodic potential due to all other silicon nuclei on the motion of the orbital electron is taken into account by giving the electron an *effective mass*, which for phosphorous doped silicon is $0.2\,m_e$. Use the Bohr model to show that the binding energy of the outermost phosphorous electron is about $-0.02\,eV$, taking into account that the Coulomb force within the crystal is diminished by the dielectric constant, which for silicon is 11.7.

Problem 7.7   Silicon has a band gap energy of $1.1\,eV$. Find the minimum photon frequency and corresponding wavelength that can be detected with intrinsic silicon.

Problem 7.8   The longest wavelength of electromagnetic radiation that is absorbed in gallium arsenide is $890\,nm$. What is the bandgap of this semiconductor?

Problem 7.9   If a crystal of germanium is doped with the following impurity atoms, determine whether the resulting semiconductor is n or p. Explain your reasoning carefully.

    a) boron

    b) indium

    c) antimony

7.3   STIMULATED EMISSION AND LIGHT AMPLIFICATION

The processes by which photons of light interact with energy levels of a material are illustrated in Figure 7.9. The interaction must conserve energy by involving specific energy levels that are separated by the photon energy hf. Figure 7.9(a) represents *absorption*, where an absorbed photon transfers its energy to a material by causing a transition from a lower energy state to a higher energy state. If the material is then left undisturbed, the excited state will eventually decay, perhaps by *spontaneous emission* of a photon. The time interval for spontaneous emission is random, as is the propagation direction of the spontaneously emitted photon. Quantum mechanics can be applied to give *probabilities* for absorption and the most probable spontaneous lifetime of the excited state. In *stimulated emission*, an incident photon stimulates an excited state to emit, as illustrated in Figure 7.9(c). In order for stimulated emission to occur, the incident stimulating photon must again have energy equal to the energy difference between the upper and lower energy level. Furthermore, the stimulating photon must interact with the excited state before spontaneous emission occurs. When stimulated emission does occur, an *exact copy* of the incident photon is emitted, identical in both frequency, phase, and propagation direction. In other words, the stimulated and stimulating photons are both spatially and temporally *coherent*. It is useful to think of stimulated emission as the time reversal of absorption (sometimes also referred to as stimulated absorption).

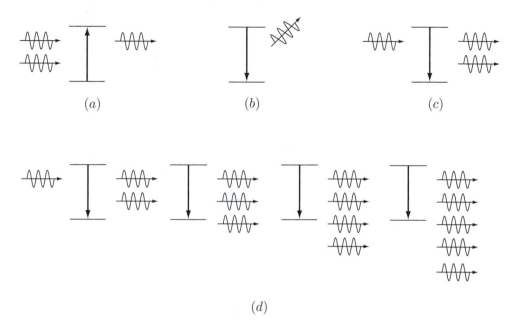

Figure 7.9 (a) Absorption. An incident coherent beam consists of a number of phase coherent photons with energy equal to the transition energy. A transition is excited from a lower level to a higher level, and in the process one photon is absorbed. (b) Spontaneous emission. A photon is emitted with random phase and propagation direction as the transition decays spontaneously from higher to lower energy level. (c) Stimulated emission. An incident photon field with energy equal to the transition energy stimulates a transition from higher level to lower level, and in the process a photon is added coherently to the incident photon field. (d) Amplification. Each stimulated emission contributes an additional photon.

The necessity of including stimulated emission in a *quantum theory of radiation* was demonstrated by Einstein in 1917. His simple and elegant argument derives the Planck blackbody radiation law of Section 4.8.4, but only when all three processes of Figure 7.9(a)—(c) are included. To follow his argument, we let the system consist of N absorbing atoms or molecules (for simplicity, a gas) occupying lower energy level E_1 and upper energy level E_2, with $E_2 > E_1$. A photon with energy $hf = E_2 - E_1$ can be absorbed by a state with energy E_1 leaving the state in E_2, and it can stimulate a state of energy E_2 to emit, causing a transition to E_1. An excited state in E_2 can also spontaneously decay to E_1.

Consider a blackbody source with internal cavity as illustrated in Figure 4.35. The cavity is filled with a photon gas of energy density $u(f)$ given by Equation 4.89, and it also contains the gas with E_1 and E_2 energy levels. The probabilities for emission and absorption are given by the *Einstein A and B coefficients*. The probabilities for stimulated emission and absorption depend on the photon energy density, but since spontaneous emission proceeds independently of any incident photon, the corresponding probability is independent of $u(f)$:

Probability of absorption of a photon: $B_{12} \, u(f)$
Probability of stimulated emission of a photon: $B_{21} \, u(f)$
Probability of spontaneous emission of a photon: A_{21}

Equilibrium is determined by the *transition rates*, which in turn are determined by the transition probabilities given above. Let N_1 be the number of states with energy E_1 and N_2 be the number of states with energy E_2. The transition rate for emission R_{21} is given by the number of states in E_2 times the net probability for emission:

$$R_{21} = [A_{21} + B_{21} \, u(f)] \, N_2 \tag{7.8}$$

Similarly, the transition rate for absorption is

$$R_{12} = [B_{12} \, u(f)] \, N_1 \tag{7.9}$$

At *equilibrium*, these rates must be equal. Thus

$$\frac{N_1}{N_2} = \frac{A_{21} + B_{21} \, u(f)}{B_{12} \, u(f)} \tag{7.10}$$

In *thermal equilibrium*, the population numbers N_1 and N_2 are given by the *Maxwell-Boltzmann distribution* discussed in Section 4.84

$$N_i = N_0 e^{-\frac{E_i}{k_B T}} \tag{7.11}$$

Since $E_2 - E_1 = hf$,

$$\frac{N_1}{N_2} = e^{-\frac{E_1}{k_B T}} e^{\frac{E_2}{k_B T}} = e^{\frac{hf}{k_B T}} \tag{7.12}$$

where k_B is Boltzmann's constant, and T is the equilibrium temperature. Combining the last two results gives

$$e^{\frac{hf}{k_B T}} = \frac{A_{21} + B_{21} \, u(f)}{B_{12} \, u(f)}$$

Thus

$$u(f) = \frac{A_{21}}{B_{12} e^{\frac{hf}{k_B T}} - B_{21}} \tag{7.13}$$

As the energy density $u(f)$ becomes large, that rate of spontaneous emission becomes negligible; thus it must be the case that $B_{12} = B_{21}$, and Equation 7.13 becomes

$$u(f) = \frac{\frac{A_{21}}{B_{21}}}{e^{\frac{hf}{k_B T}} - 1} \tag{7.14}$$

It remains only to identify the ratio $\frac{A_{21}}{B_{21}}$ with the *Jeans number* of Equation 4.76:

$$\frac{A_{21}}{B_{21}} = \frac{8\pi h f^3}{c^3} \tag{7.15}$$

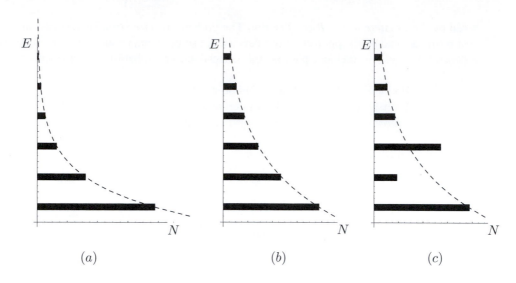

Figure 7.10   Plots of energy E vs. population number N for a system characterized by discrete energy levels. (a), (b) Equilibrium populations described by the Maxwell Boltzmann distribution (dashed line). System (a) is at a lower temperature. (c) Population inversion. A non-equilibrium condition is maintained that causes a higher energy level to have a larger population than a lower energy level.

Light amplification can occur in *non-equilibrium conditions*. Figures 7.10(a) and (b) illustrate a series of energy levels in thermal equilibrium; that is, with populations determined by Maxwell-Boltzmann distribution of Equation 7.11. The exponential dependence of Equation 7.11 causes the population to decrease as the energy increases. Figure 7.10(c) illustrates a *population inversion*, where the population of a higher state is somehow made larger than that of a lower energy state. Consider a photon traveling through a material in such non-equilibrium state, and let the photon energy equal the difference of the inverted levels. As the photon travels, more emissions are stimulated than absorptions, and the number of photons increases as illustrated in Figure 7.9(d); in other words, the photon beam is amplified.

It is curious that Einstein stopped short of suggesting the possibility of light amplification by stimulated emission in his 1917 paper. It is possible that in these early days of statistical thermodynamics, he considered it unlikely that the necessary conditions for thermal non-equilibrium could ever be reached in practice. It took another forty years for this to be achieved in the laboratory of Charles Townes[7].

Problem 7.10   Redo the derivation above without allowing for stimulated emission. Show that at a given temperature, the result obtained is only valid at high frequencies. Why is it sufficient to only include spontaneous emission in this limiting case?

---

[7]Charles H. Townes: 1915-. American physicist and corecipient of the 1964 Nobel Prize for the invention of the microwave laser (MASER).

Problem 7.11   In this problem, we show that gas molecules or atoms in thermal equilibrium at temperature T have an average thermal energy of $k_B T$. Assume that the gas particles are described by a Maxwell Boltzmann distribution of *energies*:

$$f(E) = Ce^{-\frac{E}{k_B T}}$$

where C is a normalization constant.

   a) Use the normalization condition given by $\int_0^\infty f(E)dE = 1$ to determine the constant C that normalizes the distribution.

   b) The average energy is given by

$$\langle E \rangle = \int_0^\infty E\, f\,(E)\, dE$$

   Using this, along with the normalized distribution, show that the average energy is $k_B T$.

   c) Find the average energy in electron volts of a collection of gas molecules at room temperature.

7.4 LASER SYSTEMS

A laser system can be designed around any available *gain medium*[8] that is capable of sustaining a population inversion that leads to stimulated emission. Gain mediums can consist of atoms or molecules in a gas (or plasma), liquid or solid phase. Figure 7.11(a) illustrates a typical laser process. A *pumping process* creates excited states in the gain medium, but there must also be a strategy for achieving the nonequilibruim population inversion. In the case illustrated, the initial excited state decays quickly to a *metastable state* that has a very long lifetime. This is the upper laser level; it fills quickly and drains slowly, resulting in the large population in the upper laser level necessary for an inversion. Furthermore, the lower laser level *depopulates* quickly, further enhancing and sustaining the population inversion. Examples of pumping processes include collisions in an electric discharge (discharge pumping) and photon absorption from an incident beam of light (optical pumping).

Imagine a gain medium held within a container through which a beam of light can pass, and let the beam consist of photons of energy equal to the difference between the upper and lower laser levels. Each photon that encounters an upper level state stimulates a decay to the lower level, creating a new photon identical in both energy and phase. At any point along the beam, the number of photons created by stimulated emission is proportional to the number of photons present in the beam:

$$\frac{dI}{dz} = G\,I \tag{7.16}$$

where z measures distance along the beam direction, I is the beam irradiance, and G is the *gain* with units of m^{-1}. The solution to this differential equation is

$$I(z) = I_0 e^{Gz} \tag{7.17}$$

---

[8]The gain medium is sometimes also referred to as an *active medium*, or as a *laser medium*.

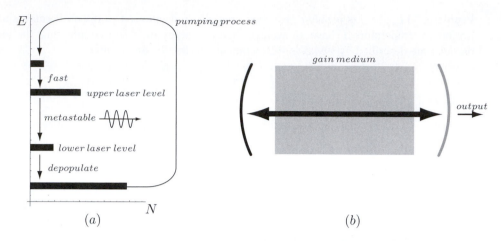

Figure 7.11   A typical laser process. (a) A pumping process selectively populates an excited state that decays very rapidly to a long-lived metastable state, creating a population inversion between the two levels responsible for the laser emission (upper laser level and lower laser level). The population inversion is enhanced if the lower laser level is rapidly depopulated by a relaxation process. (b) The gain medium consists of a collection of atoms or molecules excited by the pumping process illustrated in (a). Placing the gain medium within a Fabry-Perot resonator increases the photon density, making stimulated emission much more probable. The partially reflecting mirror on the right of the cavity transmits the output beam.

as can be verified by substitution. Thus, the beam irradiance varies exponentially with distance along the gain path. Since stimulated emission increases the photon flux, the corresponding gain term is positive. However, if a photon encounters a state in the lower laser level, it is absorbed; thus, the gain for absorption is negative. The net gain is the sum of all available processes that can add or subtract photons from the beam, and this must be positive if the beam is to be amplified. We must also include losses from optical reflections and defects in the beam path.

Laser systems are often designed around a *Fabry-Perot resonator*, as illustrated in Figure 7.11(b). The *laser cavity* typically consists of one highly reflecting mirror and a partially reflecting *output coupler* that transmits the output laser beam. Both mirrors are usually slightly curved, with a radius of curvature that is several times the mirror separation; this creates a cavity that is *stable*, meaning that a photon can multiply reflect between the two mirrors many times without walking out of the cavity. The laser beam begins as a spontaneous emission in a direction that can be contained in the cavity, then builds as amplification occurs. The amplification is enhanced by the fact that photons oscillate back and forth many times before finally passing through the output coupler. The irradiance within the laser cavity is usually much higher than that of the output beam. For example, a cavity that utilizes a highly reflecting back mirror and a 90%-reflecting output coupler will have a photon flux within the cavity that is about ten times that of the output beam. A high photon flux within the cavity makes it more probable that stimulated emission will occur before the upper laser level has time to spontaneously decay. Of course, the power exiting as the output beam must be included as a negative contribution to the overall system gain. The cavity and output power finally saturates at a steady state level that is limited by the pumping process, and perhaps also by effects such as heat generated in the gain medium and associated optics by the large cavity irradiance.

7.4.1 Atomic Gas Lasers

Atomic gas laser utilize atomic energy levels discussed in Section 7.2.1. In the process illustrated in Figure 7.11, the bottom laser level is not the ground state. A process that uses the ground state as the lower laser level makes it much more difficult to sustain a population inversion since it is difficult to sufficiently depopulate the ground state. Typically, this can only be done by very rapid pumping for very short periods of time, giving output beams that consist of pulses of short duration. Laser systems that give *continuous wave* (CW) output beams of constant irradiance are usually built around a process where the lower laser level is far enough above the ground state so that it is not thermally populated to any significant degree.

The energy level diagrams of most atoms are similar to that of hydrogen in that the transition energy from the ground state to the first excited state is much greater than transition energies between higher energy levels. This results in a low *quantum efficiency*. For example, suppose we wish to design a laser to emit the red line of the Balmer series. For each output photon of energy $1.89\,eV$, $10.2\,eV$ must be wasted as the lower laser level is depopulated back to the ground state. In this case, we say that the quantum efficiency is 18.5%. This means that the most you could ever expect for an output power is 18.5% of the power delivered by the laser power supply. In practice, working laser systems operate at an efficiency that is far less than the quantum efficiency. The point is that the quantum efficiency of atomic gas lasers is typically rather small.

7.4.1.1 Helium-Neon Laser
The ubiquitous *helium-neon* ($HeNe$) laser is illustrated in Figure 7.12. The gain medium consists of a low pressure mixture of helium and neon. The pumping process uses electron impact from an electric discharge to excite the helium atoms, as illustrated in Figure 7.12(a). The excited states of helium quickly decay to a metastable state where a large population is established. This metastable helium energy level has almost the same energy as the upper laser level of the neon atom, and so excited helium atoms can transfer energy to ground state neon during collisions. This results in a very direct population of the upper laser level. The lower laser level of neon is depopulated thermally. Despite the effective pumping process, helium-neon is a relatively low gain medium. In practice, the mirrors on both ends of the cavity typically have highly reflective dielectric coatings (see Section 5.11); this is necessary to keep the negative gain associated with extracting a beam from making the system gain negative.

Figure 7.12(b) illustrates a typical HeNe laser design. A DC voltage sustains an electric discharge, and the discharge current is limited by a series *ballast resistor*. In the design illustrated, the cavity mirrors are *internal*, meaning that the reflective surface resides within the same volume as the gain medium. It is also common to seal the gain medium with windows oriented at Brewster's angle, and to use externally mounted cavity optics. For an example of such a design, see the discussion of the carbon dioxide laser that follows. When Brewster windows are used, the laser output is polarized. The design of Figure 7.12(b) gives a randomly polarized output beam.

The energy diagram of Figure 7.12(a) gives the familiar red output of wavelength $632.8\,nm$. The neon atom has a rich spectrum that can be used for laser emission at other wavelengths such as $3.39\,\mu m$, $1.15\,\mu m$, $594\,nm$, $612\,nm$, and $543.5\,nm$, which is sometimes referred to as a GreeNe. For more details, see Siegman [19].

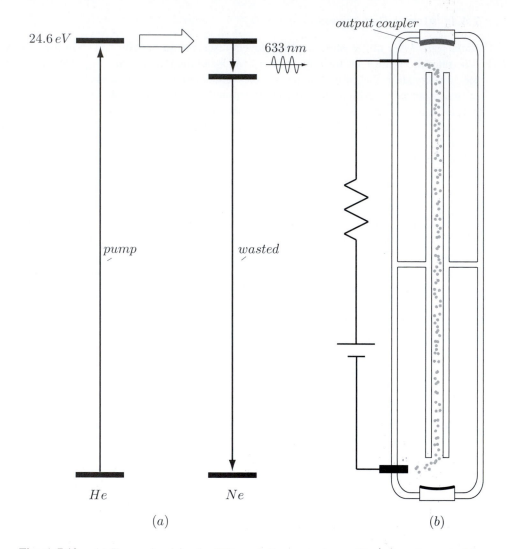

Figure 7.12 (a) Energy level for the $633\,nm$ helium-neon laser. The helium is pumped by an electric discharge, and transfers energy via collision to neon atoms. (b) A typical helium-neon laser cavity.

7.4.2   Molecular Gas Lasers

Molecular gas lasers use molecular energy levels, discussed in Section 7.2.2. Typically, these are vibration-rotation lines, illustrated in Figure 7.1(b). Since the harmonic oscillator energy levels are equally spaced, the lower laser level can be much closer to the ground state than is the case for atomic levels, and thus the quantum efficiency can be much higher. The vibration-rotation spectra of most molecules occurs in the near to mid-infrared.

7.4.2.1 *Carbon Dioxide Laser*   The carbon dioxide laser is a very efficient laser system, and can produce CW output beams of irradiance sufficient to cut metal. The gain medium typically consists of a mixture of about 10% carbon dioxide, 10% nitrogen, and 80% helium at pressures that can range from low pressure to a few atmospheres. The

pumping process is similar to helium-neon in that an electric discharge excites the nitrogen vibrational levels, which then decay quickly to a metastable state that is the first vibrational level above the ground state. This highly populated nitrogen state has almost the same energy as the upper laser level of CO_2, so a *vibration-to-vibration* energy transfer can occur that very effectively creates a population inversion, as illustrated in Figure 7.13(a).

Because CO_2 is a *triatomic* molecule, it has a more complicated spectrum. In a simple linear model, the carbon is at the center with each oxygen attached by a Hooke's law spring. Because there are two coupled springs, there are three distinct *vibrational modes*: symmetric stretch where the oxygen molecule vibrates along the internuclear axis symmetrically, asymmetric stretch, and a bending mode where the oscillation of the atoms is perpendicular to the internuclear axis. Each mode gives a distinct set of equally spaced vibrational levels, each with rotational levels superimposed, as discussed in Section 7.2.2. As illustrated in Figure 7.13(b), there are two lower laser levels, each corresponding to different vibrational modes of CO_2. This allows laser output in two distinct wavelength *bands*: one centered at $9.4\ \mu m$ and the other at $10.6\ \mu m$.

Each wavelength band consists of a discrete spectrum of R-branch and P-branch as discussed in Section 7.2.2. The output spectrum of each wavelength band has a profile similar to that shown in Figure 7.14. The center of each band corresponds to the energy spacing of the upper and lower vibrational levels that give the laser output in Figure 7.13. Lines in the R-branch result from the selection rule that gives a higher energy than the band center, and P-branch lines result from the selection rule that gives a lower energy than the band center. The profile shape comes from two effects. There is a degeneracy of $(2J + 1)$ associated with each rotational level. Also, the rotational levels are populated thermally, and so have a Boltzmann energy distribution. The profile shape comes from the product of these two effects: states with low values of J have a greater thermal population but lower degeneracy, while the high J lines (those away from the band center on either side) have higher degeneracy but lower thermal population. For most electrical discharges, the peak of the profile occurs at a J of around 18 in each branch.

A typical laser design for CO_2 is illustrated in Figure 7.13(b). The cavity includes a *blazed diffraction grating* (see Section 6.3.8 and Application Note 6.1) that allows laser output on any of the available vibration rotation lines. Recall that a blazed diffraction grating diffracts most of the incident energy into the first diffraction order. Intuitively, the grating behaves as a mirror with perfect alignment for only one of the available laser lines; for the other lines, the mirror is misaligned, and the gain is spoiled. By adjusting the grating orientation, different R- and P-branch lines may be selected for output. The other end of the cavity is terminated by a mirror that, if partially reflecting, can transmit an output beam. Note that the grating also reflects a zeroth-order beam which can also be used for output, but the direction of this beam changes as the wavelength is adjusted, so using this output is often less convenient.

In the energy diagram of Figure 7.13(a), the lower laser level is depopulated thermally, principally by collisions with the fast-moving helium atoms and by transfer to the cavity walls. For this reason, most CO_2 designs incorporate some cooling strategy, which for higher power designs usually involves water cooling. The design illustrated in Figure 7.13(b) uses *Brewster windows* to terminate each end of the discharge region. Recall that at Brewster's angle, the reflectivity is zero for polarizations parallel to the plane of incidence (see Section 3.5.2). The orthogonal polarization is reflected at each window, spoiling the gain; thus the output of a cavity that uses Brewster windows is polarized.

In many designs, one end of the cavity is controlled by a *piezoelectric transducer* (PZT). Applying a voltage to the PZT causes it to change length, typically by a few tens of microns.

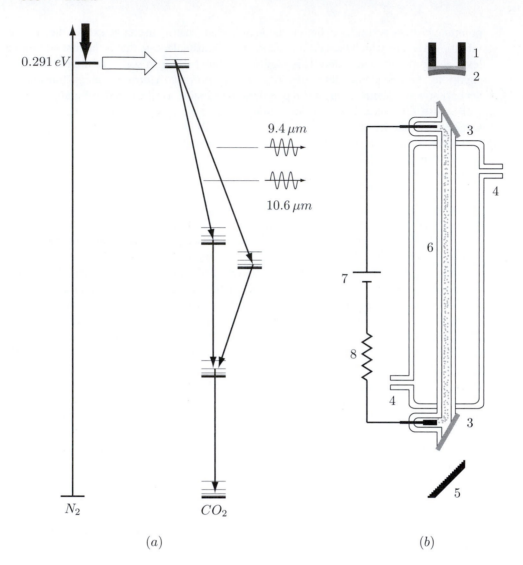

0.291 eV

9.4 µm

10.6 µm

N_2

CO_2

(a)

1

2

3

4

6

7

8

4

3

5

(b)

Figure 7.13   Carbon dioxide laser. (a) Energy level diagram. The upper laser levels of CO_2 are selectively pumped by nitrogen molecules in an excited metastable state. (b) A typical design. The cavity length is controlled by a PZT (1) attached to the output coupler (2). Brewster windows (3) allow the cavity beam to enter and exit the gain medium (6). Cooling water enters and exits (4) to remove heat generated by the discharge current. The discharge is maintained by the power supply voltage (7) and the discharge current is limited by a series ballast resistor (8). A blazed diffraction grating (5) allows specific vibration-rotation transition to be used for the laser output.

The PZT can make small changes in the length of the cavity to maintain the *standing wave condition* that gives maximum laser output (see Section 5.10) from the Fabry-Perot cavity.

A typical CO_2 laser design can provide output on more than 100 R- and P-branch lines in the two wavelength bands. The actual number of lines available is determined by the *gain-limit*, illustrated in Figure 7.14. The working gain limit is determined by many factors, including cooling effectiveness, optical alignment, and so on.

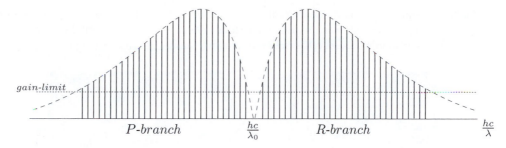

Figure 7.14   Output energy spectrum of a single band of the carbon dioxide laser. There is one band centered at $\lambda_0 = 9.4\,\mu m$ and another at $\lambda_0 = 10.6\,\mu m$. Each band consists of a spectrum of vibration rotation lines.

7.4.3 Solid-State Lasers

Solid-state lasers use a gain medium that is solid. Presumably, this would lead to the greatest quantum efficiencies, since in a solid the atoms or molecules are much more closely spaced than in a gas. However, this is not always the case; the carbon dioxide laser can operate with an efficiency that exceeds that of most solid-state lasers.

Insulators such as aluminum oxide have a large band gap, and when intrinsic they are transparent in the visible. However, when impurities are added, energy levels result that can be used as laser levels. For example, adding chromium atoms to aluminum oxide creates *ruby*, with the familiar red color that results when white light has the blue and green removed by absorption. The first laser to emit in the visible utilized ruby and was built by Maiman[9] in 1960. His design used an intense burst of white light from a xenon flash lamp to optically pump excited states in a ruby rod. The absorption is broad, and populates a very complicated and dense set of energy levels that fortunately decay quickly to a fairly sharp level that is about $1.8\,eV$ above the ground state. The ruby laser is unusual in that it uses the ground state as the bottom laser level. Thus, a population inversion may only be achieved by very intense optical pumping that almost completely depopulates the ground state. This cannot be achieved in steady state; thus the ruby laser has a pulsed output of a very short duration but with a very high intensity that can burn though a thin sheet of metal. The wavelength is deep red: $694\,nm$.

A similar gain medium uses yttrium aluminum garnet with neodinium impurities; this is commonly referred to as *neodinium YAG* (Nd:YAG). A typical Nd:YAG design uses a flashlamp shaped as a rod that is place along the focus of an elliptical cavity with reflective sides. A Nd:YAG rod with polished ends that are also coated for enhanced reflectivity is placed at the other focus. Light emitted by the flash lamp is refocused onto the gain medium to produce a pulse of output. The output wavelength of a Nd:YAG laser is $1064\,nm$, which is particularly dangerous; this wavelength is beyond the range of human vision and is thus invisible, but is still small enough to pass through the lens of the eye to focus on and damage the retina.

7.4.3.1 Diode Lasers
A *diode laser* is formed by placing a light-emitting diode (see Section 7.2.4) within a Fabry-Perot cavity. Stimulated emission occurs at the depletion layer of the p-n junction where electrons and holes recombine radiatively. The output wavelength of a diode laser can usually be varied continuously over a small wavelength

[9]T.H. Maiman: 1927-2007. American physicist and inventor of the ruby laser.

range by controlling the diode current and sometimes also with an in-cavity diffraction grating. This is referred to as *tunability*, and it facilitates many experimental opportunities that are not available to lasers that only output a discrete spectrum or a single line.

Figure 7.15 Diode lasers in a TO-220 package. (GIPhotoStock/Photo Researchers, Inc.)

■ **APPLICATION NOTE 7.2**

Diode Pumped Solid-state Lasers

Nd:YAG has a strong absorption at $808\,nm$ that is used to optically pump the upper laser level of the $1064\,nm$ transition. Diode laser sources have been developed with output at this wavelength at power levels in excess of one watt. A laser constructed with Nd:YAG and a $808\,nm$ diode laser pump is an example of a *diode pumped solid state* (DPSS) laser.

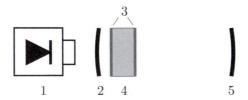

Figure 7.16 Diode-pumped solid-state laser: (1) $808\,nm$ diode laser; (2) cavity mirror, AR coated for $808\,nm$, 99.9% reflecting for $1064\,nm$; (3) AR coatings for $1064\,nm$; (4) Nd-doped laser crystal; (5) cavity mirror, partially reflecting for $1064\,nm$.

A typical DPSS laser design is illustrated in Figure 7.16. The Nd:YAG crystal is placed within a Fabry-Perot cavity. The $808\,nm$ diode laser beam enters the cavity by transmitting through one cavity mirror that is coated for high reflectivity at $1064\,nm$

and high transmission at $808\,nm$. It is absorbed by the Nd:YAG crystal, creating a population inversion for the $1064\,nm$ transition. Stimulated emission occurs as the $1064\,nm$ radiation oscillates within the cavity.

The DPSS laser output has a much better beam quality than that of the diode laser. The gain medium of a diode laser typically exists within a thin rectangular depletion layer region that surrounds the semiconductor junction, giving an output beam that can have a non-circular beam spot. Since the Nd:YAG is inside a Fabry-Perot cavity, the $1064\,nm$ beam is more monochromatic, has better coherence, and has a circular beam spot. The DPSS Fabry-Perot cavity with Nd:YAG converts poor beam quality at $808\,nm$ to a much better beam quality at $1064\,nm$.

A similar material, neodymium doped yttrium orthvanadate (Nd:YVO4) can also be pumped at $808\,nm$, and also provides DPSS output at $1064\,nm$. Both can be *frequency doubled* to provide a green output beam. See Application Note 9.5.

7.4.4 Other Laser Systems

Ion lasers use an ionized gas as a laser medium. For example, in the *argon ion laser*, the noble gas argon is striped of an outer electron, creating energy levels of the ionized atom that can provide laser levels. The quantum efficiency of argon ion lasers is quite low since the discharge must provide both excitation energy and ionization energy that is fairly high for an electron in a filled subshell. Nonetheless, argon ion lasers can be designed to produce CW beams of 50 watts or more. Argon ion lasers produce output at a number of wavelengths across the visible spectrum. *Krypton ion* lasers operate in a similar way, and also produce a number of output wavelengths across the visible.

Excimer lasers use *excited dimers* as the gain medium. A dimer is a short-lived molecule that only exists in an excited state; for example, xenon-chloride. Excimer lasers typically emit in the ultraviolet.

Dye lasers typically use a transparent liquid gain medium that contains dye molecules that provide energy levels. The pumping mechanism is usually optical, using a flashlamp or high powered ion laser. The energy level structure of a liquid dye solution is very complicated and usually allows for a high degree of tunability across a fairly large wavelength range.

Problem 7.12   Using the information in Figure 7.12, find the quantum efficiency of the helium-neon laser.

Problem 7.13   Using the information in Figure 7.13, estimate the quantum efficiency of the carbon dioxide laser in both wavelength bands.

Problem 7.14   Describe the output polarization of the laser design of Figure 7.13(b). Explain your reasoning carefully, and include a discussion of Brewster's angle.

Problem 7.15   Carbon dioxide laser lines in the P-branch of the 10.6 micron wavelength band are spaced apart by about $50\,GHz$. Estimate the grating resolution that is necessary to discriminate against adjacent lines. If the grating operates in first order, estimate the number of rulings that must be illuminated. If the cavity beam is $3\,mm$ across at the grating, estimate the number of lines per millimeter that would be sufficient.

Problem 7.16   Zinc selenide is a material that is transparent across the mid-infrared and thus is commonly used in applications that involve carbon dioxide lasers. It has a relatively high index of refraction of 2.42. Find Brewster's angle for Zinc selenide, and identify this angle in the design of Figure 7.13(b).

7.5 LONGITUDINAL CAVITY MODES

While discussing the Fabry-Perot interferometer in Section 5.10, we identified the *free spectral range* over which the interferometer may be tuned. For a laser cavity of optical length L, the free spectral range is given by

$$\Delta f_{fsr} = \frac{c}{2L} \tag{7.18}$$

Attaching one end of the cavity to a length transducer such as the PZT of Figure 7.13(b) allows L to be varied, which can vary the output frequency.

■ **EXAMPLE 7.2**

Find the free spectral range for a CO_2 laser with a cavity length of $1.5\,m$, a $HeNe$ laser with a $10\,cm$-long cavity, and a red diode laser with a $1.0\,cm$ (optical length) cavity. Estimate the wavelength range in each case.

Solution

Using Equation 7.18, we find for the $1.5\,m$ cavity:

$$\Delta f_{fsr} = \frac{3.00 \times 10^8\,\frac{m}{s}}{2(1.5\,m)} = 100\,MHz$$

which is a very small fraction of the $50\,GHz$ or so P-branch spacing. Similarly, the $10\,cm$ long cavity has a free spectral range of $1.5 GHz$, and that of the $1.0\,cm$ cavity is $15 GHz$.

To find the wavelength ranges, use Equation 5.63:

$$\Delta \lambda_{fsr} = \lambda \frac{\Delta f_{fsr}}{f} = \lambda^2 \frac{\Delta f_{fsr}}{c}$$

For carbon dioxide at $10.6\,\mu m$,

$$\Delta \lambda_{fsr} = \left(10.6 \times 10^{-6} m\right)^2 \frac{10^8\,Hz}{3.00 \times 10^8\,\frac{m}{s}} = 3.75 \times 10^{-5} \mu m$$

which again, is a very small tuning range. Similarly, we find that Δf_{fsr} is about $0.02\,nm$ for the $HeNe$ laser, and about $0.2\,nm$ for the red diode laser.

Each laser system will have a characteristic *gain profile* that for gases depends on the Doppler and collisionally broadened absorption profile, and in general also upon the diffraction grating resolution, or any similar in-cavity device for controlling the output wavelength. A longitudinal mode with a frequency that falls within the gain profile can provide laser output. A laser designed for *single longitudinal mode operation* will have positive gain over a frequency range that is roughly as wide as the free spectral range, as illustrated in Figure 7.17(a). For example, a carbon dioxide laser that uses a pressure of about 50 Torr has a gain profile that is about $100\,MHz$ wide, and in this case single-mode operation can be obtained with a cavity length of about $1.5\,m$. Designs using higher pressures are possible with gain profiles up to $1\,GHz$ wide; in this case, a cavity length of $15\,cm$ would optimize the system gain while still providing single-mode operation.

A Fabry-Perot laser cavity can also have *transverse cavity modes*; these will be discussed in Section 7.9.6.

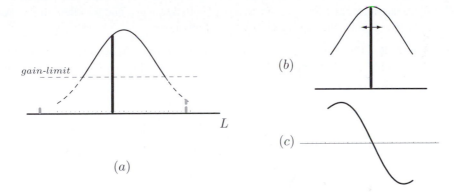

(a)

Figure 7.17   (a) The gain profile as a function of laser cavity length L, varied by a length transducer such as a PZT. The solid portion above the gain limit gives a positive system gain. The solid output line falls inside the gain region. The frequencies that are one free spectral range to each side cannot provide output because they do not have sufficient gain. (b) The laser output is dithered to give a small AC component on the laser output. (c) Measuring the AC component gives a derivative signal that can be used to feedback control the cavity length.

7.6 FREQUENCY STABILITY

According to Section 5.10, the beam that exits a Fabry-Perot cavity has a frequency profile that is equal to the free spectral range divided by the finesse. For example, a $HeNe$ laser that uses 99% reflectors has a finesse in excess of 300 (see Problem 7.17). Using the results of Example 7.2, we estimate that the minimum laser linewidth is about $6.7 \times 10^{-5}\,nm$. In practice, most lasers have linewidths that are much larger than this estimate due to external influences. For example, vibrations that couple into the cavity cause length fluctuations at acoustic frequencies that lead to laser *jitter* over the corresponding fraction of the free spectral range. Temperature changes also cause cavity length changes that lead to slower variations over the entire free spectral range of the cavity. If the cavity is not single-mode, multiple modes can beat causing additional frequency modulation.

It is possible to *feedback stabilize* a laser cavity to improve the frequency stability. For cavities that include PZT transducers (see Figure 7.13), the PZT voltage can be *dithered* by adding an AC component to the DC control voltage. The dither causes a small frequency modulation in the output beam, which if swept across the free spectral range sweeps out the *derivative* of the gain profile, as illustrated in Figure 7.17(b) and (c). This derivative signal (measured as the AC component in the laser output) can be used as a *discriminator* for *negative feedback control*. When the output is at the peak of the gain profile, the derivative is zero and the laser power is maximum. If the frequency drifts away from the peak, the derivative signal can be amplified and applied to the PZT in the proper polarity to drive the laser frequency back to gain center.

Problem 7.17   Find the coefficient of finesse and the finesse of a laser cavity designed to operate at $530\,nm$ that is terminated on each end with mirrors of reflectivity equal to 99%. Use this to estimate the minimum line width in both frequency and wavelength if the cavity length is $10\,cm$. Find the free spectral range of this laser.

Problem 7.18   Find the cavity finesse that gives a minimum frequency spread of $100\,kHz$ in the output for a $25\,cm$-long $HeNe$ laser cavity designed to operate at $633\,nm$. What is the free spectral range of this cavity?

7.7   INTRODUCTION TO GAUSSIAN BEAMS

Laser beams emitted from well-designed and well-adjusted Fabry-Perot cavities have beam profiles that are *Gaussian* in shape. Intuitively, we can understand why as follows. The curved mirrors in a stable cavity continually re-image the beam as it passes back and forth. On each traversal, the beam passes through or at least towards a Fraunhofer plane that contains a far-field diffraction pattern. At this point, any structure in the beam tends to evolve towards outer diffraction orders that are less paraxial and thus tend to be removed from the cavity. In a well-designed cavity which may include an aperture to remove unwanted diffraction orders (as discussed presently, see Figure 7.30), only the central diffraction order survives. If the cavity has cylindrical symmetry, the central diffraction order is very close to Gaussian in profile. The steady-state beam that emerges from an ideal Fabry-Perot cavity is very close to Gaussian.

Since most lasers emit Gaussian beams, it is important to know how to work with them. As we will see, the propagation of a Gaussian beam through a series of lenses and mirrors is not always accurately described by the methods of geometric optics. In particular, focal points may not be well-located by the thin-lens equation, and the focal spot has a finite size that geometric optics cannot determine. Gaussian beam propagation is affected by *diffraction*, which by definition is not included in geometrical optics.

7.7.1   Overview of Gaussian Beam Properties

We begin with a short overview of some of the more important features of Gaussian beam propagation. Derivations will follow in the next section.

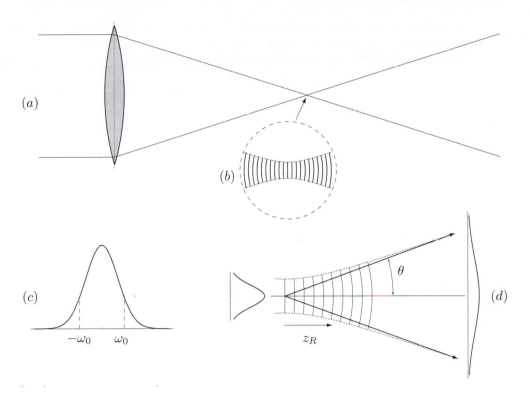

Figure 7.18   (a) A lens focuses a Gaussian beam. (b) Close-up of the waist region. The spot size is a few wavelengths across. The solid curves indicate the wavefront curvature at arbitrary points along the beam path; they are not spaced by the beam wavelength. (c) A Gaussian profile. (d) The beam divergence approaches the far field limit for distances greater than the Rayleigh range.

Gaussian beams have amplitude profiles described by the *Gaussian function*:

$$E\left(r, z\right) = E_0 \left(\frac{\omega_0}{\omega\left(z\right)}\right) e^{-\frac{r^2}{\omega^2(z)}} \qquad (7.19)$$

where r is the radial distance measured from the center of the profile, E_0 is the electric field amplitude at the center of the profile, and *Gaussian beam radius* $\omega(z)$ is given by

$$\omega\left(z\right) = \omega_0 \sqrt{1 + \left(\frac{z}{z_R}\right)^2} \qquad (7.20)$$

It important not to confuse this symbol with that used for angular frequency; the notation can be confusing but it is standard. The points $r = \pm\omega(z)$ locate the $\frac{1}{e}$-points of the amplitude profile. The distance z_R is called the *Rayleigh range*:

$$z_R = \frac{\pi\omega_0^2}{\lambda} \qquad (7.21)$$

where the quantity ω_0 represents the minimum value of Equation 7.20, and thus is called the *beam waist*. The value of ω at a point of focus is given by ω_0, as indicated in Figures 7.18(a) and (b).

It is useful to think of $w(z)$ as the *spread* of the Gaussian profile. The Rayleigh range determines the rate of change of this spread as the beam passes through a focus.

The irradiance profile is given by the square of the amplitude:

$$I(r, z) = I_0 \left(\frac{w_0}{w(z)} \right)^2 e^{-\frac{2r^2}{w^2(z)}} \tag{7.22}$$

The points $r = \pm w(z)$ locate the $\left(\frac{1}{e} \right)^2$-points of the irradiance profile, and enclose about 90% of the beam power. Intuitively, the spot that results from illuminating a white card with a laser beam has a diameter of about $2w(z)$.

As the Gaussian beam propagates, the radius of curvature of the beam wavefront varies according to

$$R(z) = z + \frac{z_R^2}{z} \tag{7.23}$$

The wavefronts are plane at the beam waist ($z = 0$). In the *far field* where $z \gg z_R$, the wavefront curvature increases in proportion to z. The minimum wavefront curvature occurs at $z = z_R$ where it has the value $R = 2z_R$ (see Problem 7.19).

When $z \gg z_R$, the *far field beam divergence* approaches the angle

$$\theta = \frac{w(z)}{z} \rightarrow \theta_f = \frac{\lambda}{\pi w_0} \tag{7.24}$$

as illustrated in Figure 7.18. Within the Rayleigh range ($|z| < z_R$), the beam remains more or less collimated. According to Equation 7.20, $w(z_R) = \sqrt{2}w_0$. When irradiance is observed, the *spot size* $s = 2w(z)$ is twice the diameter measured at the beam waist. Note that beams with a small waist diverge more quickly, as illustrated in Figure 7.19.

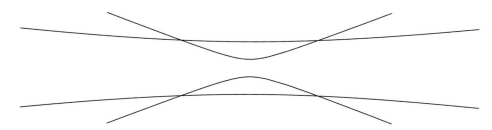

Figure 7.19   The far field divergence of a Gaussian beam is inversely proportional to the beam waist.

■ **EXAMPLE 7.3**

Find the Rayleigh range, the far field divergence angle, and the spot size at the moon for a helium-neon laser beam ($633\,nm$) with initial waist sizes of $0.5\,mm$ and $0.5\,m$. Repeat for a carbon dioxide laser with wavelength $10.0\,\mu m$. Use an Earth-Moon distance of $4.00 \times 10^5\,km$.

Solution

For the helium-neon laser with a $1.00\,mm$ minimum spot size, the Rayleigh range is

$$z_R = \frac{\pi \omega_0^2}{\lambda} = \frac{\pi \left(5.00 \times 10^{-4}m\right)^2}{633 \times 10^{-9}m} = 1.24\,m$$

The far field divergence angle is

$$\theta_f = \frac{\lambda}{\pi \omega_0} = \frac{633 \times 10^{-9}m}{\pi \left(5.00 \times 10^{-4}m\right)} = 4.03 \times 10^{-4}\,rad$$

which gives a spot size on the Moon equal to

$$s = 2x_m\theta_f = 2\left(4.00 \times 10^8 m\right)\left(4.03 \times 10^{-4}rad\right) = 3.22 \times 10^5 m = 322\,km$$

When this laser beam is expanded to an initial spot size of one meter, the Raleigh range is 10^6 times larger, and the far field divergence angle and Moon-spot size are 1000 times smaller:

$$z_R = \frac{\pi \omega_0^2}{\lambda} = \frac{\pi \left(5.00 \times 10^{-1}m\right)^2}{633 \times 10^{-9}m} = 1.24 \times 10^6\,m$$

$$\theta_f = \frac{\lambda}{\pi \omega_0} = \frac{633 \times 10^{-9}m}{\pi \left(5.00 \times 10^{-1}m\right)} = 4.03 \times 10^{-7} rad$$

$$s = 2x_m\theta_f = 2\left(4.00 \times 10^8 m\right)\left(4.03 \times 10^{-7}rad\right) = 3.22 \times 10^2 m = 322\,m$$

The carbon dioxide laser has a wavelength that is 15.8 times larger than helium-neon, so for a given waist size, the Rayleigh range is smaller by a factor of 0.0633, and the far field divergence and far field spot size are 15.8 times larger.

Problem 7.19   Show that the minimum wavefront curvature in a Gaussian beam occurs at $z = z_R$, where it has the value $R = 2z_R$.

Problem 7.20   Repeat Example 7.3 for an X-ray laser of wavelength $14.7\,nm$.

Problem 7.21   A communications satellite operating at x-band ($8.40\,GHz$) communicates with a base station on Earth using a 2.0-meter wide reflective dish. Assuming that the satellite emits a Gaussian beam with a $1.00\,m$ beam waist at the dish, find the Rayleigh range and the minimum spot size on Earth. The altitude of a geosynchronous orbit is $22,000\,km$.

7.8   DERIVATION OF GAUSSIAN BEAM PROPERTIES

In this section, we use the wave equation to determine the important properties of Gaussian beams. We begin by considering the special case of a *paraxial spherical beam*.

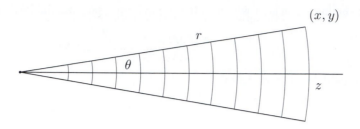

Figure 7.20   A paraxial beam with spherical wavefronts of constant amplitude.

A *spherical wave* travels outward from a source point, or converges inward towards a focus, as shown in Figure 7.20. In Section 1.7.2, we verified that such waves are described by

$$\Psi(r, t) = \frac{A}{r} e^{ikr} e^{\pm i\omega t} \tag{7.25}$$

where $r = \sqrt{x^2 + y^2 + z^2}$, A is the amplitude and in this case is constant, and importantly, ω as used here is angular frequency. The reader must be careful, especially in this section, to interpret the use of the symbol ω within the context of the discussion.

In Section 1.7.2 we showed that Equation 7.25 was a solution to the three-dimensional wave equation. Each *wavefront* is defined by a specific value of r and A, and thus is constant for a given value of r. To form a *beam* from a spherical wave, each wavefront is limited to a maximum *divergence angle* θ, as illustrated. For a well-directed beam, this divergence angle is small, and the beam satisfies appropriate *paraxial approximations* (see Section 4.3.0.1); in this case, we refer to the beam as a paraxial spherical beam.

It is useful to apply the *Fresnel approximation* of Chapter 6. The paraxial spherical beam of Figure 7.20 propagates along z, so we factor r as follows:

$$r = \sqrt{x^2 + y^2 + z^2}$$
$$= z\sqrt{1 + \left(\frac{x^2 + y^2}{z^2}\right)} \tag{7.26}$$

For a paraxial beam, the term within parentheses is small, and we may apply the binomial expansion:

$$r \approx z \left[1 + \frac{1}{2}\left(\frac{x^2 + y^2}{z^2}\right)\right]$$
$$= z + \frac{x^2 + y^2}{2z} \tag{7.27}$$

Thus, the paraxial spherical beam is approximated by

$$\Psi(x, y, z, t) = \frac{A}{\sqrt{x^2 + y^2 + z^2}} e^{ikz} e^{ik\frac{x^2+y^2}{2z}} e^{\pm i\omega t} \tag{7.28}$$

7.8.1   Approximate Solutions to the Wave Equation

Here, we investigate solutions to the three-dimensional wave equation by approximations similar to Equation 7.28. Such solutions must satisfy

$$\nabla^2 \Psi - \frac{1}{v^2}\frac{\partial^2 \Psi}{\partial x^2} = 0 \tag{7.29}$$

with $v = \frac{\omega}{k}$. For a traveling wave, the space and time dependence may be separated as follows:

$$\Psi(\vec{r}, t) = u(\vec{r}) e^{\pm i\omega t} \tag{7.30}$$

where, as discussed in Section 1.7, the notation $\vec{r}$ above represents the coordinates in any three-dimensional coordinate system. Substituting Equation 7.30 into Equation 7.29 gives, after cancellation,

$$\nabla^2 u + k^2 u = 0 \tag{7.31}$$

The wave equation expressed in this way is called the *Helmholtz equation* (see Problem 1.26).

For a beam that propagates along z as in Figure 7.20, the spatial function $u(\vec{r})$ may be expressed in Cartesian coordinates by

$$u(\vec{r}) = u(x, y, z) = \psi(x, y, z) e^{ikz} \tag{7.32}$$

We wish to substitute Equation 7.32 into the Helmholtz equation. Begin with the following derivatives:

$$\frac{\partial u}{\partial z} = \frac{\partial \psi}{\partial z} e^{ikz} + iku$$

$$\frac{\partial^2 u}{\partial z^2} = \frac{\partial^2 \psi}{\partial z^2} e^{ikz} + ik\frac{\partial u}{\partial z} = \frac{\partial^2 \psi}{\partial z^2} e^{ikz} + 2ik\frac{\partial \psi}{\partial z} e^{ikz} - k^2 u$$

$$\frac{\partial^2 u}{\partial x^2} = \frac{\partial^2 \psi}{\partial x^2} e^{ikz}$$

$$\frac{\partial^2 u}{\partial y^2} = \frac{\partial^2 \psi}{\partial y^2} e^{ikz}$$

With these results, Equation 7.31 gives

$$\left[\frac{\partial^2 \psi}{\partial x^2} + \frac{\partial^2 \psi}{\partial y^2} + \frac{\partial^2 \psi}{\partial z^2} + 2ik\frac{\partial \psi}{\partial z} \right] e^{ikz} = 0 \tag{7.33}$$

The longitudinal variation of ψ along z should be small in comparison with the transverse variation along x and y, so we may neglect any contribution due to the $\frac{\partial^2 \psi}{\partial z^2}$ term in Equation 7.33. Cancellation gives

$$\frac{\partial^2 \psi}{\partial x^2} + \frac{\partial^2 \psi}{\partial y^2} + 2ik\frac{\partial \psi}{\partial z} = 0 \tag{7.34}$$

Thus, paraxial beams represented by Equations 7.30 and 7.32 must have amplitude functions $\psi(x, y, z)$ that satisfy Equation 7.34.

To investigate solutions of Equation 7.34, we begin with a trial solution of the form:

$$\psi = \psi_0 e^{i\left(P(z) + \frac{k}{2q(z)}(x^2 + y^2)\right)} \tag{7.35}$$

In this equation, $P(z)$ is the *complex phase shift*, and $q(z)$ is the *complex beam curvature*. In preparation for substitution into Equation 7.34, we evaluate the following derivatives:

$$\frac{\partial \psi}{\partial x} = \frac{ik}{q} x\, \psi$$

$$\frac{\partial^2 \psi}{\partial x^2} = \left(\frac{ik}{q} - \frac{k^2 x^2}{q^2} \right) \psi$$

$$\frac{\partial^2 \psi}{\partial y^2} = \left(\frac{ik}{q} - \frac{k^2 y^2}{q^2} \right) \psi$$

$$\frac{\partial \psi}{\partial z} = i \left(\frac{\partial P}{\partial z} - \frac{k}{2q^2} \frac{\partial q}{\partial z} \left(x^2 + y^2 \right) \right) \psi$$

With these results, Equation 7.34 gives

$$\left\{ \frac{2ik}{q} - \frac{k^2}{q^2} \left(x^2 + y^2 \right) - 2k \left[\frac{\partial P}{\partial z} - \frac{k}{2q^2} \frac{\partial q}{\partial z} \left(x^2 + y^2 \right) \right] \right\} \psi = 0 \qquad (7.36)$$

For this to vanish at all points, the coefficients of the constant and quadratic terms must vanish separately, giving

$$\frac{2ik}{q} - 2k \frac{\partial P}{\partial z} = 0$$

Thus,

$$\frac{\partial P}{\partial z} = \frac{i}{q} \qquad (7.37)$$

Similarly

$$\frac{k^2}{q^2} - \frac{k^2}{q^2} \frac{\partial q}{\partial z} = 0$$

giving

$$\frac{\partial q}{\partial z} = 1 \qquad (7.38)$$

To summarize our progress so far, we have shown that the trial solution of Equation 7.35 represents a valid paraxial beam provided that the complex phase shift $P(z)$ and complex beam curvature $q(z)$ satisfy the relations above.

7.8.2 Paraxial Spherical Gaussian Beams

The properties of spherical Gaussian beams, introduced in Section 7.7, can be determined from the trial solution of Equation 7.35. According to Equation 7.38,

$$q = q_0 + z \qquad (7.39)$$

Consider a complex curvature defined by

$$\frac{1}{q} = \frac{1}{R(z)} + i \frac{\lambda}{\pi \omega(z)^2} \qquad (7.40)$$

where $R(z)$ and $\omega(z)$ are parameters that will be interpreted presently. Substitution of Equation 7.40 into the trial solution gives

$$\psi = \psi_0\, e^{-\frac{\left(x^2 + y^2 \right)}{\omega^2}}\, e^{i \left(P(z) + \frac{k}{2R} \left(x^2 + y^2 \right) \right)} \qquad (7.41)$$

Values of x and y with $\sqrt{x^2 + y^2} = \omega$ locate the $\frac{1}{e}$ points of amplitude ($\frac{1}{e^2}$ points of intensity); thus ω is the *beam radius*. Comparison of Equation 7.41 with the spherical beam of Equation 7.28 verifies that Equation 7.41 represents a beam with spherical wavefront of radius R with Gaussian intensity profile.

We now determine values for $R(z)$ and $\omega(z)$ that will satisfy the condition of Equation 7.39. Let

$$\frac{1}{q_0} = i\frac{\lambda}{\pi\omega_0^2}$$

where we refer to ω_0 as the *beam waist*. Inverting this gives

$$q_0 = -i\frac{\pi\omega_0^2}{\lambda} \tag{7.42}$$

We measure the complex curvature q relative to the beam waist:

$$q = q_0 + z = z - i\frac{\pi\omega_0^2}{\lambda} \tag{7.43}$$

Combining this with Equation 7.40 gives

$$\frac{1}{q} = \frac{1}{R} + i\frac{\lambda}{\pi\omega^2} = \frac{1}{z - i\frac{\pi\omega_0^2}{\lambda}}$$

$$= \frac{z + i\frac{\pi\omega_0^2}{\lambda}}{z^2 + \left(\frac{\pi\omega_0^2}{\lambda}\right)^2}$$

Equating real parts gives

$$\frac{1}{R} = \frac{z}{z^2 + \left(\frac{\pi\omega_0^2}{\lambda}\right)^2}$$

Solving this for R gives

$$R = z\left[1 + \left(\frac{\pi\omega_0^2}{\lambda z}\right)^2\right] \tag{7.44}$$

As in Section 7.7.1, we define the *Rayleigh range* z_R:

$$z_R = \frac{\pi\omega_0^2}{\lambda} \tag{7.45}$$

Using this, Equation 7.44 can be expressed as in Equation 7.23

$$R(z) = z + \frac{z_R^2}{z} \tag{7.46}$$

Similarly, equating imaginary parts above gives

$$\frac{\lambda}{\pi\omega^2} = \frac{\frac{\pi\omega_0^2}{\lambda}}{z^2 + \left(\frac{\pi\omega_0^2}{\lambda}\right)^2}$$

which, when solved for ω gives

$$\omega = \omega_0\sqrt{1 + \left(\frac{\lambda z}{\pi\omega_0^2}\right)^2} \tag{7.47}$$

Expressing this in terms of the Rayleigh range gives Equation 7.20

$$\omega(z) = \omega_0 \sqrt{1 + \left(\frac{z}{z_R}\right)^2} \tag{7.48}$$

The parameters just derived are discussed in more detail in Section 7.7.1 and are illustrated in Figure 7.18. The beam divergence approaches the *far field* value that is given by

$$\theta = \frac{\lambda}{\pi \omega_0} \tag{7.49}$$

The *complex phase shift* determined by Equation 7.37 affects the amplitude of the Gaussian beam, and results in additional phase shift relative to that of a plane wave. For more details, see Siegman [19] and Problem 7.24.

Problem 7.22   Derive Equation 7.49.

Problem 7.23   Use Equations 7.44 and 7.47 to show that

$$\frac{\lambda z}{\pi \omega_0^2} = \frac{\pi \omega^2}{\lambda R}$$

Use this to derive the following useful relations:

$$\omega_0 = \frac{\omega}{\sqrt{1 + \left(\frac{\pi \omega^2}{\lambda R}\right)^2}}$$

$$z = \frac{R}{\left[1 + \left(\frac{\lambda R}{\pi \omega^2}\right)^2\right]}$$

Problem 7.24   Show that Equation 7.37 is satisfied when

$$P(z) = -i \ln\left[1 + i\left(\frac{\lambda z}{\pi \omega_0^2}\right)\right]$$

7.8.3 Gaussian Beam Focusing

According to the analysis just discussed, Gaussian beams differ from geometric beams in that they may be expressed with a complex radius of curvature, defined by Equations 7.39 and 7.40. Using the Rayleigh range, Equation 7.43 becomes

$$q = z - i\,z_R \tag{7.50}$$

The thin-lens equation found in Section 4.3 (Equation 4.15) is written in terms of image and object distances s_i and s_o. Since a point source object on the optical axis is imaged to another point at s_i also on the optical axis, we may re-express Equation 4.15 in terms of wavefront curvatures. Assuming propagation from left to right, the usual sign conventions for wavefront radii assign positive values for radii of diverging waves (R_1 in Figure 7.21(a)) and negative values for wavefronts converging to a point (R_2 in Figure 7.21(a)). Thus, the thin-lens equation becomes

$$\frac{1}{R_1} - \frac{1}{R_2} = \frac{1}{f}$$

or, equivalently,

$$\frac{1}{R_2} = \frac{1}{R_1} - \frac{1}{f} \tag{7.51}$$

If truncation of the wings in the output beam can be neglected, the input and output beams have identical values of $\omega(z)$ at the lens. Adding a factor of $i\frac{\lambda}{\pi\omega^2}$ to both sides of Equation 7.51 gives

$$\frac{1}{q_2} = \frac{1}{q_1} - \frac{1}{f} \tag{7.52}$$

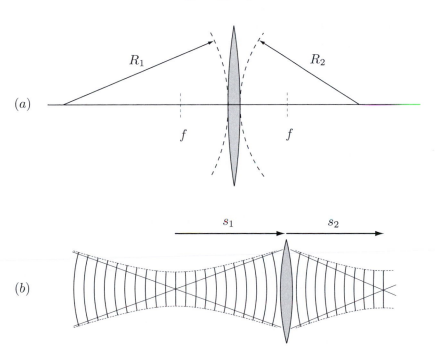

(a)

(b)

Figure 7.21 (a) A converging lens transforms the curvature of a spherical wavefront according to geometrical optics. (b) An incoming Gaussian beam with waist located at s_1 is transformed by a thin converging lens into another Gaussian beam with waist located at s_2. The sign conventions for s_1 and s_2 are identical to those for image and object distances in geometrical optics.

■ **EXAMPLE 7.4**

Find the output waist when the input waist is located one focal length behind a thin converging lens.

Solution

Using Equations 7.52 and 7.50 with $z_1 = f$ gives

$$\frac{1}{q_2} = \frac{1}{f - i z_{R1}} - \frac{1}{f} = \frac{i z_{R1}}{(f - i z_{R1}) f}$$

where z_{R1} is the Rayleigh range of the incident Gaussian beam. Solving for q_2 gives

$$q_2 = \frac{(f - i z_{R1}) f}{i z_R} = \frac{-i \left(f^2 - i f z_{R1} \right)}{z_{R1}}$$

$$= -f - i \frac{f^2}{z_{R1}}$$

$$= z_2 - i z_{R2}$$

Equating the real parts of the last result gives

$$z_2 = -f$$

Thus, an input beam with waist one focal length in front of the lens is focused to a waist one focal length behind the lens, *independent of the incident Rayleigh range*! This result may seem surprising given the corresponding case in geometrical optics where a point source one focal length behind a converging lens is transmitted as a collimated output beam. Equating the imaginary parts of the result above gives

$$z_{R2} = \frac{f^2}{z_{R1}}$$

or, equivalently

$$\omega_{01} \omega_{02} = \frac{\lambda f}{\pi} \tag{7.53}$$

Thus, a very small beam waist located f behind the lens gives an output beam with a much larger beam radius and correspondingly larger Rayleigh range.

To solve the more general case where the input waist is located at an arbitrary distance z_1 behind the lens, begin with

$$\frac{1}{q_2} = \frac{1}{z_1 - i z_{R1}} - \frac{1}{f} = \frac{f - z_1 + i z_{R1}}{f (z_1 - i z_{R1})} \tag{7.54}$$

giving

$$q_2 = \frac{f (z_1 - i z_{R1})}{-(z_1 - f) + i z_{R1}}$$

$$= \frac{f (z_1 - i z_{R1}) [-(z_1 - f) - i z_{R1}]}{(z_1 - f)^2 + z_{R1}^2} \tag{7.55}$$

$$= z_2 - i z_{R2}$$

where $z_2 < 0$ for a converging lens, since in this case the waist is formed behind the lens. Equating real and imaginary parts gives

$$z_2 = -\frac{f \left[z_1 (z_1 - f) + z_{R1}^2 \right]}{(z_1 - f)^2 + z_{R1}^2} \tag{7.56}$$

$$z_{R2} = \frac{f^2 z_{R1}}{(z_1 - f)^2 + z_{R1}^2} \qquad (7.57)$$

It is instructive to cast the result of Equation 7.56 into a form that is similar to the thin-lens equation of geometrical optics. To do so, we let define $s_1 = z_1$ but $s_2 = -z_2$ since, in geometric optics, the image distance is positive if the image is to the right of the lens as in Figure 7.54(b). Thus, by Equation 7.56,

$$\frac{1}{s_2} = \frac{(s_1 - f)^2 + z_{R1}^2}{f[s_1(s_1 - f) + z_{R1}^2]}$$

$$= \frac{1}{f}\left[\frac{s_1(s_1 - f) - f(s_1 - f) + z_{R1}^2}{s_1(s_1 - f) + z_{R1}^2}\right]$$

$$= \frac{1}{f} - \frac{1}{s_1 + \frac{z_{R1}^2}{s_1 - f}}$$

or, equivalently,

$$\frac{1}{s_1 + \frac{z_{R1}^2}{s_1 - f}} + \frac{1}{s_2} = \frac{1}{f} \qquad (7.58)$$

This result agrees with the thin-lens formula of geometric optics only as the input Rayleigh range approaches zero, as illustrated in Figure 7.22(a).

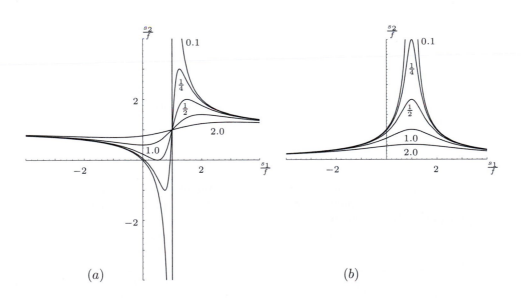

(a) $\qquad\qquad\qquad\qquad\qquad\qquad$ (b)

Figure 7.22 (a) Normalized output waist location and (b) magnification vs. input waist distance $\frac{s_1}{f}$ when the normalized input Rayleigh range is $\frac{z_{R1}}{f} = 0.1, 0.25, 0.5, 1.0,$ and 2.0.

We define the *waist magnification* as the ratio of the output to input waist value, which according to Equation 7.57 is (see Problem 7.27)

$$m = \frac{\omega_2}{\omega_1} = \frac{f}{\sqrt{(s_1 - f)^2 + z_{R1}^2}} \tag{7.59}$$

Figure 7.22 illustrates the waist magnification vs. normalized input waist location for a variety of input Rayleigh ranges. As this figure suggests, the waist magnification has a maximum when $s_1 = f$, which also gives the maximum output Rayleigh range (see Problem 7.29).

7.8.4   Matrix Methods and the ABCD Law

Transmission of Gaussian beams through arbitrary optical systems can often be described more easily using the ray transfer matrices developed in Section 4.11. In this approach, paraxial spherical rays are parameterized by column vectors, and optical elements within the optical train are represented by 2×2 matrices. Thus, for an arbitrary optical system

$$\begin{pmatrix} x_2 \\ \theta_2 \end{pmatrix} = \begin{bmatrix} A & B \\ C & D \end{bmatrix} \begin{pmatrix} x_1 \\ \theta_1 \end{pmatrix} \tag{7.60}$$

where the ray (x_1, θ_1) is the input ray, (x_2, θ_2) is the output ray, x_1 and x_2 are the transverse distances of the intersection of each ray with a reference plane, and θ_1 and θ_2 are the ray slopes relative to the optical axis. The matrix represents the combined optical system through which the input ray traverses, and is the result of matrix multiplication of the individual component matrices as described in Section 4.11. According to Equation 7.60

$$\begin{aligned} x_2 &= A\,x_1 + B\,\theta_1 \\ \theta_2 &= C\,x_1 + D\,\theta_1 \end{aligned} \tag{7.61}$$

Substituting $\theta_1 = x_1/R_1$ and $\theta_2 = x_2/R_2$ into Equation 7.60 gives (Problem 7.33)

$$R_2 = \frac{A\,R_1 + B}{C\,R_1 + D} \tag{7.62}$$

By the same argument used for Equation 7.52, we may generalize this result to spherical Gaussian beams by using the complex radius of curvature:

$$q_2 = \frac{A\,q_1 + B}{C\,q_1 + D} \tag{7.63}$$

where $q_1 = z_1 + i\,z_{R1}$ is the complex radius of curvature of the input Gaussian beam, and $q_2 = z_2 + i\,z_{R2}$ is the complex radius of curvature of the output Gaussian beam. The result is sometimes referred to as the "ABCD Law".

Problem 7.25   A Gaussian beam with a $1.00\,mm$ waist is located $10.0\,cm$ away from a thin converging lens of focal length $10.0\,cm$. If the wavelength is $633\,nm$, find the size and location of the output waist. Repeat for a wavelength of $10.6\,\mu m$.

Problem 7.26   A Gaussian beam with a $1.00\,mm$ waist is located $20.0\,cm$ away from a thin converging lens of focal length $10.0\,cm$. If the wavelength is $633\,nm$, find the size and location of the output waist. Repeat for a wavelength of $10.6\,\mu m$.

Problem 7.27   Show that the waist magnification is given by Equation 7.59.

Problem 7.28   Show that Equations 7.58 and 7.59 can be expressed in normalized form as follows:

$$\frac{s_2}{f} = 1 + \frac{\frac{s_1}{f} - 1}{\left(\frac{s_1}{f} - 1\right)^2 + \left(\frac{z_{R1}}{f}\right)^2}$$

$$m = \frac{\omega_2}{\omega_1} = \frac{1}{\sqrt{\left(\frac{s_1}{f} - 1\right)^2 + \left(\frac{z_{R1}}{f}\right)^2}}$$

Problem 7.29   (a) Show that $z_{R2} = m^2 z_{R1}$. (b) Show that the magnification is maximum when $s_1 = f$, and that this condition gives the maximum output Rayleigh range.

Problem 7.30   Show that the maximum of the normalized output waist location is given by

$$\left(\frac{s_2}{f}\right)_{max} = 1 + \frac{1}{2}\left(\frac{f}{z_{R1}}\right)$$

at

$$\frac{s_1}{f} = 1 + \frac{z_{R1}}{f}$$

Similarly, show that the minimum is given by

$$\left(\frac{s_2}{f}\right)_{min} = 1 - \frac{1}{2}\left(\frac{f}{z_{R1}}\right)$$

at

$$\frac{s_1}{f} = 1 - \frac{z_{R1}}{f}$$

Problem 7.31   Using the results of Problem 7.29, show that a thin converging lens will always form an output waist at a positive value of s_2.

Problem 7.32   Show that Equation 7.59 reduces to Equation 7.53 when the input waist is displaced from the front focal plane of a thin lens by an amount that is small in relation to the Rayleigh range.

Problem 7.33   Derive Equation 7.62.

7.9 LASER CAVITIES

A Fabry-Perot laser cavity is usually designed so that the steady-state radiation field within the cavity and in the output have Gaussian profiles. Using the results of Gaussian beam propagation just developed, we are now in a position to determine the effects of cavity design on the output beam characteristics.

7.9.1   Laser Cavity with Equal Mirror Curvatures

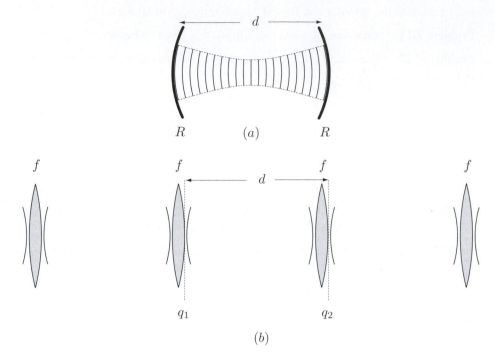

Figure 7.23   (a) A laser cavity formed by two concave reflecting surfaces of equal radius of curvature R. The mirrors are separated by distance d. (b) Equivalent sequence of lenses spaced by d. Complex beam parameters q_1 and q_2 are located just to the right of adjacent lenses.

Figure 7.23(a) shows a laser cavity formed by two converging spherical reflectors of equal radius of curvature R, separated by distance d. In order to form a *stable cavity mode*, the radiation field must reproduce itself exactly after each cycle through the cavity. Figure 7.23(b) shows a linear array of lenses that is equivalent to the cavity in (a) in that a mode that cycles back and forth with stability in (a) will propagate stably in a single direction through (b). To investigate the properties of Gaussian beam modes of such a cavity, we let q_1 and q_2 be the complex beam curvatures just to the right of adjacent lenses of Figure 7.23(b). According to Equation 7.38, translating q_1 to the position of q_2 gives $q_1 \rightarrow q_1 + d$, so by Equation 7.52

$$\frac{1}{q_2} = \frac{1}{q_1 + d} - \frac{1}{f} \tag{7.64}$$

In order to represent a stable cavity mode, $q_1 = q_2 = q$, giving

$$\frac{1}{q} = \frac{1}{q + d} - \frac{1}{f} \tag{7.65}$$

This gives the quadratic equation

$$q^2 + qd + f = 0$$

or equivalently

$$\frac{1}{q^2} + \frac{1}{fq} + \frac{1}{fd} = 0 \tag{7.66}$$

The roots of Equation 7.66 are given by

$$\frac{1}{q} = -\frac{1}{2f} \pm \frac{1}{2}\sqrt{\frac{1}{f^2} - \frac{4}{fd}}$$

or equivalently,

$$\frac{1}{q} = -\frac{1}{2f} \pm i\sqrt{\frac{1}{fd} - \frac{1}{4f^2}} \qquad (7.67)$$

This last form facilitates the identification of the beam radius and wavefront curvature by Equation 7.40. Noting that the wavefront curvature of a converging Gaussian beam is negative, we find by inspection that

$$f = \frac{R}{2} \qquad (7.68)$$

This result is significant in that it expressed the relation between focal length and curvature of a spherical reflecting surface, but utilizing the wavefront of the Gaussian beam located at the reflecting surface. Thus, *a stable cavity mode has wavefront curvature equal to the mirror curvature at each end of the resonant cavity*. Similarly, by taking the positive root of Equation 7.67 we identify

$$\frac{\lambda}{\pi\omega^2} = \sqrt{\frac{1}{fd} - \frac{1}{4f^2}}$$

giving a beam radius at each end of the cavity given by

$$\omega^2 = \frac{\frac{\lambda R}{\pi}}{\sqrt{\frac{2R}{d} - 1}} \qquad (7.69)$$

This last expression must be real, which determines the possible values of R and d that support stable cavity modes. Using the result of Problem 7.23

$$\frac{\lambda z}{\pi\omega_0^2} = \frac{\pi\omega^2}{\lambda R} \qquad (7.70)$$

where, by symmetry, $z = \frac{d}{2}$. Thus, the beam waist at the center of the cavity is given by

$$\omega_0^2 = \frac{\lambda^2 R}{\pi\omega^2}\left(\frac{d}{2}\right) = \frac{\lambda}{2\pi}\sqrt{d\left(2R - d\right)} \qquad (7.71)$$

■ **EXAMPLE 7.5**

Find waist, output beam radius, and Rayleigh range for a laser cavity that uses $10\,m$ radius of curvature mirrors separated by $1.5\,m$. Assume a wavelength of $10.6\,\mu m$.

Solution

Use Equation 7.71 to find the beam waist:

$$\omega_0^2 = \frac{\lambda}{2\pi}\sqrt{d\left(2R - d\right)}$$

$$= \frac{10.6 \times 10^{-6}m}{2\pi}\sqrt{(1.5)\,(20 - 1.5)} = 8.99 \times 10^{-6}m^2$$

giving

$$\omega_0 = 2.98\,mm$$

Use Equation 7.70 (or equivalently, Equation 7.69) to find the output beam radius:

$$\omega = \frac{\lambda \sqrt{R\frac{d}{2}}}{\pi \omega_0} = \frac{10.6 \times 10^{-6}m \sqrt{10 \left(\frac{1.5}{2}\right)}}{\pi \left(2.98 \times 10^{-3}m\right)} = 3.10\,mm$$

The output Rayleigh range is given by

$$z_R = \frac{\pi \omega_0^2}{\lambda} = \frac{\pi \left(2.98 \times 10^{-3}m\right)^2}{10.6 \times 10^{-6}m} = 2.63\,m$$

7.9.2 Laser Cavity with Unequal Mirror Curvatures

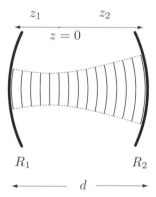

Figure 7.24   A laser cavity formed by two reflecting surfaces with radii of curvature R_1 and R_2. The mirrors are separated by distance d. The beam waist is located at $z = 0$. According to the usual sign convention, $z_1 < 0$, as are the wavefront radii of the converging wavefronts.

To find the Gaussian beam parameters for a cavity that uses mirrors with different radii of curvature, we begin with the requirement that the beam wavefronts coincide with the mirror curvatures at each end of the cavity. Figure 7.24 illustrates such a cavity formed by two converging spherical mirrors, where the left mirror has radius R_1 and the right mirror has radius R_2. According to Equation 7.46,

$$R_2 = z_2 + \frac{z_R^2}{z_2} \tag{7.72}$$

where the distance z_2 is measured relative to an origin that is located at the beam waist, as illustrated. Since the wavefront at R_2 is diverging, the radius of the Gaussian beam at this location is positive. However, the Gaussian beam radius at R_1 is converging, so it is negative. If we let the parameter R_1 represent the curvature of the mirror that, because it is concave must be positive, Equation 7.46 gives in this case

$$-R_1 = z_1 + \frac{z_R^2}{z_1} \tag{7.73}$$

Finally, since $z_1 < 0$,

$$z_2 - z_1 = d \tag{7.74}$$

Solving Equations 7.72—7.74 (see Problem 7.37) gives

$$z_2 = \frac{d(R_1 - d)}{R_1 + R_2 - 2d} \tag{7.75}$$

with

$$z_R = \sqrt{z_2(R_2 - z_2)} \tag{7.76}$$

If $0 < z_2 < d$, then the waist is formed inside the cavity. If the Rayleigh range evaluates as complex, the cavity is not stable.

Because the wavefront of the cavity beam matches the mirror curvature at both ends of the cavity, placing a flat mirror at either end gives a beam waist at the flat end, and a corresponding value of z_1 or z_2 that is zero. In this case, the output Rayleigh range must be calculated with the choice from Equations 7.72 or 7.73 that does not involve dividing by zero. In either case, the output Rayleigh range is given by (see Problem 7.35)

$$z_R = \sqrt{d(R - d)} \tag{7.77}$$

where R is the radius of curvature of the mirror that is not flat.

■ **EXAMPLE 7.6**

Find the waist size and location, beam radius at each end, and output Rayleigh range for a laser cavity that uses mirrors with radii $R_1 = 20\,m$ and $R_2 = 10\,m$. Assume a wavelength of $10.6\,\mu m$ and a mirror separation of $1.5\,m$.

Solution

Equation 7.75 gives

$$z_2 = \frac{1.5(20 - 1.5)}{30 - 2(1.5)} = 1.03\,m$$

Thus, the beam waist is within the cavity, $1.03\,m$ behind the $10\,m$ mirror. Equation 7.76 gives the Rayleigh range:

$$z_R = \frac{\pi\omega_0^2}{\lambda} = \sqrt{z_2(R_2 - z_2)} = \sqrt{(1.03)(10 - 1.03)} = 3.04\,m$$

with a beam waist equal to

$$\omega_0 = \sqrt{\frac{\lambda z_R}{\pi}} = \sqrt{\frac{(10.6 \times 10^{-6}m)(3.04\,m)}{\pi}} = 3.20\,mm$$

Equation 7.70 gives the beam radii at each mirror. At R_2,

$$\omega_2 = \frac{\lambda\sqrt{R\,z_2}}{\pi\omega_0} = \frac{(10.6 \times 10^{-6}m)\sqrt{10(1.03\,m)}}{\pi(3.20 \times 10^{-3}m)} = 3.38\,mm$$

The waist is $1.5 - 1.03 = 0.47\,m$ from mirror R_1. Equation 7.70 gives

$$\omega_1 = \frac{\lambda\sqrt{R\,|z_1|}}{\pi\omega_0} = \frac{\lambda\sqrt{R\,(d - z_2)}}{\pi\omega_0} = \frac{(10.6 \times 10^{-6}m)\sqrt{10(0.47\,m)}}{\pi(3.20 \times 10^{-3}m)} = 2.28\,mm$$

7.9.3 Stable Resonators

The approach of the previous section can be used to obtain formulas for the cavity waist, Rayleigh range, and waist at the two cavity mirrors. Before doing so, it is customary to define the following two dimensionless parameters, sometimes referred to as the *g-parameters*:

$$g_1 = 1 - \frac{d}{R_1}$$
$$g_2 = 1 - \frac{d}{R_2} \tag{7.78}$$

Using these parameters, we re-express Equation 7.75 as

$$z_2 = d \left[\frac{g_1 (1 - g_2)}{g_1 + g_2 - 2g_1 g_2} \right] \tag{7.79}$$

Similarly, Equation 7.76 gives for the Rayleigh range

$$z_R = d \sqrt{ \frac{g_1 g_2 (1 - g_1 g_2)}{(g_1 + g_2 - 2g_1 g_2)^2} } \tag{7.80}$$

with a corresponding beam waist

$$\omega_0 = \sqrt{\frac{\lambda d}{\pi}} \left[\frac{g_1 g_2 (1 - g_1 g_2)}{(g_1 + g_2 - 2g_1 g_2)^2} \right]^{\frac{1}{4}} \tag{7.81}$$

The beam radii at each end of the cavity are given by

$$\omega_2 = \omega (z_2) = \sqrt{\frac{\lambda d}{\pi}} \left[\frac{g_1}{g_2 (1 - g_1 g_2)} \right]^{\frac{1}{4}} \tag{7.82}$$

$$\omega_1 = \omega (z_1) = \sqrt{\frac{g_2}{g_1}} \, \omega_2 \tag{7.83}$$

Equations 7.80 and 7.81 give real values for the Rayleigh range and beam waist only if

$$0 \le g_1 g_2 \le 1 \tag{7.84}$$

Laser cavities with g-values that satisfy Equation 7.84 are said to have *stable resonators*. The condition of Equation 7.84 can be visualized with a *stability diagram*, as illustrated in Figure 7.25. Regions of stability are bounded by hyperbolas defined by $g_2 = \pm \frac{1}{g_1}$, as shaded in the diagram. Lasers built with stable resonator cavities are more immune to misalignments and optical imperfections.

■ **EXAMPLE 7.7**

Find the g-values for the cavity of Example 7.6 and verify that it is stable. Use these to verify the beam parameters using Equations 7.79—7.81.

Solution

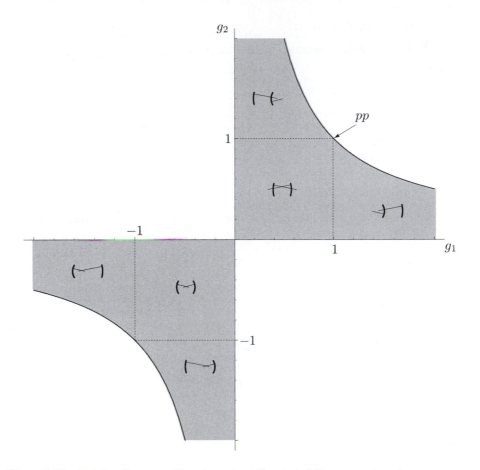

Figure 7.25   Stability diagram, with regions of stability shaded. Representative mirror arrangements in various stability regions are indicated. The special case of plane parallel mirrors (*pp*) lies on the boundary of stability and thus requires perfect mirror alignment.

Using $R_1 = 20\,m$, $R_2 = 10\,m$, $\lambda = 10.6\,\mu m$ and $d = 1.5\,m$, we find

$$g_1 = 1 - \frac{d}{R_1} = 1 - \frac{1.5}{20} = 0.925$$

$$g_2 = 1 - \frac{d}{R_2} = 1 - \frac{1.5}{10} = 0.85$$

with $g_1 g_2 = 0.786$. Thus, according to Equation 7.84, the cavity is stable. Using the above g-values with Equations 7.79—7.81 gives

$$z_2 = d\left[\frac{g_1\,(1-g_2)}{g_1 + g_2 - 2g_1 g_2}\right] = (1.5\,m)\left[\frac{(0.925)\,(1-0.850)}{0.925 + 0.850 - 2\,(0.786)}\right] = 1.03\,m$$

$$z_R = d\sqrt{\frac{g_1 g_2\,(1 - g_1 g_2)}{(g_1 + g_2 - 2g_1 g_2)^2}} = (1.5\,m)\sqrt{\frac{(0.786)\,(1 - 0.786)}{(0.203)^2}} = 3.04\,m$$

$$\omega_0 = \sqrt{\frac{\lambda d}{\pi} \left[\frac{g_1 g_2 \left(1 - g_1 g_2 \right)}{\left(g_1 + g_2 - 2 g_1 g_2 \right)^2} \right]^{\frac{1}{4}}} = \sqrt{\frac{\left(10.6 \times 10^{-6} m \right) \left(1.5\, m \right)}{\pi}} \left[4.99 \right]^{\frac{1}{4}} = 3.20\, mm$$

as before.

The *half-symmetric resonator* of Figure 7.26 is a very common design. Systems that utilize blazed diffraction gratings are usually built this way, since gratings ruled on curved surfaces are much more expensive than those ruled on plane substrates.

■ **EXAMPLE 7.8**

Find the beam parameters for the half-symmetric resonator. Locate such designs on the stability diagram.

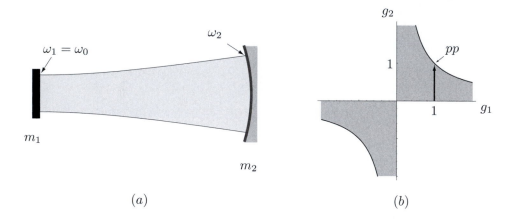

Figure 7.26   (a) A half-symmetric resonator. (b) Stability diagram.

Solution

In Figure 7.26(a), mirror M_1 is flat so $R_1 = \infty$ and $g_1 = 1$. Using Equations 7.79-7.82, we find

$$z_2 = d \left[\frac{g_1 \left(1 - g_2 \right)}{g_1 + g_2 - 2 g_1 g_2} \right] = d \left[\frac{\left(1 - g_2 \right)}{1 + g_2 - 2 g_2} \right] = d$$

In other words, the waist is located at M_1, since this is where the curvatures match. Continuing, we find

$$z_R = d \sqrt{\frac{g_2 \left(1 - g_2 \right)}{\left(1 - g_2 \right)^2}} = d \sqrt{\frac{g_2}{\left(1 - g_2 \right)}}$$

$$\omega_1 = \omega_0 = \sqrt{\frac{\lambda d}{\pi} \left[\frac{g_2 \left(1 - g_2 \right)}{\left(1 - g_2 \right)^2} \right]^{\frac{1}{4}}} = \sqrt{\frac{\lambda d}{\pi} \left[\frac{g_2}{\left(1 - g_2 \right)} \right]^{\frac{1}{4}}}$$

$$w_2 = \sqrt{\frac{\lambda d}{\pi} \left[\frac{1}{g_2 (1 - g_2)}\right]^{\frac{1}{4}}}$$

Stable designs lie along the line $g_1 = 1$ within the shaded region of Figure 7.26. As $g_2 \to 1$, the design approaches the plane parallel case.

It is usually the case that one or more of the mirrors is partially transmitting to function as an output coupler. Typically, this is achieved using dielectric coatings on substrates with the appropriate curvature. Of course, the substrate and coating materials must be transparent at the laser frequency. If the output coupler has curved surfaces, it provides refraction, which can affect the beam parameters of the laser output.

■ **EXAMPLE 7.9**

Use the design of Example 7.8 with $R_2 = 5.00\,m$, $d = 1.5\,m$, and $\lambda = 10.6\,\mu m$. Assume an output coupler made of zinc-selenide, with index of refraction 2.42 at the operating wavelength. Assume that the output coupler has one flat side that faces away from the cavity.

Solution

The zinc-selenide substrate acts like a diverging lens. Using the thin-lens formula of Equation 4.25, we find

$$\frac{1}{f} = (n - 1)\left[\frac{1}{R_1} - \frac{1}{R_2}\right] = (2.42 - 1)\left[\frac{1}{-5.00} - 0\right] = -0.284$$

Thus, the output coupler acts as a diverging lens with $f = -3.52\,m$. We use the results of Example 7.8 to find the beam parameters within the cavity:

$$g_2 = 1 - \frac{1.50\,m}{5.00\,m} = 0.7$$

$$z_R = d\sqrt{\frac{g_2}{(1 - g_2)}} = (1.50\,m)\sqrt{\frac{0.7}{0.3}} = 2.29\,m$$

$$\omega_1 = \omega_0 = \sqrt{\frac{\lambda d}{\pi}\left[\frac{g_2}{(1 - g_2)}\right]^{\frac{1}{4}}} = \sqrt{\frac{(10.6 \times 10^{-6}m)(1.50\,m)}{\pi}}\sqrt[4]{1.53} = 2.78\,mm$$

Use Equation 7.58 to find the output waist location:

$$\frac{1}{s_2} = \frac{1}{f} - \frac{1}{s_1 + \frac{z_{R1}^2}{s_1 - f}} = (-0.284) - \frac{1}{1.50\,m + \frac{(2.29\,m)^2}{1.50\,m + 3.52\,m}} = -0.677$$

or, $s_2 = -1.48\,m$, about $2\,cm$ in front of mirror M_1. To find the output waist corrected for the effect of the diverging output coupler, use Equation 7.59:

$$m = \frac{f}{\sqrt{(s_1 - f)^2 + z_{R1}^2}} = \frac{(-3.52)}{\sqrt{(1.50 + 3.52)^2 + (2.29)^2}} = -0.638$$

Ignoring the minus sign, this gives corrected output waist of $\omega_0 = (0.638)(2.78\,mm) = 1.77\,mm$ and a corrected Rayleigh range of only $z_R = \frac{\pi\omega_0^2}{\lambda} = 0.932\,m$. Thus, the diverging effect of the output coupler was substantial in this case.

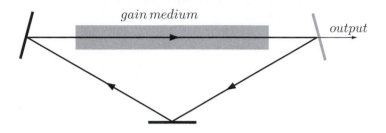

Figure 7.27   Traveling wave resonator.

7.9.4 Traveling Wave Resonators

In a *traveling wave laser*, sometimes also referred to as a *ring laser*, a beam circulates through a gain medium while traveling in the same direction, as illustrated in Figure 7.27. Of course, the beam that traverses the gain medium in the opposite direction is also equally probable, and in general both beams are present. in the *ring laser gyroscope*, both beams are aligned and combined on a detector for coherent detection (see Section 5.3.6.1). If the ring cavity rotates about an axis out of the page of the figure, the mirrors move in the time taken for a complete photon traversal, and thus one component will have a slightly longer round-trip path, resulting in a frequency difference that leads to a beat signal with frequency proportional to the rate of rotation. The ring laser gyroscope is very accurate and has no moving parts, which is an advantage over mechanical gyroscopes.

7.9.5 Unstable Resonators

If the condition of Equation 7.84 is not satisfied, then the cavity is unstable, and at least one of the end mirrors is *overfilled*. Although lossy, such a design can offer advantages in very high gain laser systems, where regions of high intensity can damage optics. Figure 7.28 illustrates a *confocal unstable resonator* design that consists of a converging mirror at the position of M_1, and a smaller diameter diverging mirror for M_2, which is mounted on a transmitting window. Since the mirrors are confocal, spherical waves reflected by M_2 are collimated by M_1, illuminating the annular opening around M_2 with plane waves, as illustrated. The output beam is decidedly non-Gaussian. In the *near field*, the output is determined by the Fresnel diffraction pattern of an annular opening (see Sections 6.4.1.2 and 6.4.3, Figure 6.30), complete with the *spot of Arago*. In the *far field*, the beam profile is that of the Fraunhofer diffraction pattern of an annular aperture (Problem 6.41). For more information on laser design with unstable resonators, see Siegman [19].

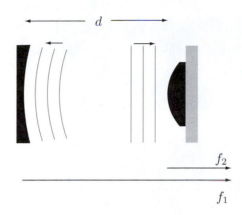

Figure 7.28 Confocal unstable resonator.

7.9.6 Transverse Cavity Modes

In a Fabry-Perot resonator, the cavity radiation consists of a collection of photons that bounce back and forth between the cavity mirrors. The radiation field in the cavity and in the output reaches a steady state that can only be achieved if the cavity field is *cyclic*. If we visualize the cavity field as a beam that propagates back and forth, then this is equivalent to saying that the beam must reproduce itself exactly at a given point in each round trip. In more precise terms, possible output beams are *eigenmodes* of the cavity. The details are beyond the scope of our discussion, but may be found in references such as Siegman [19].

The possible eigenmodes of a cavity are commonly referred to as *transverse cavity modes* (or, for simplicity, just transverse modes) to distinguish them from the *longitudinal modes* discussed in Section 7.5. Recall that longitudinal modes have frequencies that satisfy a standing wave condition and that fall within the gain limit of the laser medium. Similarly, each transverse mode has a unique frequency. Figure 7.29 illustrates a few of the lower order transverse modes. Figure 7.29(a) illustrates transverse modes with Cartesian symmetry. Oddly, these are more common, since diffraction gratings and Brewster windows tend to break the axial symmetry of the cavity. Figure 7.29(b) illustrates transverse modes with circular symmetry. Each plot of Figure 7.29 represents a possible profile for an output beam generated by a laser with a Fabry-Perot cavity.

Transverse modes are typically denoted with the acronym TEM (Transverse Electro-Magnetic) subscripted with two integers. The lowest and simplest mode is TEM_{00}, which for both circular and Cartesian symmetry is the Gaussian beam discussed previously, and is typically called the *fundamental mode*. The higher modes have more complicated profiles and in general are not desirable. Transverse modes can be difficult to eliminate since they tend to occupy a greater volume of the gain medium, and thus sometimes have more gain than the fundamental Gaussian mode. Figure 7.30 illustrates a cavity with a *mode scraper*. A circular aperture placed at the calculated beam waist of the fundamental mode will spoil the gain of higher order transverse modes.

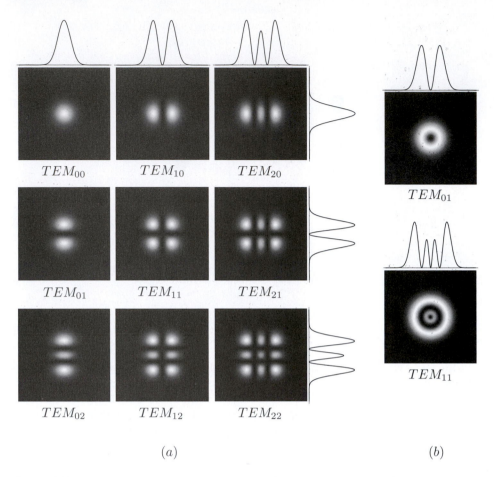

Figure 7.29 Transverse laser modes. (a) Cartesian symmetry. Each profile is generated by multiplication of the line profiles shown, each a function of an independent Cartesian coordinate. (b) Circular symmetry. Each profile is generated by rotating the line profile shown about the axis of symmetry.

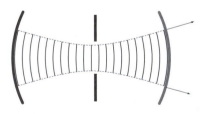

Figure 7.30 A mode scraper placed at the calculated beam waist of the fundamental cavity mode prevents higher order transverse modes from lasing.

Problem 7.34 Find the output beam radius and Rayleigh range for a laser cavity designed for helium-neon at $633\,nm$ that uses two $2.00\,m$ radius of curvature mirrors separated by $25.0\,cm$. You may ignore refraction at the output coupler.

Problem 7.35   Derive Equation 7.77 for the output Rayleigh range for a cavity with a flat mirror on one end.

Problem 7.36   Find the output beam radius and Rayleigh range for a laser cavity designed for helium-neon at $633\,nm$ that uses a $2.00\,m$ radius of curvature mirror for R_1 and a flat mirror at the output. Use a separation of $25.0\,cm$, and do not ignore refraction within the output coupler.

Problem 7.37   Derive Equation 7.75

Problem 7.38   Verify Equations 7.79—7.82.

Problem 7.39   Find the stability parameters g_1 and g_2 for a laser cavity designed for helium-neon at $633\,nm$ that uses two $2.00\,m$ radius of curvature mirrors separated by $25.0\,cm$. Locate this design on the stability diagram, and state whether or not the design is stable.

Problem 7.40   A certain laser cavity has stability parameters given by $g_1 = 0.900$ and $g_2 = 0.800$. Determine the beam waist size and location, and the Rayleigh range of the output beam for this design. Assume an operating wavelength of $633\,nm$.

Problem 7.41   Find g_1, g_2, and $g_1 g_2$ for the confocal unstable resonator of Figure 7.28 if M_1 has a focal length of $+3.00\,m$, M_2 has a focal length of $-1.00\,m$, and the cavity length is $2.00\,m$. Locate this design on the stability diagram of Figure 7.25.

Problem 7.42   In a *symmetric resonator*, $R_1 = R_2 = R$. Show that in this case, $g_1 = g_2 = g$, and that

$$\omega_0 = \sqrt{\frac{\lambda d}{\pi}} \left[\frac{1+g}{4\,(1-g)} \right]^{\frac{1}{4}}$$

with

$$\omega_1 = \omega_2 = \sqrt{\frac{\lambda d}{\pi}} \left[\frac{1}{1-g^2} \right]$$

Where do such systems lie within the stability diagram of Figure 7.25?

Problem 7.43   In a *symmetric confocal resonator*, $R_1 = R_2 = d$. Find the beam parameters, and locate this design on the stability diagram. This is in fact a very stable design.

Problem 7.44   Find the cavity waist size and location, beam radius at each end of the cavity, and output waist size, location, and Rayleigh range for a $HeNe$ laser cavity designed for $633\,nm$ that uses mirrors with $R_1 = R_2 = 3.00\,m$ separated by $25\,cm$. Assume that the output coupler has an index of refraction of 1.55 and has one flat side that faces away from the cavity.

Problem 7.45   A $10.6\,\mu m$ $1.5\,m$-long half-symmetric resonator uses a $10.0\,m$ radius of curvature output coupler made of $ZnSe$ with index of refraction 2.42 and one flat side that faces away from the cavity. Locate the design on the stability diagram, and find the size and location of the output beam waist, the beam radius as it exits the output coupler, and the output Rayleigh range.

Additional Problems

Problem 7.46   Derive all details of the Bohr model of the hydrogen-like atoms with nuclear charge Z. Begin with a careful articulation of all assumptions, and include expressions for the quantized orbits.

Problem 7.47   Find the first six wavelengths of the *Lyman series* of hydrogen that have the $n = 1$ state as the final state.

Problem 7.48   Diatomic molecules composed of different atoms of mass m_1 and m_2 oscillate about the center of mass with a natural frequency given by

$$f_0 = \frac{1}{2\pi}\sqrt{\frac{k}{\mu}}$$

where k is the spring constant, and μ is the *reduced mass*:

$$\mu = \frac{m_1 m_2}{m_1 + m_2}$$

Calculate the effective spring constant for the NO molecule if the vibrational frequency is $5.63 \times 10^{13}\, Hz$.

Problem 7.49   Find the average thermal energy in electron-volts of gas molecules in thermal equilibrium at $1000\, K$.

Problem 7.50   A *beam expander* consists of two lenses of focal lengths f_1 and f_2 separated by a positive distance d, with $d = f_1 + f_2$. If the input waist is located a distance f_1 to the left of the first lens, show that the output waist is expanded by an amount equal to the ratio of the focal lengths and is located a distance f_2 to the right of the second lens. Assume that both lenses are converging, and that $f_2 > f_1$. Comment on the appropriate lens diameters for f_1 and f_2.

Problem 7.51   A Gaussian beam with a $0.14\, mm$ waist located $20.0\, cm$ away from a thin converging lens of focal length $10.0\, cm$. If the wavelength is $633\, nm$, find the size and location of the output waist, and the far field beam divergence of the output beam.

Problem 7.52   According to the matrix formulation of geometric optics (Section 4.11), the ray transfer matrix for a thin lens is given by

$$\begin{bmatrix} A & B \\ C & D \end{bmatrix} = \begin{bmatrix} 1 & 0 \\ -\frac{1}{f} & 1 \end{bmatrix}$$

Use this matrix to verify the results of Example 7.4 using the matrix method outlined in Section 7.8.4.

Problem 7.53   In a *symmetric near-planar resonator*, $R_1 = R_2 = R$ with $R \gg L$. Show that in this case,

$$\omega_0 \approx \omega_1 \approx \omega_2 = \sqrt{\frac{\lambda d}{\pi}}\left[\frac{R}{2d}\right]^{\frac{1}{4}}$$

Thus, in this design, the resonator length is a very small fraction of the Rayleigh range. Locate such designs on the stability diagram. Comment on the challenges of mirror alignment and mirror fabrication.

CHAPTER 8

OPTICAL IMAGING

"For the limits to which our thoughts are confined, are small in respect of the vast extent of Nature itself; some parts of it are *too large* to be comprehended and some *too little* to be perceived. And from thence it must follow, that not having a full sensation of the Object, we must be very lame and imperfect in our conceptions about it, and in all the propositions which we build upon it; hence we often take the *shadow* of things for the substance, *small* appearances for *good* similitudes, *similitudes* for *definitions*; and even many of those which we think to be the most solid definitions, are rather expressions of our own misguided apprehensions than of the true nature of the things themselves."

—Robert Hooke, in *Micrographica*, 1665

Contents

8.1   INTRODUCTION

In this chapter, we integrate our previous discussions of geometrical optics and diffraction to describe the process of image formation in more detail. We begin with the Abbe theory of coherent image formation in a microscope, idealized as two stages of Fraunhofer diffraction. The image process is extended to include cameras and telescopes via definition of the point spread function. Resolving power is defined and discussed, and applied to the recording of images on both analog and digital media. A discussion of spatial filters provides an introduction to techniques of image processing. The chapter concludes with an introduction to adaptive optics techniques for removing image artifacts caused by atmospheric fluctuations.

8.2   ABBE THEORY OF IMAGE FORMATION

In the mid 1800s, Carl Zeiss founded a microscope business, and in 1866, hired Ernst Abbe[1] to study the process of image formation in order to improve the trial and error approaches of the day with firm scientific principles. Through this effort, Abbe developed a theory of *coherent image formation* that ultimately led to the modern field of *coherent optical signal processing*.

Abbe realized that the formation of a microscope image involves a *double diffraction process*. Figure 8.1 illustrates the formation of an image of a slit aperture by a single objective lens. Light passing through the slit diffracts as Huygens wavelets to form a Fraunhofer diffraction pattern at the back focal plane of the objective, as discussed in Section 6.3. Conceptually, we may resolve the light diffracted from the slit into components each with an angular direction that illuminates specific diffraction orders in a distant Fraunhofer diffraction pattern. Since the Fraunhofer diffraction pattern is infinitely far away, these angular components must be *plane parallel*, as illustrated. The objective lens focuses these plane wave components to form the Fraunhofer diffraction pattern at the back focal plane of the lens, which we referred to in Chapter 6 as the *Fraunhofer plane*. The evolution of waves that travel from the object to the Fraunhofer plane is the first diffraction process in the formation of the image. As discussed in Section 6.3.9, this evolution is described mathematically by the Fraunhofer diffraction integral of Equation 6.11:

$$E_p = C \int_{\infty} g(Y) e^{ikY \sin \theta} dY \tag{8.1}$$

where C is a constant, Y measures a point on the object (aperture) $g(Y)$ is the *aperture function*, k is the wavenumber of the monochromatic illumination, and θ locates the observation point p on the Fraunhofer plane. For the single slit of Figure 8.1, $g(Y)$ is unity within the aperture, and zero outside of the aperture opening. The Fraunhofer diffraction pattern for the single slit was found in Section 6.3.1:

$$E_p = E_0 \left(\frac{\sin \beta}{\beta} \right) \tag{8.2}$$

where E_0 is the electric field at $\theta = 0$ and $\beta = \frac{1}{2} kb \sin \theta$ with slit width b.

---

[1] Ernst Abbe: 1840-1905. German physicist, entrepreneur, and social reformer.

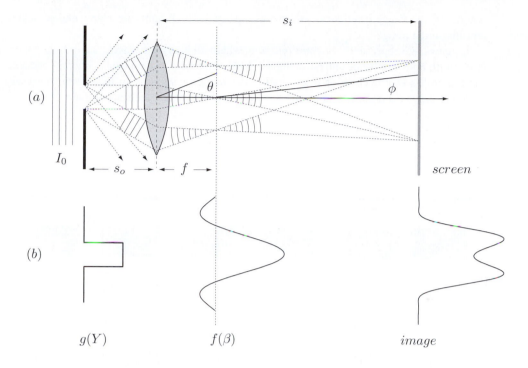

I_0

s_o f screen

$g(Y)$ $f(\beta)$ image

Figure 8.1   Abbe image formation as a two-stage diffraction process. (a) Light diffracting from a single-slit object is partially collected by an objective lens to form a truncated diffraction pattern at the Fraunhofer plane. Diffraction orders passed by the objective recombine at the image plane to form the image. (b) The aperture function $g(Y)$, the filter function $f(\beta)$, and the final image. In the case illustrated, only the first and zero diffraction orders are passed by the lens.

Because the objective lens has a finite size, Equation 8.2 does not represent the exact field distribution at the Fraunhofer plane. Diffraction structure at angles that exceed the input cone of the objective lens is not focused; this gives a diffraction pattern that is modified according to the *spatial filter function* $f(\beta)$. When the object is a single slit, this modified diffraction pattern is given by

$$f(\beta) = \begin{cases} \frac{\sin \beta}{\beta} & \beta \leq \beta_{\max} \\ 0 & \beta > \beta_{\max} \end{cases}$$

where β_{max} is determined by the maximum angle accepted by the objective lens.

The second diffraction process begins at the Fraunhofer plane. Each point on the Fraunhofer diffraction pattern represents a point of focus that re-expands as a source of Huygens wavelets, overlapping at the image plane, as illustrated in Figure 8.1. The angular distribution of propagation directions at the Fraunhofer plane is precisely that which achieves optimal overlap at the image plane, as the ray diagrams of Figure 8.1(a) illustrate. The second diffraction process includes the spatial filter function:

$$I = I_0 \left| \int_{\infty} f(\beta) \, e^{iky \sin \phi} d\beta \right|^2 \tag{8.3}$$

where I is the image irradiance, y measures a point on the Fraunhofer plane, and ϕ locates a point on the image.

Equation 8.3 represents *coherent imaging*, where all Huygens wavelets combine coherently, with both amplitude and phase intact. Mathematically, the image is coherent because the integral is evaluated prior to evaluating the modulus-square. In *incoherent imaging*, Huygens wavelets do not combine with complete coherence at the image plane.

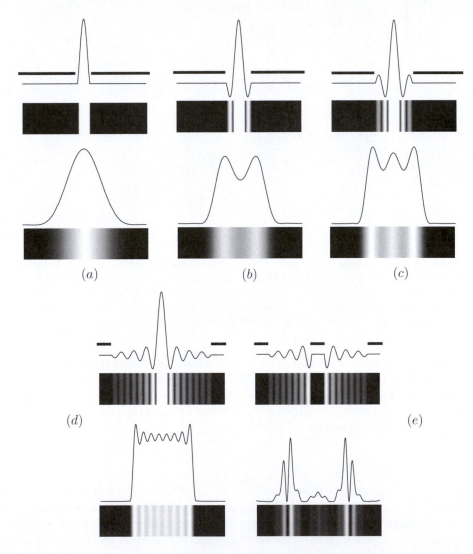

Figure 8.2   Images formed by different spatial filter functions for the case of a single-slit object. In each set, the top image is that of the filtered diffracted *field* in the Fraunhofer plane, followed by a pattern plot of the corresponding irradiance. Next, there is an irradiance plot of the corresponding image, along with a pattern plot. (a) Only the central diffraction order is passed, giving an image with no edge detail. (b) The zeroth and first orders are passed, giving an image that begins to show edge detail. Passing more diffraction orders increases the edge detail, as shown in (c) through 3rd order and (d) through order 8. (e) The central order is blocked, and nonzero orders through 8 are passed, giving an image that consists only of edge detail.

Figure 8.2 shows the result of Equation 8.3 for several examples of spatial filter function $f(\beta)$ when the object is a single slit. Each figure grouping shows a profile and fringe pattern in the Fraunhofer plane along with the corresponding image profile and pattern in the image plane. In Figure 8.2(a), only the central diffraction order is passed by the objective. In this case, $f(\beta)$ is very nearly Gaussian. Each diffraction process can be approximated as a Fourier transform, and as we found in Example 5.6, the Fourier transform of a Gaussian is another Gaussian with inverse spread. Thus, the image also has a smooth profile that lacks any edge detail. In Figure 8.2(b), the first order diffraction orders are included by $f(\beta)$, and the image begins to reveal some edge structure. Figure 8.2(c) includes the second order diffraction orders, and in Figure 8.2(d), eight orders of diffraction are passed by the objective lens, producing an image profile and pattern that closely approximates the sharp edges of the aperture function $g(Y)$. In Figure 8.2(e), the central diffraction order is blocked, giving an image profile that consists only of the edge detail. Clearly, the edge detail is contained in the nonzero diffraction orders.

In his original analysis, Abbe represented specimen detail with diffraction gratings of varying periodicity. Figure 8.3 illustrates images formed of an object that consists of five parallel slits with slit separation d equal to twice the slit width b (Figure 8.3(a)). The diffraction pattern formed by such an aperture was found in Equation 6.35:

$$E_p = E_0 \frac{\sin \beta}{\beta} \left[\frac{\sin N\gamma}{N \sin \gamma} \right] \tag{8.4}$$

where $\beta = \frac{1}{2}kb \sin\theta$, $\gamma = \frac{1}{2}kd \sin\theta$, and N is the number of slits, in this case equal to 5. This pattern consists of principal maxima determined by the slit separation d, modulated by a $\sin(\beta)/\beta$ envelope that is determined by the slit width b. In the grating illustrated $d = 2b$, so all even nonzero orders are missing as discussed in Section 6.3.7. Figure 8.3(b) illustrates the image formed when the spatial filter function passes only the central envelope order; in this case, the image shows the grating periodicity, but there is no edge detail on any of the slit images (compare Figure 8.3(a) with Figure 8.2(a)). Notice that the spatial filter function of Figure 8.3(b) passes both zeroth- and first-order principal maxima. If only the central principle maximum were passed, then the image would consist of a broad, essentially Gaussian profile that represents almost constant illumination across the entire image field. In other words, unless the first-order principal maxima are passed by the objective lens, the grating periodicity is not resolved. As Figures 8.3(c)—(e) illustrate, including more diffraction orders increases the edge detail of the image.

Abbe used the spatial filter function of Figure 8.3(f) to estimate the absolute resolution limit. In this case, the spatial filter function truncates the diffraction pattern at the center of the first-order principal maxima:

$$d \sin\theta = \lambda = \frac{\lambda_0}{n}$$

where λ_0 is the vacuum wavelength, and n is the medium that surrounds the object. Rearranging gives

$$\frac{n \sin\theta}{\lambda_0} = \frac{1}{d} \tag{8.5}$$

where the quantity $\frac{1}{d}$ is the *spatial frequency* of the periodic grating. In this expression, θ is one-half of the acceptance angle for the objective lens. Abbe called the quantity $n \sin\theta$ the *numerical aperture* of the objective lens, understanding the central role of this quantity in determining the resolution limit of the microscope. The higher the spatial frequency, the

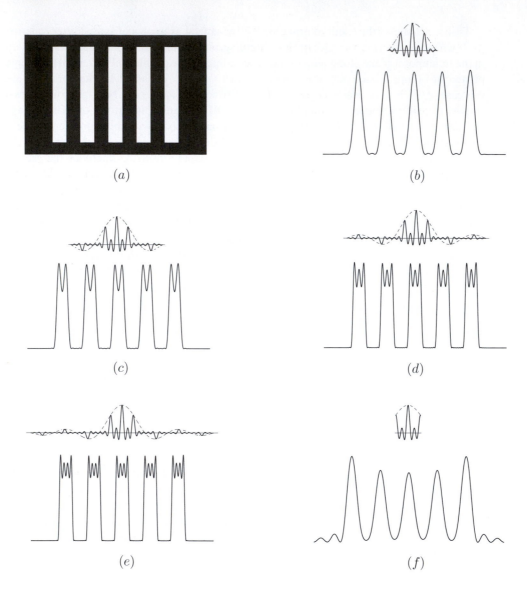

Figure 8.3 (a) Abbe grating object. (b) Only the central envelope orders are collected. (c)—(e) Progressively more envelope orders are collected. (f) Abbe's resolution limit.

larger the angular deviation for nonzero diffraction orders away from the optical axis. In order to resolve object detail with high spatial frequency, it is necessary to use an objective lens with a large acceptance angle, and a correspondingly large numerical aperture.

■ **EXAMPLE 8.1**

Estimate the maximum spatial frequency that is resolvable by an oil immersion objective lens (see Section 4.28 and Figure 4.29) which has an acceptance cone of 134° and uses an immersion oil with index of refraction 1.515. Assume green light of wavelength $550\,nm$. Estimate the smallest object that is just resolvable with this microscope.

Solution

The numerical aperture of this lens is

$$NA = n\sin\theta = (1.515)\sin(67°) = 1.39$$

The maximum resolvable spatial frequency is

$$\frac{1}{d} = \frac{NA}{\lambda_0} = \frac{1.39}{550\times 10^{-9}m} = 2.54\times 10^6\frac{lines}{m} = 2,540\frac{lines}{mm}$$

The minimum resolvable object size δ is *half* the spatial period:

$$\delta = \frac{d}{2} = \frac{1}{2\,(2540)}mm = 394\,nm$$

In the last example, notice that the resolvable structure has a periodicity that is on the order of the wavelength of light used to form the image. Edge detail would begin to appear on objects with about one third of this spatial frequency (see Figure 8.3(c)), or three times the minimum resolvable object size.

8.2.1 Phase Contrast Microscope

Biological specimens are often transparent and colorless, producing only a *phase shift* in the object plane with very little change in irradiance. The image of such a specimen shows a similar phase variation, also with almost constant irradiance. Such an image is invisible because there is no *contrast*. Zernike[2] developed *phase contrast microscopy* as a means of introducing contrast into such an image, thereby making it visible.

When discussing the scalar diffraction theory of Fresnel and Kirchhoff in Section 6.2, we noted the factor of i in Equation 6.5. This indicates a 90° phase shift in the diffracted field relative to the specimen illumination. If there are no further *relative* phase shifts in the diffracted or undiffracted waves, these two sets of waves cannot interfere at the image plane. This can be seen most clearly by noting that when complex numbers are added, the imaginary and real parts combine independently, or alternatively, that waves with a 90° relative phase shift add as scalars without interference. Zernike realized the need for a phase shift in either the illumination or diffracted fields to allow these fields to interfere, producing contrast that reveals the structure of the transparent specimen. A microscope engineered to provide such a phase shift is referred to as a *phase contrast microscope*.

A typical design is illustrated in Figure 8.4. The specimen illumination is applied via an *illumination annulus*. A focusing lens concentrates this illumination into a hollow cone with focus at the position of the specimen. Undiffracted light expands into another hollow cone, and the light diffracted from the specimen expands toward the Fraunhofer field. Both beams are collected by the second lens of Figure 8.4, and since they are for the most

---

[2]Fritz Zernike: 1888-1966. Dutch physicist who received the Nobel Prize in 1953 for the invention of the phase contrast microscope.

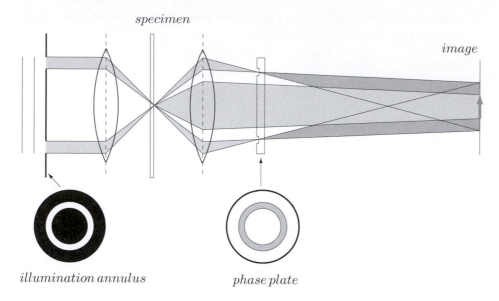

Figure 8.4   Schematic of a phase contrast microscope.

part spatially distinct, they may be manipulated separately. A *phase plate* has an annular region that has been etched to decrease the plate thickness to give a phase shift for the illumination cone of about 90°. When the phase shifted illumination combines with the diffracted field, interference occurs that causes large irradiance variations corresponding with the specimen structure, and the resulting sharp contrast causes the specimen to now be visible. In recognition of the many biological advances enabled by this innovation, Zernike received the Nobel Prize in 1953.

In *dark field microscopy*, the undiffracted illumination is blocked altogether, so that only the diffracted beam reaches the image plane. In such images, small highly diffracting objects appear as bright structure against dark background.

Problem 8.1   Image plane profiles such as those of Figures 8.1 and 8.2 can be generated easily using symbolic programming tools. Use the tool of your choice to plot a spatially filtered image that only uses the central order in the Fraunhofer plane:

$$I(x) = \left| \int_{-1}^{1} \left(\frac{\sin(\pi y)}{\pi y} \right) e^{i\pi yx} dy \right|^2$$

with n an integer. For example, Mathematica evaluates this integral directly, so there is no need for numerical analysis.

Problem 8.2   Using the approach of the preceding problem, generate the image profile of Figure 8.2(e) using two integrals, each over limits appropriate for the spatial filter function illustrated.

Problem 8.3   Find the maximum spatial frequency resolvable with a microscope objective that has a numerical aperture of 0.65. Assume $550\,nm$ illumination, and estimate the minimum resolvable object size.

Problem 8.4   Using Figure 8.4 as a guide, draw a design for a dark field microscope.

8.3   THE POINT SPREAD FUNCTION

In principle, the Abbe theory of image formation is also applicable to telescopes and cameras. For the case of a telescope, light from the disk of a distant star diffracts into the Airy pattern for a circular aperture as discussed in Section 6.3.3; even though the star may be quite large, astronomical distances are large enough so that the first diffraction process in the Abbe model has already occurred before the light enters the telescope aperture. Since the object distance is large, the image is formed at the Fraunhofer plane; thus only the second diffraction process actually occurs within the telescope. Furthermore, the Airy pattern for a distant star is much larger than the input aperture of any telescope, so no edge information is revealed in the image. As discussed in Section 6.3.4, a star is essentially a point source that illuminates the telescope with spatially coherent and essentially plane wave light that diffracts to the Airy pattern of the *telescope aperture* at the Fraunhofer plane, as illustrated in Figure 8.5. This diffraction pattern is referred to as the *point spread function*.

The same idea can be applied to any optical imaging system. For microscopes and cameras, rays from a point source located at a finite object distance are reassembled at the conjugate point in the image plane. As with the telescope, the input aperture of the imaging system produces diffraction that causes the image of the source point to become a point spread function rather than a distinct conjugate point. If the object is taken as an infinite collection of unresolvable source points, the image becomes that of a superposition of the corresponding conjugate point spread functions.

8.3.1   Coherent vs. Incoherent Images

In the Abbe model, it is assumed that the image is illuminated *coherently*. The filtered Fraunhofer diffraction patterns of the slit aperture in Figure 8.2 are only obtained in practice if the aperture is illuminated with light that is spatially coherent over the entire aperture opening. The Fourier integrals that represent the two diffraction processes are relevant only if the Fourier components of the integrand combine with the appropriate phase.

As discussed in the previous section, we may also represent an image as the superposition of point spread functions centered on the conjugate points of the object. Figure 8.6(a) illustrates this model for an object that consists of three parallel-slit apertures. In this figure, it is assumed that the object is illuminated with spatially coherent light. Each point on the object contributes a mutually coherent point spread function in the image; in the figure, only three of these are shown to emphasize the point that they combine and *interfere* at the image plane. The top figure in Figure 8.6(a) shows the result of replacing each point on the object with the corresponding coherent point spread function; the resulting image should be compared with that of the Abbe model illustrated in Figures 8.2(b) and 8.3(c).

Mathematically, the process of replacing points on the object with point spread functions on the image is referred to as a *convolution*. Assuming one dimension for simplicity, we

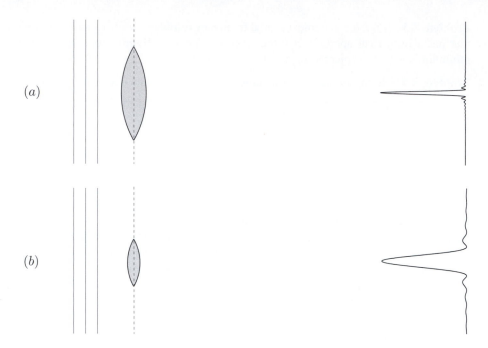

Figure 8.5   Point spread function produced at the focal plane as the image of a distant star. The central diffraction order of the star is far larger than the telescope aperture, and thus illuminates the telescope aperture as a plane wave. The diffraction patterns of the telescope aperture are greatly magnified for purposes of illustration. The size of the point spread function is inversely proportional to the objective diameter, and thus is small for a large aperture (Figure (a)) and large for a small aperture (Figure (b)).

let the coordinate X locate a point on the object function $O(X)$, and x locate a point on the image function $I(x)$. Let $p(x)$ be the point spread function at the image plane. The convolution $O * p$ of the object and the point spread function is given by

$$O * p = \int_{-\infty}^{+\infty} O(X)\, p(x - X)\, dX \tag{8.6}$$

Intuitively, the image amplitude $I(x)$ is obtained by summing the contributions of all point spread functions centered at X with amplitude $O(X)$. The image irradiance is proportional to the modulus-squared of Equation 8.6; for simplicity, we may choose the constant of proportionality to be unity since we are only interested in relative irradiance across the image profile:

$$I(x) = |O * p|^2 = \left| \int_{-\infty}^{+\infty} O(X)\, p(x - X)\, dX \right|^2 \tag{8.7}$$

Since the modulus-squared is taken after the integration, Equation 8.7 represents a model of *coherent image formation*. The top profile of Figure 8.6(a) shows the result of applying Equation 8.7 to the indicated object.

Incoherent images are formed when point spread functions centered at different objects points are mutually incoherent. For example, consider a telescope that forms an image

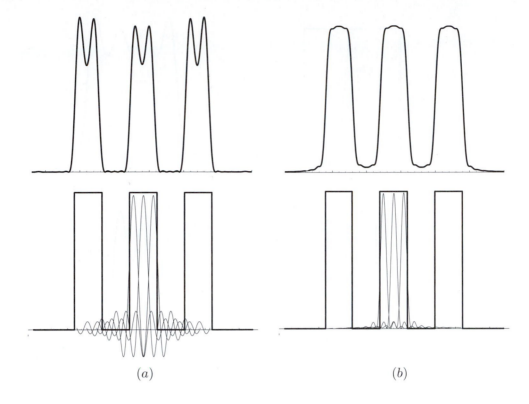

(a) (b)

Figure 8.6 (a) A coherent image as a convolution of point spread function amplitudes. (b) Incoherent image as a convolution of point spread function irradiance.

of a distant group of stars. Each star forms an independent point spread function that is uncorrelated with those of the other stars; thus each point spread function adds as irradiance instead of field amplitude. For a circular objective aperture, the *diffraction-limited* point spread function is the Airy pattern of Equations 6.26 and 6.27, illustrated in Figure 6.9. Two stars are just resolved when they produce point spread functions that overlap incoherently with the minimum angular separation indicated by Equation 6.29.

As another example, consider a camera that forms an image of an object that is illuminated by sunlight. In this case, the image is formed by the point spread function irradiance convolved across the object profile:

$$I(x) = \int\limits_{-\infty}^{+\infty} O(X) \, |p(x - X)|^2 \, dX \qquad (8.8)$$

Figure 8.6(b) illustrates the result of applying Equation 8.8. Since the modulus-square destroys all phase information in the point spread function, the interference structure in part (a) of the figure is missing.

■ **EXAMPLE 8.2**

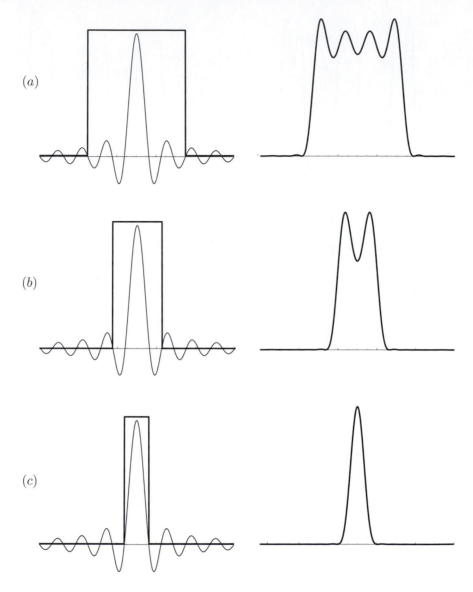

Figure 8.7 Coherent images (right) formed by convolution of the indicated object and diffraction-limited point spread functions on the left. In (a) through (c), the object size decreases with a corresponding loss of edge detail in the image. In (c), the object width is equal to the central fringe of the Airy function.

Find the diffraction-limited width of the point spread function for an $f/4$ optical system. Assume spatially incoherent illumination wavelength of $550\,nm$, and assume that the image is formed at the Fraunhofer plane.

Solution

The angular diameter measured from the first zero of intensity is given by

$$\Delta\theta = 2.44\frac{\lambda}{D_{ob}}$$

On the image plane, assumed to be at the focal plane, this corresponds to a physical size Δy given by

$$\Delta y = f\Delta\theta = 2.44\lambda\frac{f}{D}$$

Thus, the diffraction-limited point spread function has a diameter of 2.44 times the f-number in wavelengths. For an $f/4$ system, this is about ten wavelengths, or about five microns.

As the image detail becomes comparable to the size of the point spread function, edge detail begins to disappear, as illustrated by Figure 8.7 for the case of coherent image formation. In the Abbe model, illustrated for a similar object in Figure 8.2, the image is represented by a second Fourier transform over finite limits of integration. The convolution model gives the results of Figure 8.7. Comparison of these two figures shows that the two models give similar predictions.

A *diffraction-limited* image is given by the convolution of the point spread function determined by diffraction through the input aperture of the optical instrument. *Aberrations* in the optical elements (see Section 4.6) increase the size of the point spread function, increasing the size of the minimum resolvable detail.

■ **EXAMPLE 8.3**

Using Equation 8.6, compute the convolution of the functions $O(X)$ and $p(x)$, where

$$O(X) = \begin{cases} 1 & |X| \le 1 \\ 0 & |X| > 1 \end{cases} \qquad \text{and} \qquad p(x) = \begin{cases} 1 & |x| \le \frac{1}{2} \\ 0 & |x| > \frac{1}{2} \end{cases}$$

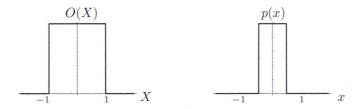

Figure 8.8   Functions used for Example 8.3.

Solution

The functions $O(X)$ and $p(x)$ are plotted in Figure 8.8. In this case, the integration of Equation 8.6 can be evaluated graphically. Figure 8.9 illustrates the result. Since

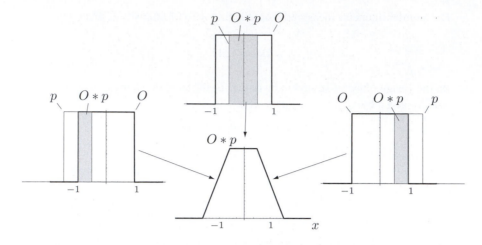

Figure 8.9 Convolution of Example 8.3.

p is more narrow than O, there is a region of complete overlap and a resulting integration area of 1. At each end, the overlap is incomplete, and the integration area decreases linearly to zero for points where there is no overlap. The convolution has the shape of the trapezoid illustrated with zero amplitude at ± 1.5.

8.3.2 Speckle

Image speckle is produced when the light collected from an object by an imaging system is spatially coherent. Figure 8.10 shows a photograph of a diffusely reflecting surface illuminated by a spatially coherent laser beam. Light reflecting from peaks and valleys in the rough surface interferes to produce the random irradiance fluctuations in the image.

Star twinkle is a speckle effect. Light that enters the eye from a distant star is composed entirely of the spatially coherent central diffraction order from the stellar disk. Atmospheric effects distort and fold the wavefronts, producing time intervals of constructive and destructive interference that we refer to as the twinkle.

Problem 8.5 Find the width of the central maximum of the diffraction-limited point spread function for a telescope with a $10.0\,cm$ objective diameter and a focal length of $1.0\,m$.

Problem 8.6 Compute the convolution of the functions $O(X)$ and $p(x)$, when both function are identical:

$$O(X) = \begin{cases} 1 & |X| \leq \frac{1}{2} \\ 0 & |X| > \frac{1}{2} \end{cases} \quad \text{and} \quad p(x) = \begin{cases} 1 & |x| \leq \frac{1}{2} \\ 0 & |x| > \frac{1}{2} \end{cases}$$

You may evaluate the convolution graphically, as in Example 8.3.

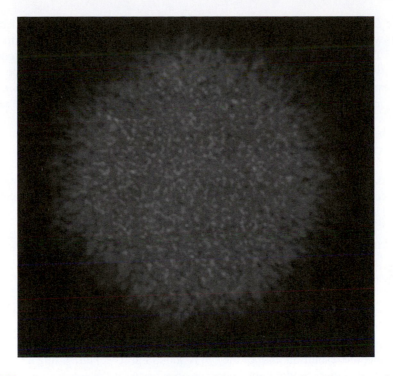

Figure 8.10   Image speckle produced by diffuse reflection of a spatially coherent laser beam.

Problem 8.7   Compute the convolution of the functions $O(X)$ and $p(x)$, where

$$O(X) = \begin{cases} 1 & |X| \leq \frac{1}{4} \\ 0 & |X| > \frac{1}{4} \end{cases} \qquad \text{and} \qquad p(x) = \begin{cases} 1 & |x| \leq \frac{1}{2} \\ 0 & |x| > \frac{1}{2} \end{cases}$$

You may evaluate the convolution graphically, as in Example 8.3.

Problem 8.8   When observing a speckle pattern such as that of Figure 8.10, the speckle pattern usually moves as you move your head. If you are nearsighted, the speckle pattern moves to the left as you move your head to the right, and vice versa. If you are farsighted, the speckle pattern moves in the same direction as your head. Explain this effect, and include detailed ray diagrams. If your eyesight was perfect, what would you see as you move your head? If an eye doctor were to use this as a quick test of your vision, what color laser beam would be best, and why?

8.4   RESOLVING POWER OF OPTICAL INSTRUMENTS

In Section 6.3.4, the *diffraction-limited resolution limit* of an optical instrument with a circular objective aperture was determined to be (See Equation 6.29)

$$\Delta\theta \cong \frac{1.22\lambda}{D_{ob}} \tag{8.9}$$

where D_{ob} is the diameter of the objective aperture, and $\Delta\theta$ is the angular separation of two points within the field of view that can just barely be resolved with the instrument. In other words, object detail that is smaller than the resolution limit cannot be detected. In Section 4.7, it was found that the magnification of telescopes and microscopes can be increased to any desired power simply by changing the eyepiece. The *maximum usable magnification* M_{max} is the magnification that causes detail at the resolution limit to just be resolvable by the human eye.

■ **EXAMPLE 8.4**

Find the diffraction limited resolution of the eye, and the corresponding minimum detectable object size. Assume a pupil diameter of $5.00\,mm$ and a wavelength of $550\,nm$.

Solution

Equation 8.9 gives

$$\Delta\theta_{eye} = 1.22\frac{\lambda}{D} = 1.22\frac{5.50 \times 10^{-7}m}{5.00 \times 10^{-3}m} = 1.34 \times 10^{-4}rad = 0.134\,mrad$$

In reality, the human eye cannot perform at this theoretical maximum. For most people, the resolution limit of the eye is about $3 \times 10^{-4}\,rad$. Using a near point of $25\,cm$, this gives a minimum detectable object size for the eye of

$$\delta_{eye} = \left(3 \times 10^{-4}rad\right)(0.25\,m) = 75\,\mu m$$

which is about the size of a human hair.

For the telescope, we estimate the maximum usable magnification as the ratio of the angular size of the smallest object the eye can detect, divided by the angular size specified by the resolution limit of Equation 8.9. Using the results of Example 8.4, we find this to be

$$M_{\max} = \frac{\Delta\theta_{eye}}{1.22\lambda}D_{ob} = \frac{3 \times 10^{-4}rad}{1.22\left(5.50 \times 10^{-7}m\right)}D_{ob} \approx 450D_{ob} \tag{8.10}$$

where the last result is to be interpreted as 450 times the objective diameter in meters. Thus, the maximum usable magnification for a telescope with $10\,in$ ($25.4\,cm$) objective is 114. Magnifications larger than this do not appreciably increase the image detail.

The resolution limit of a microscope is determined by Equation 8.5. Since the object size δ is half the spatial period d, we obtain

$$n \sin\theta_{max} = \frac{\lambda_0}{2\delta}$$

where λ_0 is the vacuum wavelength of the illuminating light. Solving for the object size, we find

$$\delta = \frac{\lambda_0}{2NA} \tag{8.11}$$

This equation represents the minimum object size that can be imaged by a microscope with a given numerical aperture NA.

To find the maximum usable magnification M_{max} of a microscope, we again allow object detail at the resolution limit to be just detectable by the eye

$$M_{\max}\delta = \delta_{eye}$$

giving

$$M_{\max} = \frac{\delta_e}{\delta} = 2NA\frac{\delta_e}{\lambda_0} \tag{8.12}$$

Using the results of Example 8.4, we find that $M_{\max} \approx 275NA$. Air immersion objectives have numerical apertures that range from about 0.1 to 1.0, and oil immersion objectives have numerical apertures as high as 1.4. Of course, the microscopist may have good reason to use magnifications in excess of M_{max}, but as a rule, exceeding the value of Equation 8.12 leads to *empty magnification*.

Problem 8.9   Find the minimum angular separation that can just be resolved by a telescope with a $10.0\,cm$ objective diameter. Repeat when the objective diameter is $10.0\,m$. In each case, express your answer in arc-seconds, where one arc-second is one degree divided by 3600. Assume an illumination wavelength of $550\,nm$.

Problem 8.10   Verify that the maximum usable magnification for a microscope is given by $275NA$, as stated in the text.

Problem 8.11   Find the maximum usable magnification and minimum detectable object diameter when $\lambda = 550\,nm$ and again when $\lambda = 450\,nm$ for the following microscope objectives: (a) air immersion with $NA = 0.4$, (b) air immersion with $NA = 0.95$, and (c) oil immersion with $NA = 1.4$.

Problem 8.12   Find the maximum usable magnification for the 200 inch Hale telescope of the Mt. Palomar observatory.

Problem 8.13   Find the maximum usable magnification for both telescopes and microscopes in terms of the numerical aperture using the diffraction-limited resolution of the eye. Assume an iris diameter of $5.00\,mm$ and a wavelength of $550\,nm$.

Problem 8.14   Estimate the maximum usable magnification for binoculars with $50\,mm$ diameter objectives used during daylight when the pupil of the eye has decreased to $2.0\,mm$.

8.5   IMAGE RECORDING

Images formed by optical instruments can be recorded in an *analog* format using photographic film, and in a *digital* format using detector arrays. Both have advantages; however, digital recording is becoming more common as this technology improves.

8.5.1   Photographic Film

Photographic film uses a transparent *emulsion* that randomly suspends tiny photosensitive silver halide *photographic grains*, as illustrated in Figure 8.11. The grains become strongly absorbing only after *exposure* and *development* (each of these processes will be discussed below). Regions on the developed film that receive more exposure have more absorbing grains, thus the film records an image *negative* where bright areas in the image appear darker in the negative. A positive image can be obtained by re-recording the negative.

The developed film has an *irradiance transmittance T* that is defined as the ratio of transmitted irradiance to incident irradiance measured at each point on the film. The *photographic density D* is defined as the logarithm of the reciprocal of the transmittance:

$$D(x, y) = \log_{10}\left(\frac{1}{T(x, y)}\right) \tag{8.13}$$

where the coordinates (x, y) locate a point on the photographic film. The *exposure* is the product of the incident irradiance I and the *exposure time t*:

$$E(x, y) = I(x, y)\, t \tag{8.14}$$

The *Hurter-Driffield curve*[3] is a plot of density versus exposure, as illustrated in Figure 8.11(b). Typical photographic film is *linear* only over a range of densities in the neighborhood of 1.0, as illustrated. The slope of the linear region is referred to as the film *gamma* (γ). *High-contrast film* has a high value of γ, whereas films with low values of γ are referred to as *low-contrast*.

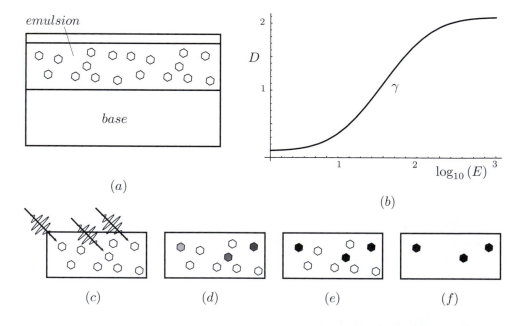

Figure 8.11   (a) Photographic film. (b) HD curve. (c)—(d) Image capture and development in photographic film (see text).

[3]F. Hurter: 1844-1898, V. C. Driffield: 1848-1915; published a classic analysis of photographic film.

After exposure, film must be *developed* in a process that is illustrated in Figures 8.11(c)—(f). In Figure 8.11(c), photons in the image field illuminate the unexposed emulsion grains, typically composed of a silver halide salt such as silver bromide ($AgBr$). Absorbed photons create single metallic silver atoms within the grain at the point where the absorption occurs. The exposed film is now immersed in a *developer* which transforms exposed grains completely into metallic silver in a chemical process referred to as *reduction*. In practice, a given grain must absorb more than the *threshold number* of photons in order for reduction to proceed completely within a given development time. The exposed grains in the undeveloped emulsion record the *latent image* illustrated in Figure 8.11(d), whereas the developed grains in Figure 8.11(e) record the final image with the full range of photographic density. A final *fixing* step washes away unreduced silver halide salt grains as illustrated in Figure 8.11(f).

The *film speed* and *film resolution* are determined largely by the *grain size*. The film speed determines the exposure required to achieve a given photographic density; high-speed film requires less exposure than film with low speed. The film speed determines the horizontal scale of the HD curve of Figure 8.11(b). Large grains present a larger cross section for absorption, and thus more quickly absorb the threshold number of photons; thus, high speed is generally achieved by making the grain size large. On the other hand, small grains give better resolution. In general, the higher the resolution, the lower the film speed. Film grains typically range from a few nanometers for high resolution holographic film, up to a few microns for high-speed film.

8.5.2 Digital Detector Arrays

A *detector array* consists of small photo-detectors arranged in a two-dimensional matrix, as illustrated in Figure 8.12(a). If the detectors are *linear*, each detector registers a voltage level over some finite time interval that is proportional to the number of photons detected. This voltage is subsequently converted into a *digital level* by the *analog to digital* circuitry of the detector array.

■ **EXAMPLE 8.5**

Find the voltage spacing between adjacent digital levels for a 16-bit detector with a voltage output that varies between 0 and 5 volts.

Solution

The number of digital levels available for a 16-bit detection system is given by 2 raised to the 16th power:

$$2^{16} = 65\ 536$$

Each of these 65 536 digital levels are spaced equally across the entire operating range - in this case 5.00 volts. The voltage spacing ΔV between adjacent digital levels is given by

$$\Delta V = \frac{5.00\ V}{65\ 536} = 7.63 \times 10^{-5}\ V$$

This represents the smallest voltage change than can be detected with this detection system.

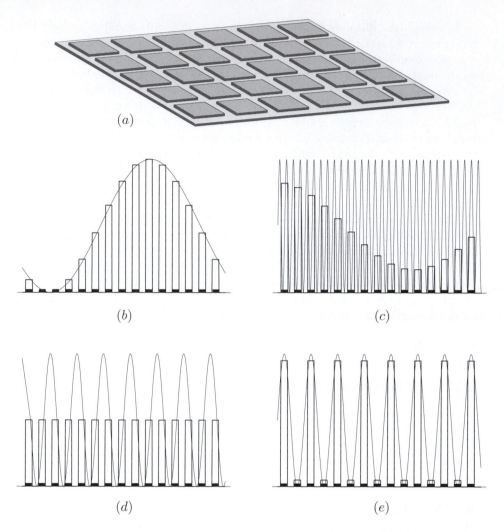

Figure 8.12   (a) A focal plane detector array. In plots (b)-(e), the bar above each detector element indicates the measured signal, calculated as the irradiance average over the detector surface. In each case, the image is a sinusoid with the illustrated spatial frequency. (b) The image spatial frequency is lower than the Nyquist limit, resulting in a digital recording with the same spatial frequency as the object. In (c), the spatial frequency exceeds the Nyquist limit, giving an aliased signal modulation. In (d) and (e) the sampled spatial frequency is at the Nyquist limit. In (d), each detector gives a constant output. In (e), the sampled image is shifted slightly and now produces a maximum signal modulation.

The regular spacing of a digital detector array can produce image artifacts that result from *aliasing*. Figure 8.12(b) illustrates the measurement of a sinusoidal image of spatial period that is much longer than the detector spacing. In this case, there are several detector readings over each cycle of the image, and the recorded image (indicated by the bars above each detector element, computed as an average over the detector element) has a spatial frequency equal to that of the image. In Figure 8.12(c), the spatial period of the image is less than the detector spacing. In this case the recorded image, again computed as an average of irradiance over each detector element, has a spatial frequency that is *less* than

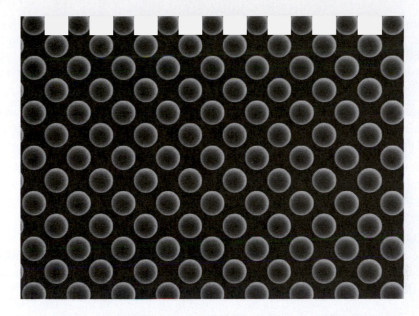

Figure 8.13   Scanning electron micrograph (2,275x) of a CCD detector array. (David Scharf/Peter Arnold, Inc.)

that of the image, resulting in false detail that is referred to as aliasing. In order to avoid aliasing, it is necessary that the spatial frequency of the image be less than the *Nyquist frequency*, defined as *twice* the spatial frequency defined by the detector spacing. Strictly, the image frequency must exceed the Nyquist limit. Figures 8.12(d) and (e) illustrate two possibilities for recording an image with frequency equal to the Nyquist limit. In Figure 8.12(d) each detector produces a constant output, and with a small lateral adjustment, the same image produces a large signal amplitude. As long as the frequency of the image exceeds the Nyquist limit, the recorded image has some modulation at the image frequency.

In a well-designed digital image recording system, the resolution limit of the system optics serves as a low-pass filter that functions as an *anti-aliasing filter*. In such a system, blurring from aberrations and diffraction remove spatial frequencies in the image that exceed the Nyquist limit. As with photographic film, the smaller the detector element, the fewer photons it receives for a given image irradiance; thus detector arrays that significantly *oversample* require longer integration times.

Most modern digital cameras utilize *charge-coupled devices* (CCD) as the individual image sensor in the detector array. Each element in a CCD array can be thought of as a "charge bucket" that increases the amount of charge stored with each new photon detection. The signal is integrated for some length of time, then each bucket is emptied as the detector reading is recorded. Individual detectors are referred to as *pixels* which is short for "picture element." Electronic noise in the detector circuitry can result in a background signal that is referred to as "dark signal." The dark signal is usually generated as thermal noise, and can be reduced substantially by cooling the detector.

Problem 8.15   A *neutral density filter* has an irradiance transmittance defined by Equation 8.13. For example, a ND 2.0 filter has a photographic density of 2.0. The term "neutral" describes the fact that this density is constant over a wide optical spectrum. Find the transmittance of ND 2.0 and ND 1.0 filters. What density gives a transmittance of one-half?

Problem 8.16   A certain CCD array is constructed from individual 24-bit detectors that provide a 0V—10V voltage output. Find the spacing between adjacent voltage levels for this detector.

Problem 8.17   A certain digital camera uses a CCD array that is $22.7\,mm$ wide. Each line in the detector array contains 2160 pixels. If the detector width is equal to the detector spacing, what is the width of each detector? What is the Nyquist limit for this detector array? Express your answer in cycles per millimeter.

Problem 8.18   You are designing a digital camera, and want each detector element to have a width equal to the distance between the first zeros of the diffraction-limited point spread function. If the optical system is $f/4$, what is the detector width? If the detectors are spaced by the detector width, how many pixels are contained in a line that is $22.7\,mm$ wide? How many square pixels are contained in an array with this spacing, width, and a height of $15.1\,mm$? Assume a wavelength of $550\,nm$.

8.6 CONTRAST TRANSFER FUNCTION

The resolution performance of a real optical system is typically dominated by aberrations, falling well short of the diffraction limit. Figure 8.14(a) is an example of a *resolution target* that can provide a useful resolution benchmark. The target illustrated consists of small periodic *square wave* structures of variable spatial frequency. As discussed above, the final image is a convolution of the magnified object function and the system point spread function. As the size of the point spread function becomes comparable to a single cycle of the magnified object, the image becomes *blurred*, and no longer resembles a square wave.

The *contrast* is defined as

$$C = \frac{I_{\max} - I_{\min}}{I_{\max} + I_{\min}} \tag{8.15}$$

where I_{max} is maximum image intensity, and I_{min} is the minimum image intensity. Figure 8.14(b) illustrates these quantities in an image profile for which there is no blurring. The contrast approaches unity when $I_{max} << I_{min}$.

The *contrast transfer function* is obtained by plotting the image contrast versus spatial frequency of the magnified resolution target. A typical plot is illustrated in Figure 8.15(f). The spatial frequency is typically measured in line-pairs per millimeter ($\frac{lp}{mm}$), where a line-pair is a bright/dark pair. This is sometimes also referred to as cycles/mm. The measurement is not straightforward, and can be affected by edge effects as illustrated in Figures 8.15(a)—(d).

The *modulation transfer function* is obtained by using a resolution target with a sinusoidal profile rather than the step profile of Figure 8.14(a). The modulation transfer function is the modulus of the Fourier transform of the point spread function. For more details, see Goodman [7].

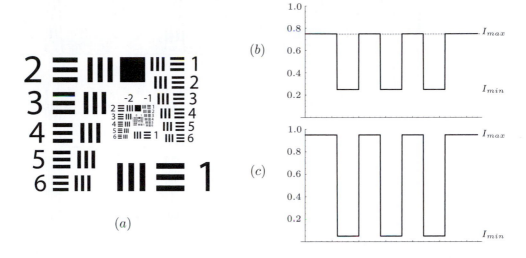

Figure 8.14 (a) A standard resolution target. (b) Image of (a) recorded by an optical system with point spread function much smaller than the spatial period of the image. (c) Object intensity levels for the target in (a).

Problem 8.19 What object distance would give a focused image with a spatial frequency of $50\,lp/mm$ when the object consists of a bar chart with black bars $5.00\,mm$ wide separated by $5.00\,mm$? Assume an objective lens with focal length $50.0\,mm$.

Problem 8.20 Find the contrast of the recorded image, the target and the image contrast for the data illustrated in Figures 8.14 (b) and (c).

Problem 8.21 Find the contrast for the CCD image of a bar target when the maximum pixel value is 20 times that of the minimum pixel value.

8.7 SPATIAL FILTERING

Spatial filtering is a technique of *optical signal processing* where the irradiance content in the Fraunhofer plane is manipulated to control the irradiance structure in the image plane. We have seen an example of this in Application Note 6.3 that describes a *spatial filter* that is used to remove spatial noise content from a laser beam. The spatial filter of Figure 6.20 can be thought of as a *low-pass filter* that passes only the central diffraction order that is very nearly Gaussian in profile.

As another example, consider Figure 8.2(e), where the filter function has been modified to block the central diffraction order. This can be achieved by placing an opaque object in the Fraunhofer plane that blocks the zeroth diffraction order while passing the higher

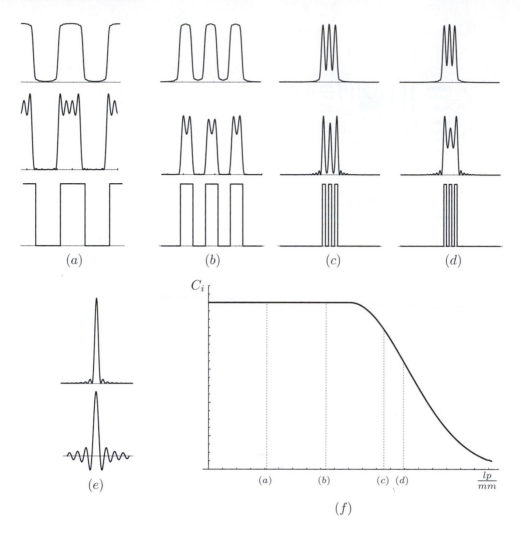

Figure 8.15 (a)-(d) Coherent and incoherent images formed of a three-bar object. The bottom figure is the object, the middle figure is the coherent image, and the top figure is the incoherent image. Each image is the convolution of the object and the points spread functions shown in (e) The bottom point spread function generates the coherent images and the top point spread function generates incoherent images. (f) Modulation transfer function. The curve shown is for illustration only and does not correspond analytically to either of the image sets of (a)-(d).

diffraction orders. Since edge detail is contain in the nonzero orders, the edges of the image are enhanced. This can often reveal detail in the image that would not otherwise be visible.

A *spatial light modulator* (SLM) consists of an array of pixel elements each capable of controlling both amplitude and phase of an illumination wave. SLM's employ several different technologies to achieve modulation, including liquid crystals and deformable mirrors. Liquid crystals control the amplitude and phase of transmitted light, and are also commonly used for laptop and digital projector displays. SLM's that use deformable mirrors adjust the amplitude and phase of reflected light by adjusting the tilt and displacement of an array of small mirrors using MEMS (Micro Electro-Mechanical Systems) technology.

Other approaches are also commonly used; see Goodman[7] and references therein for more information.

Figure 8.16   Spatial filtering with an electrically addressable spatial light modulator. (Courtesy Michael K. Giles.)

When placed in the Fraunhofer plane, SLM's can be used as real-time spatial filters with very fast temporal response. An optical system can be designed to respond preferentially to objects in a given size range by placing a spatial filter in the Fraunhofer plane that only passes the far-field diffraction pattern of such objects, resulting in an image where objects in that size range have enhanced brightness. Since this can be done in real time, the size range can be scanned as the image is observed. Figure 8.18 shows an example.

■ **EXAMPLE 8.6**

Design a spatial filter to be placed at the Fraunhofer plane of a lens with focal length f that will enhance the brightness of objects with the shape of a vertical slit of width b.

Solution

According to Section 6.3.1, the far field diffraction pattern of a single slit is given by Equation 6.16

$$E = E_0 \frac{\sin \beta}{\beta}$$

where

$$\beta = \tfrac{1}{2} kb \sin \theta$$

where k is the wavenumber, b is the slit width and θ is defined in Figure 8.17.

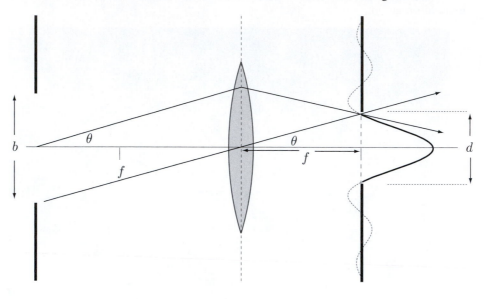

Figure 8.17   A spatial filter.

A spatial filter that just passes the central order of the single-slit diffraction pattern has edges located by the angle θ that is given by

$$\frac{1}{2}\frac{2\pi}{\lambda}b\sin\theta = \pi$$

Using the small angle approximation gives

$$\frac{\lambda}{b} = \sin\theta \simeq \theta \simeq \frac{d}{2f}$$

where d is the width of the spatial filter. Solving for d gives

$$d = \frac{2\lambda f}{b}$$

A vertical filter of this width centered on the optical axis in the Fraunhofer plane will pass the central plus possibly higher orders for vertical slits of width greater than or equal to b. Vertical slits of width less than b will have a larger diffraction pattern; in this case only a portion of the central order passes the filter, and the resulting image will have diminished brightness.

With sufficient resolution, SLM's can also be used to project real-tine holograms that produce three-dimensional images via wavefront reconstruction, as discussed in Section 6.4.2.

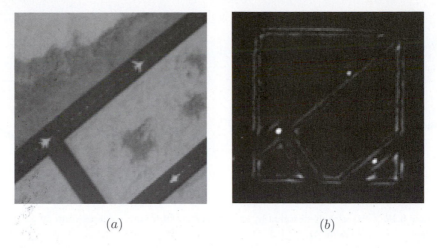

(a) (b)

Figure 8.18   The image in (a) is spatially filtered in (b) to enhance objects in the size of interest. (Courtesy Michael K. Giles.)

Problem 8.22   Design a spatial filter to be placed at the Fraunhofer plane of a $50.0\,mm$ focal length lens if the optical system is designed to preferentially detect horizontal slit apertures of width greater than $0.100\,mm$. Assume an illumination wavelength of $550\,nm$.

Problem 8.23   Design a spatial filter to be placed at the Fraunhofer plane of a lens of focal length f if the optical system is designed to preferentially detect rectangular slit apertures of minimum width w and minimum height h. Assume an illumination wavelength of $550\,nm$.

Problem 8.24   Design a spatial filter to be placed at the Fraunhofer plane of a lens with focal length f if the optical system is designed to preferentially detect objects of circular cross-section with minimum radius R. Assume an illumination wavelength of $550\,nm$.

8.8   ADAPTIVE OPTICS

An *adaptive optics* (AO) system removes image aberrations by adjusting one or more optical surfaces in the optical train. For example, wavefronts diffracted from distant astronomical objects are distorted as they pass through density fluctuations in the Earth's atmosphere, causing distortion in images formed by ground-based telescopes. Figure 8.19 illustrates a distorted wavefront that reflects from a mirror with a surface adjusted to reflect this particular wavefront as a plane wave. Adaptive optics systems designed for astronomy can sense the arriving wavefront and make the necessary adjustments to an *active mirror* surface to remove atmospheric image aberrations in real time.

In Figure 8.20, plane wavefronts from a distant star are distorted by the atmosphere before collection by the AO system. Light reflected from a static mirror is divided by a beam splitter before reflecting from the active mirror. The split beam is analyzed by a *waveform sensor* that measures the wavefront distortion. The result of this measurement is passed to the active mirror control electronics which make adjustments to the mirror surface that remove the distortion. The corrected beam then passes through additional optics that form the final undistorted image.

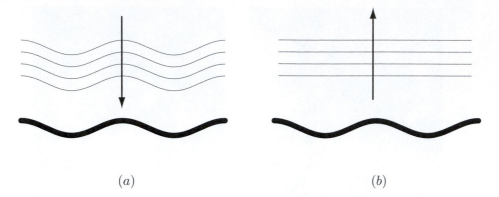

(a) (b)

Figure 8.19   Wavefronts distorted by the atmosphere in (a) are reflected from an adaptive optics surface as plane wves in (b).

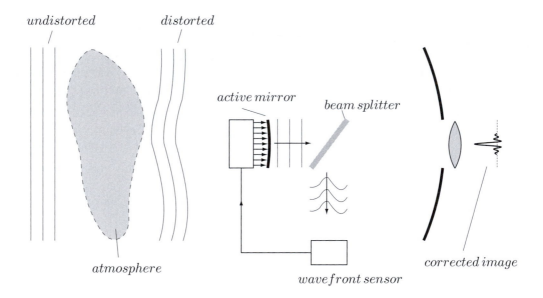

Figure 8.20   An AO system design for astronomy. Plane waves from a distant star are distorted by the atmosphere. A large static mirror focuses the distorted wavefronts onto an active mirror whose shape can be adjusted to remove the distortion. In this design, a beams plitter sends a portion of the collected signal to a wavefront sensor that measures the distortion, and sends the necessary control signals to the active mirror. The corrected signal is processed by the imaging system to produce to corrected image.

Figure 8.21 illustrates the workings of a typical deformable mirror. In this design, *actuators* are connected to discrete points behind the reflecting surface. The length of each actuator is controlled to achieve the desired mirror profile. Actuators are commonly made from *piezo-electric* materials whose length can be controlled by application of a voltage. Atmospheric fluctuations tend to occur on time scales of about $10\,ms$, and piezo-electric actuators are fast enough to correct these in *real time* (i.e., as they occur).

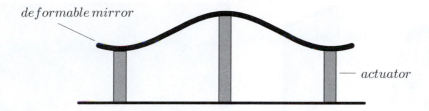

Figure 8.21   Active mirror design that uses mechanical actuators to bend a flexible mirror surface.

A typical wavefront sensor is illustrated in Figure 8.22. A two-dimensional array of lenses focuses portions of the distorted wavefront from a *guide star* onto independent CCD arrays. An undistorted plane wave focuses on the optical axis, as illustrated in Figure 8.22(a). Curvature in a distorted wave causes the image formed by each lens to be formed away form the optical axis, as in Figure 8.22(b). Measurement of the lateral displacement allows the distorted curvature to be measured and then corrected by the active mirror.

The guide star must be bright enough to allow accurate wavefront sensing, and should also have a small angular separation from the object to be imaged. Ground-based lasers are often used as *laser guide stars*. Sodium lasers have beams that are resonantly scattered by sodium atoms, which happen to be concentrated in the atmospheric *sodium layer* at an altitude of about $90\ km$. Blue and UV lasers are Rayleigh scattered (Section 3.8) at lower altitudes. Laser guide stars are brighter with a position that can be adjusted.

Figure 8.22   A lens array focuses portions of the wavefront from a guide star onto independent CCD arrays. (a) If the wavefront is undistorted, the guide star is focused on the optical axis of each lens. (b) Wavefront distortion causes each image of the guide star to be displaced. Measurement of this displacement allows the curvature of the wavefront to be determined. (c) Sodium laser guidestar being used at the Gemini North Telescope atop Mauna Kea, Hawaii. (Stephen & Donna O'Meara/Photo Researchers, Inc.)

Additional Problems

Problem 8.25   Estimate the minimum resolvable object size for a microscope of numerical aperture of 1.15. Assume $550\,nm$ illumination.

Problem 8.26   Find the width of the central maximum of the diffraction-limited point spread function for a telescope with a $10.0\,m$ objective diameter and a focal length of $12.0\,m$.

Problem 8.27   Find the normalized irradiance profile for the convolution found in Example 8.3.

Problem 8.28   What object distance would give a focused image with a spatial frequency of $20\,lp/mm$ when the object consists of a bar chart with black bars $0.500\,mm$ wide separated by $0.500\,mm$? Assume an objective lens with focal length $200\,mm$.

Problem 8.29   Use the programming language of your choice to verify the coherent image profiles of Figure 8.2. See Problem 8.1.

Problem 8.30   Use the programming language of your choice to verify the coherent image profiles of Figure 8.3. See Problem 8.1.

Problem 8.31   Design a spatial filter to be placed at the Fraunhofer plane of a $1.00\, m$ focal length lens if the optical system is designed to preferentially detect objects of diameter greater than $10.0\, m$. Assume $550\, nm$ illumination.

CHAPTER 9

POLARIZATION AND NONLINEAR OPTICS

The only laws of matter are those that our minds must fabricate and the only laws of mind are fabricated for it by matter.

—Maxwell

Contents

9.1   INTRODUCTION

In this chapter, we continue the discussion of electromagnetic wave polarization that began in Chapter 2. After reviewing linear polarization and discussing linear polarizers, we discuss birefringence and how this may be used to construct waveplates and retarders that produce states of elliptical and circular polarization. The Jones calculus is then introduced to facilitate and simplify calculations that involve polarizers and waveplates. Methods for actively controlling the optical properties of materials properties via electro-optic and acousto-optic effects, optical activity, and Faraday rotation are introduced. This discussion is extended in an introduction to nonlinear optics, which includes discussion of harmonic generation and frequency mixing.

9.2   LINEAR POLARIZATION

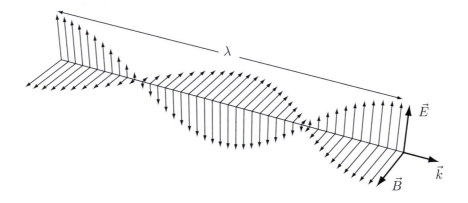

Figure 9.1   A linearly polarized transverse electromagnetic wave.

In Chapter 2, it was shown that *linearly polarized* transverse electromagnetic waves (see Figure 9.1) were solutions to the electromagnetic wave equation, and thus were solutions to Maxwell's equations. Using Maxwell's equations, we also determined several important properties of transverse electromagnetic waves. For example, we showed that such a wave has both an electric field $\vec{E}$ and a magnetic field $\vec{B}$, that $\vec{E}$ and $\vec{B}$ are mutually perpendicular, and that both fields are oriented transversely to the direction of propagation, determined by the wave vector $\vec{k}$. Moreover, the fields are aligned so that $\vec{E} \times \vec{B}$ points in the direction of $\vec{k}$, with the Poynting vector $\vec{s} = \frac{1}{\mu_0} \vec{E} \times \vec{B}$ giving the flux of energy transmitted by the wave. The fields $\vec{E}$ and $\vec{B}$ oscillate in phase, with amplitudes determined by $E = \frac{c}{n} B$. Finally, we showed that it was redundant to specify both $\vec{E}$ and $\vec{B}$ for any given electromagnetic wave, since either field is specified completely once the other field is given. We will always use the electric field to specify the complete electromagnetic wave.

An electromagnetic wave with an electric field vector $\vec{E}$ that oscillates back and forth along a fixed direction is said to be linearly polarized. In this case, the electric field oscillates within the *plane of polarization*, as shown in Figure 9.2. The plane of polarization for the wave illustrated is inclined between the x-z and y-z planes. At any instant, the electric

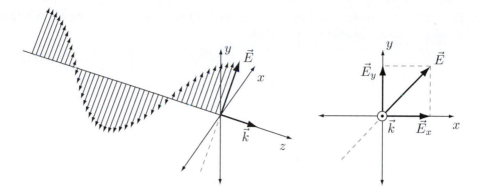

Figure 9.2 Linear polarization in the x-y plane. The electric field vector may be resolved into components along the x and y axes.

field vector has x and y components given by E_{0x} and E_{0y}. The corresponding forward traveling electromagnetic wave is determined by

$$\vec{E} = \vec{E}_0 e^{i(kz - \omega t + \varphi)} = \left(E_{0x}\hat{i} + E_{0y}\hat{j} \right) e^{i(kz - \omega t + \varphi)} \tag{9.1}$$

where we have used the complex representation discussed in Section 1.6. Equation 9.1 represents a linearly polarized electromagnetic wave traveling in the positive-z direction, as illustrated in Figure 9.2. The initial phase angle ϕ represents the phase of the wave when z and t both equal zero.

9.2.1 Linear Polarizers

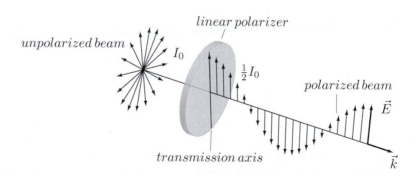

Figure 9.3 A linear polarizer is illuminated by unpolarized light. The transmitted beam is linearly polarized parallel to the transmission axis of the linear polarizer. The transmitted irradiance is one half that of the unpolarized beam.

A *linear polarizer* selectively removes light that is linearly polarized along a direction that is perpendicular to its *transmission axis*, as illustrated in Figure 9.3. In an *ideal* linear polarizer, the transmission is zero for electric field components perpendicular to the transmission axis, and 100% for electric field components parallel to the transmission axis.

The light that is transmitted by the polarizer is linearly polarized along an axis that is parallel to the transmission axis, as shown.

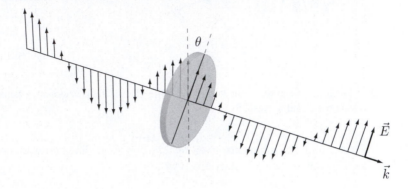

Figure 9.4   A linear polarizer is illuminated by a linearly polarized beam. The linear polarizer rotates incident polarization, giving a transmitted beam that is linearly polarized parallel to the transmission axis. The transmitted irradiance is given by Malus's law.

The linear polarizer in Figure 9.3 is illuminated with *unpolarized light*, illustrated by the bundle of arrows in the figure. By unpolarized, we mean that the beam is a mixture of many polarization states, each random in both orientation and phase. Unpolarized light is emitted, for example, by thermal sources such as the filament of an incandescent lightbulb. On a microscopic scale, this emission can be visualized as resulting from the random motions of independent radiators. It is always possible to visualize each independent emission as a transverse electromagnetic wave that is linearly polarized along an arbitrary axis. A typical emission process for visible light lasts for only a short time (for visible wavelengths, typically about 10^{-8} s), so an unpolarized beam can be thought of as consisting of many short-lived randomly oriented linear polarization states that we refer to as *wavepackets*. In general, an incident wavepacket will be polarized along an axis that is inclined at some random angle θ to the transmission axis of the linear polarizer. If the incident wavepacket has amplitude E_0, the transmitted wavepacket will have amplitude $E_0 \cos\theta$. As shown in Section 2.3.3, irradiance is proportional to the square of the amplitude, so the irradiance of the transmitted wavepacket is proportional to $E_0^2 \cos^2\theta$. Since the orientation of each wavepacket is random, we must average over all possible orientations to obtain the average transmission. As shown in Example 2.4, the average of $\cos^2\theta$ over many cycles of θ is $1/2$. Thus, if an unpolarized beam of irradiance I_0 is incident on an ideal linear polarizer, the transmitted beam consists of wavepackets that are linearly polarized parallel to the transmission axis. The average irradiance of the transmitted beam is $\frac{1}{2}I_0$, as shown in Figure 9.3.

A beam that has been previously polarized transmits through a linear polarizer according to *Malus's law*, as illustrated in Figure 9.4. If the incident beam has amplitude E_0, the transmitted beam has amplitude $E_0 \cos\theta$. Similarly, if the incident beam has irradiance I_0, the transmitted beam has irradiance

$$I = I_0 \cos^2\theta \tag{9.2}$$

where θ is the angle between the incident polarization axis and the transmission axis of the linear polarizer. It is important to note that a linear polarizer *rotates* the polarization state, transmitting an output beam that is linearly polarized in a direction that is parallel to the polarizer's transmission axis.

■ **EXAMPLE 9.1**

Unpolarized light is incident upon two coaxial ideal linear polarizers whose transmission angles are inclined at an angle θ, as shown in Figure 9.5. If the irradiance of the unpolarized beam is I_0, find the transmitted irradiance and describe the transmitted polarization state when θ is $0°$, $45°$, $90°$, $135°$, and $180°$.

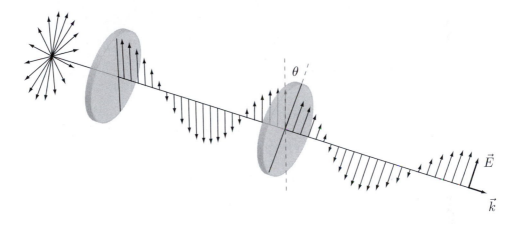

Figure 9.5   Two coaxial linear polarizers. The first polarizer is illuminated by an unpolarized beam.

Solution

The first polarizer transmits half the unpolarized irradiance, and the second transmits according to Malus' law. If the unpolarized irradiance is I_0, the irradiance transmitted through the second polarizer is

$$I = \frac{1}{2} I_0 \cos^2 \theta$$

When $\theta = 0$ and $180°$, the second ideal polarizer has no effect. When $\theta = 45°$ and $135°$, the transmitted irradiance is $0.25\, I_0$ with the beam polarized parallel to the *second* polarizer transmission axis. When $\theta = 90°$, the two polarizers are *crossed*, and the transmission is zero.

The second linear polarizer rotates the polarization axis of the beam that is linearly polarized by the first polarizer. A rotation of $90°$ cannot be achieved with this arrangement.

A linear polarization state can also be rotated by simple reflection, as shown in Figure 9.6. This is especially useful for high-power lasers whose beams are already polarized. The arrangement shown in Figure 9.6 allows the polarization to be rotated by 90° without rotating the laser. This approach has the disadvantage of changing the beam height and beam direction.

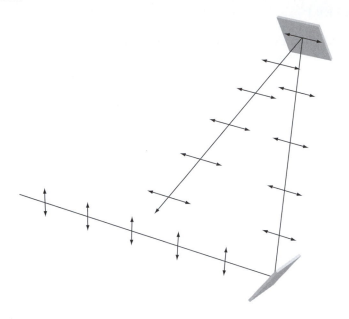

Figure 9.6   Vertical linear polarization becomes horizontal linear polarization after two mirror reflections.

9.2.2   Linear Polarizer Design

A *wire grid* linear polarizer is made of an arrangement of small closely spaced parallel wires, as shown in Figure 9.7. A linearly polarized wave with electric field $\vec{E}$ parallel to the wires can initiate currents in the wires, and the energy of the wave dissipated as heat generated by the wire's resistance. A wave linearly polarized along an axis perpendicular to the wires cannot initiate currents and so is not absorbed. Thus, the transmission axis of the wire grid polarizer is *perpendicular* to the wires, as shown in Figure 9.7. The wires will not appreciably scatter the beam if they can be spaced apart by half a wavelength or less of the illuminating light; however, this becomes difficult for wavelengths as small as that of visible light. For this reason, wire grid polarizers are more useful in the infrared region of the electromagnetic spectrum.

Linear polarizers for the visible region can be made from materials that are *dichroic*. A dichroic material selectively absorbs only one of two perpendicular linear polarization states. Dichroic materials typically contain long, thin molecules or crystals that can be aligned mechanically or electromagnetically. When aligned, they are similar to a wiregrid polarizer, and have greater absorption for waves that are linearly polarized parallel to the alignment. This is the process used for Polaroid© material, invented by Edwin Land[1]. Polaroid H-sheet material is made by a process that results in long needle-shaped

---

[1] Edwin H. Land: 1909-1991. American physicist and founder of the Polaroid Corporation.

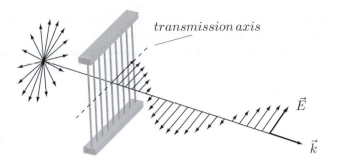

Figure 9.7   A wiregrid polarizer. The transmission axis is perpendicular to the wires.

iodine crystals aligned within a transparent substrate at an average spacing of only a few Angstroms. Polaroid material can be made in thin sheets that are ideal for *Polaroid sunglasses* which selectively absorbs $\vec{E}_{\parallel}$ glare from horizontal surfaces, as discussed in Section 3.5.

It is also possible to utilize Brewster's angle to polarize by reflection using a pile-of-plates linear polarizer as shown in Figure 9.8(a). This device consists of an array of dielectric slabs oriented so that the input beam is incident at Brewster's angle. As noted in Section 3.5, light externally incident at Brewster's angle refracts so that it is also incident internally at the internal Brewster angle. At each interface, the reflection is zero for the $\vec{E}_{\parallel}$ polarization, which is to say that the reflection is completely $\vec{E}_{\perp}$. Thus, the proportion of $\vec{E}_{\perp}$ becomes successively less, giving a transmitted beam that is almost completely $E_{\parallel}$. In Figure 9.8(a), $\vec{E}_{\perp}$ points out of the page, giving a transmission axis that lies in the plane of the page, as shown. Orienting two sets of plates as shown in Figure 9.8 gives a transmitted beam that travels along the same optical axis as the incident beam.

Figure 9.8(b) illustrates *polarizing beam splitter*. The space between the two wedge halves is filled with alternating layers of high and low-index films. A typical design uses two identical right angle prisms, giving a reflected beam at 90°. From the formula for Brewster angle ($\theta_B = \tan^{-1}\left(\frac{n_t}{n_i}\right)$) it is evident that it is not possible to have $\theta_B = 45°$. Thus, three indices of refraction are used. A typical design might use $n_p = 1.62$ for the right angle prisms, $n_l = 1.38$ for the low-index layer, and $n_h = 2.03$ for the high-index layer. According to Snell's law, a ray incident normally on the n_p-n_l layer transmits into the low-index layer at 56°. This becomes the incident angle for the first low-high index interface, and for $n_h = 2.03$, this incident angle is also Brewster's angle. Since each layer is a slab, the transmitted ray encounters all subsequent interfaces at Brewster's angle except for the last n_l-n_p interface where it emerges parallel to the original direction. Since the layers are thin, they act much like a uniform reflecting surface, giving a reflected $\vec{E}_{\perp}$ beam and a transmitted $\vec{E}_{\parallel}$ beam. Reflections of $\vec{E}_{\parallel}$ from the first and last interfaces adds to the reflected beam, while multiple reflections within the layers can introduce some of the $\vec{E}_{\perp}$ light back into the transmitted beam.

The *degree of linear polarization* V_p is a measure of how completely a beam of light is linearly polarized. Suppose a polarizer can be adjusted in one orientation to pass a maximum irradiance I_{max}, and in another orientation a minimum irradiance I_{min}. In this case, the degree of polarization is

$$V_p = \frac{I_{\max} - I_{\min}}{I_{\max} + I_{\min}} \tag{9.3}$$

V_p typically depends upon wavelength. For example, inexpensive Polaroid sheets typically transmit violet light when crossed. V_p can also depend upon the incident angle. Clearly, the pile-of-plates polarizer will be most effective for collimated beams incident exactly at θ_B.

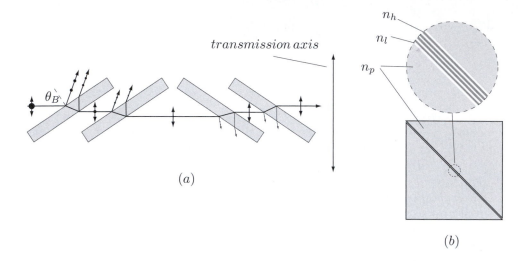

(a)

(b)

Figure 9.8 (a) A pile-of-plates polarizer. At each interface, reflection removes a portion of the $\vec{E}_\perp$ polarization. After several such reflections, the transmitted beam is almost entirely $\vec{E}_\parallel$. (b) A polarizing beam splitter. The circle shows a zoom image of the narrow region between the two wedge halves, which is filled with alternating layers of high- and low-index of refraction layers. A beam incident normally on the cube face encounters each of these layer interfaces at Brewster's angle.

■ **EXAMPLE 9.2**

Design a pile-of-plates polarizer similar to the one shown in Figure 9.8(a) using glass with index $n = 1.5$. Estimate the fraction of horizontally polarized light passed in a collimated beam incident exactly at Brewster's angle.

Solution

According to Section 3.5, the external Brewster's angle this case is

$$\theta_B = \tan^{-1}\left(\frac{n_t}{n_i}\right) = 56.3°$$

The angle between the plate surfaces and the optical axis should be the compliment of this angle.

According to Equation 3.58, the reflection coefficient for $\vec{E}_\perp$ polarization is

$$R_\perp = r_\perp^2 = \left(\frac{n_i \cos\theta_i - n_t \cos\theta_t}{n_i \cos\theta_i + n_t \cos\theta_t}\right)^2$$

For external incidence, $\theta_i = 56.3°$ and $\theta_t = 33.7°$, giving

$$R_\perp = \left(\frac{\cos(56.3°) - 1.5\cos(33.7°)}{\cos(56.3°) + 1.5\cos(33.7°)} \right)^2 = 0.148$$

For internal incidence, $\theta_i = 33.7°$ and $\theta_t = 56.3°$, giving

$$R_\perp = \left(\frac{1.5\cos(33.7°) - \cos(56.3°)}{1.5\cos(33.7°) + \cos(56.3°)} \right)^2 = 0.148$$

The polarizer shown in Figure 9.8 has eight interfaces, each reflecting 14.8% of the $\vec{E}_\perp$. After eight reflections, the fraction of $\vec{E}_\perp$ polarization left in the transmitted beam is only $(1 - 0.148)^8 = 0.278$. Using more plates with a higher index of refraction can decrease this fraction.

This type of polarizer is commonly used with high-power lasers. The reflected beams are absorbed by a water-cooled absorbing enclosure.

Problem 9.1 Find the transmitted irradiance and describe the transmitted polarization state when an unpolarized beam of irradiance I_0 is incident on three coaxial ideal linear polarizers, where the first transmission axis is vertical, the second is inclined at $45°$ to the vertical, and the third is horizontal. Repeat when the middle polarizer is removed without disturbing the first and third. If possible, find three linear polarizers and verify your conclusions.

Problem 9.2 For what angle of external incidence will light reflected from sapphire $(n = 1.76)$ be completely polarized? Repeat for internal incidence.

Problem 9.3 Calculate the degree of polarization for the output beam of Example 9.2 if the input beam is unpolarized.

Problem 9.4 Zinc selenide (ZnSe) is optically transparent with very low absorption in the infrared region. The index of refraction of ZnSe at $10.6\,\mu m$ is 2.42. Find the appropriate plate orientation and ideal extinction ratio for the design of Example 9.2. Include a design sketch that shows the plate orientation.

Problem 9.5 Suppose you have a piece of Polaroid material that has no indication of its transmission axis direction. Describe how you might use glare from a horizontal surface to determine the transmission axis for this polarizer.

9.3 BIREFRINGENCE

Crystalline materials often exhibit *birefringence*, otherwise known as *double refraction*. Crystals possess an ordered atomic structure that causes the index of refraction to be *anisotropic*, having values that depend on both propagation and polarization direction. Double-refraction can be used to separate an incident beam into two beams that are linearly

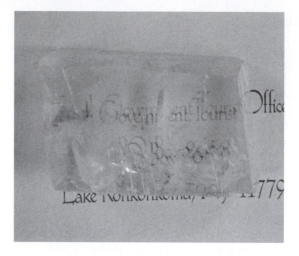

Figure 9.9   Double refraction through a calcite crystal. (Leonard Lessin/Peter Arnold, Inc.)

polarized and mutually orthogonal. If either beam can be isolated, the material can be used as a linear polarizer. Calcite ($CaCO_3$) crystals exhibit a very large double refraction effect, and have long been used in this way.

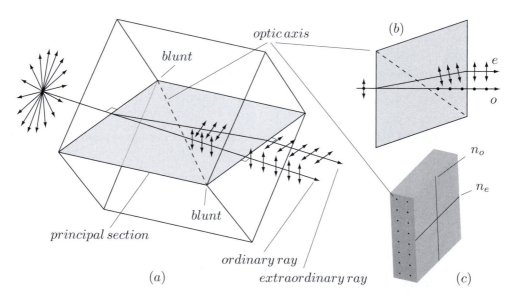

Figure 9.10   (a) A calcite crystal with sides determined by cleavage planes. An unpolarized beam incident normally on one cleavage plane is doubly-refracted. An optic axis passes through the center of symmetry of both blunt ends. (b) A principal section contains the ordinary and extraordinary rays and an optic axis. (c) A square slab of calcite cut with each square face parallel to an optic axis.

Many crystalline materials are birefringent, but some are not. Materials such as rock salt (NaCl) with a cubic crystal lattice are highly symmetric and do not exhibit double refraction. Calcite ($CaCO_3$) has a hexagonal crystal structure with three *cleavage planes*, as illustrated in Figure 9.10(a). Cleavage planes for calcite intersect at two *blunt ends* with three edges inclined equally at $101.55°$. A line that passes through the line of symmetry at either blunt end identifies an *optic axis* of the calcite crystal. The particular calcite crystal illustrated in Figure 9.10(a) has sides determined by cleavage planes, and is shaped symmetrically so that a single optic axis passes through both blunt ends. An optic axis is not unique; any line parallel to a given optic axis is also an optic axis of the crystal. In other words, an optic axis defines a direction of symmetry within the crystal. Calcite is an example of a *uniaxial* crystal with only one such direction.

Experiments on double refraction in calcite crystals lead to the following observations[2]:

1. An unpolarized ray incident normally on a cleavage plane is refracted into two rays, as shown in Figures 9.10(a) and (b). The *ordinary ray* propagates according to Snell's law for normal incidence, and for all other angles of incidence. The index of refraction for the ordinary ray is given by n_o. For calcite, $n_o = 1.6584$ at $\lambda = 589 \, nm$.

2. The second refracted ray obviously does *not* propagate according to Snell's law (see Figure 9.10(b)), and thus is referred to as the *extraordinary ray*. When the incident ray is normal to a cleavage plane, both rays lie within a *principal section* of the crystal, which also contains an optic axis. There are three sets of cleavage planes, and three corresponding families of principal sections. Placing a calcite crystal over a page of text results in two images; when viewed normally, rotating the crystal causes the image produced by the extraordinary ray to move in a circle about the stationary image produced by the ordinary ray, as indicated by Figure 9.10(b).

3. The ordinary ray is polarized perpendicular to its principle section, and the polarization axis of the extraordinary ray lies within its principle section, as illustrated in Figures 9.10(a) and (b).

4. There is no double refraction for a ray traveling along an optical axis. The index of refraction for this direction of propagation is n_o.

5. A slab cut so that the optic axis is parallel to the slab face, as shown in Figure 9.10(c), is characterized by two indices of refraction. For normal incidence, a ray polarized parallel to the optic axis travels with index of refraction n_o, and a ray polarized perpendicular to the optic axis propagates with index of refraction n_e. For calcite, $n_e = 1.4849$ at $\lambda = 589 \, nm$. In this orientation, the ordinary and extraordinary rays do not refract into separate beams. Normally incident light with arbitrary linear polarization may be resolved into components, with the component parallel to the optic axis propagating with velocity determined by n_e, and the component perpendicular to the optic axis propagating with velocity determined by n_o.

■ **APPLICATION NOTE 9.1**

Birefringent Polarizers

To use double refraction in calcite as a linear polarizer, we must isolate at least one of the polarized rays. Two ways of doing this are illustrated in Figure 9.11. Figure

---

[2]These observations may be verified analytically by deriving a set of Fresnel equations that do not assume isotropic dielectric properties. In crystals, the dielectric constant depends on direction, and is described by a tensor rather than a single constant number. Details may be found in Guenther [10].

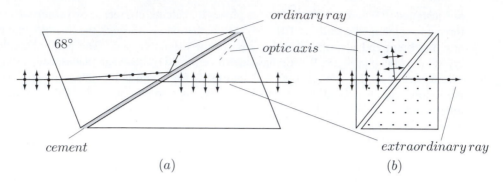

Figure 9.11   (a) Nicol prism. (b) Glan-Foucault prism.

9.11(a) illustrates a *Nicol prism*, which uses total internal reflection to remove the ordinary ray. The figure shows a principal section of a calcite crystal that has been cut into two sections. The two sections are then glued back together with Canada balsam cement, which has an index of refraction of 1.55. The ordinary ray, with $n_o = 1.66$, totally reflects at the calcite-glue interface. The extraordinary ray transmits through the glue and into the second calcite section. The ends of the prism are trimmed to an angle of $68°$ which gives a linearly-polarized transmitted ray that is parallel to the incident ray. Figure 9.11(b) shows a *Glan-Foucault* prism prepared from two sections of calcite that are separated by an air gap. The optic axis of both sections is perpendicular to the page. The angle of incidence at the internal calcite-air interface is such that the ordinary ray is totally reflected, but the extraordinary ray is not. The second section of calcite causes the transmitted ray to propagate parallel to the incident ray. A similar design where the two sections are cemented together is known as a *Glan-Thompson* prism. The non-cemented version is typically more useful for applications involving high-power lasers. Because they are based on total internal reflection, the extinction ratios for both prisms illustrated in Figure 9.11 can be very large.

Problem 9.6   Find the critical angle for total internal reflection for the ordinary ray at an interface between calcite and Canada balsam cement. Repeat when the cement layer is replaced by air.

Problem 9.7   Estimate the fraction of the extraordinary ray that is lost by both internal and external reflection at the calcite-cement interface of a Nichol prism. Repeat for the air interface of a Glan-Foucault prism.

9.4 CIRCULAR AND ELLIPTICAL POLARIZATION

Circular and elliptical polarization results from introducing a relative phase shift between the components $\vec{E}_x$ and $\vec{E}_y$ shown in Figure 9.2. Such phase shifts are very common, and can occur, for example, when a linearly polarized wave propagates through a birefringent material. When a wave *circularly polarized*, the tip of the electric field vector $\vec{E}$ sweeps out a circle over each wavelength of travel, as illustrated in Figure 9.13. For an *elliptically polarized* wave, this becomes an ellipse. It is important to note that circular and elliptical polarization is just as fundamental, just as important, and just as common as linear polarization. The capability to measure and manipulate these additional polarization states can create opportunities for experiment and application.

Suppose that we somehow manage to introduce a phase shift for $\vec{E}_x$ that is different from that for $\vec{E}_y$. In this case, Equation 9.1 becomes

$$\vec{E} = \hat{i}E_{0x}e^{i(kz-\omega t+\phi_x)} + \hat{j}E_{0y}e^{i(kz-\omega t+\phi_y)}$$

Factoring out the phase term ϕ_y from both terms gives

$$\vec{E} = e^{i\phi_x}\left(\hat{i}E_{0x}e^{i(kz-\omega t)} + \hat{j}E_{0y}e^{i(kz-\omega t+\Delta\phi)}\right)$$

where $\Delta\phi = \phi_y - \phi_x$. The term $e^{i\phi_x}$ that multiplies both components of $\vec{E}$ represents an absolute phase that in practice is not measured, so we can safely ignore it. Thus, the electromagnetic wave given by

$$\vec{E} = \hat{i}E_{0x}e^{i(kz-\omega t)} + \hat{j}E_{0y}e^{i(kz-\omega t+\Delta\phi)} \tag{9.4}$$

represents a wave where $\vec{E}_y$ has a phase difference of $\Delta\phi$ relative to $\vec{E}_x$.

Consider the case where $\Delta\phi = \frac{\pi}{2}$ and where the amplitudes E_{0x} and E_{0y} are equal, with $E_{0x} = E_{0y} = E_0$. In this case, Equation 9.4 becomes

$$\vec{E} = \hat{i}E_0e^{i(kz-\omega t)} + \hat{j}E_0e^{i\left(kz-\omega t+\frac{\pi}{2}\right)} \tag{9.5}$$

We will now show that the tip of vector $\vec{E}$ sweeps out a complete *circle* as the wave of Equation 9.5 travels through a complete wavelength. To show this, we take the real part of Equation 9.5, which by the Euler relation (see Section 1.6) is proportional to a cosine:

$$\vec{E} = \hat{i}E_0 \cos(kz - \omega t) + \hat{j}E_0 \cos\left(kz - \omega t + \frac{\pi}{2}\right) \tag{9.6}$$

Figure 9.12 illustrates the orientation of $\vec{E}$ at two times t_1 and t_2 that are short compared to the period of the wave. In this figure, the propagation vector $\vec{k}$ points out of the page, along positive-z. If we let $z = 0$ at the origin, then Equation 9.6 becomes

$$\vec{E} = \hat{i}E_0 \cos(-\omega t) + \hat{j}E_0 \cos\left(-\omega t + \frac{\pi}{2}\right) \tag{9.7}$$

Note that the time term in Equation 9.7 is negative. At $t = 0$, the component $E_{0x} = E_0$ and $E_{0y} = 0$. A short time t_1 later, E_{0x} has decreased, and E_{0y} is small and positive. Still later at $t = t_2$, the vector $\vec{E}$ has a smaller x-component and a larger y-component. Since the magnitude of $\vec{E}$ is E_0 at all times, the vector $\vec{E}$ sweeps out a circle of radius E_0 as the tip of $\vec{E}$ rotates *counter-clockwise*.

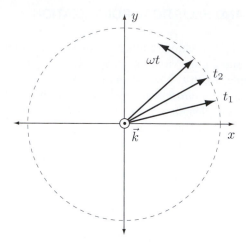

Figure 9.12   The electric field vector $\vec{E}$ in a left-circularly polarized electromagnetic wave rotates counter-clockwise as the wave propagates out of the page.

Figure 9.13(a) shows the wave of Equation 9.5 from another perspective. Figure 9.13(c) shows the corresponding counterclockwise electric field rotation for a wave that emerges from the plane of the page. Circularly polarized light with an electric field vector that rotates as in Figure 9.13(a) is said to be *left-circularly polarized*[3].

If the value of $\Delta\phi$ in Equation 9.4 is equal to $-\frac{\pi}{2}$, again with $E_{0x} = E_{0y} = E_0$, the wave is given by

$$\vec{E} = \hat{i}E_0 e^{i(kz-\omega t)} + \hat{j}E_0 e^{i\left(kz-\omega t-\frac{\pi}{2}\right)} \tag{9.8}$$

In this case, the electric field rotates as in Figure 9.13(b). Such a wave is said to be *right-circularly polarized*. As this wave emerges from the plane of the page, the electric field vector rotates *clockwise*, as shown in Figure 9.13(d).

If E_{0x} and E_{0y} are not equal, or if $\Delta\phi$ is anything other than $\pm\frac{\pi}{2}$, the wave can be *elliptically polarized*. There is left and right elliptical polarization that follows the same convention as the circular polarization of Figure 9.13. Figure 9.14 illustrates elliptical polarization states for various values of $\Delta\phi$. In Figure 9.13(a), a value of $\Delta\phi = 0$ gives linear polarization at an angle of 60° from the horizontal. As $0 < \Delta\phi < \pi$, the wave assumes various states of left elliptical polarization. Similarly, as $0 > \Delta\phi > -\pi$, the wave is right elliptically polarized. A value of $\Delta\phi = \pm\pi$ results in a linear polarization state that is rotated by 60° from the $\Delta\phi = 0$ state. In Figure 9.13(b) a value of $\Delta\phi = 0$ gives linear polarization at an angle of 45° from the horizontal, with $E_{0x} = E_{0y}$. Thus, $\Delta\phi = \frac{\pi}{2}$ gives left-circular polarization and $\Delta\phi = -\frac{\pi}{2}$ gives right-circular polarization, as illustrated. For other values of $\Delta\phi$ the polarization is left and right elliptical, as illustrated. When $\Delta\phi = \pm\pi$, the polarization is rotated by 90°.

9.4.1   Wave Plates and Circular Polarizers

A *wave plate* (also known as a *retarder*) may be used to introduce the $\Delta\phi$ phase shift necessary to produce elliptically and circularly polarized light. An example of a wave

---

[3]The terminology seems opposite to the right-hand rule, but the convention used here is common. The choice is arbitrary, but once made it is important to be consistent.

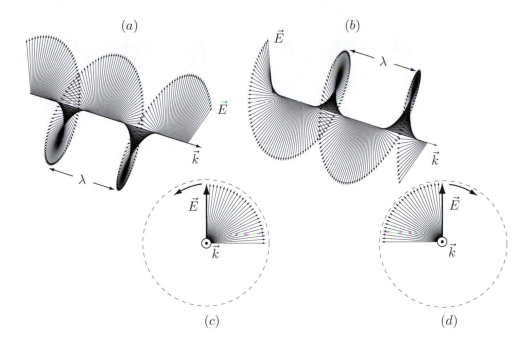

Figure 9.13 (a) Left circular polarization. (b) Right-circular polarization. (c) Rotation of $\vec{E}$ as a left-circular wave propagates out of the page. (d) Rotation of $\vec{E}$ as a right-circular wave propagates out of the page.

plate is illustrated by Figure 9.10(c), which shows a birefringent slab characterized by two indices of refraction n_o and n_e. According to observation 5 in Section 9.2.1, a linearly polarized beam incident normally on such a slab may resolved into components that each travel a different optical distance (see Section 4.2) through the slab. If the thickness of the slab is d, then the n_e ray travels optical distance $n_e d$ and the n_o ray travels optical distance $n_o d$. Since a wave changes phase by $k \Delta z$ when traveling a distance Δz, the relative phase shift between the n_o and n_e components after traveling through the slab is

$$\Delta\phi = k_o d - k_e d = \frac{2\pi n_e}{\lambda_0} d - \frac{2\pi n_o}{\lambda_0} d = \frac{2\pi d}{\lambda_0} (n_o - n_e) \tag{9.9}$$

where λ_0 is the vacuum wavelength. According to Equation 9.4, we defined $\Delta\phi$ as $\Delta\phi = \phi_y - \phi_x$. Using this convention, Equation 9.9 becomes

$$\Delta\phi = \frac{2\pi d}{\lambda_0} (n_y - n_x) \tag{9.10}$$

In calcite, $n_e = 1.4849$ and $n_0 = 1.6584$ at $\lambda = 589 nm$. The axis with the smallest value of n is called the *fast axis*, since light with linear polarization along this axis propagates with the greater speed. In Figure 9.10(c), the fast axis is horizontal.

A *quarter-wave plate* can be used to produce circularly polarized light. It is designed so that

$$d |n_e - n_o| = (4m + 1) \frac{\lambda_0}{4} \tag{9.11}$$

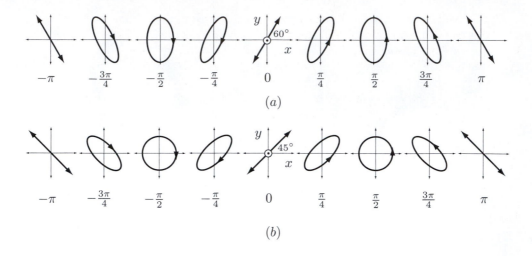

Figure 9.14 Polarization states for various phase differences between E_x and E_y. In (a), $E_x \neq E_y$ and in (b) they are equal.

where $m = 1, 2, 3, \ldots$ Notice that this design is only applicable at the target vacuum wavelength λ_0. The parameter m allows for a phase shift of $m(2\pi) \pm \pi/2$. A quarter-wave plate illuminated with light linearly polarized at $45°$ transmits circularly polarized light. If $n_y > n_x$ (i.e. if the fast axis is horizontal), then $\Delta\phi = +\frac{\pi}{2}$.[4] If such a quarter-wave plate is illuminated with linearly polarized light inclined a $45°$ to the horizontal fast axis, the transmitted beam is left-circularly polarized, as described by Equation 9.5 (see Figure 9.15(a)). Rotating the quarter-wave plate so that its fast axis is vertical gives $\Delta\phi = -\frac{\pi}{2}$, which according to Equation 9.8 gives right-circular polarization with the same illumination, as illustrated in Figure 9.15(b).

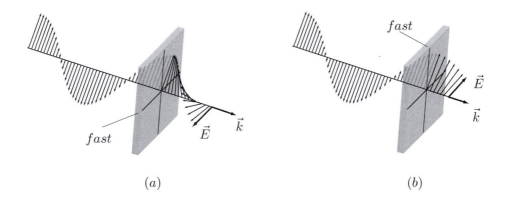

Figure 9.15 A quarter-wave plate is illuminated by light linearly polarized at $45°$. In (a), the fast axis of the wave plate is horizontal, giving a left-circular output beam. In (b), the fast axis is vertical, producing right-circular polarization.

[4]We ignore the factor of 2π, since it produces no observable effect.

Calcite is brittle and soft, and is not commonly used for wave plates. Crystalline quartz has a much smaller double refraction, with $n_o = 1.5442$ and $n_e = 1.5533$[5], which allows for smaller values of m. Many plastics become birefringent when stretched, as you can easily verify by placing a stretched piece of plastic wrap between two crossed polarizers. A very controlled stretch applied to a plastic sheet while monitoring with a probe beam at the target wavelength is often used to produce wave plates with somewhat less optical quality than those made of crystalline quartz.

■ EXAMPLE 9.3

Find the minimum thickness of a quarter-wave plate for use at 589 nm using quartz. Find the thickness when $m = 200$.

Solution

The minimum thickness uses $m = 0$. According to Equation 9.11

$$d = \frac{\lambda_0}{4\,|n_e - n_o|} = \frac{5.89 \times 10^{-7} m}{4\,(1.5533 - 1.5442)} = 16.2\,\mu m$$

If $m = 200$,

$$d = \frac{\lambda_0}{4\,|n_e - n_o|} = \frac{(201)\,5.89 \times 10^{-7} m}{4\,(1.5533 - 1.5442)} = 3.25\,mm$$

The second result gives a plate thickness of about $1/8$ inch, which is easier to fabricate and is also more durable.

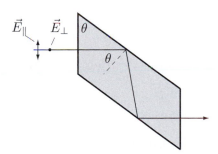

Figure 9.16   A Fresnel rhomb. In this orientation, the transmitted beam is left-circular. Rotating the rhomb by $90°$ gives right-circular.

■ APPLICATION NOTE 9.2

Fresnel Rhomb

It is possible to construct a quarter-wave retarder using total internal reflection. In Section 3.5.5, we saw that the total internal reflection phase shift for $\vec{E}_\perp$ was larger

---

[5]Both indices are measured at 589 nm. Since $n_e > n_o$, quartz is called a *positive* uniaxial crystal. Calcite has a negative uniaxial crystal.

than that for $\vec{E}_{\parallel}$, and we defined the relative phase shift $\Delta\phi = \phi_{\perp} - \phi_{\parallel}$, which is plotted for a glass-air interface in Figure 3.16. The data indicate a $\Delta\phi$ of 45° for an incident angle about 53°. A *Fresnel rhomb* (see Figure 9.16) uses two such reflections to give a net phase shift of 90°.

_____

A *half-wave plate* can be used to rotate a linear polarization state. It is designed so that

$$d\left|n_e - n_o\right| = (4m + 1)\frac{\lambda_0}{2} \tag{9.12}$$

where $m = 1, 2, 3, \ldots$ and λ_0 is the target wavelength. Consider a half-wave plate illuminated with light linearly polarized along an axis inclined at θ to the horizontal. If the fast axis is horizontal, the y-component is phase-shifted by π and if the fast axis is vertical, the y-component is phase-shifted by $-\pi$. In either case, the sign on the y-component reverses, and the polarization is rotated by 2θ. Using $\theta = 45°$ rotates the polarization by 90°, which cannot be done with a single linear polarizer. It should be obvious from Equations 9.11 and 9.12 that two quarter-wave plates can be combined to give a half-wave plate.

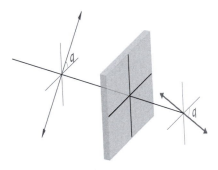

Figure 9.17   A half-wave plate.

■ **EXAMPLE 9.4**

A monochromatic beam of left-circularly polarized light is incident normally on a quarter-wave plate with fast axis horizontal. Describe the transmitted beam.

Solution

The incident beam is given by Equation 9.5:

$$\vec{E} = \hat{i}E_0 e^{i(kz-\omega t)} + \hat{j}E_0 e^{i\left(kz-\omega t+\frac{\pi}{2}\right)}$$

Since the fast axis is horizontal, the vertical axis receives an extra $\pi/2$ phase shift after passing through the quarter-wave plate. This phase shift adds to the one already there, resulting in an output beam given by

$$\vec{E} = \hat{i}E_0 e^{i(kz-\omega t)} + \hat{j}E_0 e^{i(kz-\omega t+\pi)} = \hat{i}E_0 e^{i(kz-\omega t)} - \hat{j}E_0 e^{i(kz-\omega t)}$$

This is a linearly-polarized beam along an axis at $-45°$.

Problem 9.8　In Example 9.3, what value of m would be required if calcite were used to make a quarter-wave plate with a thickness of about $3.25\,mm$.

Problem 9.9　Repeat Example 9.4 when the fast axis is vertical.

Problem 9.10　Repeat Example 9.4 when the quarter-wave plate is replaced by a half-wave plate with fast axis horizontal, and again when the fast axis is vertical.

Problem 9.11　Describe the effect of combining two half-wave plates with their fast axes parallel, and again when their fast axes are perpendicular.

9.5 JONES VECTORS AND MATRICES

Jones vectors and matrices can make it easier to calculate the effect of optical elements such as linear polarizers and wave plates, especially when the optical train consists of several of these elements combined coaxially in series. We begin by defining a Jones vector for linearly polarized light. According to Equation 9.1,

$$\vec{E} = \vec{E}_0 e^{i(kz-\omega t+\varphi)} = \left(E_{0x}\hat{i} + E_{0y}\hat{j} \right) e^{i(kz-\omega t+\varphi)}$$

$$= E_0 \left(\cos\theta\hat{i} + \sin\theta\hat{j} \right) e^{i(kz-\omega t+\varphi)}$$

In the Jones representation, the $e^{i(kz-\omega t+\varphi)}$ is understood. Thus, the Jones vector for linearly polarized light is given by

$$\tilde{E} = \left[\begin{array}{c} E_{0x} \\ E_{0y} \end{array} \right] = E_0 \left[\begin{array}{c} \cos\theta \\ \sin\theta \end{array} \right] \tag{9.13}$$

where the notation $\tilde{E}$ simply means that we are representing the vector in this abbreviated way. Similarly, the elliptically polarized light of Equation 9.4 has a Jones vector given by

$$\tilde{E} = \left[\begin{array}{c} E_{0x} \\ E_{0y}e^{i\Delta\phi} \end{array} \right] = E_0 \left[\begin{array}{c} \cos\theta \\ \sin\theta e^{i\Delta\phi} \end{array} \right] \tag{9.14}$$

Jones vectors can be *normalized* so that the magnitude of the vector is one. Thus, the normalized version of Equation 9.13 is

$$\tilde{E} = \left[\begin{array}{c} \cos\theta \\ \sin\theta \end{array} \right] \tag{9.15}$$

This is normalized since $\cos^2\theta + \sin^2\theta = 1$. If the vector is complex, the *magnitude-squared* of both components must sum to one. For example, the normalized version of Equation 9.14 is

$$\tilde{E} = \left[\begin{array}{c} \cos\theta \\ \sin\theta e^{i\Delta\phi} \end{array} \right] \tag{9.16}$$

This has magnitude one since $\left|e^{i\Delta\phi}\right| = \left(e^{i\Delta\phi}\right)\left(e^{i\Delta\phi}\right)^* = \left(e^{i\Delta\phi}\right)\left(e^{-i\Delta\phi}\right) = 1$.

In Section 2.3.3, we found that the irradiance of an electromagnetic wave is given by

$$I = \frac{\epsilon v}{2} E_0^2 \tag{9.17}$$

In the Jones notation, a normalized Jones vector represents "one unit" of irradiance. Thus, the Jones vector

$$\tilde{E} = \frac{1}{2}\begin{bmatrix} 1 \\ 1 \end{bmatrix} = \frac{1}{\sqrt{2}}\left(\frac{1}{\sqrt{2}}\begin{bmatrix} 1 \\ 1 \end{bmatrix}\right)$$

represents a wave of one-half unit irradiance, since $\left(\frac{1}{\sqrt{2}}\right)^2 = \frac{1}{2}$.

We may think of the normalized Jones vectors for horizontal and vertical linear polarization as *unit vectors* $\hat{e}_h$ and $\hat{e}_v$:

$$\hat{e}_h = \begin{bmatrix} 1 \\ 0 \end{bmatrix} \qquad \hat{e}_v = \begin{bmatrix} 0 \\ 1 \end{bmatrix} \tag{9.18}$$

Using this notation, the arbitrary state of elliptical polarization given by Equation 9.14 can be represented as a linear combination of linear polarization states:

$$\tilde{E} = \begin{bmatrix} \cos\theta \\ \sin\theta\, e^{i\Delta\phi} \end{bmatrix} = (\cos\theta)\,\hat{e}_h + \left(\sin\theta\, e^{i\Delta\phi}\right)\hat{e}_v \tag{9.19}$$

In the Jones notation, left-circular polarization given by Equation 9.7 becomes

$$\tilde{E} = \begin{bmatrix} E_0 \\ E_0 e^{i\frac{\pi}{2}} \end{bmatrix} = \begin{bmatrix} E_0 \\ iE_0 \end{bmatrix} = E_0 \begin{bmatrix} 1 \\ i \end{bmatrix}$$

since $e^{i\frac{\pi}{2}} = i$. Similarly, right-circular polarization of Equation 9.8 is given by

$$\tilde{E} = E_0 \begin{bmatrix} 1 \\ -i \end{bmatrix}$$

We may also define unit vectors $\hat{e}_r$ and $\hat{e}_l$ given by the normalized Jones vectors for right and left-circular polarization:

$$\hat{e}_r = \frac{1}{\sqrt{2}}\begin{bmatrix} 1 \\ -i \end{bmatrix} \qquad \hat{e}_l = \frac{1}{\sqrt{2}}\begin{bmatrix} 1 \\ i \end{bmatrix} \tag{9.20}$$

Notice that $\hat{e}_r + \hat{e}_l$ gives a *linear* polarization state:

$$\hat{e}_r + \hat{e}_l = \frac{1}{\sqrt{2}}\begin{bmatrix} 1 \\ -i \end{bmatrix} + \frac{1}{\sqrt{2}}\begin{bmatrix} 1 \\ i \end{bmatrix} = \frac{1}{\sqrt{2}}\begin{bmatrix} 2 \\ 0 \end{bmatrix} = \sqrt{2}\hat{e}_h$$

Thus,

$$\hat{e}_h = \frac{\hat{e}_l + \hat{e}_r}{\sqrt{2}} \tag{9.21}$$

$$\hat{e}_v = \frac{\hat{e}_l - \hat{e}_r}{\sqrt{2}i} \tag{9.22}$$

Thus, any arbitrary state of linear polarization may be represented as a linear combination of $\hat{e}_r$ and $\hat{e}_l$. In fact, both sets of unit vectors are equally fundamental, and may be used to represent any state of linear or elliptical polarization. In the language of vector algebra, both sets of unit vectors *span the space* of all possible polarization states.

Table 9.1. Jones Matrices

Linear Polarizers

Transmission Axis Horizontal $\begin{bmatrix} 1 & 0 \\ 0 & 0 \end{bmatrix}$ Transmission Axis Vertical $\begin{bmatrix} 0 & 0 \\ 0 & 1 \end{bmatrix}$

Transmission Axis $+45°$ $\frac{1}{2}\begin{bmatrix} 1 & 1 \\ 1 & 1 \end{bmatrix}$ Transmission Axis $-45°$ $\frac{1}{2}\begin{bmatrix} 1 & -1 \\ -1 & 1 \end{bmatrix}$

Quarter-wave Plates

Fast Axis Horizontal $\begin{bmatrix} 1 & 0 \\ 0 & i \end{bmatrix}$ Fast Axis Vertical $\begin{bmatrix} 1 & 0 \\ 0 & -i \end{bmatrix}$

Rotator

$\theta \rightarrow \theta + \alpha$ $\begin{bmatrix} \cos\alpha & -\sin\alpha \\ \sin\alpha & \cos\alpha \end{bmatrix}$

Jones Matrices

Optical elements such as linear polarizers and wave plates are represented in the Jones notation by 2×2 matrices. They are used according to the following calculation convention. Suppose an incident beam $\tilde{E}_i$ illuminates an optical element represented by the matrix M. The beam $\tilde{E}_t$ that is transmitted by this element is given by

$$\tilde{E}_t = M\,\tilde{E}_i \qquad (9.23)$$

It is important to note that matrix multiplication is *not* commutative, so the order is important in Equation 9.23. Jones matrices for a variety of optical elements are given in Table 9.1..

■ **EXAMPLE 9.5**

Show that the Jones matrix for rotation is given by the matrix shown in Table 9.1..

Solution

The rotation matrix should increment the angle in each component of an incident wave by α:

$$\begin{bmatrix} a & b \\ c & d \end{bmatrix} E_0 \begin{bmatrix} \cos\theta \\ \sin\theta \end{bmatrix} = E_0 \begin{bmatrix} \cos(\theta+\alpha) \\ \sin(\theta+\alpha) \end{bmatrix}$$

where we have used the symbols a—d to represent the elements of the rotation matrix. Multiplying and canceling the factor of E_0 gives

$$a\cos\theta + b\sin\theta = \cos(\theta+\alpha)$$
$$c\cos\theta + d\sin\theta = \sin(\theta+\alpha)$$

Use the trigonometry identities for angle addition:

$$\cos (\theta + \alpha) = \cos \theta \cos \alpha + \sin \theta \sin \alpha$$
$$\sin (\theta + \alpha) = \sin \theta \cos \alpha + \cos \theta \sin \alpha$$

Solve the two sets of equations simultaneously to give

$$a = \cos \alpha \qquad b = -\sin \alpha$$
$$c = \sin \alpha \qquad d = \cos \alpha$$

giving the matrix indicated in Table 9.1..

When $\tilde{E}_i$ is a normalized Jones vector, the magnitude-squared of $\tilde{E}_t$ represents the fraction of incident intensity that is passed by the optical element. The matrices in Table 9.1. are easily derived from Equation 9.23, but it is sufficient to simply verify that they have the intended effect.

■ **EXAMPLE 9.6**

Find the Jones vector of the transmitted beam and the fraction of intensity transmitted for the following cases: (a) Linearly polarized light at $45°$ incident on a linear polarizer with transmission axis vertical and at $45°$, and (b) light polarized at $45°$ incident on a quarter-wave plate with horizontal fast axis.

Solution

(a) Using a normalized Jones vector and the Jones matrix for a linear polarizer with vertical transmission axis, we obtain

$$\tilde{E}_t = \begin{bmatrix} 0 & 0 \\ 0 & 1 \end{bmatrix} \frac{1}{\sqrt{2}} \begin{bmatrix} 1 \\ 1 \end{bmatrix} = \frac{1}{\sqrt{2}} \begin{bmatrix} 0 \\ 1 \end{bmatrix} = \frac{1}{\sqrt{2}} \hat{e}_v$$

Note that the polarizer has passed the vertical component, but not the horizontal component, as required. The magnitude squared of the transmitted Jones vector is one half, so half of the incident intensity is transmitted.

For a linear polarizer with transmission axis at $45°$,

$$\tilde{E}_t = \frac{1}{2} \begin{bmatrix} 1 & 1 \\ 1 & 1 \end{bmatrix} \frac{1}{\sqrt{2}} \begin{bmatrix} 1 \\ 1 \end{bmatrix} = \frac{1}{2\sqrt{2}} \begin{bmatrix} 2 \\ 2 \end{bmatrix} = \frac{1}{\sqrt{2}} \begin{bmatrix} 1 \\ 1 \end{bmatrix}$$

Of course, in this case, the entire beam is transmitted since the polarization is aligned with the transmission axis. Similarly, the magnitude squared of the transmitted beam is one, so all of the intensity is transmitted.

(b) Using a normalized Jones vector for the incident beam and the Jones matrix for a quarter-wave plate with horizontal fast axis gives

$$\tilde{E}_t = \begin{bmatrix} 1 & 0 \\ 0 & i \end{bmatrix} \frac{1}{\sqrt{2}} \begin{bmatrix} 1 \\ 1 \end{bmatrix} = \frac{1}{\sqrt{2}} \begin{bmatrix} 1 \\ i \end{bmatrix}$$

giving the normalized Jones vector for a left-circularly polarized transmitted beam. Thus, none of the beam power was removed by the quarter-wave plate.

The real power of the Jones approach lies in the ability to treat more than one optical element. Equation 9.23 may be extended to include any number of optical elements in the following way:

$$\tilde{E}_t = M_N \dots M_2 M_1 \tilde{E}_i \qquad (9.24)$$

where M_1 represents the element the incident light encounters first, and M_N represents the element the transmitted beam emerges from. The elements are assumed to be arranged coaxially so that the beam transmitting from one falls normally on the center of the next.

■ **EXAMPLE 9.7**

Find the transmitted beam when light polarized at $45°$ is incident on a half-wave plate with fast axis horizontal, followed by a quarter-wave plate with fast axis vertical, followed by a linear polarizer with transmission axis at $-45°$.

Solution

There is no half-wave plate in Table 9.1., but we can represent one with a group of two quarter-wave plates in series. The transmitted beam is given by

$$\tilde{E}_t = \frac{1}{2}\begin{bmatrix} 1 & -1 \\ -1 & 1 \end{bmatrix}\begin{bmatrix} 1 & 0 \\ 0 & -i \end{bmatrix}\begin{bmatrix} 1 & 0 \\ 0 & i \end{bmatrix}\begin{bmatrix} 1 & 0 \\ 0 & i \end{bmatrix}\frac{1}{\sqrt{2}}\begin{bmatrix} 1 \\ 1 \end{bmatrix}$$

The matrices must be multiplied in groups of two. Begin with the last two on the right (M_1 and M_2) to obtain the Jones matrix for a half-wave plate with horizontal fast axis:

$$\tilde{E}_t = \frac{1}{2}\begin{bmatrix} 1 & -1 \\ -1 & 1 \end{bmatrix}\begin{bmatrix} 1 & 0 \\ 0 & -i \end{bmatrix}\begin{bmatrix} 1 & 0 \\ 0 & -1 \end{bmatrix}\frac{1}{\sqrt{2}}\begin{bmatrix} 1 \\ 1 \end{bmatrix}$$

Now multiply that last two matrices on the right above to obtain the combination of a half-wave plate with horizontal fast axis and a quarter-wave plate with vertical fast axis:

$$\tilde{E}_t = \frac{1}{2}\begin{bmatrix} 1 & -1 \\ -1 & 1 \end{bmatrix}\begin{bmatrix} 1 & 0 \\ 0 & i \end{bmatrix}\frac{1}{\sqrt{2}}\begin{bmatrix} 1 \\ 1 \end{bmatrix}$$

You should recognize the last matrix above as that of a quarter-wave plate with horizontal fast axis. The quarter-wave plate removed half of the phase shift introduced by the half-wave plate, giving the effective combination of a quarter-wave plate with horizontal fast axis. The remaining calculations can proceed in any order. It is instructive to now multiply the last matrix and the column vector representing $\tilde{E}_i$:

$$\tilde{E}_t = \frac{1}{2}\begin{bmatrix} 1 & -1 \\ -1 & 1 \end{bmatrix}\frac{1}{\sqrt{2}}\begin{bmatrix} 1 \\ i \end{bmatrix}$$

This represents left-circular light (produced by the quarter-wave plate and the incident linearly polarized beam) incident on a linear polarizer at $45°$. Note that this combination should give the same transmission regardless of orientation of the second polarizer. The final transmitted beam is given by

$$\tilde{E}_t = \frac{1}{2\sqrt{2}}\begin{bmatrix} 1-i \\ -(1-i) \end{bmatrix}$$

The complex number $z = 1 - i$ may be converted to polar form to give $z = \sqrt{2}e^{-i\frac{\pi}{4}}$. Using this, the transmitted beam becomes

$$\tilde{E}_t = \frac{1}{2}e^{-i\frac{\pi}{4}}\begin{bmatrix} 1 \\ -1 \end{bmatrix} = \frac{1}{\sqrt{2}}e^{-i\frac{\pi}{4}}\left(\frac{1}{\sqrt{2}}\begin{bmatrix} 1 \\ -1 \end{bmatrix}\right)$$

where we have grouped the result in terms of the normalized Jones vector for linear polarization at $-45°$. The phase term $e^{-i\frac{\pi}{4}}$ can usually be ignored, since it represents a phase shift that occurs for both components. Since $\left|e^{-i\frac{\pi}{4}}\right| = 1$, this term does not affect the transmitted irradiance, which is one-half the incident irradiance.

The Jones matrix for rotation can be used to rotate a Jones vector as illustrated in Example 9.5. Using the rotation matrix to rotate a Jones matrix can be determined in the following way. Consider a Jones matrix M that acts upon $\tilde{E}_i$ to give $\tilde{E}_t$

$$\tilde{E}_t = M\tilde{E}_i$$

Now consider a rotation matrix $R(\alpha)$ that acts upon $\tilde{E}_i$ to give $\tilde{E}_i'$, and also acts upon $\tilde{E}_t$ to give rotated Jones vector $\tilde{E}_t'$. Noting that the inverse of $R(\alpha)$ is given by $R(-\alpha)$ (see Problem 9.16), we obtain

$$\begin{aligned} \tilde{E}_t' &= R(\alpha)\tilde{E}_t \\ &= R(\alpha)\left(M\tilde{E}_i\right) \\ &= R(\alpha)M(R(-\alpha)R(\alpha))\tilde{E}_i \\ &= R(\alpha)M(R(-\alpha)R(\alpha))\tilde{E}_i \end{aligned}$$

where in the last step the quantity within the parentheses is the identity matrix. Rearranging gives

$$\tilde{E}_t' = (R(\alpha)MR(-\alpha))\tilde{E}_i'$$

Thus, the matrix $R(\alpha)MR(-\alpha)$ produces the same transformation as M acting on the unrotated Jones vectors. The rotation of a Jones matrix is therefore given by

$$M' = R(\alpha)MR(-\alpha) \tag{9.25}$$

Problem 9.12   Describe the transmitted beam for the following cases: (a) Vertically polarized light incident on a linear polarizer with transmission axis horizontal and at $-45°$, and (b) light polarized at $45°$ incident on a quarter-wave plate with vertical fast axis.

Problem 9.13   Find the transmitted Jones vector when light polarized at $45°$ is incident on a half-wave plate with fast axis vertical, followed by a quarter-wave plate with fast axis vertical, followed by a linear polarizer with vertical transmission axis. Determine the irradiance of the transmitted beam relative to the incident beam.

Problem 9.14   Use Table 9.1. to show that the Jones matrix for the following optical elements are given by the expressions indicated.

a) Left-circular polarizer:

$$\frac{1}{2} \begin{bmatrix} 1 & -i \\ i & 1 \end{bmatrix}$$

b) Right-circular polarizer:

$$\frac{1}{2} \begin{bmatrix} 1 & i \\ -i & 1 \end{bmatrix}$$

c) Half-wave plate, fast axis horizontal:

$$\begin{bmatrix} 1 & 0 \\ 0 & -1 \end{bmatrix}$$

d) Show that the Jones matrix for a half-wave plate with fast axis vertical is given by the expression above, and explain why this makes sense.

Problem 9.15   Show that the combination of two quarter-wave plates, one with fast axis vertical and one with fast axis horizontal, produces no change in the transmitted wave. Explain why this is so.

Problem 9.16   Show that if a rotation matrix rotates a Jones vector through an angle α, then the rotation matrix that rotates the same Jones vector through $-\alpha$ is its inverse. In other words, show that

$$R(\alpha) R(-\alpha) = \begin{bmatrix} 1 & 0 \\ 0 & 1 \end{bmatrix}$$

and thus, $R(-\alpha)$ is the inverse of $R(\alpha)$:

$$R(-\alpha) = R^{-1}(\alpha)$$

Show also that this result is commutative:

$$R(-\alpha) R(\alpha) = \begin{bmatrix} 1 & 0 \\ 0 & 1 \end{bmatrix}$$

Problem 9.17   Use the rotation matrix to rotate the Jones matrix for a linear polarizer with vertical transmission axis by $90°$, and show that it gives the result for a linear polarizer with horizontal transmission axis.

Problem 9.18   Use the rotation matrix to verify the Jones matrices given in Table 9.1. for linear polarizers with transmission axes at $\pm 45°$.

9.5.1 Birefringent Colors

Many plastics consist of long, randomly oriented polymer molecules, and stretching or stressing them can cause a partial alignment that can cause a *stress birefringence*. You can easily observe this effect by stretching a piece of plastic wrap and placing it between two polarizers. In most cases, you will see a quite beautiful display of colors that change as you rotate one of the polarizers or the plastic material.

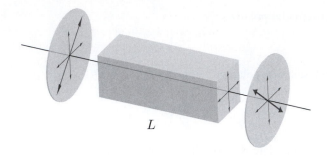

Figure 9.18 A retarder placed between two crossed linear polarizers.

The colors result from the wavelength dependent phase shifts that result from the birefringence of the plastic material. Consider the retarder illustrated in Figure 9.18. The incident light is polarized at $45°$, and this light illuminates the retarder characterized by indices n_x and n_y. Light emerging from the retarder passes through a second linear polarizer with transmission axis at $-45°$. The Jones matrix for the retarder is given by

$$\begin{bmatrix} 1 & 0 \\ 0 & e^{i\delta} \end{bmatrix}$$

where the phase shift δ is given by

$$\delta = \frac{2\pi\left(n_y - n_x\right)L}{\lambda_0} \tag{9.26}$$

To find the transmitted Jones vector, we must replace the first linear polarizer with an incident beam linearly polarized at $45°$ (there is no Jones vector for unpolarized light). The transmitted Jones vector is then given by

$$\begin{aligned}
\tilde{E}_t &= \frac{1}{2}\begin{bmatrix} 1 & -1 \\ -1 & 1 \end{bmatrix}\begin{bmatrix} 1 & 0 \\ 0 & e^{i\delta} \end{bmatrix}\frac{1}{\sqrt{2}}\begin{bmatrix} 1 \\ 1 \end{bmatrix} \\
&= \frac{1}{2}\begin{bmatrix} 1 & -1 \\ -1 & 1 \end{bmatrix}\frac{1}{\sqrt{2}}\begin{bmatrix} 1 \\ e^{i\delta} \end{bmatrix} \\
&= \frac{1}{2}\left(1 - e^{i\delta}\right)\frac{1}{\sqrt{2}}\begin{bmatrix} 1 \\ -1 \end{bmatrix}
\end{aligned} \tag{9.27}$$

We define the transmission of the optical system according to

$$\begin{aligned}
T &= \frac{\tilde{E}_t \cdot \tilde{E}_t^*}{\tilde{E}_i \cdot \tilde{E}_i^*} = \left|\frac{1}{2}\left(1 - e^{i\delta}\right)\right|^2 \\
&= \left|ie^{-i\frac{\delta}{2}}\left(\frac{e^{i\frac{\delta}{2}} - e^{-i\frac{\delta}{2}}}{2i}\right)\right|^2 \\
&= \sin^2\left(\frac{\delta}{2}\right)
\end{aligned} \tag{9.28}$$

The transmission is maximum for wavelengths where δ is an odd-integer multiple of π; in this case the retarder acts as a half-wave plate, rotating the initial linear polarization by $90°$ so that it passes without loss through the second linear polarizer. Similarly, wavelengths where δ is an integer multiple of 2π cause the retarder to function as a full-wave plate, so the initial linear polarization state is unchanged and so does not pass through the second crossed polarizer. Illumination with white light results in a transmitted beam with a mixture of colors, the brightest of which individually give a δ that is an odd-integer multiple of π.

Rotating either polarizer or the retarder will cause the colors of the transmitted light to change. In general, the transmitted beam will contain polarization states that are highly elliptical along the transmission axis of the second polarizer for the brightest colors transmitted.

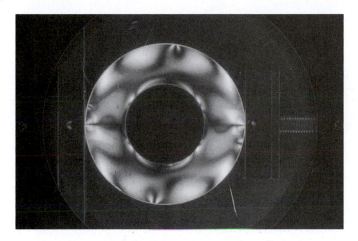

Figure 9.19 A compressed plastic ring viewed between crossed polarizers. Stress induced birefringence in the plastic creates elliptical polarization and corresponding variable transmission through the front polarizer. (Loren Winters/Visuals Unlimited)

■ EXAMPLE 9.8

A certain retarder with a length of $1.00\,cm$ has $n_y - n_x = 1.00 \times 10^{-2}$. Find δ if the retarder is illuminated by light of vacuum wavelength $500\,nm$. What new wavelength will cause δ to increase by π?

Solution

According to Equation 9.26, δ for this case is

$$\delta = 2\pi \left(1.00 \times 10^{-2}\right) \frac{\left(1.00 \times 10^{-2}\,m\right)}{\left(5.00 \times 10^{-7}\,m\right)} = 400\pi$$

To find a new wavelength λ' that will cause δ to increase by π, we proceed as follows:

$$\Delta\delta = 2\pi \left(n_y - n_x\right) L \left(\frac{1}{\lambda_0'} - \frac{1}{\lambda_0}\right)$$

$$= 2\pi \left(n_y - n_x\right) L \left(\frac{1}{\lambda_0 - \Delta\lambda} - \frac{1}{\lambda_0}\right)$$

$$= 2\pi \left(n_y - n_x\right) \frac{L}{\lambda_0} \left(\frac{1}{1 - \frac{\Delta\lambda}{\lambda_0}} - 1\right)$$

where we have let the new wavelength decrease in order to obtain a larger value of δ. Using the approximation that $\frac{1}{1-x} \approx 1 + x$ for small x, we obtain

$$\pi = \Delta\delta = 2\pi \left(n_y - n_x\right) \frac{L}{\lambda_0} \left(\frac{\Delta\lambda}{\lambda_0}\right)$$

$$= 400\pi \left(\frac{\Delta\lambda}{\lambda_0}\right)$$

Solving for $\Delta\lambda$ gives

$$\Delta\lambda = \frac{\lambda_0}{400}$$

Problem 9.19   Design a quartz retarder with $\delta = 201\pi$ using the indices given in Section 9.4.1, using a vacuum wavelength of $589\,nm$. Find wavelength values on either side of this that will give maximum transmission through the arrangement of Figure 9.18.

9.6   THE ELECTRO-OPTIC EFFECT

The *electro-optic effect* causes a change in the index of refraction of a dielectric material by applying an external electric field. Intuitively, this is caused when the applied electric field induces or aligns dipole moments within the material. Since the electric field has a particular direction, the change in refractive index is also directional, producing birefringence. Thus, the electro-optic effect allows the birefringence of materials to be controlled with an external voltage.

Some materials exhibit a *linear electro-optic effect*, also referred to as the *Pockels*[6] *effect*. The *quadratic electro-optic effect*, also referred to as the *Kerr*[7] *effect* that depends on the square of the applied electric field. The Pockels effect, if present, is usually much stronger than the Kerr effect. Materials must be crystalline and without a center of symmetry to exhibit a Pockels effect. The Kerr effect is useful in glasses and liquids as well as crystalline materials.

An *electro-optic modulator* uses an electro-optically controlled birefringent phase retarder in an arrangement similar to that of Figure 9.18. By application of a control voltage, transmission through the second polarizer can be made maximum or zero. Such changes can be achieved in a few nanoseconds, enabling very high amplitude modulation rates. Electro-optic modulators that utilize the Pockels effect are sometimes referred to as *Pockels cells*, and those that use the Kerr effect are sometimes referred to as Kerr cells.

9.6.1   Pockels Cells

For many materials used in Pockels cells, it is necessary to apply the external electric field *longitudinally*, as illustrated in Figure 9.20. This arrangement induces a birefringence

---

[6]Friedrich Pockels: 1865-1913. German physicist who discovered the effect that bears his name.
[7]Rev. John Kerr: 1824-1907. Scottish physicist and minister.

characterized by refractive indices give by

$$n_x = n_0 - \frac{1}{2}n_0^3 \, r \, E_z$$
$$n_y = n_0 + \frac{1}{2}n_0^3 \, r \, E_z \tag{9.29}$$

where n_0 is the index at zero applied field, E_z is the applied longitudinal electric field, and r is the electro-optic constant of the material (unit: $\left(\frac{m}{V}\right)$), which because of the crystalline state of the Pockels material is the component of a tensor. The derivation of this result is beyond the scope of our discussion, but may be found in Boyd [3]. The corresponding retardation of Equation 9.26 is given by

$$\delta = \frac{2\pi n_0^3 \, r \, E_z L}{\lambda_0} \tag{9.30}$$

If we take the electric field to be uniform, the last two terms in the numerator may be combined as the applied voltage V:

$$\delta = \frac{2\pi n_0^3 \, r \, V}{\lambda_0} \tag{9.31}$$

See Table 9.2. for values of r and n_0 for selected materials. Full modulation occurs when the Pockels material becomes a half-wave retarder.

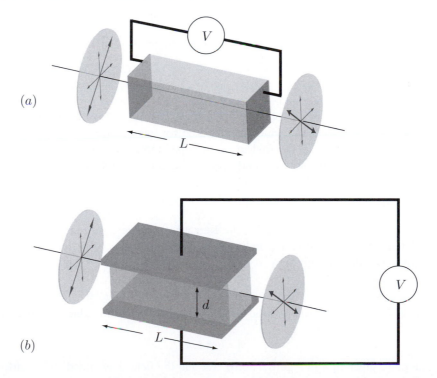

Figure 9.20   Electro-optic modulators. In (a) the applied electric field is longitudinal, and in (b) it is transverse.

Table 9.2.   Linear electro-optic constants for selected materials.

Material	$r(\frac{m}{V})$	n_0
$NH_4H_2PO_4$ (ADP)	8.5×10^{-12}	1.52
KH_2PO_4 (KDP)	10.6×10^{-12}	1.51
KH_2AsO_4 (KDA)	13×10^{-12}	1.57
KD_2PO_4 (KD*P)	23.3×10^{-12}	1.52

■ EXAMPLE 9.9

Find the voltage required for full modulation of a Pockels cell constructed with KDP for use at $550\,nm$.

Solution

Setting Equation 9.31 equal to π and solving for V gives

$$V = \frac{\lambda_0}{2n_0^3 r}$$

Substitution using values from Table 9.2. gives

$$V = \frac{5.50 \times 10^{-7}\,m}{2\,(1.51)^3\,\left(10.6 \times 10^{-12}\frac{m}{V}\right)} = 7.4\,kV$$

Notice that this result is independent of the length of the KDP crystal.

In the longitudinal design of Figure 9.20(a), the beam must pass through conducting surfaces that also provide the cell electrode surfaces. There are in fact conducting materials that are also transparent and which can be used for this purpose. For example: indium tin oxide (ITO), otherwise referred to as tin-doped indium oxide. Thin coatings of this material are transparent and sufficiently conducting for this application.

9.6.2   Kerr Cells

Electro-optic modulators that utilize the Kerr effect utilize a transverse applied field as illustrated in Figure 9.20(b). The induced refractive index change is along the direction of the applied field, and is given by

$$\Delta n = \lambda_0 K E^2 \tag{9.32}$$

where K is the *Kerr constant* of the material. The corresponding phase change δ is given by

$$\delta = 2\pi K L E^2 \tag{9.33}$$

If the plate separation in Figure 9.20(b) is d and the field is assumed to be uniform, the retardation can be expressed as a function of applied voltage V

$$\delta = 2\pi K L \frac{V^2}{d^2} \tag{9.34}$$

The electro-optic retarder will function as a half-wave plate to give full modulation when the applied voltage is

$$V = \frac{d}{\sqrt{2\,K\,L}} \qquad (9.35)$$

The full modulation voltage for a Kerr cell is typically much higher that for a Pockels cell.

Table 9.3. Kerr constants for selected materials.

	Material	$K \left(\frac{m}{V^2} \right)$
Benzene	C_6H_6	1.2×10^{-14}
Carbon disulfide	CS_2	6.4×10^{-14}
Chloroform	$CHCl_3$	-7.0×10^{-14}
Water	H_2O	9.4×10^{-14}
Nitrotoluene	$C_5H_7NO_2$	2.46×10^{-12}
Nitrobenzene	$C_6H_5NO_2$	4.4×10^{-12}

Problem 9.20   Find the full modulation voltage for a Pockels cells constructed with deuterated KDP (KD*P) when the wavelength is $546.1\ nm$.

Problem 9.21   Some electro-optic materials are expensive, so some designs use a quarter-wave electro-optic retarder in series with a quarter-wave plate that is not electro-optic. Design such a device. Include a diagram that clearly indicates the modulator indices of refraction relative to the fast and slow axis of the quarter-wave plate.

Problem 9.22   If the voltage source for a Pockels cell is modulated according to $V = V_m \sin \omega t$, show that the transmitted beam is given by

$$T = \frac{1}{2} \left(1 - \cos \left[\frac{2\pi n_0^3\, r\, V_m \sin \omega t}{\lambda_0} \right] \right)$$

When plotted vs. time, is the amplitude modulation sinusoidal?

Problem 9.23   Show that the Kerr retarder gives full modulation for the voltage given by Equation 9.35.

Problem 9.24   A $3.00\ cm$-long Kerr cell has a plate separation of $5.00\ mm$. What voltage will give full modulation if the cell is filled with carbon disulfide? Repeat if the cell is filled with nitrobenzene.

9.7   OPTICAL ACTIVITY

Linearly polarized light passing through an *optically active* medium experiences a rotation of the linear polarization state, as illustrated in Figure 9.21. Optically active materials are classified according to whether, as looking through the material at the light source, the polarization is rotated counter-clockwise (*levorotary*) or clockwise (*dextrorotary*).

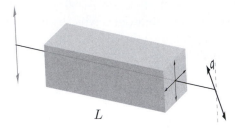

Figure 9.21   An optically active material rotates a linear polarization state. The rotation illustrated is counter-clockwise (levorotary) when viewed from the plane of emergence. The opposite rotation would be dextrorotary.

The rotation of the polarization state in an optically active material is not the same as circular polarization. In circularly polarized light, $\vec{E}$ rotates a full 2π radians in each wavelength of travel. In an optically active material, $\vec{E}$ might only rotate a small fraction of a radian in a few centimeters of travel.

Optical activity is due to *circular birefringence*, in which a material has a different index of refraction for left and right-circular polarizations. Examples of such materials include the sugar glucose, which is dextrorotary, and the sugar fructose, which is levorotary. Let n_l be the index of refraction for left-circular light, and n_r be the index of refraction for right-circular polarization. The phase shift between left and right-circular components of a beam after passing through a material of length L is given by

$$\Delta\phi = \frac{2\pi L\left(n_l - n_r\right)}{\lambda_0} \tag{9.36}$$

where λ_0 is the vacuum wavelength, and $\Delta\phi$ is the phase difference between the left- and right-circular components of the wave.

As noted in Section 9.5, a linear polarization state can be represented as a combination of left and right polarization states. Consider a horizontally polarized beam of unit irradiance. According to Equation 9.21,

$$\hat{e}_h = \frac{1}{\sqrt{2}}\left(\hat{e}_r + \hat{e}_l\right) = \frac{1}{\sqrt{2}}\left(\frac{1}{\sqrt{2}}\begin{bmatrix} 1 \\ -i \end{bmatrix} + \frac{1}{\sqrt{2}}\begin{bmatrix} 1 \\ i \end{bmatrix}\right)$$

After passing through a length L of optically active material, the emerging wave has Jones vector

$$\tilde{E}_t = \frac{1}{\sqrt{2}}\left(\frac{1}{\sqrt{2}}\begin{bmatrix} 1 \\ -i \end{bmatrix} + \frac{1}{\sqrt{2}}\begin{bmatrix} e^{i\Delta\phi} \\ ie^{i\Delta\phi} \end{bmatrix}\right) = \frac{1}{2}\left(\begin{bmatrix} 1 + e^{i\Delta\phi} \\ -i\left(1 - e^{i\Delta\phi}\right) \end{bmatrix}\right)$$

$$= e^{i\frac{\Delta\phi}{2}}\begin{bmatrix} \frac{e^{i\frac{\Delta\phi}{2}} + e^{-i\frac{\Delta\phi}{2}}}{2} \\ -\left(\frac{e^{i\frac{\Delta\phi}{2}} - e^{-i\frac{\Delta\phi}{2}}}{2i}\right) \end{bmatrix} = e^{i\frac{\Delta\phi}{2}}\begin{bmatrix} \cos\left(\frac{\Delta\phi}{2}\right) \\ -\sin\left(\frac{\Delta\phi}{2}\right) \end{bmatrix}$$

As usual, we ignore the phase term that is the same for both components. The phase shift $\beta = \frac{\Delta\phi}{2}$ is the *observed rotation*.

$$\beta = \frac{\pi L \left(n_l - n_r \right)}{\lambda_0} \tag{9.37}$$

The transmitted beam has a Jones vector given by

$$\hat{E}_t = \left[\begin{array}{c} \cos\beta \\ -\sin\beta \end{array} \right] \tag{9.38}$$

The *specific rotation* ρ is a standard measure of optical activity. For solids, ρ is defined according to a $1\,mm$ thick sample

$$\rho = \frac{\beta}{1\,mm} \tag{9.39}$$

For example, the specific rotation of quartz is $21.684°/mm$.

9.8 FARADAY ROTATION

An external magnetic field can cause rotation of a linear polarization state that is very similar to that caused by optical activity. This is usually referred to as *Faraday rotation* in honor of the discoverer[8]. The effect is caused by a magnetic field directed along the light propagation path, as illustrated in Figure 9.22. In the Faraday effect, the observed rotation of Equation 9.37 is given by

$$\beta = V\,B\,L \tag{9.40}$$

where B is the magnetic field magnitude, L is the light travel length, and V is the *Verdet*[9] *constant*. The Verdet constant, which is a strong function of wavelength, is on the order of $5\,\frac{rad}{Tm}$ for solids and $5 \times 10^{-3}\,\frac{rad}{Tm}$ for gases. *Terbium gallium garnet* (TGG) has a very large Verdet constant of $-134\,\frac{rad}{Tm}$ at $632.8\,nm$.

The Faraday effect can be used to construct a *Faraday modulator* as illustrated in Figure 9.22(a). Faraday modulators are not as widely used as Pockels cells due to the difficulty of modulating large magnetic fields. More commonly, the Faraday effect is used to construct a *Faraday isolator*, illustrated in Figure 9.22(b). An *isolator* is a device that blocks reflected beams. Here, it is important to note that materials with a positive Verdet constant rotate the polarization in a sense determined by the current in a solenoid that would produce the applied magnetic field, *for both forward and backward propagation paths*. Thus, for the field indicated in the figure, the rotation is counter-clockwise when viewed from the plane where $\vec{B}$ points out of the page. In this design, the first polarizer has a vertical transmission axis, and the second has a transmission axis that is rotated by $45°$. The magnetic field is adjusted to give maximum transmission in the forward direction, which begins with vertical polarization. The return reflection begins with linear polarization inclined at $45°$, and ends *horizontal* and thus is blocked by the first polarizer.

---

[8]Michael Faraday: 1791-1867. English physicist who made many remarkable discoveries in electromagnetism.
[9]Emile Verdet: 1824-1866. French physicist.

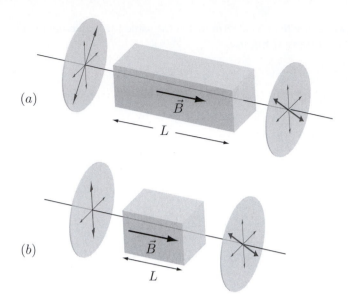

Figure 9.22   Magneto-optic devices. (a) Faraday modulator. (b) Faraday isolator.

Problem 9.25   Find the observed rotation for a material with $n_\ell = 1.5582$ and $n_r = 1.5581$, measured at a wavelength of $633\,nm$ with a sample thickness of $1.00\,mm$. Repeat when $n_\ell = 1.5581$ and $n_r = 1.5582$. Which is the levorotary case?

Problem 9.26   Modify the design of the Faraday isolator of Figure 9.22(b) to use a material such as TGG which has a negative Verdet constant.

Problem 9.27   What length of material is needed to construct a Faraday isolator if the Verdet constant is $5.00\,\frac{rad}{Tm}$ and the magnetic field has magnitude $0.100\,T$. Repeat if the isolator is constructed of TGG for use at $633\,nm$.

9.9   ACOUSTO-OPTIC EFFECT

Figure 9.23(a) illustrates an *acousto-optic* (AO) device, where an acoustic transducer attached to a material launches a traveling acoustic wave within the material. In the device illustrated, the acoustic transducer is a piezo-electric transducer (PZT) with a thickness that changes in proportion to an applied voltage. By modulating this voltage at an acoustic frequency, an acoustic wave is launched within the material that travels upward in the figure. An acoustic absorber at the other end eliminates reflections that would otherwise occur.

The acoustic wave causes *photoelastic* changes in the density and hence the index of refraction of the material. These index fluctuations, which have a spatial period that is identical to the wavelength λ_s of the acoustic wave, behave as a *diffraction grating* (see Section 6.3.7). The PZT is typically modulated at radio-frequencies (e.g. $40\,MHz$), producing closely spaced index fluctuations. A light beam that illuminates the material will thus typically overlap many acoustic wavelengths. The diffraction will be strong whenever reflections from all acoustic wavefront combine in phase. According to Figure 9.23, the

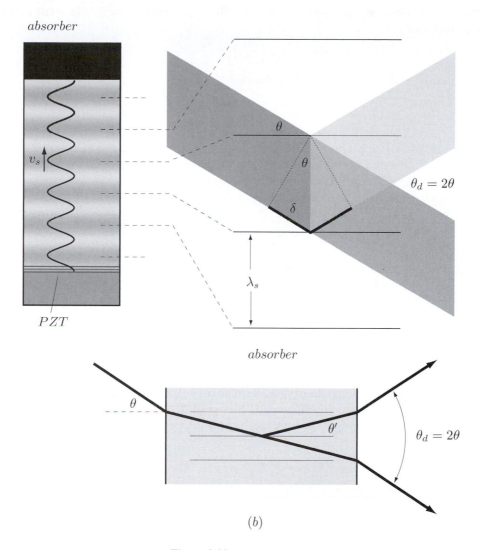

absorber

PZT

(b)

Figure 9.23 Acousto-optic device.

path difference between adjacent acoustic wavefronts is 2δ, where $\delta = \lambda_s \sin \theta$ and θ is the angle that the incident beam makes with the acoustic wavefronts, as illustrated. If 2δ is equal to an integral number of *light* wavelengths λ, then light reflected from many acoustic wavefronts will constructively interfere to produce a diffracted beam:

$$2\delta = 2\lambda_s \sin \theta = m\lambda$$

Solving for $\sin \theta$ gives

$$\sin \theta = m\frac{\lambda}{2\lambda_s} \qquad (9.41)$$

The first-order diffracted beam ($m = 1$) is deflected according to the law of reflection, and is thus deviated through a diffracted angle of $\theta_d = 2\theta$ with respect to the incident beam direction, as illustrated. Furthermore, the acoustic wavelength is usually much larger than the illumination wavelength, so it is usually appropriate to apply the small angle approximation:

$$\sin\theta = \sin\left(\frac{\theta_d}{2}\right) \approx \frac{\theta_d}{2} = \frac{\lambda}{2\lambda_s}$$

The first-order diffracted angle is thus given by

$$\theta_d = \frac{\lambda}{\lambda_s} \qquad (9.42)$$

If the medium is isotropic, has parallel sides, and these sides are parallel to the acoustic wave propagation direction, Equation 9.42 uses external incidence and transmission angles, and vacuum illumination wavelengths. Figure 9.23(b) illustrates this at each surface. According to Snell's law and again using the small angle approximation,

$$\sin\theta = n\sin\theta' = n\left(\frac{\lambda}{2n\lambda_s}\right) = \frac{\lambda}{2\lambda_s}$$

where θ' is the refracted incidence angle. The beam that transmits through the medium is parallel to the incident beam (see Example 3.1), so Equation 9.42 is valid using external angle θ and vacuum illumination wavelength λ. The refraction caused by the index fluctuations of the acoustic wave is negligible.

■ **EXAMPLE 9.10**

An acousto-optic device is attached to a voltage source that drives the PZT at $40\,MHz$, and is illuminated with light from a HeNe laser ($\lambda = 632.8\,nm$). If AO device is constructed from a material for which the speed of $40\,MHz$ acoustic waves is $4,000\,m/s$, find the deviation angle for the first-order diffracted beam.

Solution

The acoustic wavelength within the material is

$$\lambda_s = \frac{v_s}{f} = \frac{4000\frac{m}{s}}{40\times10^6\frac{1}{s}} = 10^{-4}m$$

According to Equation 9.42,

$$\theta_d = \frac{\lambda}{\lambda_s} = \frac{6.328\times10^{-7}m}{10^{-4}m} = 0.36°$$

Note the smallness of this deflection angle.

The photoelastic index variations result in a *phase grating*, and since these variations have a sinusoidal profile, the grating typically couples a large percentage (80% or more) of the incident beam into first-order diffraction. Amplitude modulating the PZT voltage can modulate the diffracted beam; in this case the AO device is referred to as an *AO modulator*. The PZT voltage can also be frequency modulated to give a raster-scanned diffracted beam.

The device illustrated in Figure 9.23, with the acoustic absorber opposite the PZT, produces a *traveling* acoustic wave and associated index variations that move at the acoustic speed v_s. The incident beam thus diffracts from a *moving* grating, giving a first-order diffracted beam that is Doppler-shifted by *exactly* the PZT modulation frequency[10]. The frequency-shifted diffracted beam is useful for *optical heterodyne detection*, as described in Section 5.3.6.1.

Problem 9.28   Using the parameters of Example 9.10, find the PZT modulation frequency that will produce a diffracted angle of $1.00°$.

Problem 9.29   Find the deviation angle for $633\,nm$ if the AO medium is fused quartz with an acoustic wave speed of $5960\,m/s$ at the modulation frequency of $40\,MHz$.

Problem 9.30   The *rise time* t_r for an AO modulator is given by

$$t_r = \frac{d}{v_s}$$

where d is the beam diameter and v_s is the acoustic wave speed. Intuitively, the acoustic wave must interact with the entire illumination beam before it can be fully deflected. Find the rise time for an illumination beam that is focused to a $100\,\mu m$-diameter spot, using the parameters of Example 9.10, and again if the medium is fused quartz with an acoustic wave speed is $5960\,m/s$. If the AO device were to be used as an amplitude modulator, estimate the maximum beam modulation frequency in both cases. Describe the signal to be applied to the PZT to achieve this.

■ **APPLICATION NOTE 9.3**

Laser Q-Switching

Fast modulators can be used to *Q-switch* a laser. In this application, the modulator is placed within the laser cavity. When the modulator is adjusted for zero transmission, the cavity cannot oscillate. In this case, the pumping mechanism (see Section 7.4) creates a much larger population inversion than would be possible if oscillation were permitted. A voltage is then applied to the modulator that quickly adjusts it for maximum transmission, allowing the large population inversion to produce laser output by stimulated emission. The Q-switched output is typically a short pulse with peak irradiance that is many orders of magnitude larger than that obtained without a Q-switch.

A Q-switch that uses a modulator is sometimes referred to as an *active Q-switch*. A *passive Q-switch* uses a *saturable absorber*. In this nonlinear application, an absorbing material is placed within the laser cavity. As the population inversion becomes large, the irradiance within the cavity can be large enough to excite every ground state within the absorber into an excited state. When this happens, the absorber becomes transparent, and the absorber is said to be *bleached*. The Q-switched pulsed

---

[10]This is an interesting result that is not immediately obvious. For a derivation that obtains this by a conservation of energy argument, see Pedrotti [16].

is released at the instant of transparency. As the pulse completes, the population inversion decreases, excited states decay back to the ground state of the absorber, and the process starts again. The laser output consists of a series of pulses with a pulse rate that is determined by the density of absorbing states within the beam path.

■ **APPLICATION NOTE 9.4**

Laser Mode Locking

Mode locking is a technique for obtaining a series of pulses with extremely short pulse duration. The cavity of a mode-locked laser typically contains a very fast shutter that only allows a single large pulse to oscillate within the cavity. In *active mode locking*, the shutter is a fast EO or AO modulator that is opened and closed with an applied voltage, and in *passive mode locking* the shutter is a saturable absorber that is bleached by the mode-locked pulse. In either case, the shutter opens with a timing that only allows the mode-locked pulse to pass.

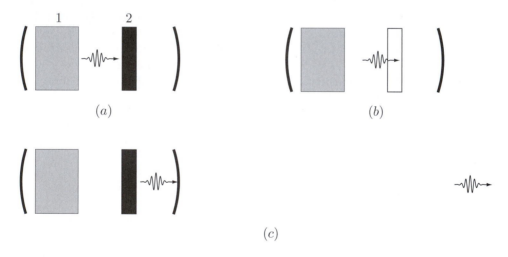

Figure 9.24   The cavity of a model locked laser contains a gain medium (1) and a fast modulator or saturable absorber (2) that is timed to allow a single large pulse to oscillate within the cavity. (a) The modulator or saturable absorber is opaque when the pulse does not propagate through it. (b) The modulator switches to transparent, allowing the pulse to pass. If a saturable absorber is used, the pulse intensity is large enough to bleach it. (c) The modulator or saturable absorber returns to opaque when the pulse passes. The mode-locked output is a series of short pulses spaced by twice the cavity length.

Pulse durations on the order of $10^{-15}\,s$ (femtoseconds) are achievable. As discussed in Section 5.4, amplitude modulating a monochromatic beam introduces new frequency components, making it less monochromatic. Lasers that are capable of being mode locked must have frequency components available with which to form the short mode-locked pulses. These frequencies come from available longitudinal cavity modes (see Section 7.5); thus, lasers capable of forming mode-locked pulses of very short duration must have a rich longitudinal mode structure. The solid state titanium sapphire (Ti:sapphire) has such a mode structure, and is frequently used for mode locking.

Problem 9.31   Ti:sapphire lasers are commonly mode locked to output short, intense pulses with a duration of about ten femtoseconds. If the output of the laser is $800\ nm$, how long is the pulse in wavelengths? If the laser cavity length is $100\ cm$, what is the repetition rate of the laser output?

9.10 NONLINEAR OPTICS

An optical material becomes *nonlinear* when a beam passes through it with sufficient irradiance to cause changes in its optical properties. The saturable absorber of Application Note 9.3 is one example: when the beam irradiance is small, the absorber is opaque, but for a large irradiance the absorber becomes transparent. As another example, focusing the output of a pulsed laser can create electric fields at the beam waist that are large enough to cause air breakdown (see Problem 9.32). Superposition (Section 1.5) is no longer valid when a material becomes nonlinear. For example, two beams that individually pass through air can cause breakdown when the beam paths overlap.

By far, the most useful nonlinear materials are *dielectrics*. In this case, the electric field of a passing electromagnetic wave is strong enough to change the dielectric properties of the material. *Linear isotropic* dielectric materials are characterized by a *dielectric constant* K_E, as described in Section 2.2, and an associated constant index of refraction $n = \sqrt{K_E}$ (Section 2.3). Dielectric materials consist of molecules that either have a *permanent dipole moment*, or a dipole moment that is *induced* by an external electric field. Molecules with dipole moments can be thought of as oblong, where one end is on the average positive and the other end negative, as illustrated in Figure 9.25(a). A *dipole* consists of equal and opposite charges of magnitude Q separated by a displacement vector $\vec{s}$. The *dipole moment* is defined to be

$$\vec{p} = Q\vec{s} \tag{9.43}$$

The direction of $\vec{s}$ and hence $\vec{p}$ is from $-Q$ to $+Q$, and indicated in Figure 9.25(a).

In order to describe the dielectric properties of bulk materials, we define the *dipole moment per unit volume* $\vec{P}$:

$$\vec{P} = N\vec{p} \tag{9.44}$$

where N is the number of dipole moments (i.e. molecules) per unit volume. Like $\vec{E}$, $\vec{P}$ is a vector field that is commonly referred to as the *electric polarization*, or simply the *polarization*[11]. Ordinarily, the individual molecules are randomly oriented and $\vec{P}$ is *zero*, since in this case $\sum \vec{p} = 0$. However, an external electric field will orient the individual $\vec{p}$ dipoles to give a nonzero $\vec{P}$, as illustrated in Figure 9.25. The degree of alignment determines the value of $\vec{P}$. It is customary to define the *electric susceptibility* X, defined

---

[11] This must not be confused with the electric field direction of a transverse electromagnetic wave.

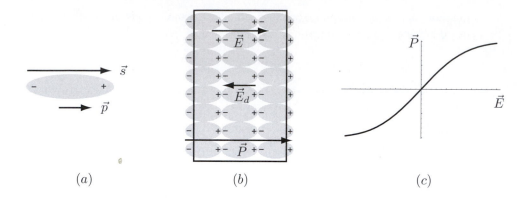

Figure 9.25   (a) A molecule with dipole moment $\vec{p}$. (b) A dielectric slab with net dipole moment per unit volume $\vec{P}$, induced by an external electric field $\vec{E}$. The external field is partly canceled by the dielectric field $\vec{E}_d$. (c) Nonlinear effects can occur when $\vec{P}$ becomes saturated.

by

$$\vec{P} = \epsilon_0 \mathrm{X} \vec{E} \tag{9.45}$$

The dielectric constant and index of refraction are related to X according to (see Section 3.7.1)

$$K_E = \mathrm{X} + 1 \tag{9.46}$$

$$n = \sqrt{\mathrm{X} + 1} \tag{9.47}$$

In many cases, $\vec{P}$ varies linearly with the external field $\vec{E}$, in which case X is a constant. This dependence can become *nonlinear* when, for example, *saturation* occurs as illustrated by Figure 9.25(c). Saturation occurs for strong electric fields, where alignment is so nearly complete that additional field $\vec{E}$ produces a relatively small change in $\vec{P}$. In this case, X is a function of E, and we may expand it as a polynomial series:

$$\mathrm{X} = \mathrm{X}_1 + \mathrm{X}_2 \mathrm{E} + \mathrm{X}_3 \mathrm{E}^2 + \dots \tag{9.48}$$

where the X_i are constants of the polynomial expansion. Combining Equations 9.48 and 9.45 gives

$$\begin{aligned} P &= \epsilon_0 \left(\mathrm{X}_1 \mathrm{E} + \mathrm{X}_2 \mathrm{E}^2 + \mathrm{X}_3 \mathrm{E}^3 + \dots \right) \\ &= P_1 + P_2 + P_3 + \dots \end{aligned} \tag{9.49}$$

In Equation 9.49, P_1 is the linear term and the rest are nonlinear. From Equation 9.46, it is evident that X_1 is unitless. Units for X_2 are m/V, for X_3 $(m/V)^2$, and so on.

A *nonlinear optical effect* can occur whenever the electric field $\vec{E}$ of an electromagnetic wave is sufficiently large to induce nonlinearity in $\vec{P}$. Materials commonly used for nonlinear optics applications have values of X_2 on the order of $5 \times 10^{-12} \, m/V = 5 \, pm/V$.

A dielectric material that is *isotropic* cannot have a non-zero second-order susceptibility. Consider a material for which $\mathrm{X} \neq 0$, so that

$$P_2 = \epsilon_0 \mathrm{X}_2 E^2$$

An isotropic material is one for which effects of polarization are independent of direction. Hence, reversing the electric field direction should produce the opposite polarization, giving

$$-P_2 = \epsilon_0 X_2 \left(-E\right)^2$$

Since the electric field is squared, the last two results give $P_2 = -P_2$, which can only be true if $P_2 = 0..$ Thus, second-order nonlinear effects may only be observed in materials such as crystals that are non-isotropic.

9.11 HARMONIC GENERATION

Consider a material with a non-zero second-order susceptibility that is illuminated with an electromagnetic wave having a sinusoidally oscillating electric field

$$E = E_m \cos \omega t$$

This produces a second-order polarization term given by

$$P_2 = \varepsilon_0 X_2 E^2$$
$$= \varepsilon_0 X_2 E_m^2 \cos^2 \omega t$$
$$= \varepsilon_0 X_2 E_m^2 \frac{1}{2} \left(1 + \cos 2\omega t\right)$$

where we have used the double-angle identity to substitute for $\cos^2 \omega t$. Rearranging gives

$$P_2 = \frac{1}{2}\varepsilon_0 X_2 E_m^2 \ + \ \frac{1}{2}\varepsilon_0 X_2 E_m^2 \cos 2\omega t \tag{9.50}$$

The second term in Equation 9.50 represents a sinusoidally varying polarization term that oscillates at *twice* the frequency of the illuminating radiation. This oscillating dipole moment radiates a new electromagnetic wave that is *frequency doubled* compared to the illuminating light in a process referred to as *second harmonic generation* (SHG).

It is useful to develop a simple physical picture of electromagnetic wave propagation within a nonlinear optical material. In Figure 9.26(a), an illuminating electromagnetic wave induces an oscillating dipole. The energy absorbed by the dipole is then re-radiated into a pattern that is isotropic over all azimuthal angles as illustrated. If the induced dipole depends linearly on the illuminating field amplitude, the re-radiated light only contain the frequency f of the illuminating field, but if there is a quadratic dependence then there will be a portion of the re-radiated light of frequency $2f$. Figure 9.26(b) illustrates a bulk dielectric which is composed of an enormous number of individual radiating dipoles. Since each dipole is driven by the same illuminating field, the dipoles radiate as a *phased array*, with a corresponding radiation pattern that only emits in the extreme forward direction. Thus, as a beam of light travels through a dielectric, the beam profile is maintained as many dipole radiators absorb and re-emit the incident radiation.

In a typical SHG application, the illuminating wave is focused into a nonlinear material, as illustrated in Figure 9.27(a). For Gaussian beam focusing (see Section 7.7), the beam spot size is $\pi \omega_0^2$, where ω_0 is the Gaussian beam waist radius. If the beam power is P_b, then the irradiance at the focus is given by (see Section 2.3.3)

$$I = \frac{P_b}{\pi \omega_0^2} = \frac{1}{2}\varepsilon_0 c\, n\, E_m^2 \tag{9.51}$$

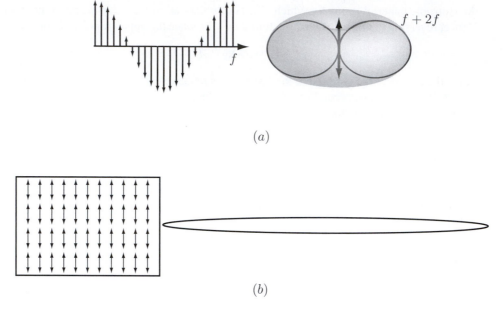

(a)

(b)

Figure 9.26   (a) The electric field in an illuminating wave induces an oscillating dipole that re-radiates at the illuminating frequency as well as at the frequency-doubled SHG frequency. The radiation pattern for a single oscillating dipole is azimuthally isotropic.   (b) A large number of oscillating dipole moments acting as phased array to give a radiation pattern that is extremely lobed in the forward direction.

Let us focus so that the nonlinear material has a length L of twice the Rayleigh range within the material:

$$L = \frac{2\pi\omega_0^2 n}{\lambda} \tag{9.52}$$

where λ is the vacuum wavelength of the illuminating radiation, and n is the index of refraction of the nonlinear material. Using the last two results, we obtain

$$E_m = 2\sqrt{\frac{P_b}{\varepsilon_0 c \lambda L}} \tag{9.53}$$

In order to achieve SHG, the field E_m must be large enough to induce second-order effects within the nonlinear material.

■ **EXAMPLE 9.11**

The material potassium titanyl phosphate, $KTiOPO_4$, (KTP) has a large second-order nonlinear susceptibility X_2 of $16.6 \times 10^{-12}\,m/V$, and an index of refraction of 1.8. What beam power P_b at a wavelength of $1.064\,nm$ focused into a $1.0\,cm$-long KTP crystal will cause a second-order nonlinear polarization P_2 that is 0.1% of the linear term P_1? Assume that the KTP crystal has a length that is twice the Rayleigh range of the illuminating beam.

Solution

By Equation 9.47,
$$X_1 = n^2 - 1 = (1.8)^2 - 1 = 2.24$$

We wish to find an irradiance that gives a electric field so that

$$X_2 E^2 = (0.001) X_1 E$$

giving

$$E = \frac{(0.001) X_1}{X_2} = \frac{(0.001)(2.24)}{16.6 \times 10^{-12} \frac{m}{V}} = 1.35 \times 10^8 \frac{V}{m}$$

Solve Equation 9.53 for the beam power to give

$$P_b = \frac{\varepsilon_0 c L \lambda}{4} E_m^2$$

$$= \frac{\left(8.85 \times 10^{-12} \frac{C^2}{Nm^2}\right) \left(3.00 \times 10^8 \frac{m}{s}\right) \left(10^{-2} m\right) \left(1.064 \times 10^{-6} m\right)}{4} \left(1.35 \times 10^8 \frac{V}{m}\right)^2$$

$$= 1.29 \times 10^5 W$$

This is a beam power that can be achieved with pulsed laser sources.

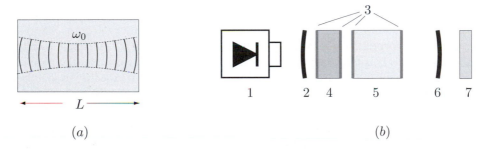

(a) $\qquad\qquad\qquad\qquad\qquad\qquad$ (b)

Figure 9.27 (a) Gaussian beam focusing into a nonlinear material. The length L of the material is twice the Rayleigh range. (b) Diode-pumped solid state laser: (1) $808\,nm$ diode laser; (2) cavity mirror, AR coated for $808\,nm$, 99.9% reflecting for $1064\,nm$; (3) AR coatings for $1064\,nm$ and $532\,nm$; (4) Nd-doped laser crystal; (5) KTP nonlinear crystal; (6) cavity mirror, AR coating for $532\,nm$, 99.9% reflecting for $1064\,nm$; (7) infrared filter to block any leakage of $808\,nm$ or $1064\,nm$.

Recent technological advances have resulted in a generation of low-cost *diode-pumped solid state* (DPSS) lasers. A typical design is illustrated in Figure 9.27(b). The process begins with a diode laser that outputs a $808\,nm$ optical pump[12] beam with a beam power that is typically hundreds of milliwatts. This beam illuminates a crystal of neodymium-doped yttrium orthvanadate ($Nd : YVO_4$) or neodymium-doped yttrium aluminum garnet ($Nd : YAG$), both of which are very efficiently pumped by $808\,nm$ for laser output at $1064\,nm$. The optically pumped laser crystal resides within a resonant Fabry-Perot cavity, which also contains a SHG material that is typically KTP. Placing the KTP within the resonant cavity exposes it to a much higher irradiance, increasing SHG. The mirror

---

[12]for a discussion of optical pumping, see Section 7.4.

that is illuminated by the diode laser is coated for high reflectivity at $1064\,nm$ and high transmission at $808\,nm$. The output coupler at the opposite end of the cavity is coated for high reflectivity at $1064\,nm$ and high transmission at the frequency-doubled $532\,nm$. All optical surfaces within the cavity are coated for zero reflection at both $532\,nm$ and $1064\,nm$. The $1064\,nm$ beam oscillates back and forth within the KTP, generating more $532\,nm$ on each pass.

■ **APPLICATION NOTE 9.5**

The Green Laser Pointer

It is currently possible to purchase green laser pointers that utilize the technology illustrated in Figure 9.27(b) and described above for under $100. The optical pump diode typically emits a few hundred mW at $808\,nm$, yielding several milliwatts in a green $532\,nm$ beam.

Extreme caution should be used when working with high-power infrared laser sources such as the $808\,nm$ diode, $Nd:YAG$, and $Nd:YVO_4$ sources of Figure 9.27(b). These wavelengths are too long to be detected by the eye, but still short enough to be focused by the lens onto the retina. Such beams are completely invisible, but are more than powerful enough to cause permanent eye damage. The last optical element of Figure 9.27(b) is an infrared filter whose job it is to remove any remaining infrared in the output. Disassembling such a laser and/or removing the IR filter can be very dangerous.

The nonlinear term P_3 in Equation 9.49 can produce *frequency-tripled* radiation, sometimes referred to as *third harmonic generation* (THG). As above, we let the illuminating field be given by $E = E_m \cos \omega t$. Using a similar trigonometric identity to that used for Equation 9.50, we obtain (see Problem 9.36)

$$P = \frac{3\varepsilon_0 \mathrm{X}_3}{4} E_m^3 \cos \omega t + \frac{\varepsilon_0 \mathrm{X}_3}{4} E_m^3 \cos(3\omega t) \tag{9.54}$$

The second term in Equation 9.54 represents THG. The first term contributes to the linear polarization effect, and if X_2 is positive *increases* the index of refraction within regions of higher irradiance. An Gaussian beam traveling through such a material will encounter a higher refractive closer to the beam axis, and in this case the material acts as a focusing lens that leads to the phenomena of *self focusing*. Self focusing can result in optical damage within the material. It can be reduced or eliminated by *thermal blooming*, where increased irradiance and absorption causes heating, which decreases the density and index closer to the beam axis, causing the material as a diverging lens.

In the photon picture, SHG is a *two photon process*. In other words, the oscillating dipole of the nonlinear medium absorbs *two* photons of the illuminating light, re-emitting one photon of the frequency-doubled light. Similarly, in THG, three photons are absorbed for every one photon of frequency-tripled light emitted.

■ **APPLICATION NOTE 9.6**

The National Ignition Facility (NIF)

The *National Ignition Facility* (Figure 9.28) at Lawrence Livermore National Laboratory currently houses the world's largest laser. The facility exists to study controlled thermonuclear fusion initiated by the radiation pressure of 192 UV laser beams that compress small hydrogen- and deuterium-rich targets to temperatures and pressures similar to that within the core of a star.

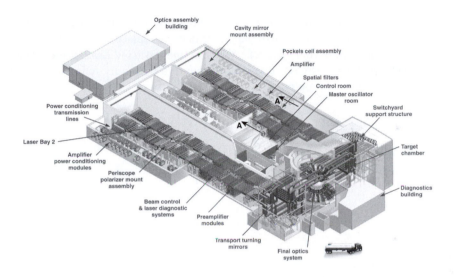

Figure 9.28   The National Ignition Facility.

The beam begins in the master oscillator that outputs a small nanojoule pulse at $1064\,nm$ which is then amplified by a factor of about 10^6 in a preamplifier module. This beam is amplified further by making exactly four passes through an amplifier constructed of flashlamp-pumped Nd-doped glass. A deformable Adaptive Optics mirror optimizes the beam profile. The timing for this portion of the beam path is controlled by a optical switch that uses a Pockels cell to rapidly rotate the polarization; after switching, a polarizing beam splitter directs the amplified beam, now with a pulse energy of about $10\,J$, into the remaining beam path. Final amplification of the $1064\,nm$ beam is achieved by a power amplification section consisting of 3074 $40\,kg$ slabs of flashlamp-pumped Nd-doped glass mounted at Brewster's angle. At this point, the pulse energy at $1064\,nm$ is in excess of $20\,kJ$.

The nonlinear material KDP is used for SHG and THG conversion. This material was chosen due to the capability to grow very large high quality crystals, as illustrated in the figure. Current single-beam performance for the green SHG is $11.4\,kJ$ and at the UV THG it is $10.4\,kJ$. With all 192 beams in operation, the fusion target can be illuminated by $2.2\,MJ$ in SHG, and $2.0MJ$ with THG.

9.11.1 Phase Conjugation Reflection by Degenerate Four-Wave Mixing

In *phase conjugate reflection*, an incident wave is reflected as a wave proportional its *complex conjugate*. As we will show, this is quite different from ordinary reflection. Consider a polarized forward traveling wave given by

$$\vec{E}(\vec{r}, t) = \vec{E}_m e^{i\vec{k}\cdot\vec{r}} e^{-i\omega t} \tag{9.55}$$

The conjugated wave is given by

$$\vec{E}^*(\vec{r}, t) = \vec{E}_m^* e^{-i\vec{k}\cdot\vec{r}} e^{+i\omega t} \tag{9.56}$$

In order for the conjugate wave to be a *reflection* of the forward traveling incident wave, it must be backward traveling, and this will only be true if we replace t by $-t$ in Equation 9.56. In other words, a phase conjugate reflection is a *time-reversed copy* of the incident wave, except perhaps for a decrease in amplitude due to a non-unity reflectivity. Even polarization is preserved (see Problem 9.39).

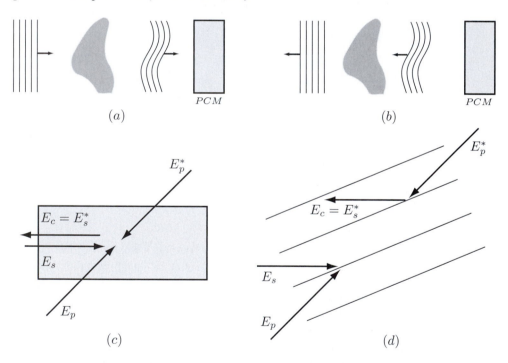

Figure 9.29   (a) A plane wave passes through a distorting medium before reflecting from a phase conjugate mirror (PCM). (b) After reflection, the time-reversed wave passes back through the same distorting medium to transmit as the original plane wave. (c) Geometry of four-wave mixing. Pump beams E_p and E_p^* are counter-propagating, and are typically obtained within a resonant laser cavity. The signal beam E_s is reflected to give a phase conjugate beam E_c. (d) An index grating is created by E_p and E_s. The conjugate pump beam E_p^* is Bragg reflected from the index grating to give the conjugate reflection E_c.

Phase conjugate reflection has some remarkable properties. For example, consider a beam that travels through a distorting medium, as illustrated in Figure 9.29(a). If the distorted wave reflects from a phase conjugate mirror, the reflection is a conjugate copy that passes back through the distorting medium as a time-reversed copy, so the distortion is removed in transmission as indicated in Figure 9.29(b). An ordinary reflection would have transmitted with further distortion.

Phase conjugate mirrors can be produced with *degenerate four-wave mixing* as illustrated in Figure 9.29(c). In this application, counter-propagating pump beams E_p and E_p^* that comprise the standing wave inside a resonant laser cavity illuminate the material. A signal beam E_s is incident as indicated in the figure, generating the conjugate reflection E_c that propagates counter to E_s. The term degenerate indicates that all four beams have the same frequency. Phase conjugate reflection will only occur if the fields are large enough to support THG. The first term in Equation 9.54 causes an index variation that produces a stationary phase diffraction grating within the nonlinear material. If we conceptualize the grating as being generated by E_p and E_s as indicated in Figure 9.29(d), then E_p^* Bragg reflects to give the conjugate reflection E_c. For more information, see Boyd [3].

Problem 9.32 The *dielectric strength* of a material is the magnitude of the largest electric field that can exist within the material before breakdown occurs. In solids, breakdown almost always results in optical damage. If a $5.00 \, cm$-diameter laser beam is focused to a spot of diameter $100 \, mum$, what beam power in the unfocused beam will cause air breakdown at the focused spot? What beam power in the unfocused beam will cause breakdown without focusing? The dielectric strength of air is $3 \times 10^6 \, V/m$. You may assume a step profile for both focused and unfocused beams.

Problem 9.33 Use the complex notation to derive Equation 9.50. With a complex E, we compute the second-order nonlinear polarization according to

$$P_2 = \varepsilon_0 X_2 \left|E\right|^2 = \varepsilon_0 X_2 E E^*$$

Show that the correct result is *not* obtained if the illuminating field is represented as $E = E_m e^{i\omega t}$. However, if we use the real part of E by adding the complex conjugate

$$E = \frac{1}{2} \left[E_m e^{i\omega t} + \left(E_m e^{i\omega t} \right)^* \right]$$

then Equation 9.50 is obtained. You may assume that the amplitude E_m is real. Note the extra factor of $\frac{1}{2}$ in the expression for E. Using the complex representation when computing products requires care. See also Problem 2.23.

Problem 9.34 The nonlinear material potassium dihydrogen phosphate KH_2PO_4 (KDP) has a second-order nonlinear susceptibility X_2 of $2.2 \times 10^{-12} \, m/V$, and an index of refraction of 1.5. What beam power P_b at a wavelength of $1.064 \, nm$ focused into a $1.0 \, cm$-long KDP crystal will cause a second-order nonlinear polarization P_2 that is 0.1% of the linear term P_1? Assume that the KDP crystal has a length that is twice the Rayleigh range of the illuminating beam.

Problem 9.35 If the cavity in Figure 9.27 is terminated on each end by mirrors with a reflectivity 99.9% at $1064 \, nm$, estimate the number of passes through the KTP crystal that a photon makes before passing through the output coupler.

Problem 9.36   Verify Equation 9.54.

Problem 9.37   Can isotropic materials be used for THG? Explain your reasoning.

Problem 9.38   A certain material has a third-order susceptibility $X_3 = 1.00 \times 10^{-20}\, m^2/V^2$ and an index of refraction of 1.79. Estimate the electric field amplitude that will produce a third-order contribution to the polarization that is 0.1% of the linear term. Find the associated beam power for a $1.064\mu m$ beam focused into a $1.00\, cm$ long crystal that is twice the Rayleigh range of the illuminating light.

Problem 9.39   Show that if the incident wave is right-circularly polarized, a phase conjugate reflection is also right-circularly polarized. Show that for normal reflection, such a reflected wave would be left-circular.

9.12   FREQUENCY MIXING

As a more general case, we consider the case of a beam of illumination consisting of two discrete frequencies with identical linear polarization $E_1 e^{i\omega_1 t} + E_2 e^{i\omega_2 t}$, using the complex notation to facilitate algebra. Let the medium have a nonzero second-order susceptibility, so that

$$P_2 = \varepsilon_0 X_2 E^2 \tag{9.57}$$

Letting the amplitudes E_1 and E_2 be real, and noting the results of Problem 9.33, we compute the second-order polarization term as follows (the symbol cc indicates the complex conjugate):

$$
\begin{aligned}
P_2 &= \frac{\varepsilon_0 X_2}{4} \left(E_1 e^{i\omega_1 t} + E_2 e^{i\omega_2 t} + cc \right)^2 \\
&= \frac{\varepsilon_0 X_2}{4} \left(E_1^2 e^{2i\omega_1 t} + E_2^2 e^{2i\omega_2 t} + 2E_1 E_2 e^{i(\omega_1+\omega_2)t} + 2E_1 E_2 e^{i(\omega_1-\omega_2)t} + cc \right) \\
&\quad + \frac{\varepsilon_0 X_2}{4} \left[2 \left(E_1^2 + E_2^2 \right) \right]
\end{aligned}
$$

Rearrangement and the Euler relations give

$$
\begin{aligned}
P_2 = {}& \frac{\varepsilon_0 X_2}{2} \left(E_1^2 + E_2^2 \right) + \underbrace{\frac{\varepsilon_0 X_2}{2} \left[E_1^2 \cos\left(2\omega_1 t\right) + E_1^2 \cos\left(2\omega_1 t\right) \right]}_{SHG} \\
& + \underbrace{\varepsilon_0 X_2 E_1 E_2 \cos\left(\omega_1 + \omega_2\right) t}_{SFG} \\
& + \underbrace{\varepsilon_0 X_2 E_1 E_2 \cos\left(\omega_1 - \omega_2\right) t}_{DFG}
\end{aligned}
\tag{9.58}
$$

The first term in Equation 9.58 represents a constant polarization that is sometimes referred to as *optical rectification* (there is a similar term in Equation 9.50). The next term represents second harmonic generation by each frequency component in the composite wave. The term labeled SFG represents *sum frequency generation* and results in radiation at the sum of the composite frequencies as illustrated in Figure 9.30(a), and the term labeled DFG represents *difference frequency generation* and results in radiation at the difference of the composite frequencies as illustrated in Figure 9.30(b).

A laser source is *tunable* if its frequency can be varied. Tunable laser sources are useful for applications such as spectroscopy; unfortunately, laser mediums that support tunability are not common. A tunable laser source can be extended into new wavelength regions by frequency mixing: into a longer wavelength region by DFG or into a region of shorter wavelengths by SFG.

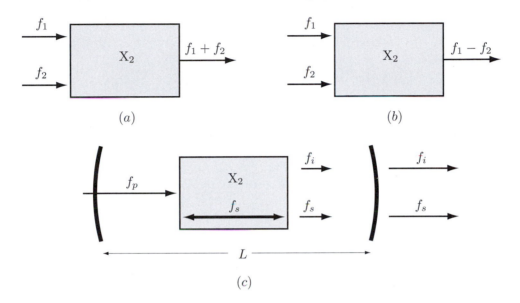

Figure 9.30 (a) Sum frequency generation (SFG). (b) Difference frequency generation (DFG). (c) Optical parametric oscillator. The signal frequency f_s is determined by the Fabry-Perot cavity resonance. The idler frequency f_i is determined by conservation of energy.

In an *optical parametric oscillator* (OPO) the nonlinear material is placed inside of a resonant cavity, as illustrated in Figure 9.30(c). A pump beam of frequency f_p enters the resonant cavity and illuminates the nonlinear material. In what is essentially the reversal of Figure 9.30(b), two new frequencies are generated. The *signal beam* of frequency f_s corresponds to a longitudinal resonance of the Fabry-Perot cavity, and thus is amplified. An *idler beam* of frequency f_i must also be generated to conserve energy; each pump photon is converted according to

$$hf_p = hf_s + hf_i \tag{9.59}$$

The OPO is tuned by adjusting the cavity length L. The idler beam is usually discarded, and the signal beam is used as output.

Problem 9.40 Verify Equation 9.58. Show that eliminating either frequency term gives Equation 9.50.

Problem 9.41   A tunable dye laser provides output over a wavelength range of $600\,nm$ to $700\,nm$. If this beam is combined with a $1064\,nm$ beam within a second-order nonlinear medium, what frequencies can be obtained with SFG and with DFG?

Problem 9.42   A $1064\,nm$ beam pumps an OPO adjusted to output at $1500\,nm$. What is the frequency of the idler? Can this OPO/pump combination be adjusted to output at $800\,nm$? If not, why?

Additional Problems

Problem 9.43   Find the percent transmission of a collimated beam of light incident at Brewster's angle on a single slab of sapphire with index of refraction 1.76 if the incident polarization is perpendicular to the plane of incidence. Repeat when the polarization if parallel to the plane of incidence.

Problem 9.44   Unpolarized light passes through three ideal linear polarizers with transmission axes at $0°$, $35°$, and $55°$. Find the fraction of incident light that is passed by this combination.

Problem 9.45   A beam of $589\,nm$ light passes through calcite in a direction perpendicular to the optic axis. Calculate the wavelengths of the ordinary and extraordinary waves.

Problem 9.46   Vertically polarized light passes through nine ideal linear polarizers. The first polarizer is inclined at $10°$ with respect to the vertical, and the remaining have transmission axes inclined at $10°$ from the previous polarizer. Find the percentage of incident irradiance transmitted and final polarization state of the beam passed by this combination.

Problem 9.47   Find the transmitted beam when light polarized at $30°$ is incident on a quarter-wave plate with fast axis horizontal. Repeat when the fast axis is vertical.

Problem 9.48   Three optical beams are prepared: one is unpolarized, one is left-circularly polarized, and the third is right-circular. Using only a quarter-wave plate and a linear polarizer, describe a method for identifying each beam.

Problem 9.49   If the specific rotation of quartz is $21.684°$, find the observed rotation from a $3.50\,mm$-thick slab illuminated with $633\,nm$ light.

Problem 9.50   Each THG beam of the NIF laser outputs a $10\,kJ$ optical pulse that lasts for about $5\,ns$. What is the average power of this beam during the time that it is on? If this beam is focused to a $100\,\mu m$ spot, what is the irradiance? What is the maximum electric field magnitude within the beam?

Problem 9.51   The nonlinear material potassium dihydrogen phosphate KH_2PO_4 (KDP) has a second-order nonlinear susceptibility X_2 of $2.2 \times 10^{-12}\,m/V$, and an index of refraction of 1.5. What electric field will give a second-order polarization P_2 that is equal to the first-order term? What would the intensity be in such a beam?

Problem 9.52   After final amplification at $1064\,nm$ the NIF laser outputs a $20\,kJ$ optical pulse that lasts for about $3.5\,ns$. If the beam profile is taken to be a step function with a $10\,cm$ diameter, what is the electric field amplitude within the beam? If this beam passes through KDP (see Problem 9.51), what is the ratio of the second-order polarization term relative to P_1?

Problem 9.53   In *third harmonic microscopy*, an infrared probe beam is focused into a biological sample, and the frequency-tripled THG light is detected point by point. Explain why THG is more useful than SHG for this application.

REFERENCES

1. M. L. Boas. *Mathematical Methods in the Physical Sciences, 2nd ed.* John Wiley and Sons, Hoboken, NJ, 1983.

2. M. Born and E. Wolf. *Principles of Optics, 7th ed.* Cambridge University Press, Cambridge, England, 1999.

3. Robert W. Boyd. *Nonlinear Optics.* Academic Press, Inc., New York, 1992.

4. R. N. Bracewell. *The Fourier Transform and its Applications, 3rd ed.* McGraw-Hill, Boston, MA, 2000.

5. R. P. Feynman. *QED: The Strange Theory of Light and Matter.* Princeton University Press, Princeton, NJ, 1985.

6. A. Gerrard and J. M. Burch. *Introduction to Matrix Methods in Optics.* John Wiley and Sons, Inc., New York, 1975.

7. J. W. Goodman. *Fourier Optics, 3rd ed.* Roberts and Company, Denver, CO, 2004.

8. G. Greenstein and A. G. Zajonc. *The Quantum Challenge.* Jones and Bartlett Publishers, Sudbury, MA, 1997.

9. D. J. Griffiths. *Introduction to Electrodynamics, 3rd ed.* Prentice-Hall, Englewood Cliffs, NJ, 1999.

10. R. D. Guenther. *Modern Optics.* John Wiley and Sons, Hoboken, NJ, 1990.

11. D. Halliday, R. R. Resnick, and J. Walker. *Fundamentals of Physics, 7th ed.* John Wiley and Sons, Hoboken, NJ, 2004.

12. J. D. Jackson. *Classical Electrodynamics, 3rd ed.* John Wiley and Sons, Inc., New York, 1999.

13. G. B. Thomas Jr., M. D. Weir, J. R. Hass, and F. R. Giordano. *Calculus, 11th ed.* Addison Wesley, Reading, MA, 1999.

14. Kenneth S. Krane. *Modern Physics, 2nd ed.* John Wiley and Sons, Inc., New York, NY, 1995.

15. T. S. Kuhn. *Black-Body Theory and the Quantum Discontinuity 1894—1912*. Clarendon Press-Oxford, New York, 1978.

16. F. L. Pedrotti and L. S. Pedrotti. *Introduction to Optics, 2nd ed.* Prentice-Hall, Englewood Cliffs, NJ, 1993.

17. B. A. E. Saleh and M. C. Teich. *Fundamentals of Photonics*. John Wiley and Sons, Hoboken, NJ, 1991.

18. D. V. Schroeder. *An Introduction to Thermal Physics*. Addison-Wesley, Reading, MA, 2000.

19. Anthony E. Siegman. *Lasers*. University Science Books, Mill Valley, CA, 1986.

20. Sherman K. Stein. *Calculus and Analytic Geometry, 4th Edition*. McGraw-Hill, New York, 1987.

21. E. F. Taylor and J. A. Wheeler. *Spacetime Physics, 2nd ed.* W. H. Freeman and Co., San Francisco, CA, 1992.

INDEX